Mike Holt's Illustrated Guide to

Understanding the
National Electrical Code
Volume 1

Based on the 2005 *NEC*©

www.*NEC*code.com
1.888.NEC®code

Mike Holt Enterprises, Inc.
7310 W. McNab Road • Suite 201• Tamarac, Florida 33321
1.888.NEC.CODE • NECcode.com • Info@NECcode.com

NOTICE TO THE READER

Mike Holt's Illustrated Guide to
Understanding the *National Electrical Code*, Volume 1
5th Edition
Graphic Illustrations: Mike Culbreath
Cover Design: Tracy Jette
Layout Design and Typesetting: Cathleen Kwas

1st Printing: February 2005.

COPYRIGHT © 2005 Charles Michael Holt Sr.
ISBN: 1-932685-16-2

For more information, contact:
Mike Holt Enterprises, Inc.
7310 West McNab Road, Suite 201
Tamarac, Florida 33321
www.MikeHolt.com

This logo is a registered trademark of Mike Holt Enterprises, Inc.

www.*NEC*code.com
1.888.NEC®code

I dedicate this book to the
Lord Jesus Christ,
my mentor and teacher.

To request examination copies of this or other
Mike Holt Publications, call or write to:

Mike Holt Enterprises, Inc.
7310 West McNab Road, Suite 201
Tamarac, Florida 33321
Phone: 1-888-NEC CODE • Fax: 1-954-720-7944

Mike Holt Online
www.NECcode.com
Sales@NECcode.com

You can download a sample PDF of all our
publications by visiting www.NECcode.com

Table of Contents

Notes

Introduction

This edition of *Mike Holt's Illustrated Guide to Understanding the National Electrical Code, Volume 1* textbook is intended to provide you with the tools necessary to understand the technical requirements of the *National Electrical Code (NEC)*®. The writing style of this textbook, and in all of Mike Holt's products, is meant to be informative, practical, useful, informal, easy to read, and applicable for today's electrical professional. Also, just like all of Mike Holt's textbooks, it contains hundreds of full-color illustrations to help you see the safety requirements of the *NEC* in practical use, as they apply to today's electrical installations.

This illustrated textbook contains advice, cautions about possible conflicts or confusing *Code* requirements, tips on proper electrical installations, and warnings of dangers related to improper electrical installations. This textbook cannot eliminate confusing, conflicting, or controversial *NEC* requirements, but it does try to put these requirements into sharper focus to help you understand their intended purpose. Sometimes a requirement is so confusing that nobody really understands its actual application. When this occurs, the textbook will be upfront and straightforward to point this out.

The *NEC* is updated every three years to accommodate new electrical products and material, emerging and dynamically changing technologies, as well as improved installation techniques. Fortunately, the *Code* allows the authority having jurisdiction, typically called the "Electrical Inspector," the flexibility to waive specific *NEC* requirements, and to permit alternative wiring methods contrary to the *Code* requirements, when he/she is assured that the completed electrical installation is equivalent in establishing and maintaining effective safety [90.4].

Keeping up with the *NEC* should be the goal of all those who are involved in electrical safety as it relates to electrical installations. This includes the electrical installer, contractor, owner, inspector, architect, engineer, instructor, and others concerned with safety as it relates to electrical installations.

Scope of *Understanding the NEC, Volume 1*

This textbook, *Understanding the National Electrical Code, Volume 1,* covers the general installation requirements contained in Articles 90 through 460 (*NEC* Chapters 1 through 4) that Mike considers to be of critical importance. But the textbook contains the following stipulations:

- **Power Systems and Voltage.** All power-supply systems are assumed to be solidly grounded and of any of the following voltages: 120V single-phase, 120/240V single-phase, 120/208V three-phase, 120/240V three-phase, or 277/480V three-phase, unless identified otherwise.

- **Electrical Calculations.** Unless the question or example specifies three-phase, the questions and examples are based on a single-phase power supply. In addition, all ampere calculations are rounded to the nearest ampere in accordance with 220.5(B).

- **Conductor Material.** All conductors are considered copper, unless aluminum is identified or specified.

- **Conductor Sizing.** All conductors are sized based on a THHN copper conductor terminating on a 75°C terminal in accordance with 110.14(C), unless the question or example identifies otherwise.

- **Protection Device.** The term "circuit protection device" refers to a molded case circuit breaker, unless identified otherwise. Where a fuse is identified, it is the single-element type, also known as a "one-time fuse," unless identified otherwise.

Workbook to Accompany Understanding the NEC, Volume 1

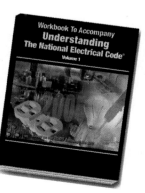

The *Workbook To Accompany Understanding the NEC, Volume 1* (2005 Edition) contains over 1,200 NEC practice questions plus seven 50-question exams which will test your knowledge and comprehension of the covered material.

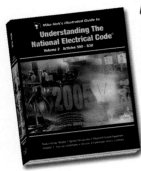

Understanding the NEC, Volume 2

To understand the entire *National Electrical Code*, you need to also study Mike's *Understanding the NEC, Volume 2* textbook, which covers requirements for wiring in special occupancies, special equipment, under special conditions, as well as communications systems requirements in Articles 500 through 830 (*NEC* Chapters 5-8).

Companion Video or DVDs

There are seven companion videos or DVDs (a total of 28 hours) available for *Understanding the National Electric Code, Volume 1* (Articles 90–460). There are three companion videos or DVDs (a total of 13.5 hours) available for *Understanding the National Electric Code, Volume 2* (Articles 500–830). Visit www.NECcode.com for details.

NEC Library

The *NEC* Library includes the Understanding the *NEC* Volumes 1 & 2 textbooks, the *NEC* Practice Questions book and ten videos or DVD's (a total of 41.5 hours). This option allows you to learn at the most cost effective price.

How to Use This Textbook

This textbook is to be used with the *NEC*, not as a replacement for the *NEC*, so be sure to have a copy of the 2005 *National Electrical Code* handy. Compare what Mike has explained in the text to your *Code* book, and discuss those topics that you find difficult to understand with others. As you read through this textbook, be sure to take the time to review the text with the outstanding graphics and examples.

Cross-References

The textbook contains thousands of *NEC* cross-references to other related *Code* requirements to help you develop a better understanding of how the *NEC* rules relate to one another. These cross-references are identified by a *Code* Section number in brackets, such as "90.4," which would look like "[90.4]."

Author's Comments

This textbook contains hundreds of "Author's Comments." These sections were written by Mike to help you better understand the *NEC* material, and to bring to your attention things he believes you should be aware of. To help you find them more easily, they are printed differently than the rest of the material. Mike's first Author's Comment is contained in 90.1(B).

Textbook Format

This textbook follows the *NEC* format, but it doesn't cover every *Code* requirement. For example, it doesn't cover every *NEC* Article, Section, Subsection, Exception, or Fine Print Note. So don't be concerned if you see that the textbook contains Exception No. 1 and Exception No. 3, but not Exception No. 2. In addition, at times, the title of an Article, Section, or Subsection might be rephrased differently.

Difficult Concepts

As you progress through this textbook, you might find that you don't understand every explanation, example, calculation, or comment. Don't get frustrated, and don't get down on yourself. Remember, this is the *National Electrical Code* and sometimes the best attempt to explain a concept isn't enough to make it per-fectly clear. When this happens to you, just make it a point to highlight the section that is causing you difficulty. If you can, take the textbook to someone you feel can provide additional insight, possibly your boss, the electrical inspector, a co-worker, your instructor, etc.

Not an *NEC* Replacement

This textbook is intended to explain the requirements of the *NEC*. It isn't intended to be a replacement for the *NEC*, so be sure you have a copy of the current *Code* book, and always compare Mike's explanation, comments, and graphics to the actual language contained in the *NEC*.

You'll sometimes notice that the titles of a few Articles and Sections are different than they appear in the actual *Code*. This only occurs when Mike feels it's easier to understand the content of the rule, so please keep this in mind when comparing the two documents. For example, 250.8 in the *NEC* is titled "Connection of Grounding and Bonding Equipment," but Mike used the title "Termination of Grounding (Earthing) and Bonding Conductors," because he felt his title better reflects the content of the rule.

Textbook Errors and Corrections

Humans develop the text, graphics, and layout of this textbook, and since currently none of us is perfect, there may be a few errors. This could occur because the *NEC* is dramatically changed each *Code* cycle; new Articles are added, some deleted, some relocated, and many renumbered. In addition, this textbook must be written within a very narrow window of opportunity; after the *NEC* has been published (September), yet before it's enforceable (January).

You can be sure we work a tremendous number of hours and use all of our available resources to produce the finest product with the fewest errors. We take great care in researching the *Code* requirements to ensure this textbook is correct. If you feel there's an error of any type in this textbook (typo, grammar, or technical), no matter how insignificant, please let us know.

Any errors found after printing are listed on our Website, so if you find an error, first check to see if it has already been corrected. Go to www.MikeHolt.com, click on the "Books" link, and then the "Corrections" link (www.MikeHolt.com/bookcorrections.htm).

If you do not find the error listed on the Website, contact us by E-mailing corrections@MikeHolt.com, calling 1.888.NEC.CODE (1.888.632.2633), or faxing 954.720.7944. Be sure to include the book title, page number, and any other pertinent information.

Internet

Today as never before, you can get your technical questions answered by posting them to Mike Holt's *Code* Forum. Just visit www.MikeHolt.com and click on the "*Code* Forum" link.

Different Interpretations

Some electricians, contractors, instructors, inspectors, engineers, and others enjoy the challenge of discussing the *Code* requirements, hopefully in a positive and a productive manner. This action of challenging each other is important to the process of better understanding the *NEC*'s requirements and its intended application. However, if you're going to get into an *NEC* discussion, please do not spout out what you think without having the actual *Code* in your hand. The professional way of discussing an *NEC* requirement is by referring to a specific section, rather than by talking in vague generalities.

The *National Electrical Code*

The *National Electrical Code (NEC)* is written for persons who understand electrical terms, theory, safety procedures, and electrical trade practices. These individuals include electricians, electrical contractors, electrical inspectors, electrical engineers, designers, and other qualified persons. The *Code* was not written to serve as an instructive or teaching manual for untrained individuals [90.1(C)].

Learning to use the *NEC* is somewhat like learning to play the game of chess; it's a great game if you enjoy mental warfare. You must first learn the names of the game pieces, how the pieces are placed on the board, and how each piece moves.

In the electrical world, this is equivalent to completing a comprehensive course on basic electrical theory, such as:

- What electricity is and how is it produced
- Dangers of electrical potential: fire, arc blast, arc fault, and electric shock
- Direct current
- Series and parallel circuits
- Electrical formulas
- Alternating current
- Induction, motors, generators, and transformers

Once you understand the fundamentals of the game of chess, you're ready to start playing the game. Unfortunately, at this point all you can do is make crude moves, because you really do not understand how all the information works together. To play chess well, you will need to learn how to use your knowledge by working on subtle strategies before you can work your way up to the more intriguing and complicated moves.

Again, back to the electrical world, this is equivalent to completing a course on the basics of electrical theory. You have the foundation upon which to build, but now you need to take it to the next level, which you can do by reading this textbook, watching the companion video or DVD, and answering the *NEC* practice questions in the *Workbook to Accompany Understanding the National Electric Code, Volume 1.*

Not a Game

Electrical work isn't a game, and it must be taken very seriously. Learning the basics of electricity, important terms and concepts, as well as the basic layout of the *NEC* gives you just enough knowledge to be dangerous. There are thousands of specific and unique applications of electrical installations, and the *Code* doesn't cover every one of them. To safely apply the *NEC*, you must understand the purpose of a rule and how it affects the safety aspects of the installation.

NEC Terms and Concepts

The *NEC* contains many technical terms, so it's crucial that *Code* users understand their meanings and their applications. If you do not understand a term used in a *Code* rule, it will be impossible to properly apply the *NEC* requirement. Be sure you understand that Article 100 defines the terms that apply to *two or more* Articles. For example, the term "Dwelling Unit" applies to many Articles. If you do not know what a Dwelling Unit is, how can you possibly apply the *Code* requirements for it?

In addition, many Articles have terms that are unique for that specific Article. This means that the definition of those terms is only applicable for that given Article. For example, Article 250 Grounding and Bonding has the definitions of a few terms that are only to be used within Article 250.

Small Words, Grammar, and Punctuation

It's not only the technical words that require close attention, because even the simplest of words can make a big difference to the intent of a rule. The word "or" can imply alternate choices for equipment wiring methods, while "and" can mean an additional requirement. Let's not forget about grammar and punctuation. The location of a comma "," can dramatically change the requirement of a rule.

Slang Terms or Technical Jargon

Electricians, engineers, and other trade-related professionals use slang terms or technical jargon that isn't shared by all. This makes it very difficult to communicate because not everybody understands the intent or application of those slang terms. So where possible, be sure you use the proper word, and do not use a word if you do not understand its definition and application. For example, lots of electricians use the term "pigtail" when describing the short conductor for the connection of a receptacle, switch, luminaire, or equipment. Although they may understand it, not everyone does. **Figure 1**

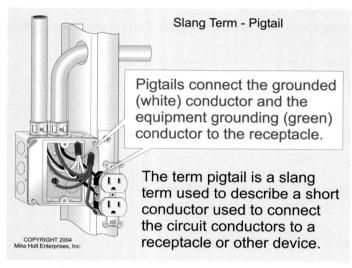

Slang Term - Pigtail

Pigtails connect the grounded (white) conductor and the equipment grounding (green) conductor to the receptacle.

The term pigtail is a slang term used to describe a short conductor used to connect the circuit conductors to a receptacle or other device.

COPYRIGHT 2004
Mike Holt Enterprises, Inc.

Figure 1

NEC Style and Layout

Before we get into the details of the *NEC*, we need to take a few moments to understand its style and layout. Understanding the structure and writing style of the *Code* is very important before it can be used effectively. If you think about it, how can you use something if you don't know how it works? Okay, let's get started. The *National Electrical Code* is organized into nine components.

- Table of Contents
- Chapters 1 through 9 (major categories)
- Articles 90 through 830 (individual subjects)
- Parts (divisions of an Article)
- Sections and Tables (*Code* requirements)
- Exceptions (*Code* permissions)
- Fine Print Notes (explanatory material)
- Index
- Annexes (information)

1. Table of Contents. The Table of Contents displays the layout of the Chapters, Articles, and Parts as well as the page numbers. It's an excellent resource and should be referred to periodically to observe the interrelationship of the various *NEC* components. When attempting to locate the rules for a particular situation, knowledgeable *Code* users often go first to the Table of Contents to quickly find the specific *NEC* section that applies.

2. Chapters. There are nine Chapters, each of which is divided into Articles. The Articles fall into one of four groupings: General Requirements (Chapters 1 through 4), Specific Requirements (Chapters 5 through 7), Communications Systems (Chapter 8), and Tables (Chapter 9).

- Chapter 1 General
- Chapter 2 Wiring and Protection
- Chapter 3 Wiring Methods and Materials
- Chapter 4 Equipment for General Use
- Chapter 5 Special Occupancies
- Chapter 6 Special Equipment
- Chapter 7 Special Conditions
- Chapter 8 Communications Systems (Telephone, Data, Satellite, and Cable TV)
- Chapter 9 Tables—Conductor and Raceway Specifications

3. Articles. The *NEC* contains approximately 140 Articles, each of which covers a specific subject. For example:

- Article 110 General Requirements
- Article 250 Grounding
- Article 300 Wiring Methods
- Article 430 Motors
- Article 500 Hazardous (Classified) Locations
- Article 680 Swimming Pools, Spas, Hot Tubs, and Fountains
- Article 725 Remote-Control, Signaling, and Power-Limited Circuits
- Article 800 Communications Systems

4. Parts. Larger Articles are subdivided into Parts. For example, Article 110 has been divided into multiple parts:

- Part I. General (Sections 110.1—110.23)
- Part II. 600 Volts, Nominal, or Less (110.26—110.27)
- Part III. Over 600 Volts, Nominal (110.30—110.59)

Note: Because the Parts of a *Code* Article aren't included in the Section numbers, we have a tendency to forget what "Part" the *NEC* rule is relating to. For example, Table 110.34(A) contains the working space clearances for electrical equipment. If we aren't careful, we might think this table applies to all electrical installations, but Table 110.34(A) is located in Part III, which contains the requirements for Over 600 Volts, Nominal installations. The rules for working clearances for electrical equipment for systems 600V or less are contained in Table 110.26(A)(1), which is located in Part II. 600 Volts, Nominal, or Less.

5. Sections and Tables.

Sections: Each *NEC* rule is called a *Code* Section. A *Code* Section may be broken down into subsections by letters in parentheses (A), (B), etc. Numbers in parentheses (1), (2), etc., may further break down a subsection, and lower-case letters (a), (b), etc., further breaks the rule down to the third level. For example, the rule requiring all receptacles in a dwelling unit bathroom to be GFCI protected is contained in Section 210.8(A)(1). Section 210.8(A)(1) is located in Chapter 2, Article 210, Section 8, subsection (A), sub-subsection (1).

Many in the industry incorrectly use the term "Article" when referring to a *Code* Section. For example, they say "Article 210.8," when they should say "Section 210.8."

Tables: Many *Code* requirements are contained within Tables, which are lists of *NEC* requirements placed in a systematic arrangement. The titles of the Tables are extremely important; they must be carefully read in order to understand the contents, applications, limitations, etc., of each Table in the *Code*. Many times notes are provided in a table; be sure to read them as well, since they are also part of the requirement. For example, Note 1 for Table 300.5 explains how to measure the cover when burying cables and raceways, and Note 5 explains what to do if solid rock is encountered.

6. Exceptions. Exceptions are *Code* requirements that provide an alternative method to a specific requirement. There are two types of exceptions—mandatory and permissive. When a rule has several exceptions, those exceptions with mandatory requirements are listed before the permissive exceptions.

Mandatory Exception: A mandatory exception uses the words "shall" or "shall not." The word "shall" in an exception means that if you're using the exception, you're required to do it in a particular way. The term "shall not" means it isn't permitted.

Permissive Exception: A permissive exception uses words such as "is permitted," which means that it's acceptable to do it in this way.

7. Fine Print Note (FPN). A Fine Print Note contains explanatory material intended to clarify a rule or give assistance, but it isn't a *Code* requirement.

8. Index. The Index contained in the *NEC* is excellent and is helpful in locating a specific rule.

9. Annexes. Annexes aren't a part of the *NEC* requirements, and are included in the *Code* for informational purposes only.

- Annex A. Product Safety Standards
- Annex B. Application Information for Ampacity Calculation
- Annex C. Conduit and Tubing Fill Tables for Conductors and Fixture Wires of the Same Size
- Annex D. Examples
- Annex E. Types of Construction
- Annex F. Cross-Reference Tables (1999, 2002, and 2005 *NEC*)
- Annex G. Administration and Enforcement

Note: Changes to the *NEC*, since the previous edition(s) are identified in the margins by a vertical line (|), but rules that have been relocated aren't identified as a change. In addition, the location from which the *Code* rule was removed has no identifier.

How to Locate a Specific Requirement

How to go about finding what you're looking for in the *Code* depends, to some degree, on your experience with the *NEC*. *Code* experts typically know the requirements so well that they just go to the *NEC* rule without any outside assistance. The Table of Contents might be the only thing very experienced *Code* users need to locate their requirement. On the other hand, average *Code* users should use all of the tools at their disposal, and that includes the Table of Contents and the Index.

Table of Contents: Let's work out a simple example: What *NEC* rule specifies the maximum number of disconnects permitted for a service? If you're an experienced *Code* user, you'll know that Article 230 applies to "Services," and because this Article is so large, it's divided up into multiple parts (actually 8 parts). With this knowledge, you can quickly go to the Table of Contents (page 70-2) and see that it lists the Service Equipment Disconnecting Means requirements in Part VI, starting at page 70-77.

Note: The number 70 precedes all page numbers because the *NEC* is standard number 70 within the collection of *NFPA* standards.

Index: If you used the Index, which lists subjects in alphabetical order, to look up the term "service disconnect," you would see that there's no listing. If you tried "disconnecting means," then "services," you would find the Index specifies that the rule is located at 230, Part VI. Because the *NEC* doesn't give a page number in the Index, you'll need to use the Table of Contents to get the page number, or flip through the *Code* to Article 230, then continue to flip until you find Part VI.

As you can see, although the index is very comprehensive, it's not that easy to use if you do not understand how the index works. But if you answer the over 1,200 *NEC* practice questions or seven 50-question exams contained in the *Workbook to Accompany Understanding the National Electric Code, Volume 1,* you'll become a master at finding things in the *Code* quickly.

Many people complain that the *NEC* only confuses them by taking them in circles. As you gain experience in using the *Code* and deepen your understanding of words, terms, principles, and practices, you will find the *NEC* much easier to understand and use than you originally thought.

Customizing Your *Code* Book

One way to increase your comfort level with the *Code* is to customize it to meet your needs. You can do this by highlighting and underlining important *NEC* requirements, and by attaching tabs to important pages.

Highlighting: As you read through this textbook and answer the questions in the workbook, be sure you highlight those requirements in the *Code* that are most important to you. Use yellow for general interest and orange for important requirements you want to find quickly. Be sure to highlight terms in the Index and Table of Contents as you use them.

Because of the size of the 2005 *NEC*, I recommend you highlight in green the Parts of Articles that are important for your applications, particularly:

> Article 230 Services
> Article 250 Grounding
> Article 430 Motors

Underlining: Underline or circle key words and phrases in the *NEC* with a red pen (not a lead pencil) and use a 6-in. ruler to keep lines straight and neat. This is a very handy way to make important requirements stand out. A small 6-in. ruler also comes in handy for locating specific information in the many *Code* tables.

Tabbing the *NEC*: Placing tabs on important *Code* Articles, Sections, and Tables will make it very easy to access important *NEC* requirements. However, too many tabs will defeat the purpose. You can order a custom set of *Code* tabs, designed by Mike Holt, online at www.MikeHolt.com, or by calling us at 1.888.NEC.Code (1.888.632.2633).

Mike Holt Enterprises Team

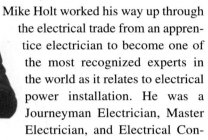

About the Author

Mike Holt worked his way up through the electrical trade from an apprentice electrician to become one of the most recognized experts in the world as it relates to electrical power installation. He was a Journeyman Electrician, Master Electrician, and Electrical Contractor. Mike came from the real world, and he has a unique understanding of how the *NEC* relates to electrical installations from a practical standpoint. You will find his writing style to be simple, nontechnical, and practical.

Did you know that Mike didn't finish high school? So if you struggled in high school or if you didn't finish it at all, don't let this get you down, you're in good company. As a matter of fact, Mike Culbreath, Master Electrician, who produces the finest electrical graphics in the history of the electrical industry, didn't finish high school either! So two high school dropouts produced the text and graphics in this textbook! However, realizing that success depends on one's continuing pursuit of education, Mike immediately attained his GED (as did Mike Culbreath) and ultimately attended the University of Miami's Graduate School for a Master's degree in Business Administration (MBA).

Mike Holt resides in Central Florida, is the father of seven children, and has many outside interests and activities. He is a former National Barefoot Waterskiing Champion (1988 and 1999), who set five barefoot water-ski records, and he continues to train year-round at a national competition level [www.barefootcentral.com].

Mike enjoys motocross racing, but at the age of 52 decided to retire from that activity (way too many broken bones, concussions, collapsed lung, etc., but what a rush). Mike also enjoys snow skiing and spending time with his family. What sets Mike apart from some is his commitment to living a balanced lifestyle; he places God first, then family, career, and self.

Educational Director

Sarina Snow was born and raised in "The Bronx" (Yankee Stadium Area), then moved to New Jersey with her husband Freddy where they raised their three children. They moved to Florida in 1979 and have totally loved the move to a warmer climate. Sarina has worked with Mike Holt for over twenty years and has literally learned the business from the ground up. She remembers typing Mike's books using carbon paper BC (Before Computers). She has developed a strong relationship with the industry by attending Mike's classes and seminars, and by accompanying Mike to many trade shows in various states to get a deeper understanding of the trade firsthand. Her love and devotion to Mike and Company (actually the original name of Mike Holt Enterprises) has given her the ability to be the "Mom" in the industry. "She Cares," sums it up.

Graphic Illustrator

Mike Culbreath devoted his career to the electrical industry and worked his way up from an apprentice electrician to master electrician. While working as a journeyman electrician, he suffered a serious on-the-job knee injury. With a keen interest in continuing education for electricians, and as part of his rehabilitation program, he completed courses at Mike Holt Enterprises, Inc. and then passed the exam to receive his Master Electrician's license.

In 1986, after attending classes at Mike Holt Enterprises, he joined the staff to update material and later studied computer graphics and began illustrating Mike Holt's textbooks and magazine articles. He's worked with Mike Holt Enterprises for over 15 years and, as Mike Holt has proudly acknowledged, has helped to transform his words and visions into lifelike graphics.

Mike Culbreath resides in northern Michigan with his wife Toni, and two children: Dawn and Mac. He is helping Toni fulfill her dream by helping her develop and build a quality horse boarding, training, and teaching facility. Mike enjoys working with children by volunteering as a leader for a 4-H archery club and assisting with the local 4-H horse club. He also enjoys fishing, gardening, and cooking.

Editorial

Toni Culbreath completed high school graduation requirements by the end of the first semester of her senior year. She went on to complete courses for computer programming at a trade school by March of that year, and then returned to participate in graduation ceremonies with her high school class.

Toni became associated with Mike Holt Enterprises in 1994 in the area of software support and training and now enjoys the challenges of editing Mike Holt's superb material. She is certified as a therapeutic riding instructor and is extensively involved in Michigan's 4-H horse programs at both the county and state level.

Barbara Parks has been working for Mike Holt Enterprises for the last several years as a Writer's Assistant. She has edited most of Mike Holt's books and various projects over this period of time. She is a retired lady, working part time at home and thoroughly enjoys "keeping busy."

Technical Editorial Director

Steve Arne has been involved in the electrical industry since 1974 working in various positions from electrician to full-time instructor and department chair in technical post secondary education. Steve has developed curriculum for many electrical training courses and has developed university business and leadership courses. Currently, Steve offers occasional exam prep and Continuing Education *Code* classes.

Steve believes that as a teacher he understands the joy of helping others as they learn and experience new insights. His goal is to help others understand more of the technological marvels that surround us. Steve thanks God for the wonders of His creation and for the opportunity to share it with others.

Steve and his lovely wife Deb live in Rapid City, South Dakota where they are both active in their church and community. They have two grown children and five grandchildren.

Cover Design

Tracy Jette has enjoyed working in the field of Graphic Design for over 10 years. She loves all aspects of design, and finds that spending time outdoors camping and hiking with her family and friends is a great inspiration. Tracy is very happy to have recently joined Mike Holt Enterprises and has found that working from home brings a harmony to her life with her 3 boys (10-year-old twins and a 7-year-old), her husband of 16 years, Mario, and her work life.

Layout Design and Production

Cathleen Kwas has been in the publishing industry for over 26 years. She's worn many hats–copy editor, desktop publisher, prepress manager, project coordinator, communications director, book designer, and graphic artist.

Cathleen is very happily married to Michael and lives in beautiful Lake Mary, Florida with their adorable Maltese-ShihTzu puppy, Bosco.

Acknowledgments

Special Acknowledgments

First I want to thank God for my life. I want to thank Him for even the most difficult of times, because this has helped me become a man that I hope honors Him in my actions. My loving Godly wife is always by my side, and there's no question that I could not have achieved any of my success without her continued support. She is a selfless mother and wife. She made the difficult decision to stay at home and support her family; the successes of her husband and her children are God's reward for her sacrifice. To my wonderful children, Belynda, Melissa, Autumn, Steven, Michael, Meghan, and Brittney—I love every moment we shared together (well, most moments). Thank you for loving me and knowing God.

I would like to thank all the people in my life that believed in me, and those who spurred me on. Thanks to the Electrical Construction & Maintenance (*EC&M*) magazine for my first "big break" in 1980, and Joe McPartland who helped and encouraged me from 1980 to 1992. Joe, I'll never forget to help others as you've helped me. I would also like to thank Joe Salimando, the former publisher of the Electrical Contractor magazine produced by the National Electrical Contractors Association (*NECA*) for my second "big break" in 1995.

A special thank you must be sent to the staff at the National Fire Protection Association (NFPA), publishers of the *NEC*—in particular Jeff Sargent for his assistance in answering my many *NEC* questions. Jeff, you're a "first class" guy, and I admire your dedication and commitment to helping others, including me, to understand the *Code*. Other former NFPA staff members I would like to thank include John Caloggero, Joe Ross, and Dick Murray for their help in the past.

Phil Simmons, former Executive Director of the International Association of Electrical Inspectors (IAEI)—you're truly a Godly man whom I admire, and I do want to thank you for your help, especially in grounding. Other people who have been important in my personal and technical development include James Stallcup, Dick Loyd, Mark Ode, DJ Clements, Morris Trimmer, Tony Silvestri, and the infamous Marvin Weiss.

A personal thank you goes to Sarina, my long-time friend and office manager. Thank you for covering the office for me while I spend so much time writing textbooks, conducting seminars, and producing videos and DVDs. Your love and concern for the customer has contributed significantly to the success of Mike Holt Enterprises, Inc., and it has been wonderful working side-by-side and nurturing this company's growth from its small beginnings. Also thank you for loving my family and me and for being there during those many difficult times.

Mike Holt Enterprises Team Acknowledgments

There are many people who played a role in the development and production of this textbook. I would like to start with Mike Culbreath, Master Electrician, who has been with me for over 15 years, helping me transform my words and thoughts into lifelike graphics.

Also, a thank you goes to Cathleen Kwas for the outstanding electronic layout of this textbook, and Tracy Jette for the amazing front and back cover.

Finally, I would like to thank the following individuals who worked tirelessly to proofread and edit the final stages of this publication: Toni Culbreath and Barbara Parks. Their attention to detail and dedication to this project is greatly appreciated.

Advisory Committee

Thanks are also in order for the following individuals who reviewed the manuscript and offered invaluable suggestions and feedback.

Victor M. Ammons, P.E.
Director of Electrical Engineering,
The Prisco Group,
Hopewell, New Jersey

Steve Arne
Technical Director, Mike Holt Enterprises, Inc.,
Rapid City, South Dakota

Mike Culbreath
Graphic Designer, Mike Holt Enterprises, Inc.
Alden, Michigan

Leo W. Moritz, P.E.
Senior Electrical Engineer,
Chicago, Illinois

Jerry Peck
Inspector and Instructor,
Inspection Services Associates, Inc.,
Pembroke Pines, Florida

Terry Schneider
Electrical Field Inspection Supervisor,
Colorado Springs, Colorado

Brooke Stauffer
Executive Director of Standards and Safety,
National Electrical Contractors Association,
Bethesda, Maryland

James Thomas
Electrical/Electronics Instructor,
James Sprunt Community College,
Kenansville, North Carolina

J. Kevin Vogel, P.E.
Design and Quotations, Crescent Electric Supply,
Coeur d'Alene, Idaho

Joseph Wages Jr.
Instructor
Siloam Springs, Arkansas

A Very Special Thank You

To my beautiful wife, Linda, and my seven children:
Belynda, Melissa, Autumn, Steven, Michael,
Meghan, and Brittney—
thank you for loving me so much.

Video Team Members

Steve Arne
Technical Editorial Director,
Mike Holt Enterprises, Inc.
Electrical Instructor, Arne Electro Tech,
Rapid City, South Dakota
http://electricalmaster.com

Steve Arne has been involved in the electrical industry since 1974, working in various positions from electrician to full-time instructor and department chair in technical post secondary education. He has a Bachelor's Degree in Technical Education and a Master's Degree in Administrative Studies with a human resources emphasis. Licenses held by Steve include Electrical Master, Electrical Inspector, Electrical Contractor, and Real Estate Home Inspector. He is a board member of the Black Hills Chapter of the SD Electrical Council and a member of the SD Real Estate Task Force on Home Inspection. He also enjoys developing his own Websites.

Steve and his lovely wife Deb have celebrated over 32 years of marriage in Rapid City, South Dakota where they are both active in their church and community. They have two grown children and five grandchildren.

Tarry L. Baker
Chief Electrical Code Compliance Officer
Broward County Board of Rules & Appeals
Fort Lauderdale, Florida

Tarry has been the Chief Electrical Code Compliance Officer for the Broward County Board of Rules and Appeals (Florida) for the last twelve years, standardizing enforcement in 30 municipalities and the unincorporated area of Broward County. He has served on the Electrical Technical Committee to the Board of Rules and Appeals for over ten years, and as an advisor to the committee for nine years. He has been a member of National Electrical Code (NEC) Code-Making Panel-13 of the National Fire Protection Association (NFPA) since 2000 for the 2002 & 2005 NEC. He helped develop the NFPA & IAEI National Certified Residential Electrical Inspector and Master Electrical Inspector examinations. Tarry is currently a member of the Electrical/Alarms Technical Advisory Committee to the Florida Building Commission, and a former member of the Education, Disciple, and Licensing Work Group for the Governor's Building Code Study Commission for the State of Florida.

Tarry has been married to his wife Beckie for 34 years. They have twin daughters and two grandchildren.

Mike Culbreath
Graphic Illustrator,
Mike Holt Enterprises, Inc.
Alden, Michigan

Mike Culbreath devoted his career to the electrical industry and worked his way up from an apprentice electrician to master electrician. While working as a journeyman electrician, he suffered a serious on-the-job knee injury. With a keen interest in continuing education for electricians, and as part of his rehabilitation program, he completed courses at Mike Holt Enterprises, Inc. and then passed the exam to receive his Master Electrician's license.

In 1986, after attending classes at Mike Holt Enterprises, he joined the staff to update material and later studied computer graphics and began illustrating Mike Holt's textbooks and magazine articles. He's worked with Mike Holt Enterprises for over 15 years and, as Mike Holt has proudly acknowledged, has helped to transform his words and visions into lifelike graphics.

Mike Culbreath resides in northern Michigan with his wife Toni, and two children: Dawn and Mac. He is helping Toni fulfill her dream by helping her develop and build a quality horse boarding, training, and teaching facility. Mike enjoys working with children by volunteering as a leader for a 4-H archery club and assisting with the local 4-H horse club. He also enjoys fishing, gardening, and cooking.

Charles Fulmer
Electrical Inspector
City of Little Rock
Little Rock, Arkansas

Charles Fulmer is currently an Electrical Inspector for the City of Little Rock, Arkansas. He is also an Electrical Instructor for the Arkansas Electrical Apprenticeship Program and a member of the International Association of Electrical Inspectors.

Before becoming an Inspector, Charles worked for 25 years in the commercial electrical trade, mostly as a supervisor over large jobs such as hospitals, high rise construction, hotels, and schools.

Charles has been married to his wife Meryl for 28 years. They have raised three sons and one daughter and now have one granddaughter who is two years old.

Charles has been a leader in a church youth scouting organization called the Royal Rangers for 20 years, reaching, teaching, and keeping boys for Christ! Charles feels it is better to build men than to repair them.

Lynn E. Palmer
Construction Supervisor
State of California
Sacramento, California

Lynn started his career as an electrician in 1979, became a licensed Electrical Contractor in 1985, and owned and operated L & B Palmer Electric. He later went to work as an Electrical Inspector for the County of Sacramento. Lynn is a member of the International Association of Electrical Inspectors, Sacramento Valley Chapter, where he also served as Chairman, and was a member of IBEW Local 340 Sacramento.

Lynn has taught a variety of apprenticeship, journeyman and *Code* training classes. Lynn currently teaches in the Building Inspection Technology Program at Cosumnes River College in Sacramento. After working for the County of Sacramento for 10 years, he went to work for the State of California as a Construction Supervisor.

Lynn currently resides in Elk Grove, California, with his wife Brenda and two children, Christopher and Jennifer.

Ray Winkel
Electrical Technology Department Chair,
North Central Kansas Technical College
Beloit, Kansas
www.ncktc.tec.ks.us/default.htm

Ray completed an electronics program at a trade school immediately after high school. He started his career in the consumer electronics industry servicing electronic products. Ray and partner Randy Paxson started their own electrical contracting business. As contractors, their company did all types of wiring: residential, commercial, and industrial.

After 10 years in the contracting business, he began teach electricity at North Central Kansas Technical College. Ray is a licensed Professional Engineer and a licensed Master Electrician and is currently department chair in the Electrical Technology Department at North Central Kansas Area Vocational Technical College (NCKAVTC) in Beloit, Kansas. Ray gets a great deal of satisfaction from teaching and seeing students succeed. Ray is currently working on a Masters degree.

Ray and his wife Ilene enjoy traveling and restoring antique jukeboxes in their spare time.

Chapter 4 Video Team Members

Larry D. Abernathy
Master Electrician/Instructor
Ford Motor Co. Research & Engineering
Dearborn, Michigan

Steve Arne
Technical Editorial Director
Mike Holt Enterprises, Inc.
www.ElectricalMaster.com

Mike Culbreath
Graphic Designer
Mike Holt Enterprises, Inc.

Ryan Jackson
Inspector/Instructor
Ryan Jackson Electrical Training
City of Draper, Utah
www.RyanJacksonElectricalTraining.com

Chuck Williams
Master Electrician/Instructor
North Idaho College
Post Falls, Idaho

David A. Williams
Inspector/Instructor
Delta Township/Lansing Community College
Lansing, Michigan

Introduction

Many *NEC* violations and misunderstandings wouldn't occur if people doing the work simply understood Article 90. For example, many people see *Code* requirements as performance standards. In fact, *NEC* requirements are the bare minimum for safety. This is exactly the stance electrical inspectors, insurance companies, and courts will take when making a decision regarding electrical design or installation.

Article 90 opens by saying the *NEC* isn't intended as a design specification or instruction manual. The *National Electrical Code* has one purpose only. That is "the practical safeguarding of persons and property from hazards arising from the use of electricity."

Article 90 then describes the scope and arrangement of the *Code*. A person who says, "I can't find anything in the *Code,*" is really saying, "I never took the time to review Article 90." The balance of Article 90 provides the reader with information essential to understanding those items you do find in the *NEC*.

Typically, electrical work requires you to understand the first four Chapters of the *NEC*, plus have a working knowledge of the Chapter 9 tables. Chapters 5, 6, 7, and 8 make up a large portion of the *NEC*, but they apply to special situations. They build on, and extend, what you must know in the first four chapters. That knowledge begins with Chapter 1.

90.1 Purpose of the *NEC.*

(A) Practical Safeguarding. The purpose of the *NEC* is to ensure that electrical systems are installed in a manner that protects people and property by minimizing the risks associated with the use of electricity.

(B) Adequacy. The *Code* contains requirements that are considered necessary for a safe electrical installation. When an electrical installation is installed in compliance with the *NEC*, it will be essentially free from electrical hazards. The *NEC* is a safety standard, not a design guide.

The *NEC* requirements aren't intended to ensure that the electrical installation will be efficient, convenient, adequate for good service, or suitable for future expansion. Specific items of concern, such as electrical energy management, maintenance, and power quality issues aren't within the scope of the *NEC*.
Figure 90–1

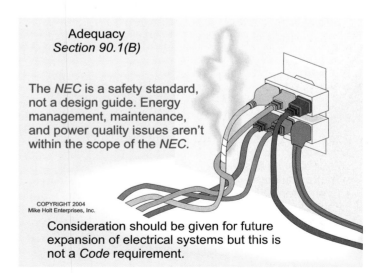

Adequacy
Section 90.1(B)

The *NEC* is a safety standard, not a design guide. Energy management, maintenance, and power quality issues aren't within the scope of the *NEC*.

COPYRIGHT 2004
Mike Holt Enterprises, Inc.

Consideration should be given for future expansion of electrical systems but this is not a *Code* requirement.

Figure 90–1

FPN: Hazards in electrical systems often occur because circuits are overloaded or not properly installed in accordance with the *NEC*. The initial wiring often did not provide reasonable provisions for system changes or for the increase in the use of electricity.

Author's Comments:

- See Article 100 for the definition of "Overload."

- The *NEC* does not require electrical systems to be designed or installed to accommodate future loads. However, the electrical designer, typically an electrical engineer, is concerned with not only ensuring electrical safety (*Code* compliance), but also ensuring that the system meets the customers' needs, both of today and in the near future. To satisfy customers' needs, electrical systems must be designed and installed above the minimum requirements contained in the *NEC*.

(C) Intention. The *Code* is to be used by those skilled and knowledgeable in electrical theory, electrical systems, construction, and the installation and operation of electrical equipment. It isn't a design specification standard or instruction manual for the untrained and unqualified.

(D) Relation to International Standards. The requirements of the *NEC* address the fundamental safety principles contained in International Electrotechnical Commission standards, including protection against electric shock, adverse thermal effects, overcurrent, fault currents, and overvoltage. Figure 90–2

Author's Comments:

- See Article 100 for the definition of "Overcurrent."

- The *NEC* is used in Chile, Ecuador, Peru, and the Philippines. It's also the *Electrical Code* for Colombia, Costa Rica, Mexico, Panama, Puerto Rico, and Venezuela. Because of these adoptions, the *NEC* is available in Spanish from the National Fire Protection Association, 1.617.770.3000.

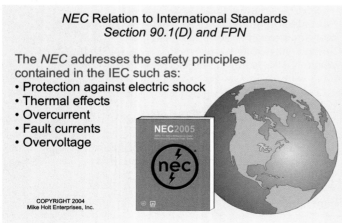

NEC Relation to International Standards
Section 90.1(D) and FPN

The *NEC* addresses the safety principles contained in the IEC such as:
- Protection against electric shock
- Thermal effects
- Overcurrent
- Fault currents
- Overvoltage

NEC2005

COPYRIGHT 2004
Mike Holt Enterprises, Inc.

Figure 90–2

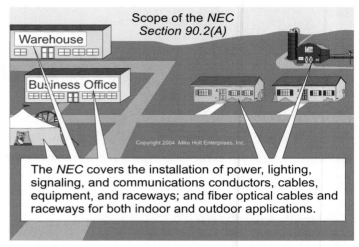

Scope of the *NEC*
Section 90.2(A)

Warehouse

Business Office

Copyright 2004. Mike Holt Enterprises, Inc.

The *NEC* covers the installation of power, lighting, signaling, and communications conductors, cables, equipment, and raceways; and fiber optical cables and raceways for both indoor and outdoor applications.

Figure 90–3

90.2 Scope of the *NEC*.

(A) What is Covered. The *NEC* contains requirements necessary for the proper electrical installation of electrical conductors, equipment, and raceways; signaling and communications conductors, equipment, and raceways; as well as fiber optic cables and raceways for the following locations: Figure 90–3

(1) Public and private premises, including buildings or structures, mobile homes, recreational vehicles, and floating buildings.

(2) Yards, lots, parking lots, carnivals, and industrial substations.

(3) Conductors and equipment that connect to the utility supply.

(4) Installations used by an electric utility, such as office buildings, warehouses, garages, machine shops, recreational buildings, and other electric utility buildings that are not an integral part of a utility's generating plant, substation, or control center. Figure 90–4

(B) What isn't Covered. The *National Electrical Code* doesn't apply to the following applications:

(1) Transportation Vehicles. Installations in cars, trucks, boats, ships and watercraft, planes, electric trains, or underground mines.

(2) Mining Equipment. Installations underground in mines and self-propelled mobile surface mining machinery and its attendant electrical trailing cables.

(3) Railways. Railway power, signaling, and communications wiring.

(4) Communications Utilities. The installation requirements of the *NEC* do not apply to communications (telephone), CATV, or network-powered broadband utility equipment located in

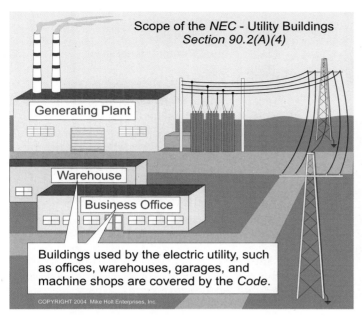

Figure 90–4

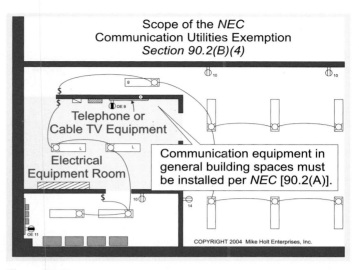

Figure 90–6

building spaces used exclusively for such use or outdoors, if the installation is under the exclusive control of the communications utility. **Figure 90–5**

Author's Comment: Interior wiring for communications systems, not in building spaces used exclusively for such use, must be installed in accordance with the following Chapter 8 requirements: **Figure 90–6**

- Phone and Data, Article 800
- CATV, Article 820
- Network-Powered Broadband, Article 830

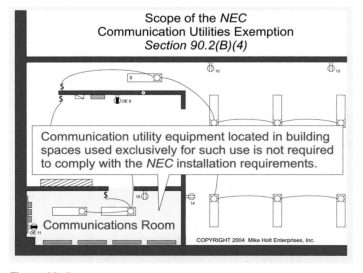

Figure 90–5

(5) Electric Utilities. The *NEC* doesn't apply to electric installations under the exclusive control of an electric utility where such installations:

    a. Consist of service drops or service laterals and associated metering. **Figure 90–07**

    b. Are located on legally established easements, rights-of-way, or by other agreements recognized by public/utility regulatory agencies, or property owned or leased by the electric utility. **Figure 90–8**

    c. Are on property owned or leased by the electric utility for the purpose of generation, transformation, transmission, distribution, or metering of electric energy. **See Figure 90–8.**

Author's Comment: Luminaires (lighting fixtures) located in legally established easements, or rights-of-way, such as at poles supporting transmission or distribution lines, are exempt from the requirements of the *NEC*. However, if the electric utility provides

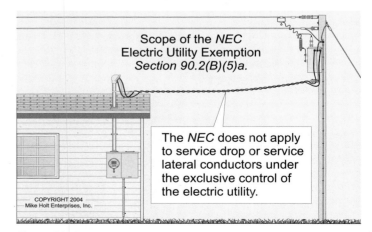

Figure 90–7

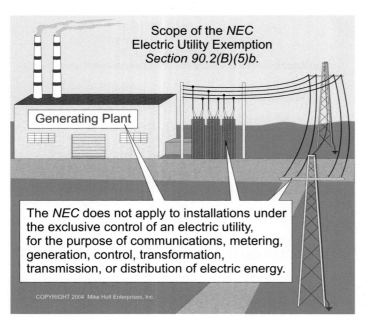

Figure 90–8

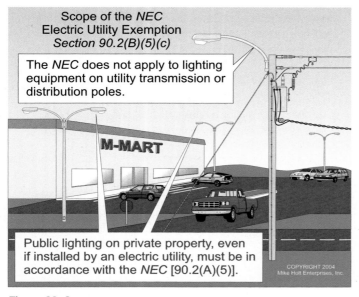

Figure 90–9

site and public lighting on private property, then the installation must comply with the *NEC* [90.2(A)(4)]. **Figure 90–9**

FPN to 90.2(B)(4) and (5): Utilities include entities that install, operate, and maintain communications systems (telephone, CATV, Internet, satellite, or data services) or electric supply systems (generation, transmission, or distribution systems) and are designated or recognized by governmental law or regulation by public service/utility commissions. Utilities may be subject to compliance with codes and standards covering their regulated activities as adopted under governmental law or regulation.

Code Arrangement
Section 90.3
General Requirements

• Chapter 1 - General
• Chapter 2 - Wiring and Protection
• Chapter 3 - Wiring Methods and Materials
• Chapter 4 - Equipment for General Use
Chapters 1 through 4 apply to all applications.

Special Requirements

• Chapter 5 - Special Occupancies
• Chapter 6 - Special Equipment
• Chapter 7 - Special Conditions
Chapters 5 through 7 can supplement or modify the general requirements of Chapters 1 through 4.

• Chapter 8 - Communications Systems
Chapter 8 requirements are not subject to requirements in Chapters 1 through 7, unless there is a specific reference in Chapter 8 to a rule in Chapters 1 through 7.

• Chapter 9 - Tables
Chapter 9 tables are used for calculating raceway sizes, conductor fill, and voltage drop.

• Annex A through F
Annexes are for information only and not enforceable.

COPYRIGHT 2004 Mike Holt Enterprises, Inc.

Figure 90–10

90.3 *Code* Arrangement. The *Code* is divided into an Introduction and nine chapters. **Figure 90–10**

General Requirements. The requirements contained in Chapters 1, 2, 3, and 4 apply to all installations.

> **Author's Comment:** The scope of this textbook includes *NEC* Chapters 1 through 4.

Special Requirements. The requirements contained in Chapters 5, 6, and 7 apply to special occupancies, special equipment, or other special conditions. They can supplement or modify the requirements in Chapters 1 through 4.

For example, the general requirement contained in 250.118 of Article 250 Grounding and Bonding states that a metal raceway, such as Electrical Metallic Tubing, is considered suitable to provide a low-impedance path to the power supply for ground-fault current. However, 517.13(B) of Article 517 Health Care Facilities doesn't consider the raceway to be sufficient. It requires an insulated copper conductor to be installed in the raceway for this purpose.

Communications Systems. Chapter 8 contains the requirements for communications systems, such as telephone, antenna wiring, CATV, and network-powered broadband systems. Communications systems aren't subject to the general requirements of Chapters 1 through 4, or the special requirements of Chapters 5 through 7, unless there's a specific reference in Chapter 8 to a rule in Chapters 1 through 7.

> **Author's Comment:** Mike Holt's *Understanding the NEC, Volume 2 [Articles 500 through 830]*, explains the wiring requirements of special occupancies, special equipment, and special conditions, as well as communications systems.

Table. Chapter 9 consists of tables necessary to calculate raceway sizing, conductor fill, and voltage drop.

Annexes. Annexes aren't part of the *Code*, but are included for informational purposes. They are:

- Annex A. Product Safety Standards
- Annex B. Conductor Ampacity Under Engineering Supervision
- Annex C. Raceway Size Tables
- Annex D. Examples
- Annex E. Types of Construction
- Annex F. Cross-Reference Tables

90.4 Enforcement. This *Code* is intended to be suitable for enforcement by governmental bodies that exercise legal jurisdiction over electrical installations for power, lighting, signaling circuits, and communications systems, such as: **Figure 90–11**

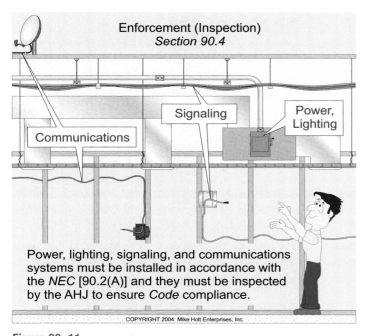

Power, lighting, signaling, and communications systems must be installed in accordance with the *NEC* [90.2(A)] and they must be inspected by the AHJ to ensure *Code* compliance.

Figure 90–11

Signaling circuits, which include:

- Article 725 Class 1, Class 2, and Class 3 Remote-Control, Signaling, and Power-Limited Circuits
- Article 760 Fire Alarm Systems
- Article 770 Optical Fiber Cables and Raceways

Communications circuits, which include:

- Article 800 Communications Circuits (twisted-pair conductors)
- Article 810 Radio and Television Equipment (satellite dish and antenna)
- Article 820 Community Antenna Television and Radio Distribution Systems (coaxial cable)
- Article 830 Network-Powered Broadband Communications Systems

> **Author's Comment:** The installation requirements for signaling circuits and communications circuits are covered in Mike's *Understanding the NEC, Volume 2* textbook.

The enforcement of the *NEC* is the responsibility of the authority having jurisdiction (AHJ), who is responsible for interpreting requirements, approving equipment and materials, waiving *Code* requirements, and ensuring that equipment is installed in accordance with listing instructions.

> **Author's Comment:** See Article 100 for the definition of "Authority Having Jurisdiction."

Interpretation of the Requirements. The authority having jurisdiction is responsible for interpreting the *NEC*, but his or her decisions must be based on a specific *Code* requirement. If an installation is rejected, the authority having jurisdiction is legally responsible for informing the installer which specific *NEC* rule was violated.

> **Author's Comment:** The art of getting along with the authority having jurisdiction consists of doing good work and knowing what the *Code* actually says (as opposed to what you only think it says). It's also useful to know how to choose your battles when the inevitable disagreement does occur.

Approval of Equipment and Materials. Only the authority having jurisdiction has authority to approve the installation of equipment and materials. Typically, the authority having jurisdiction will approve equipment listed by a product testing organization, such as Underwriters Laboratories Inc. (UL), but the *NEC* doesn't require all equipment to be listed. See 90.7, 110.2, 110.3, and the definitions in Article 100 for Approved, Identified, Labeled, and Listed. **Figure 90–12**

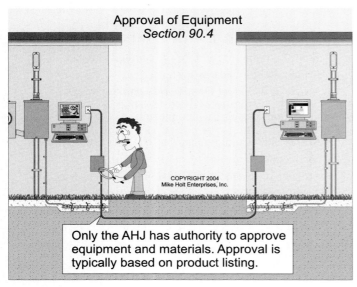

Approval of Equipment
Section 90.4

COPYRIGHT 2004
Mike Holt Enterprises, Inc.

Only the AHJ has authority to approve equipment and materials. Approval is typically based on product listing.

Figure 90–12

Author's Comment: According to the *NEC*, the authority having jurisdiction determines the approval of equipment. This means that he/she can reject an installation of listed equipment and he/she can approve the use of unlisted equipment. Given our highly litigious society, approval of unlisted equipment is becoming increasingly difficult to obtain.

Waiver of Requirements. By special permission, the authority having jurisdiction can waive specific requirements in this *Code* or permit alternative methods where it's assured that equivalent safety can be achieved and maintained.

Author's Comment: Special permission is defined in Article 100 as the written consent of the authority having jurisdiction.

Waiver of New Product Requirements. If the 2005 *NEC* requires products that aren't yet available at the time the *Code* is adopted, the authority having jurisdiction can allow products that were acceptable in the previous *Code* to continue to be used.

Author's Comment: Sometimes it takes years before testing laboratories establish product standards for new *NEC* product requirements, and then it takes time before manufacturers can design, manufacture, and distribute these products to the marketplace.

Compliance with Listing Instructions. It's the authority having jurisdiction's responsibility to ensure that electrical equipment is installed in accordance with equipment listing and/or labeling instructions [110.3(B)]. In addition, the authority having jurisdiction can reject the installation of equipment modified in the field [90.7].

Author's Comment: The *NEC* doesn't address the maintenance of electrical equipment (NFPA 70B does), because the *Code* is an installation standard, not a maintenance standard.

90.5 Mandatory Requirements and Explanatory Material.

(A) Mandatory Requirements. In the *NEC* the words "shall" or "shall not," indicate a mandatory requirement.

Author's Comment: For the ease of reading this textbook, the word "shall" has been replaced with the word "must," and the words "shall not" have been replaced with the word "cannot."

(B) Permissive Requirements. When the *Code* uses "shall be permitted" it means the identified actions are allowed but not required, and the authority having jurisdiction is not to restrict an installation from being done in that manner. A permissive rule is often an exception to the general requirement.

Author's Comment: For ease of reading, the phrase "shall be permitted" as used in the *Code*, has been replaced in this textbook with the words "is permitted."

(C) Explanatory Material. References to other standards or sections of the *NEC*, or information related to a *Code* rule, are included in the form of Fine Print Notes (FPN). Fine Print Notes are for information only and aren't intended to be enforceable.

For example, Fine Print Note No. 4 in 210.19(A)(1) <u>recommends</u> that the circuit voltage drop not exceed three percent. This isn't a requirement; it's just a recommendation.

90.6 Formal Interpretations.
To promote uniformity of interpretation and application of the provisions of the *National Electrical Code,* formal interpretation procedures have been established and are found in the NFPA Regulations Governing Committee Projects.

Author's Comment: This is rarely done because it's a very time-consuming process, and formal interpretations from the *NFPA* are not binding on the authority having jurisdiction!

90.7 Examination of Equipment for Product Safety.
Product evaluation for safety is typically performed by a testing laboratory, which publishes a list of equipment that meets a nationally recognized test standard. Products and materials listed, labeled, or identified by a testing laboratory are generally approved by the authority having jurisdiction.

Author's Comment: See Article 100 for the definition of "Approved."

Listed, factory-installed, internal wiring and construction of equipment need not be inspected at the time of installation, except to detect alterations or damage [300.1(B)]. **Figure 90–13**

90.9 Units of Measurement.

(B) Dual Systems of Units. Both the metric and inch-pound measurement systems are shown in the *NEC*, with the metric units appearing first and the inch-pound system immediately following in parentheses.

> **Author's Comment:** This is a normal practice in all *NFPA* standards, even though the U.S. construction industry uses inch-pound units of measurement.

(D) Compliance. Installing electrical systems in accordance with the metric system or the inch-pound system is considered to comply with the *Code*.

> **Author's Comment:** Since compliance with either the metric or the inch-pound system of measurement constitutes compliance with the *NEC*, this textbook uses only inch-pound units.

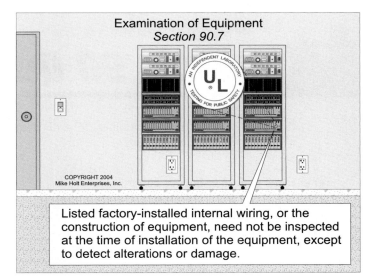

Examination of Equipment
Section 90.7

Listed factory-installed internal wiring, or the construction of equipment, need not be inspected at the time of installation of the equipment, except to detect alterations or damage.

COPYRIGHT 2004
Mike Holt Enterprises, Inc.

Figure 90–13

Article 90 Questions

(• Indicates that 75% or fewer exam takers get the question correct.)

1. The *NEC* is _____.

    (a) intended to be a design manual
    (b) meant to be used as an instruction guide for untrained persons
    (c) for the practical safeguarding of persons and property
    (d) published by the Bureau of Standards

2. •The *Code* applies to the installation of _____.

    (a) electrical conductors and equipment within or on public and private buildings
    (b) outside conductors and equipment on the premises
    (c) optical fiber cable
    (d) all of these

3. Service laterals installed by an electrical contractor must be installed in accordance with the *NEC*.

    (a) True   (b) False

4. The requirements in "Annexes" must be complied with.

    (a) True   (b) False

5. Explanatory material, such as references to other standards, references to related sections of the *NEC*, or information related to a *Code* rule, are included in the form of Fine Print Notes (FPNs).

    (a) True   (b) False

CHAPTER 1
General

Introduction

Many people skip Chapter 1 of the *NEC* because they want something prescriptive—they want something that tells them what to do, cookbook style. But electricity isn't a simple topic you can jump right into. You cannot just follow a few simple steps to get a safe installation. You need a foundation from which you can apply the *Code*.

Consider Ohm's law. Would Ohm's Law make sense to you if you did not know what an ohm was? Similarly, you must become familiar with a few basic rules, concepts, definitions, and requirements that apply to the rest of the *NEC*, and you must maintain that familiarity as you continue to apply the *Code*.

Chapter 1 consists of two main parts. Article 100 provides definitions so people can understand one another when trying to communicate on *Code* related matters. Article 110 provides general requirements that you need to know so you can correctly apply the rest of the *NEC*.

Time spent learning this general material is a great investment. After understanding Chapter 1, some of the *Code* requirements that seem confusing to other people—those who do not understand Chapter 1—will become increasingly straight forward to you. That is, they will strike you as being "common sense," because you'll have the foundation from which to understand and apply them. Because you'll understand the principles upon which many *NEC* requirements in later Chapters are based, you'll read those requirements and not be surprised at all. You'll read them and feel like you already knew them.

Article 100—Definitions. Part I of Article 100 contains the definitions of terms used throughout the *Code* for systems that operate at 600V or less. The definitions of terms in Part II apply to systems that operate at over 600V.

> **Author's Comment:** The requirements covered in this textbook apply to systems that operate at 600V or less.

Definitions of standard terms, such as volt, voltage drop, ampere, impedance, and resistance, aren't listed in Article 100. If the *NEC* doesn't define a term, then a dictionary suitable to the authority having jurisdiction should be consulted. A building code glossary might provide a better definition than a dictionary found at your home or school.

Definitions at the beginning of an article apply only to that specific article. For example, the definition of a "Swimming Pool" is contained in 680.2, because this term applies only to the requirements contained in Article 680 Swimming Pools.

Article 110—Requirements for Electrical Installations. This article contains the general requirements for electrical installations for the following:

- PART I. GENERAL
- PART II. 600V, NOMINAL, OR LESS
- PART III. OVER 600V, NOMINAL
- PART IV. TUNNEL INSTALLATIONS OVER 600V, NOMINAL
- PART V. MANHOLES AND OTHER ELECTRIC ENCLOSURES INTENDED FOR PERSONNEL ENTRY

Notes

Mike Holt Enterprises, Inc. • www.NECcode.com • 1.888.NEC.Code

Introduction

Have you ever had a conversation with someone, only to discover what you said and what he/she heard were completely different? This happens when one or more of the people in a conversation do not understand the definitions of the words being used, and that's why the definitions of key terms are located right up in the front of the *NEC,* in Article 100.

If we can all agree on important definitions, then we speak the same language and avoid misunderstandings. Because the *Code* exists to protect people and property, we can agree it's very important to know the definitions presented in Article 100.

Now, here are a couple of things you may not know about Article 100:

- Article 100 contains the definitions of many, but not all, of the terms used throughout the *NEC.* In general, only those terms used in two or more articles are defined in Article 100.
- Part I of Article 100 contains the definitions of terms used throughout the *Code*.
- Part II of Article 100 contains only terms that apply to systems that operate at over 600V.

How can you possibly learn all these definitions? There seem to be so many. Here are a few tips:

- Break the task down. Study a few words at a time, rather than trying to learn them all at one sitting.
- Review the graphics in the textbook. These will help you see how a term is applied.
- Relate them to your work. As you read a word, think of how it applies to the work you're doing. This will provide a natural reinforcement of the learning process.

Definitions.

Accessible as it Applies to Equipment. Admitting close approach and not guarded by locked doors, elevation, or other effective means.

Accessible as it Applies to Wiring Methods. Not permanently closed in by the building structure or finish and capable of being removed or exposed without damaging the building structure or finish. **Figure 100–1**

Author's Comments:

- Conductors in a concealed raceway are considered concealed, even though they may become accessible by withdrawing them. See the definition of "Concealed" in this article.
- Raceways, cables, and enclosures installed above a suspended ceiling or within a raised floor are considered accessible, because the wiring methods can be accessed without damaging the building structure. See "Concealed" and "Exposed."

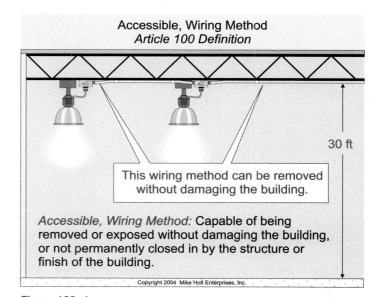

Accessible, Wiring Method
Article 100 Definition

30 ft

This wiring method can be removed without damaging the building.

Accessible, Wiring Method: Capable of being removed or exposed without damaging the building, or not permanently closed in by the structure or finish of the building.

Copyright 2004 Mike Holt Enterprises, Inc.

Figure 100–1

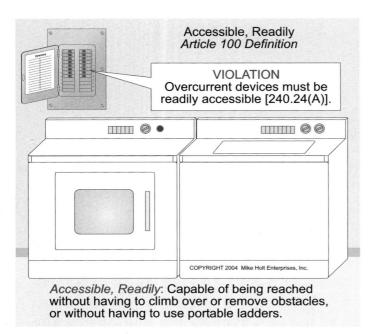

Accessible, Readily
Article 100 Definition

VIOLATION
Overcurrent devices must be
readily accessible [240.24(A)].

COPYRIGHT 2004 Mike Holt Enterprises, Inc.

Accessible, Readily: Capable of being reached
without having to climb over or remove obstacles,
or without having to use portable ladders.

Figure 100–2

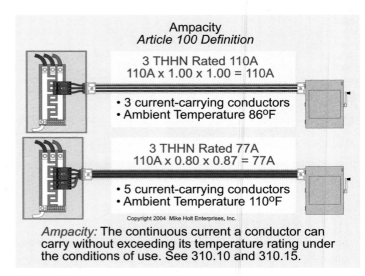

Ampacity
Article 100 Definition

3 THHN Rated 110A
110A x 1.00 x 1.00 = 110A

• 3 current-carrying conductors
• Ambient Temperature 86ºF

3 THHN Rated 77A
110A x 0.80 x 0.87 = 77A

• 5 current-carrying conductors
• Ambient Temperature 110ºF

Copyright 2004 Mike Holt Enterprises, Inc.

Ampacity: The continuous current a conductor can
carry without exceeding its temperature rating under
the conditions of use. See 310.10 and 310.15.

Figure 100–4

Accessible, Readily (Readily Accessible). Capable of being reached quickly without having to climb over or remove obstacles or resort to portable ladders. **Figures 100–2 and 100–3**

Ampacity. The current in amperes a conductor can carry continuously, where the temperature will not be raised in excess of the conductor's insulation temperature rating. See 310.10 and 310.15 for details and examples. **Figure 100–4**

Appliance [Article 424]. Electrical equipment, other than industrial equipment, built in standardized sizes, such as ranges, ovens, cooktops, refrigerators, drinking water coolers, or beverage dispensers.

Approved. Acceptable to the authority having jurisdiction, usually the electrical inspector.

Author's Comment: Product listing doesn't mean that the product is approved, but it's a basis for approval. See 90.4, 90.7, 110.2, and the definitions in Article 100 for Authority Having Jurisdiction, Identified, Labeled, and Listed.

Attachment Plug (Plug Cap)(Plug). A wiring device at the end of flexible cord intended to be inserted into a receptacle. **Figure 100–5**

Author's Comment: The use of a cord with an attachment plug is limited by 210.50(A), 400.7, 410.14, 410.30, 422.33, 590.4, and 645.5.

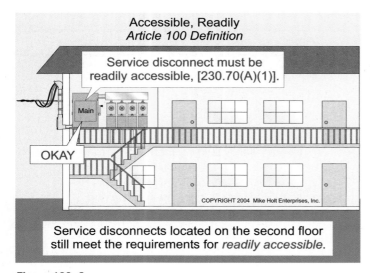

Accessible, Readily
Article 100 Definition

Service disconnect must be
readily accessible, [230.70(A)(1)].

Main

OKAY

COPYRIGHT 2004 Mike Holt Enterprises, Inc.

Service disconnects located on the second floor
still meet the requirements for *readily accessible*.

Figure 100–3

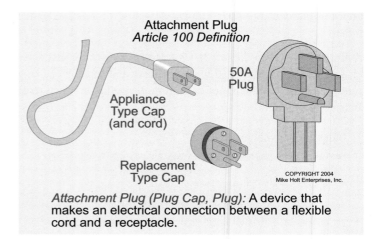

Attachment Plug
Article 100 Definition

50A
Plug

Appliance
Type Cap
(and cord)

Replacement
Type Cap

COPYRIGHT 2004
Mike Holt Enterprises, Inc.

Attachment Plug (Plug Cap, Plug): A device that
makes an electrical connection between a flexible
cord and a receptacle.

Figure 100–5

Authority Having Jurisdiction (AHJ). The organization, office, or individual that is responsible for approving equipment, materials, an installation, or a procedure. See 90.4, 90.7, and 110.2 for more information.

FPN: The authority having jurisdiction may be a federal, state, or local government, or an individual such as a fire chief, fire marshal, chief of a fire prevention bureau or labor department or health department, a building official or electrical inspector, or others having statutory authority. In some circumstances, the property owner or his/her agent assumes the role, and at government installations, the commanding officer, or departmental official may be the authority having jurisdiction.

Author's Comments:

- Typically, the authority having jurisdiction will be the electrical inspector who has legal statutory authority. In the absence of federal, state, or local regulations, the operator of the facility or his/her agent, such as an architect or engineer of the facility, can assume the role.

- Many feel that the authority having jurisdiction should have a strong background in the electrical field, such as having studied electrical engineering or having obtained an electrical contractor's license, and in a few states this is a legal requirement. Memberships, certifications, and active participation in electrical organizations, such as the IAEI (www.IAEI.org), speak to an individual's qualifications.

Bathroom. A bathroom is an area that includes a basin with a toilet, tub, or shower. **Figure 100–6**

Author's Comment: All 15A and 20A, 125V receptacles located in bathrooms must be GFCI protected [210.8].

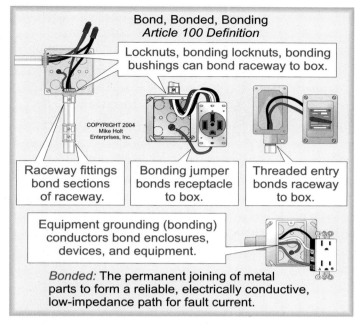

Bond, Bonded, Bonding
Article 100 Definition

Locknuts, bonding locknuts, bonding bushings can bond raceway to box.

COPYRIGHT 2004 Mike Holt Enterprises, Inc.

Raceway fittings bond sections of raceway.

Bonding jumper bonds receptacle to box.

Threaded entry bonds raceway to box.

Equipment grounding (bonding) conductors bond enclosures, devices, and equipment.

Bonded: The permanent joining of metal parts to form a reliable, electrically conductive, low-impedance path for fault current.

Figure 100–7

Bonding (Bond) (Bonded). The permanent joining of metallic parts together to form an electrically conductive path. Such a path must have the capacity to conduct safely any fault current likely to be imposed on it. **Figure 100–7**

Author's Comment: Bonding is accomplished by the use of bonding conductors, metallic raceways and cables, connectors, couplings, or other devices listed for this purpose [250.8, 250.118, and 300.10].

Bonding Jumper. A reliable conductor that is properly sized in accordance with Article 250, to ensure electrical conductivity between metal parts of the electrical installation. **Figure 100–8**

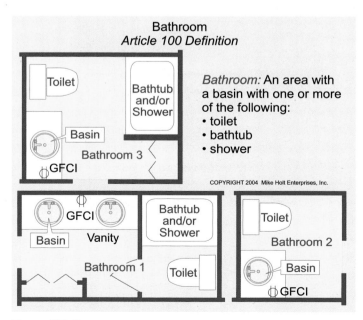

Bathroom
Article 100 Definition

Toilet

Bathtub and/or Shower

Basin

Bathroom 3

GFCI

Bathroom: An area with a basin with one or more of the following:
- toilet
- bathtub
- shower

COPYRIGHT 2004 Mike Holt Enterprises, Inc.

GFCI

Basin   Vanity

Bathtub and/or Shower

Toilet

Bathroom 2

Bathroom 1   Toilet

Basin

GFCI

Figure 100–6

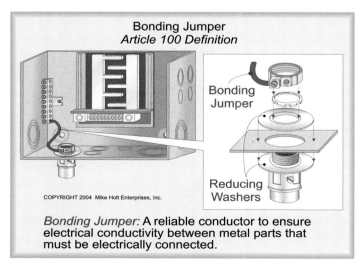

Bonding Jumper
Article 100 Definition

Bonding Jumper

Reducing Washers

COPYRIGHT 2004 Mike Holt Enterprises, Inc.

Bonding Jumper: A reliable conductor to ensure electrical conductivity between metal parts that must be electrically connected.

Figure 100–8

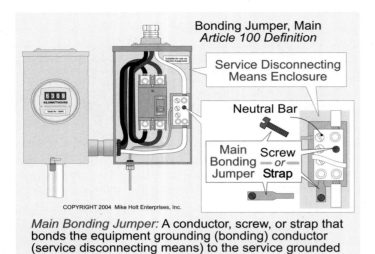

Bonding Jumper, Main
Article 100 Definition

Service Disconnecting Means Enclosure

Neutral Bar

Main Bonding Jumper

Screw *or* Strap

Main Bonding Jumper: A conductor, screw, or strap that bonds the equipment grounding (bonding) conductor (service disconnecting means) to the service grounded neutral conductor [250.24(B), 250.28].

Figure 100–9

Bonding Jumper, Main. A conductor, screw, or strap that bonds the equipment grounding (bonding) conductor (service disconnecting means) to the grounded neutral conductor in accordance with 250.24(B). For more details, see 250.24(A)(4), 250.28, and 408.3(C). **Figure 100–9**

Bonding Jumper, System. The conductor, screw, or strap that bonds the metal parts of a separately derived system to a system winding in accordance with 250.30(A)(1). **Figure 100–10**

> **Author's Comment:** The system bonding jumper provides the low-impedance fault-current path to the power source to facilitate the clearing of a ground fault by opening the circuit protection device. For more information, see 250.4(A)(5), 250.28, and 250.30(A)(1).

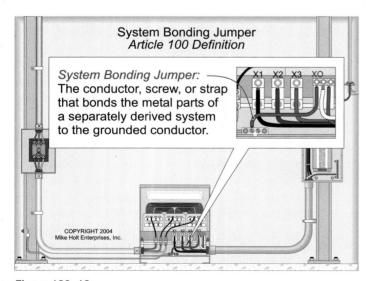

System Bonding Jumper
Article 100 Definition

System Bonding Jumper: The conductor, screw, or strap that bonds the metal parts of a separately derived system to the grounded conductor.

X1   X2   X3   XO

Figure 100–10

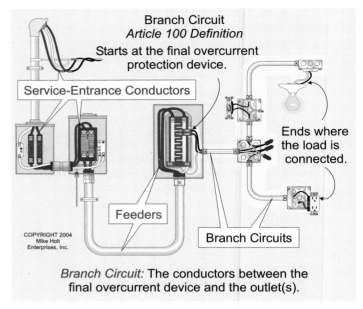

Branch Circuit
Article 100 Definition
Starts at the final overcurrent protection device.

Service-Entrance Conductors

Ends where the load is connected.

Feeders

Branch Circuits

Branch Circuit: The conductors between the final overcurrent device and the outlet(s).

Figure 100–11

Branch Circuit [Article 210]. The conductors between the final overcurrent device and the receptacle outlets, lighting outlets, or other outlets as defined in Article 100. **Figure 100–11**

Branch Circuit, Multiwire. A branch circuit that consists of two or more ungrounded circuit conductors with a common grounded neutral conductor. There must be a voltage potential between the ungrounded conductors and an equal voltage potential from each ungrounded conductor to the grounded neutral conductor. **Figure 100-12**

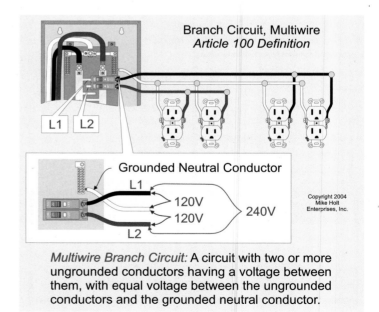

Branch Circuit, Multiwire
Article 100 Definition

L1   L2

Grounded Neutral Conductor

L1

120V
120V

240V

L2

Multiwire Branch Circuit: A circuit with two or more ungrounded conductors having a voltage between them, with equal voltage between the ungrounded conductors and the grounded neutral conductor.

Figure 100–12

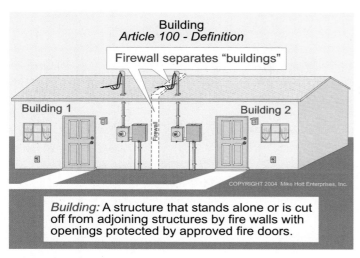

Building
Article 100 - Definition

Firewall separates "buildings"

Building 1   Building 2

COPYRIGHT 2004 Mike Holt Enterprises, Inc.

Building: A structure that stands alone or is cut off from adjoining structures by fire walls with openings protected by approved fire doors.

Figure 100–13

Author's Comment: Multiwire branch circuits offer the advantage of fewer conductors in a raceway, smaller raceway sizing, and a reduction of material and labor costs. In addition, multiwire branch circuits can reduce circuit voltage drop by as much as 50 percent. However, because of the dangers associated with multiwire branch circuits, the *NEC* contains additional requirements to ensure a safe installation. See 210.4, 300.13(B), and 408.40 for additional details.

Building. A structure that stands alone or is cut off from other structures by firewalls with all openings protected by fire doors that are approved by the authority having jurisdiction. **Figure 100–13**

Cabinet [Article 312]. An enclosure for either surface mounting or flush mounting provided with a frame in which a door can be hung. **Figure 100–14**

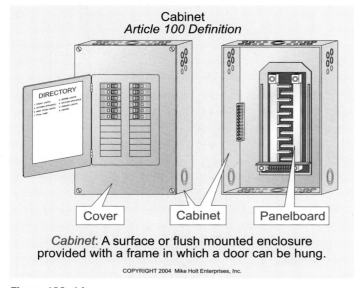

Cabinet
Article 100 Definition

DIRECTORY

Cover   Cabinet   Panelboard

Cabinet: A surface or flush mounted enclosure provided with a frame in which a door can be hung.

COPYRIGHT 2004 Mike Holt Enterprises, Inc.

Figure 100–14

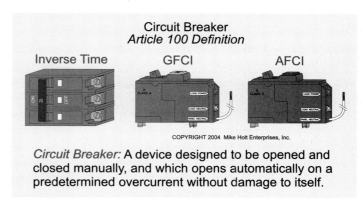

Circuit Breaker
Article 100 Definition

Inverse Time   GFCI   AFCI

COPYRIGHT 2004 Mike Holt Enterprises, Inc.

Circuit Breaker: A device designed to be opened and closed manually, and which opens automatically on a predetermined overcurrent without damage to itself.

Figure 100–15

Circuit Breaker. A device designed to be opened and closed manually, and which opens automatically on a predetermined overcurrent without damage to itself. Circuit breakers are available in different configurations, such as inverse time molded case, adjustable (electronically controlled), and instantaneous trip/motor circuit protectors. **Figure 100–15**

• *Inverse Time:* Inverse-time breakers operate on the principle that as the current increases, the time it takes for the devices to open decreases. This type of breaker provides overcurrent protection (overload, short circuit, and ground fault).

• *Adjustable Trip:* Adjustable-trip breakers permit the thermal trip setting to be adjusted. The adjustment is often necessary to coordinate the operation of the circuit breakers with other overcurrent protection devices.

Author's Comment: Coordination means that the devices with the lowest ratings, closest to the fault, operate and isolate the fault and disruption, if possible, so that the rest of the system can remain energized and functional.

• *Instantaneous Trip:* Instantaneous-trip breakers operate on the principle of electromagnetism only and are used for motors; sometimes these devices are called motor short-circuit protectors (MCPs). This type of protection device doesn't provide overload protection. It only provides short-circuit and ground-fault protection; overload protection must be provided separately.

Author's Comment: Instantaneous-trip circuit breakers have no intentional time delay and are sensitive to current inrush, and to vibration and shock. Consequently, they should not be used where these factors are known to exist.

Concealed. Rendered inaccessible by the structure or finish of the building. Conductors in a concealed raceway are considered concealed, even though they may become accessible by withdrawing them. **Figure 100–16**

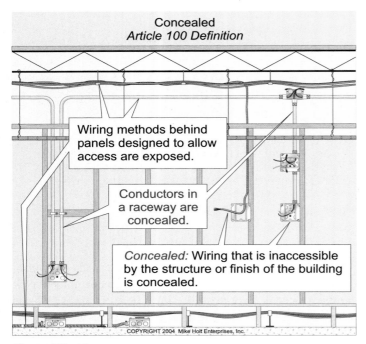

Figure 100–16

Author's Comment: Wiring behind panels that are designed to allow access is considered exposed.

Conduit Body. A fitting that provides access to conductors through a removable cover. **Figure 100–17**

Connector, Pressure (Solderless). A device that establishes a conductive connection between conductors and a terminal by the means of mechanical pressure.

Continuous Load. A load where the current is expected to exist for three hours or more, such as store or parking lot lighting.

Controller. A device that controls, in some predetermined manner, the electric power delivered to electrical equipment. This includes time clocks, lighting contactors, photocells, etc. **Figure 100–18**

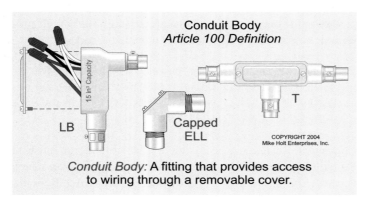

Conduit Body: A fitting that provides access to wiring through a removable cover.

Figure 100–17

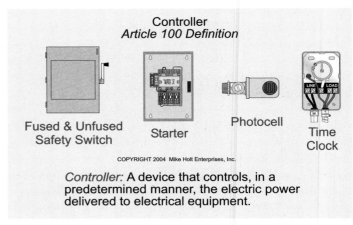

Controller: A device that controls, in a predetermined manner, the electric power delivered to electrical equipment.

Figure 100–18

Coordination (Selective). Localization of an overcurrent condition to restrict outages to the circuit or equipment affected, accomplished by the choice of overcurrent protective devices.

Author's Comment: Selective coordination is required for:

- Orderly Shutdown, 240.12
- Motors, 430.52(C)(3)
- Elevators, 620.62
- Fire Pumps, 695.5(C)(2)
- Emergency Power Systems, 700.27
- Legally Required Standby Power Systems, 701.18

Selective coordination means the circuit protection scheme confines the interruption to a particular area rather than to the whole system. For example, if someone plugs in a space heater and raises total demand on a 20A circuit to 25A, or if a short circuit or ground fault occurs with selective coordination, the only breaker/fuse that will open is the one protecting just that branch circuit. Without selective coordination, an entire floor of a building could go dark!

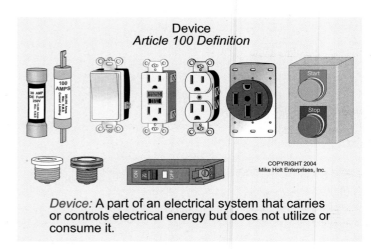

Device: A part of an electrical system that carries or controls electrical energy but does not utilize or consume it.

Figure 100–19

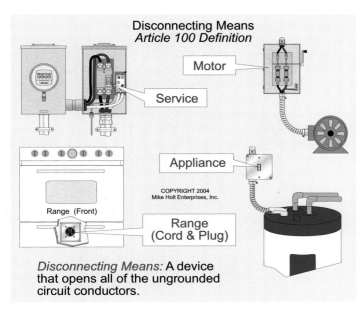

Disconnecting Means
Article 100 Definition

Motor

Service

Appliance

COPYRIGHT 2004
Mike Holt Enterprises, Inc.

Range (Front)

Range
(Cord & Plug)

Disconnecting Means: A device
that opens all of the ungrounded
circuit conductors.

Figure 100–20

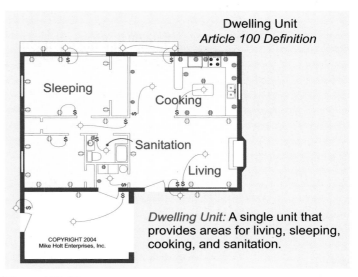

Dwelling Unit
Article 100 Definition

Sleeping

Cooking

Sanitation

Living

COPYRIGHT 2004
Mike Holt Enterprises, Inc.

Dwelling Unit: A single unit that
provides areas for living, sleeping,
cooking, and sanitation.

Figure 100–21

Cutout Box. Cutout boxes are designed for surface mounting with a swinging door.

Device. A component of an electrical installation that is intended to carry or control, but not consume electrical energy.

> **Author's Comment:** Devices include receptacles, switches, circuit breakers, fuses, time clocks, controllers, etc., but not locknuts or other mechanical fittings. **Figure 100–19**

Disconnecting Means. A device that opens all of the ungrounded circuit conductors from their power source. These include switches, attachment plugs and receptacles, and circuit breakers. **Figure 100–20**

> **Author's Comment:** Review the following for the specific requirements for equipment disconnecting means:
>
> - Air Conditioning and Refrigeration, 440.14
> - Appliances, Article 422, Part III
> - Building supplied by a feeder, Article 225, Part II
> - Electric space heating, 424.19
> - Electric duct heaters, 424.65
> - Motor control conductors, 430.74
> - Motor controllers, 430.102(A)
> - Motors, 430.102(B)
> - Refrigeration equipment, 440.14
> - Services, Article 230, Part VI
> - Swimming pool, spa, hot tub, and fountain equipment, 680.12

Dwelling Unit. A single unit that provides independent living facilities for persons, including permanent provisions for living, sleeping, cooking, and sanitation. **Figure 100–21**

Dwelling, Multifamily. A building that contains three or more dwelling units.

Energized. Electrically connected to, or is, a source of voltage.

Exposed (Wiring Methods). On, or attached to the surface of a building, or behind panels designed to allow access.

> **Author's Comment:** An example is wiring located in the space above a suspended ceiling or below a raised floor. **Figure 100–22**

Feeder. The conductors between the service equipment, the source of a separately derived system, or other power source and the final branch-circuit overcurrent device. **Figure 100–23**

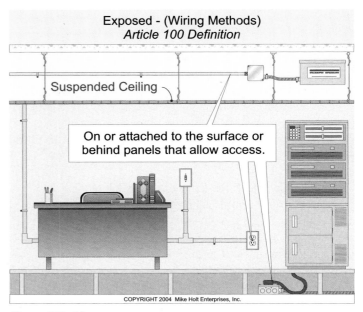

Exposed - (Wiring Methods)
Article 100 Definition

Suspended Ceiling

On or attached to the surface or
behind panels that allow access.

COPYRIGHT 2004 Mike Holt Enterprises, Inc.

Figure 100–22

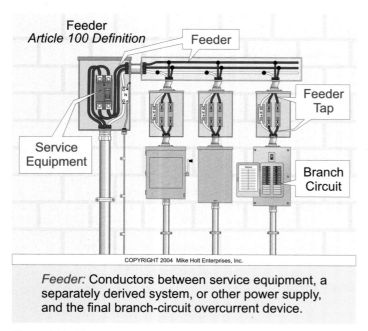

Feeder
Article 100 Definition
Feeder
Feeder Tap
Service Equipment
Branch Circuit

COPYRIGHT 2004 Mike Holt Enterprises, Inc.

Feeder: Conductors between service equipment, a separately derived system, or other power supply, and the final branch-circuit overcurrent device.

Figure 100–23

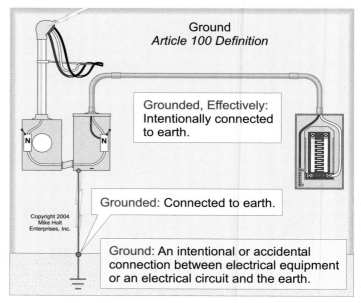

Ground
Article 100 Definition

Grounded, Effectively: Intentionally connected to earth.

Grounded: Connected to earth.

Ground: An intentional or accidental connection between electrical equipment or an electrical circuit and the earth.

Copyright 2004 Mike Holt Enterprises, Inc.

Figure 100–24

Author's Comments:

- An "other power source" would include a solar energy system (photovoltaic or PV).

- To have a better understanding of what a feeder is, be sure to review the definitions of service equipment and separately derived systems.

Fitting. An accessory, such as a locknut, that is intended to perform a mechanical function.

Garage. A building or portion of a building where self-propelled vehicles can be kept.

Author's Comment: Receptacles can be installed at any height in a dwelling unit garage, but no less than 18 in. above the floor for a commercial garage [511.3(A)(5)], unless they are listed as explosionproof.

Ground. An intentional or accidental connection to the earth. **Figure 100–24**

Author's Comment: The *NEC* also defines this term as "connection to some conducting body that serves in place of the earth," which leads to much confusion.

Grounded. Connected to earth. **See Figure 100–24.**

Author's Comment: The *NEC* also defines this term as "connected to some conducting body that serves in place of the earth." However, nobody really knows what this means.

Effectively Grounded. Intentional connection to earth through a conductor of sufficiently low impedance.

Grounded, Solidly. The intentional electrical connection of one system terminal to ground.

Author's Comment: In reality, solidly grounded means the bonding of the system to the metal case of the derived system in accordance with 250.30(A)(1). **Figure 100–25**

Grounded Neutral Conductor. The conductor that is intentionally grounded to the earth. See Article 200 in this textbook for additional details.

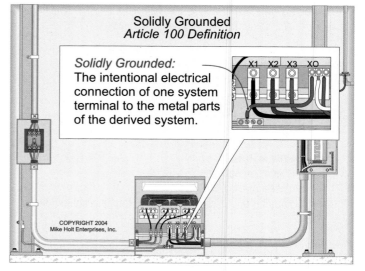

Solidly Grounded
Article 100 Definition

Solidly Grounded: The intentional electrical connection of one system terminal to the metal parts of the derived system.

X1 X2 X3 XO

COPYRIGHT 2004 Mike Holt Enterprises, Inc.

Figure 100–25

Mike Holt Enterprises, Inc. • www.NECcode.com • 1.888.NEC.Code

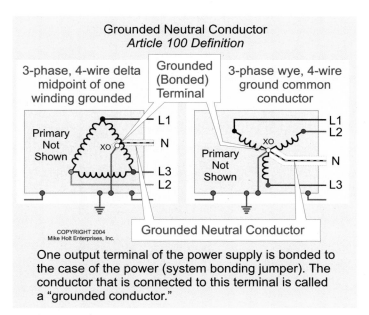

Grounded Neutral Conductor
Article 100 Definition

3-phase, 4-wire delta midpoint of one winding grounded

Grounded (Bonded) Terminal

3-phase wye, 4-wire ground common conductor

COPYRIGHT 2004
Mike Holt Enterprises, Inc.

Grounded Neutral Conductor

One output terminal of the power supply is bonded to the case of the power (system bonding jumper). The conductor that is connected to this terminal is called a "grounded conductor."

Figure 100–26

Author's Comment: In reality, one output terminal of a power supply is bonded to the case of the power supply (system bonding jumper). The conductor that is connected to this grounded terminal is called a "grounded conductor." **Figure 100–26**

Grounding (Earthing) Conductor. A conductor used to connect equipment to a grounding (earthing) electrode.

Author's Comment: A grounding (earthing) conductor is often used to connect the metal parts of electrical equipment to a supplementary grounding electrode [250.54]. This actually serves no performance purpose, but some equipment manufacturers require this connection in their installation instructions. **Figure 100–27**

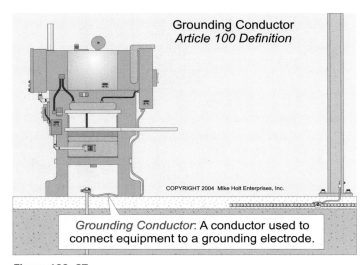

Grounding Conductor
Article 100 Definition

COPYRIGHT 2004 Mike Holt Enterprises, Inc.

Grounding Conductor: A conductor used to connect equipment to a grounding electrode.

Figure 100–27

Grounding (Bonding) Conductor, Equipment. The low-impedance fault-current path used to connect the noncurrent-carrying metal parts of equipment, raceways, and other metal enclosures to the grounded neutral conductor at service equipment or at the source of a separately derived system.

Author's Comments:

- The equipment grounding (bonding) conductor actually serves as the "effective ground-fault current path" as defined in 250.2. Its purpose is to provide the low-impedance fault-current path necessary to facilitate the operation of overcurrent protection devices, and to remove dangerous voltage potentials between conductive parts of building components and electrical systems [250.4(A)(3)]. Because this is actually a bonding conductor, not a grounding (earthing) conductor, we will identify this conductor in this textbook as the equipment grounding (bonding) conductor.

- According to 250.118, the equipment grounding (bonding) conductor must be one or a combination of the following: **Figure 100–28**

  - A bare or insulated conductor
  - Rigid Metal Conduit
  - Intermediate Metal Conduit
  - Electrical Metallic Tubing
  - Listed Flexible Metal Conduit as limited by 250.118(5)
  - Listed Liquidtight Flexible Metal Conduit as limited by 250.118(6)
  - Type AC Armored Cable
  - Copper metal sheath of Mineral Insulated Cable
  - Metal Clad Cable as limited by 250.118(10)

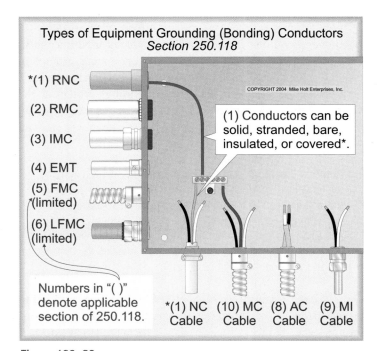

Types of Equipment Grounding (Bonding) Conductors
Section 250.118

COPYRIGHT 2004 Mike Holt Enterprises, Inc.

*(1) RNC

(2) RMC

(3) IMC

(4) EMT

(5) FMC
(limited)

(6) LFMC
(limited)

(1) Conductors can be solid, stranded, bare, insulated, or covered*.

Numbers in "( )" denote applicable section of 250.118.

*(1) NC Cable   (10) MC Cable   (8) AC Cable   (9) MI Cable

Figure 100–28

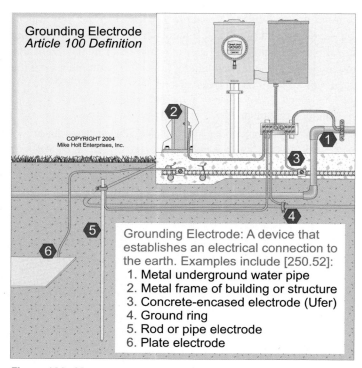

Grounding Electrode
Article 100 Definition

COPYRIGHT 2004
Mike Holt Enterprises, Inc.

Grounding Electrode: A device that establishes an electrical connection to the earth. Examples include [250.52]:
1. Metal underground water pipe
2. Metal frame of building or structure
3. Concrete-encased electrode (Ufer)
4. Ground ring
5. Rod or pipe electrode
6. Plate electrode

Figure 100–29

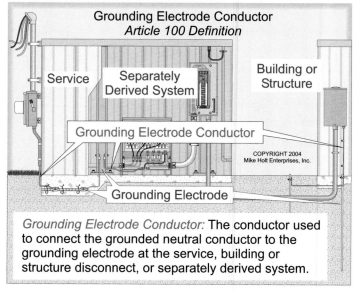

Grounding Electrode Conductor
Article 100 Definition

Service | Separately Derived System | Building or Structure

Grounding Electrode Conductor

COPYRIGHT 2004
Mike Holt Enterprises, Inc.

Grounding Electrode

Grounding Electrode Conductor: The conductor used to connect the grounded neutral conductor to the grounding electrode at the service, building or structure disconnect, or separately derived system.

Figure 100–30

- Metallic cable trays as limited by 250.118(11) and 392.7
- Electrically continuous metal raceways listed for grounding
- Surface Metal Raceways listed for grounding.

Grounding (Earthing) Electrode. A device that establishes an electrical connection to the earth. See 250.50 through 250.70. Figure 100–29

Grounding Electrode Conductor. The conductor used to connect the grounded neutral conductor or the equipment grounding (bonding) conductor (metal parts of the disconnecting means), or both, to the grounding electrode (earthing) system at the service [250.24(A)], at each building or structure supplied by feeder(s) [250.32(A)], or the source of a separately derived system [250.30(A)]. Figure 100–30

Author's Comment: At a service or separately derived system, the grounding electrode conductor connects the grounded neutral conductor and the equipment grounding (bonding) conductor to the grounding electrode (earthing) system [250.24(D) and 250.30]. At a separate building, the grounding electrode conductor connects the metal parts of the building disconnect to the grounding electrode [250.32(A)].

Ground-Fault Circuit Interrupter (GFCI). A device intended to protect people by de-energizing a circuit when the current-to-ground exceeds the value established for a "Class A" device. Figure 100–31

FPN: A "Class A" ground-fault circuit interrupter opens the circuit when the current-to-ground has a value between 4 mA and 6 mA.

Author's Comment: A GFCI operates on the principle of monitoring the unbalanced current between the ungrounded and grounded neutral conductor. GFCI protective devices are commercially available in receptacles, circuit breakers, cord sets, and other types of devices. Figure 100–32

Ground-Fault Protection of Equipment. A system intended to provide protection of equipment from damaging ground-fault currents by opening all ungrounded conductors of the faulted circuit.

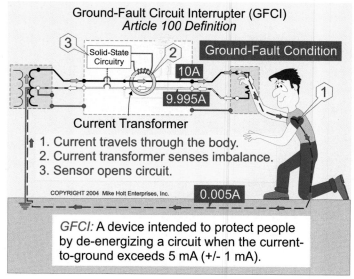

Ground-Fault Circuit Interrupter (GFCI)
Article 100 Definition

Solid-State Circuitry

Ground-Fault Condition

10A

9.995A

Current Transformer
1. Current travels through the body.
2. Current transformer senses imbalance.
3. Sensor opens circuit.

COPYRIGHT 2004 Mike Holt Enterprises, Inc.

0.005A

GFCI: A device intended to protect people by de-energizing a circuit when the current-to-ground exceeds 5 mA (+/- 1 mA).

Figure 100–31

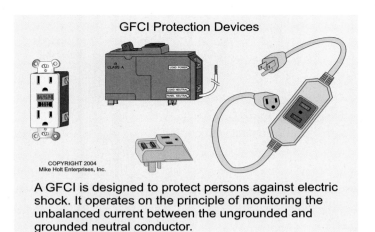

GFCI Protection Devices

A GFCI is designed to protect persons against electric shock. It operates on the principle of monitoring the unbalanced current between the ungrounded and grounded neutral conductor.

Figure 100–32

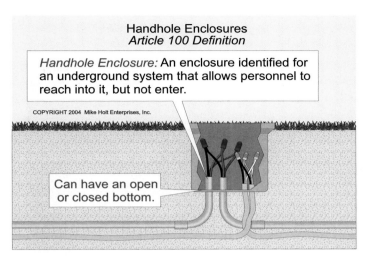

Handhole Enclosures
Article 100 Definition

Handhole Enclosure: An enclosure identified for an underground system that allows personnel to reach into it, but not enter.

Can have an open or closed bottom.

Figure 100–34

This protection is provided at current levels less than those required to protect conductors from damage through the operation of a supply circuit overcurrent device. See 215.10, 230.95, and 240.13.

Author's Comment: This type of protective device isn't intended to protect people, only connected utilization equipment.

Guest Room. An accommodation that combines living, sleeping, sanitary, and storage facilities. **Figure 100–33**

Guest Suite. An accommodation with two or more contiguous rooms comprising a compartment, with or without doors between such rooms, that provides living, sleeping, sanitary, and storage facilities.

Handhole Enclosure. An enclosure identified for underground system use, provided with an open or closed bottom, and sized to allow personnel to reach into, but not enter, for the purpose of installing or maintaining equipment or wiring. **Figure 100–34**

Author's Comment: See 314.30 for the installation requirements for handhole enclosures.

Identified Equipment. Recognized as suitable for a specific purpose, function, or environment by listing and labeling. See 90.4, 90.7, 110.3(A)(1), and the definitions in Article 100 for Approved, Labeled, and Listed.

In Sight From (Within Sight). Visible and not more than 50 ft distant from the equipment. **Figure 100–35**

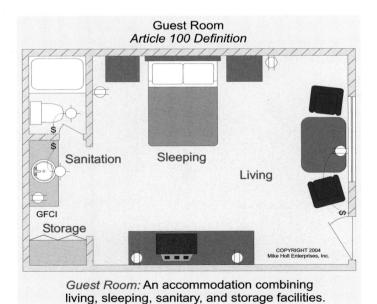

Guest Room
Article 100 Definition

Sanitation

Sleeping

Living

GFCI

Storage

Guest Room: An accommodation combining living, sleeping, sanitary, and storage facilities.

Figure 100–33

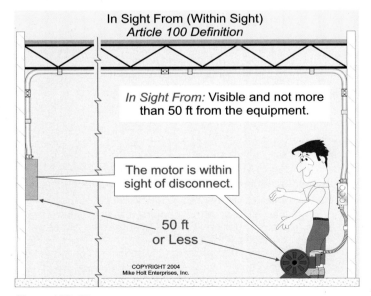

In Sight From (Within Sight)
Article 100 Definition

In Sight From: Visible and not more than 50 ft from the equipment.

The motor is within sight of disconnect.

50 ft or Less

Figure 100–35

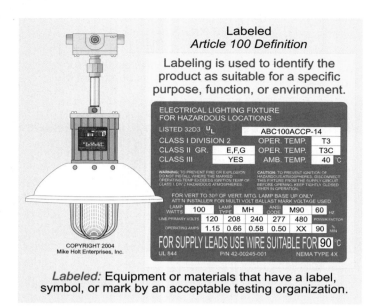

Labeled
Article 100 Definition

Labeling is used to identify the product as suitable for a specific purpose, function, or environment.

Labeled: Equipment or materials that have a label, symbol, or mark by an acceptable testing organization.

Figure 100–36

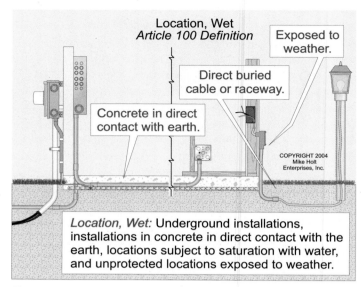

Location, Wet
Article 100 Definition

Location, Wet: Underground installations, installations in concrete in direct contact with the earth, locations subject to saturation with water, and unprotected locations exposed to weather.

Figure 100–38

Interrupting Rating. The highest short-circuit current at rated voltage that the device can safely interrupt. For more information, see 110.9 in this textbook.

Labeled. Equipment or materials that have a label, symbol, or other identifying mark in the form of a sticker, decal, or printed label, or molded or stamped into the product by a testing laboratory acceptable to the authority having jurisdiction. See Identified and Listed. Figure 100–36

> **Author's Comment:** Labeling and listing of equipment typically provides the basis for equipment approval by the authority having jurisdiction. See 90.4, 90.7, 110.2, and 110.3 for more information.

Lighting Outlet. An outlet for the connection of a lampholder, luminaire, or lampholder pendant cord. Figure 100–37

Listed. Equipment or materials included in a list published by a testing laboratory that is acceptable to the authority having jurisdiction. The listing organization must periodically inspect the production of listed equipment or material to ensure that the equipment or material meets appropriate designated standards and is suitable for a specified purpose. See Identified and Labeled.

> **Author's Comment:** The *NEC* doesn't require all electrical equipment to be listed, but some *Code* requirements do specifically require product listing. Increasingly, organizations such as OSHA require that listed equipment be used when such equipment is available. See 90.7, 110.2, and 110.3 for more information.

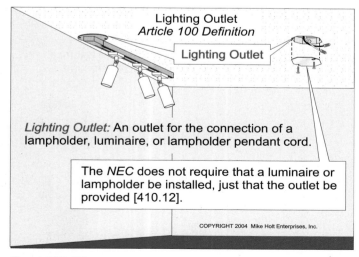

Lighting Outlet
Article 100 Definition

Lighting Outlet

Lighting Outlet: An outlet for the connection of a lampholder, luminaire, or lampholder pendant cord.

The *NEC* does not require that a luminaire or lampholder be installed, just that the outlet be provided [410.12].

Figure 100–37

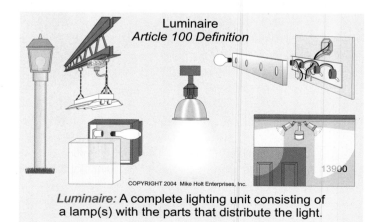

Luminaire
Article 100 Definition

Luminaire: A complete lighting unit consisting of a lamp(s) with the parts that distribute the light.

Figure 100–39

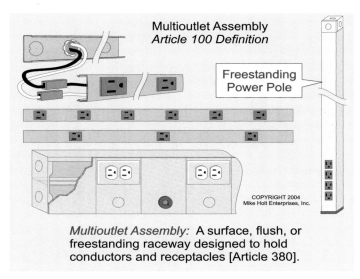

Multioutlet Assembly
Article 100 Definition

Freestanding
Power Pole

COPYRIGHT 2004
Mike Holt Enterprises, Inc.

Multioutlet Assembly: A surface, flush, or
freestanding raceway designed to hold
conductors and receptacles [Article 380].

Figure 100–40

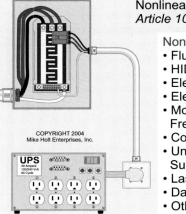

Nonlinear Load Examples
Article 100 Definition FPN

COPYRIGHT 2004
Mike Holt Enterprises, Inc.

UPS
40 Ampere
120/240 Volt
60 Cycle

Nonlinear loads can include:
• Fluorescent Lighting
• HID Lighting
• Electronic Ballasts
• Electronic Dimmers
• Motors with Variable
  Frequency Drives
• Computers
• Uninterrupted Power
  Supplies (UPS)
• Laser Printers
• Data-Processing Equipment
• Other Electronic Equipment

Figure 100–42

Location, Damp. Locations protected from weather and not subject to saturation with water or other liquids. This includes locations partially protected under canopies, marquees, roofed open porches, and interior locations subject to moderate degrees of moisture, such as some basements, barns, and cold-storage warehouses.

Location, Dry. An area not normally subjected to dampness or wetness, but which may temporarily be subject to dampness or wetness, such as a building under construction.

Location, Wet. Underground installations, installations in concrete in direct contact with the earth, locations subject to saturation with water, and unprotected locations exposed to weather. Figure 100–38

Luminaire. A complete lighting unit that consists of a lamp or lamps together with the parts designed to distribute the light. Figure 100–39

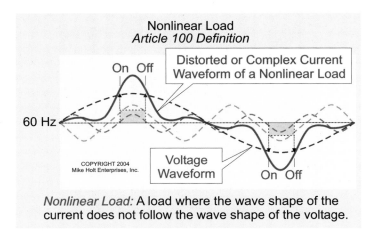

Nonlinear Load
Article 100 Definition

On Off

Distorted or Complex Current
Waveform of a Nonlinear Load

60 Hz

COPYRIGHT 2004
Mike Holt Enterprises, Inc.

Voltage
Waveform

On Off

Nonlinear Load: A load where the wave shape of the current does not follow the wave shape of the voltage.

Figure 100–41

Multioutlet Assembly. A surface, flush, or freestanding raceway designed to hold conductors and receptacles. See Article 380 for additional details. Figure 100–40

Nonlinear Load. A load where the current waveform doesn't follow the applied sinusoidal voltage waveform. See 210.4(A) FPN, 220.61(C)(2) FPN 2, 310.15(B)(4)(c), and 450.3 FPN 2. Figure 100–41

> **FPN:** Single-phase nonlinear loads include electronic equipment, such as copy machines, laser printers, and electric-discharge lighting. Three-phase nonlinear loads include uninterruptible power supplies (UPSs), induction motors, and electronic switching devices, such as variable-frequency and variable-speed drives (VFD-VSD). Figure 100–42

Author's Comment: The subject of nonlinear loads is outside the scope of this textbook. For more information on this topic, visit http://www.MikeHolt.com/news/archive/html/master/necharmonics.htm

Outlet. A point in the wiring system where electric current is taken to supply a load, such as receptacle(s), luminaire(s), and equipment. Figure 100–43

Outline Lighting. An arrangement of incandescent lamps, electric-discharge lighting, or other electrically powered light sources to outline or call attention to certain features such as the shape of a building or the decoration of a window.

Overcurrent. Current in amperes that is greater than the rated current of the equipment or conductors, resulting from an overload, short circuit, or ground fault. Figure 100–44

Author's Comment: See the definitions of "Ground Fault" in 250.2 and "Overload" in Article 100.

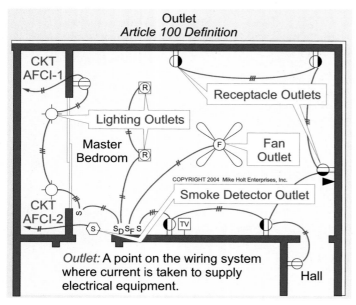

Figure 100–43

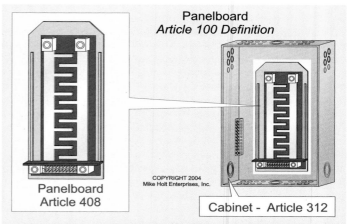

Panelboard: A distribution point containing protection devices, which is designed to be placed in a cabinet.

Figure 100–45

Author's Comments:

- See the definition of "Cabinet" in this article.
- The slang term in the electrical field for a panelboard is "the guts."

Plenum. A compartment or chamber to which one or more ducts are connected and that forms part of the air distribution system.

Premises Wiring. The interior and exterior wiring, including power, lighting, control, and signal circuits, and all associated hardware, fittings, and wiring devices, both permanently and temporarily installed. This doesn't include the internal wiring of electrical equipment and appliances, such as luminaires, dishwashers, water heaters, motors, controllers, motor control centers, A/C equipment, etc. See 90.7 and 300.1(B).

Overload. The operation of equipment above its ampere current rating or current in excess of conductor ampacity. When an overload condition persists for a sufficient length of time, it could result in equipment failure or a fire from damaging or dangerous overheating. A fault, such as a short circuit or ground fault, isn't an overload.

Panelboard [Article 408]. A distribution point containing overcurrent protection devices and designed to be installed in a cabinet. Figure 100–45

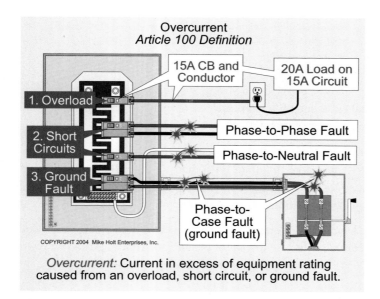

Overcurrent: Current in excess of equipment rating caused from an overload, short circuit, or ground fault.

Figure 100–44

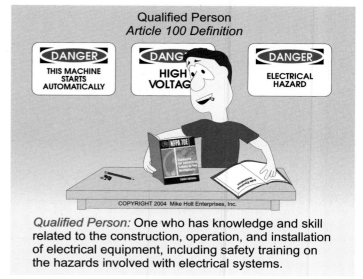

Qualified Person: One who has knowledge and skill related to the construction, operation, and installation of electrical equipment, including safety training on the hazards involved with electrical systems.

Figure 100–46

Qualified Person. A person who has the skill and knowledge related to the construction and operation of the electrical equipment and its installation. This person must have received safety training on the hazards involved with electrical systems. **Figure 100–46**

FPN: Refer to NFPA 70E, *Standard for Electrical Safety in the Workplace*, for electrical safety training requirements.

Author's Comments:

- Examples of this safety training include, but aren't limited to, training in the use of special precautionary techniques, of personal protective equipment, of insulating and shielding materials, and of using insulated tools and test equipment when working on or near exposed conductors or circuit parts that are or can become energized.

- In many parts of the United States, electricians, electrical contractors, electrical inspectors, and electrical engineers must complete from 6 to 24 hours of *NEC* review each year as a requirement to maintain licensing. This in itself doesn't make one qualified to deal with the specific hazards involved.

Raceway. An enclosure designed for the installation of conductors, cables, or busbars. Raceways in the *NEC* include:

Raceway Type	Article
• Busways	368
• Electrical Metallic Tubing	358
• Electrical Nonmetallic Tubing	362
• Flexible Metal Conduit	348
• Intermediate Metal Conduit	342
• Liquidtight Flexible Metal Conduit	350
• Liquidtight Flexible Nonmetallic Conduit	356
• Metal Wireways	376
• Multioutlet Assembly	380
• Nonmetallic Wireways	378
• Rigid Metal Conduit	344
• Rigid Nonmetallic Conduit	352
• Strut-Type Channel Raceways	384
• Surface Metal Raceways	386
• Surface Nonmetallic Raceways	388

Author's Comment: A cable tray system isn't a raceway; it's a support system for cables, raceways, and enclosures. See Article 392.

Receptacle. A contact device installed at an outlet for the connection of an attachment plug. A single receptacle contains one device on a strap (mounting yoke), and a multiple receptacle contains more than one device on a common yoke. **Figure 100–47**

Receptacle Outlet. An opening in an outlet box where one or more receptacles have been installed.

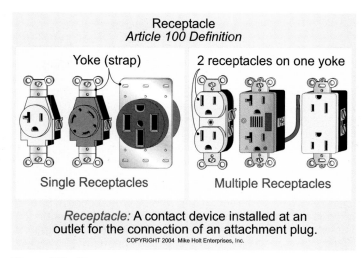

Receptacle
Article 100 Definition

Yoke (strap)   2 receptacles on one yoke

Single Receptacles   Multiple Receptacles

Receptacle: A contact device installed at an outlet for the connection of an attachment plug.
COPYRIGHT 2004 Mike Holt Enterprises, Inc.

Figure 100–47

Remote-Control Circuit [Article 725]. An electric circuit that controls another circuit by a relay or equivalent device installed in accordance with Article 725. **Figure 100–48**

Separately Derived System. A wiring system whose power is derived from a source of electric energy or equipment other than the electric utility service. This includes a battery, a solar photovoltaic system, a transformer, or a converter winding, where there's no direct electrical connection to the supply conductors of another system. **Figure 100–49**

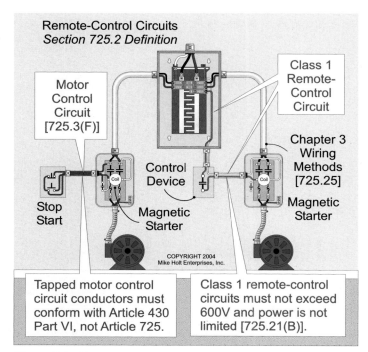

Remote-Control Circuits
Section 725.2 Definition

Motor Control Circuit [725.3(F)]

Class 1 Remote-Control Circuit

Control Device

Chapter 3 Wiring Methods [725.25]

Stop Start

Magnetic Starter

Magnetic Starter

COPYRIGHT 2004 Mike Holt Enterprises, Inc.

Tapped motor control circuit conductors must conform with Article 430 Part VI, not Article 725.

Class 1 remote-control circuits must not exceed 600V and power is not limited [725.21(B)].

Figure 100–48

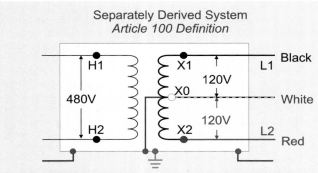

Separately Derived System: A wiring system whose power is derived from a source of electrical energy or equipment other than the electric utility service. Such systems have no direct electrical connection, including a solidly connected grounded circuit conductor to supply conductors originating in another system.

Figure 100–49

Author's Comments:

- The revised definition clarifies that a separately derived system also includes equipment such as transformers, converters, and inverters, which might not be considered a source of energy.

- Separately derived systems are actually a lot more complicated than the above definition suggests, and understanding them requires additional study. For more information, see 250.20(D) and 250.30.

Service. The conductors from the electric utility that deliver electric energy to the premises.

Author's Comment: Conductors from a UPS system, solar photovoltaic system, generator, or transformer are not identified as service conductors. See the definitions for "Feeder" and for "Service Conductors" in this article.

Service Conductors. Conductors originating from the "service point" and terminating in "service equipment." See the definition of Service Point and Service Equipment. **Figure 100–50**

Author's Comment: Conductors from a generator, UPS system, or transformers are feeder conductors, not service conductors. See the definition of Feeder.

Service Equipment. The one to six disconnects connected to the load end of service conductors intended to control and cut off the supply to the building or structure. **Figure 100–51**

Author's Comment: It's important to know where a service begins and where it ends in order to properly apply the *NEC* requirements. Sometimes the service ends before the metering equipment. **Figure 100–52**

Service Point. The point where the electrical utility conductors make contact with premises wiring. **Figure 100–53**

Author's Comments:

- See the definition of "Premises Wiring" in this article.

- The service point can be at the utility transformer, at the service weatherhead, or at the meter enclosure, depending on where the utility conductors terminate. **Figure 100–54**

Signaling Circuit [Article 725]. Any electric circuit that energizes signaling equipment.

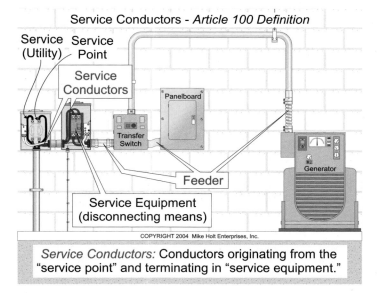

Service Conductors: Conductors originating from the "service point" and terminating in "service equipment."

Figure 100–50

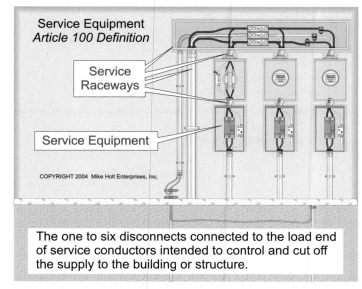

The one to six disconnects connected to the load end of service conductors intended to control and cut off the supply to the building or structure.

Figure 100–51

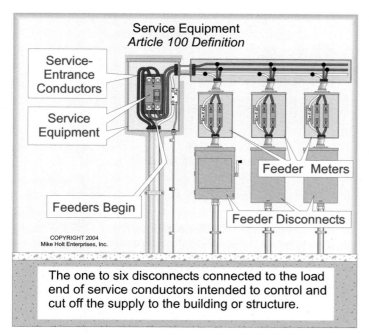

Service Equipment
Article 100 Definition

Service-Entrance Conductors

Service Equipment

Feeders Begin

Feeder Meters

Feeder Disconnects

COPYRIGHT 2004
Mike Holt Enterprises, Inc.

The one to six disconnects connected to the load end of service conductors intended to control and cut off the supply to the building or structure.

Figure 100–52

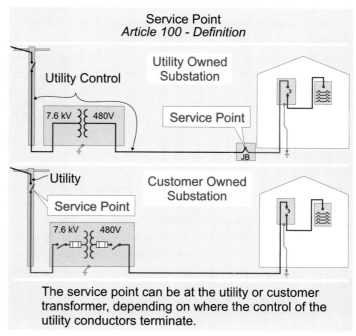

Service Point
Article 100 - Definition

Utility Control

Utility Owned Substation

7.6 kV   480V

Service Point

JB

Utility

Customer Owned Substation

Service Point

7.6 kV   480V

The service point can be at the utility or customer transformer, depending on where the control of the utility conductors terminate.

Figure 100–54

Special Permission. Written consent from the authority having jurisdiction.

> **Author's Comment:** See the definition for Authority Having Jurisdiction.

Structure. That which is built or constructed.

Supplementary Overcurrent Protective Device. A device intended to provide limited overcurrent protection for specific applications and utilization equipment, such as luminaires and appliances. This limited protection is in addition to the protection provided in the required branch circuit by the branch-circuit overcurrent protective device. **Figure 100–55**

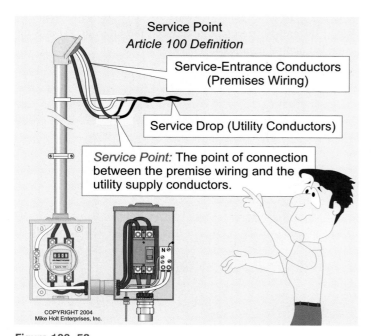

Service Point
Article 100 Definition

Service-Entrance Conductors (Premises Wiring)

Service Drop (Utility Conductors)

Service Point: The point of connection between the premise wiring and the utility supply conductors.

COPYRIGHT 2004
Mike Holt Enterprises, Inc.

Figure 100–53

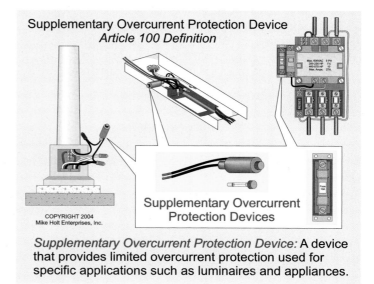

Supplementary Overcurrent Protection Device
Article 100 Definition

Supplementary Overcurrent Protection Devices

COPYRIGHT 2004
Mike Holt Enterprises, Inc.

Supplementary Overcurrent Protection Device: A device that provides limited overcurrent protection used for specific applications such as luminaires and appliances.

Figure 100–55

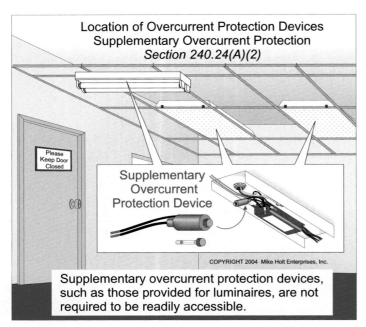

Location of Overcurrent Protection Devices
Supplementary Overcurrent Protection
Section 240.24(A)(2)

Please Keep Door Closed

Supplementary Overcurrent Protection Device

COPYRIGHT 2004 Mike Holt Enterprises, Inc.

Supplementary overcurrent protection devices, such as those provided for luminaires, are not required to be readily accessible.

Figure 100–56

Author's Comment: Supplementary overcurrent devices aren't required to be readily accessible [240.10 and 240.24(A)(2)]. **Figure 100–56**

Switch, General-Use Snap. A switch constructed to be installed in a device box or a box cover.

Voltage of a Circuit. The greatest effective root-mean-square (RMS) difference of potential between any two conductors of the circuit.

Author's Comment: Voltages can be reported in many ways: peak, peak-to-peak, average, and root-mean-square (RMS). Electrical voltage measuring equipment that provides readings in either average or RMS are commercially available. RMS meters provide the most accurate reading when used on circuits with nonlinear loads that have a high harmonic content.

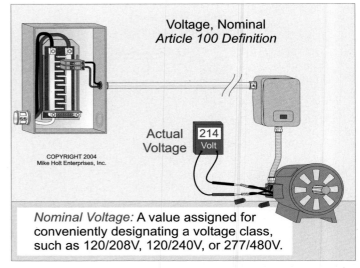

Voltage, Nominal
Article 100 Definition

COPYRIGHT 2004
Mike Holt Enterprises, Inc.

Actual Voltage 214 Volt

Nominal Voltage: A value assigned for conveniently designating a voltage class, such as 120/208V, 120/240V, or 277/480V.

Figure 100–57

Voltage, Nominal. A value assigned for the purpose of conveniently designating voltage class, such as 120/240V, 120/208V, or 277/480V. See 220.5(A). **Figure 100–57**

Author's Comment: The actual voltage at which a circuit operates can vary from the nominal within a range that permits satisfactory operation of equipment. In addition, common voltage ratings of electrical equipment are 115, 200, 208, 230, and 460. The electrical power supplied might be at the nominal voltage (240), but the voltage at the equipment will be less than this voltage value (230). Therefore, electrical equipment is rated at a value less than the nominal system voltage.

Voltage-to-Ground. The greatest effective root-mean-square (RMS) difference of potential between the ungrounded and the grounded neutral conductor.

Article 100 Questions

(• Indicates that 75% or fewer exam takers get the question correct.)

1. Admitting close approach, not guarded by locked doors, elevation, or other effective means, is commonly referred to as _____.

    (a) accessible (equipment)   (b) accessible (wiring methods)
    (c) accessible, readily   (d) all of these

2. •A device that, by insertion in a receptacle, establishes a connection between the conductors of the attached flexible cord and the conductors connected permanently to the receptacle is called a(n) _____.

    (a) attachment plug   (b) plug cap   (c) plug   (d) any of these

3. •For a circuit to be considered a multiwire branch circuit, it must have _____.

    (a) two or more ungrounded conductors with a voltage potential between them
    (b) a grounded neutral conductor having equal voltage potential between it and each ungrounded conductor of the circuit
    (c) a grounded neutral conductor connected to the grounded neutral terminal of the system
    (d) all of these

4. NM cable is considered _____ if rendered inaccessible by the structure or finish of the building.

    (a) inaccessible   (b) concealed   (c) hidden   (d) enclosed

5. A component of an electrical system that is intended to carry or control but not utilize electric energy is a(n) _____.

    (a) raceway   (b) fitting   (c) device   (d) enclosure

6. The *NEC* term to define wiring methods that are not concealed is _____.

    (a) open   (b) uncovered   (c) exposed   (d) bare

7. _____ is defined as intentionally connected to earth through a ground connection or connections of sufficiently low impedance and having sufficient current-carrying capacity, to prevent the build up of voltages that may result in undue hazards to connected equipment or to persons.

    (a) Effectively grounded   (b) A proper wiring system
    (c) A lighting rod   (d) A grounded neutral conductor

8. The grounding electrode conductor is the conductor used to connect the grounding electrode to the equipment grounding (bonding) conductor and the grounded neutral conductor at _____.

    (a) the service   (b) each building or structure supplied by feeder(s)
    (c) the source of a separately derived system   (d) all of these

9. Recognized as suitable for the specific purpose, function, use, environment, and application is the definition of _____.

    (a) labeled   (b) identified (as applied to equipment)
    (c) listed   (d) approved

10. A _____ location may be temporarily subject to dampness and wetness.

(a) dry   (b) damp   (c) moist   (d) wet

11. Outline lighting may not include light sources such as light emitting diodes (LEDs).

(a) True   (b) False

12. NFPA 70E, *Standard for Electrical Safety in the Workplace,* provides information to help determine the electrical safety training requirements expected of a "qualified person."

(a) True   (b) False

13. When one electrical circuit controls another circuit through a relay, the first circuit is called a _____.

(a) control circuit   (b) remote-control circuit   (c) signal circuit   (d) controller

14. The _____ is the necessary equipment, usually consisting of a circuit breaker(s) or switch(es) and fuse(s) and their accessories, connected to the load end of service conductors to a building or other structure, or an otherwise designated area, and intended to constitute the main control and cutoff of the supply.

(a) service equipment   (b) service
(c) service disconnect   (d) service overcurrent protection device

15. A form of general-use switch constructed so that it can be installed in device boxes or on box covers, or otherwise used in conjunction with wiring systems recognized by the *Code,* is called a _____ switch.

(a) transfer   (b) motor-circuit   (c) general-use snap   (d) bypass isolation

ARTICLE 110
Requirements for Electrical Installations

Introduction

Article 110 sets the stage for how you will implement the rest of the *NEC*. This article contains a few of the most important and yet neglected parts of the *Code*. For example:

- What do you do with unused openings in enclosures?
- What's the right working clearance for a given installation?
- How should you terminate conductors?
- What kinds of warnings, markings, and identification does a given installation require?

It's critical that you master Article 110, and that's exactly what this Illustrated Guide is designed to help you do. As you read this article, remember that doing so helps build your foundation for correctly applying much of the *NEC*. In fact, the article itself is a foundation for much of the *Code*. You may need to read something several times to understand it. The time you take to do that will be well spent. The illustrations will also help. But if you find your mind starting to wander, take a break. What matters is how well you master the material and how safe your work is—not how fast you blazed through a book.

PART I. GENERAL REQUIREMENTS

110.1 Scope. Article 110 covers the general requirements for the examination and approval, installation and use, access to and spaces about electrical equipment; as well as general requirements for enclosures intended for personnel entry (manholes, vaults, and tunnels).

110.2 Approval of Equipment. The authority having jurisdiction must approve all electrical conductors and equipment. Figure 110–1

> **Author's Comment:** For a better understanding of product approval, review 90.4, 90.7, 110.3 and the definitions for Approved, Identified, Labeled, and Listed in Article 100.

110.3 Examination, Identification, Installation, and Use of Equipment.

(A) Guidelines for Approval. The authority having jurisdiction must approve equipment, and consideration must be given to the following:

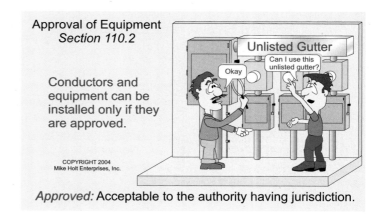

Approval of Equipment
Section 110.2

Conductors and equipment can be installed only if they are approved.

COPYRIGHT 2004
Mike Holt Enterprises, Inc.

Approved: Acceptable to the authority having jurisdiction.

Figure 110–1

(1) Listing or labeling
(2) Mechanical strength and durability
(3) Wire-bending and connection space
(4) Electrical insulation
(5) Heating effects under conditions of use
(6) Arcing effects

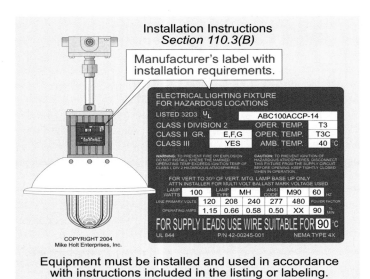

Figure 110–2

(7) Classification by voltage, current capacity, and specific use

(8) Other factors contributing to the practical safeguarding of persons using or in contact with the equipment

(B) Installation and Use. Equipment must be installed and used in accordance with any instructions included in the listing or labeling requirements.

Author's Comments:

- See Article 100 for the definitions of "Labeling" and "Listing."

- Equipment is listed for a specific condition of use, operation, or installation, and it must be installed and used in accordance with those listed instructions. Failure to follow product listing instructions, such as torquing of terminals and sizing of conductors, is a violation of this *Code* rule [110.3(B)]. **Figure 110–2**

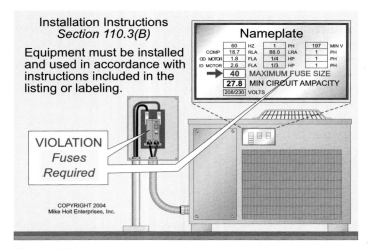

Figure 110–3

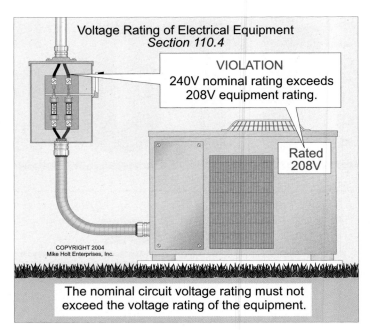

Figure 110–4

- When an air conditioner nameplate specifies "Maximum Fuse Size," one-time or dual-element fuses must be used to protect the equipment. **Figure 110–3**

110.4 Voltages. The voltage rating of electrical equipment must not be less than the nominal voltage of a circuit. Put another way, electrical equipment must be installed on a circuit where the nominal system voltage doesn't exceed the voltage rating of the equipment. This rule is intended to prohibit the installation of 208V rated motors on a 240V nominal voltage rated circuit. **Figure 110–4**

Author's Comments:

- See Article 100 for the definition of "Nominal Voltage."

- According to 110.3(B), equipment must be installed in accordance with any instructions included in the listing or labeling. Therefore, equipment must not be connected to a circuit where the nominal voltage is less than the rated voltage of the electrical equipment. For example, you cannot place a 230V rated motor on a 208V system. **Figure 110–5**

110.5 Copper Conductors. Where conductor material isn't specified in a rule, the material and the sizes given in the *Code* (and this textbook) are based on copper.

110.6 Conductor Sizes. Conductor sizes are expressed in American Wire Gage (AWG), typically from 18 AWG up to 4/0 AWG. Conductor sizes larger than 4/0 AWG are expressed in kcmil (thousand circular mils). **Figure 110–6**

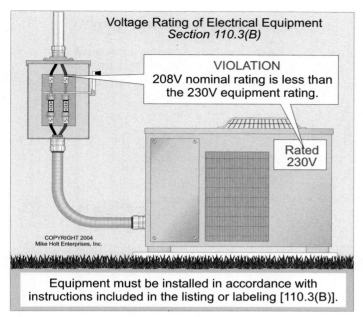

Figure 110–5

110.7 Conductor Insulation.

All wiring must be installed so as to be free from short circuits and ground faults.

Author's Comments:

- A ground fault is an unintentional electrical connection between an ungrounded or grounded neutral conductor and metallic enclosures, raceways, or equipment [250.2].

- Short circuits and ground faults often arise from insulation failure due to mishandling or improper installation. This happens when, for example, wire is dragged on a sharp edge, when insulation is scraped on boxes and enclosures, when wire is pulled hard, when insulation is nicked while being stripped, or when cable clamps and/or staples are installed too tightly.

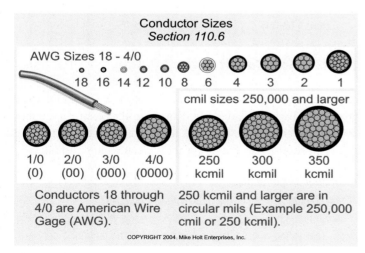

Figure 110–6

- To protect against accidental contact with energized conductors, the ends of abandoned conductors must be covered with an insulating device identified for the purpose, such as a twist-on or push-on wire connector [110.14(B)].

- See Article 100 for the definition of "Device."

110.8 Suitable Wiring Methods.

Only wiring methods recognized as suitable are included in the *NEC*, and they must be installed in accordance with the *Code*.

> **Author's Comment:** See Chapter 3 for power and lighting wiring methods, Chapter 7 for signaling circuits, and Chapter 8 for communications circuits. **Figure 110–7**

110.9 Interrupting Protection Rating.

Overcurrent protection devices such as circuit breakers and fuses are intended to interrupt the circuit, and they must have an interrupting rating sufficient for the short-circuit current available at the line terminals of the equipment. **Figure 110–8**

> **Author's Comments:**
> - See Article 100 for the definition of "Interrupting Rating."
> - Unless marked otherwise, the ampere interrupting rating for circuit breakers is 5,000A [240.83(C)], and for fuses is 10,000A [240.60(C)(3)]. **Figure 110–9**

Available Short-Circuit Current. Available short-circuit current is the current in amperes that is available at a given point in the electrical system. This available short-circuit current is first determined at the secondary terminals of the utility transformer.

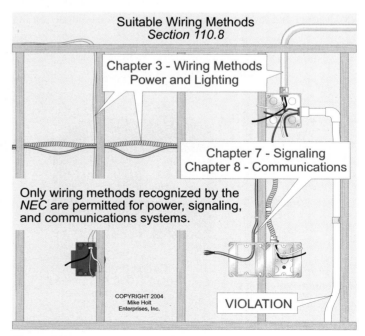

Figure 110–7

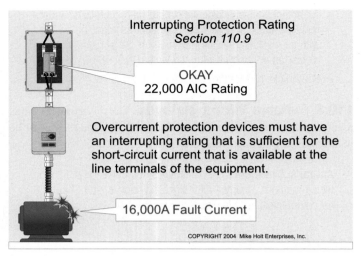

Figure 110–8

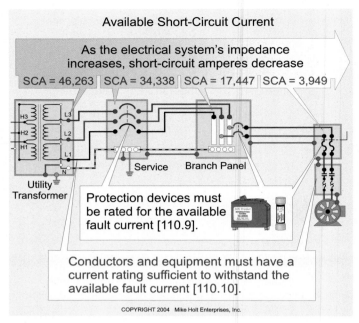

Figure 110–10

Thereafter, the available short-circuit current is calculated at the terminals of service equipment, then at branch-circuit panelboards and other equipment. The available short-circuit current is different at each point of the electrical system. It's highest at the utility transformer and lowest at the branch-circuit load.

The available short-circuit current depends on the impedance of the circuit, which increases moving downstream from the utility transformer. The greater the circuit impedance (utility transformer and the additive impedances of the circuit conductors), the lower the available short-circuit current. **Figure 110–10**

Factors that impact the available short-circuit current at the utility transformer include the system voltage, the transformer kVA rating, and its impedance (expressed in a percentage on the

equipment nameplate). Properties that impact the impedance of the circuit include the conductor material (copper versus aluminum), conductor size, and conductor length.

Author's Comment: Many in the industry describe Amperes Interrupting Rating (AIR) as "Amperes Interrupting Capacity" (AIC).

DANGER: *Extremely high values of current flow (caused by short circuits or ground faults) produce tremendously destructive thermal and magnetic forces. If the circuit overcurrent protection device isn't rated to interrupt the current at the available fault values at its listed voltage rating, it could explode while attempting to clear a fault. Naturally this can cause serious injury or death, as well as property damage.* **Figure 110–11**

110.10 Short-Circuit Current Rating. Electrical equipment must have a short-circuit current rating that permits the circuit overcurrent protection device to clear a short circuit or ground fault without extensive damage to the electrical components of the circuit. For example, a motor controller must have a sufficient short-circuit rating for the available fault-current.

Author's Comments:
- See Article 100 for the definition of "Controller."
- If the fault exceeds the controller's 5,000A short-circuit current rating, the controller could explode, endangering persons and property. **Figure 110–12**

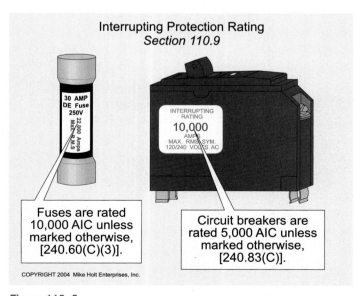

Figure 110–9

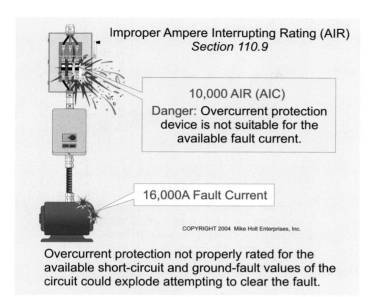

Improper Ampere Interrupting Rating (AIR)
Section 110.9

10,000 AIR (AIC)
Danger: Overcurrent protection device is not suitable for the available fault current.

16,000A Fault Current

COPYRIGHT 2004 Mike Holt Enterprises, Inc.

Overcurrent protection not properly rated for the available short-circuit and ground-fault values of the circuit could explode attempting to clear the fault.

Figure 110–11

To solve this problem, a current-limiting protection device (fast-clearing fuse) can be used to reduce the let-through current to less than 5,000A. **Figure 110–13**

Author's Comment: For more information on the application of current limiting devices, see 240.2 and 240.60(B).

110.11 Deteriorating Agents. Electrical equipment and conductors must be suitable for the environment and conditions of use, and consideration must be given to the presence

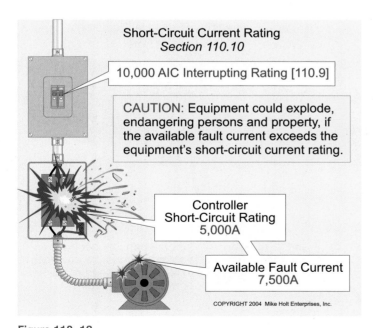

Short-Circuit Current Rating
Section 110.10

10,000 AIC Interrupting Rating [110.9]

CAUTION: Equipment could explode, endangering persons and property, if the available fault current exceeds the equipment's short-circuit current rating.

Controller
Short-Circuit Rating
5,000A

Available Fault Current
7,500A

COPYRIGHT 2004 Mike Holt Enterprises, Inc.

Figure 110–12

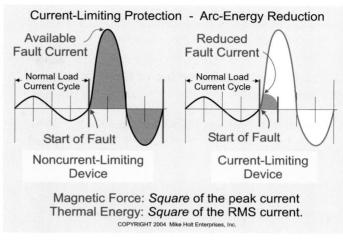

Current-Limiting Protection - Arc-Energy Reduction

Available Fault Current

Normal Load Current Cycle

Start of Fault

Noncurrent-Limiting Device

Reduced Fault Current

Normal Load Current Cycle

Start of Fault

Current-Limiting Device

Magnetic Force: *Square* of the peak current
Thermal Energy: *Square* of the RMS current.

COPYRIGHT 2004 Mike Holt Enterprises, Inc.

Figure 110–13

of corrosive gases, fumes, vapors, liquids, or chemicals that can have a deteriorating effect on the conductors or equipment. **Figure 110–14**

Author's Comment: Conductors must not be exposed to ultra-violet rays from the sun unless identified for the purpose [310.8(D)].

FPN No. 1: Raceways, cable trays, cablebus, auxiliary gutters, cable armor, boxes, cable sheathing, cabinets, elbows, couplings, fittings, supports, and support hardware must be of materials suitable for the environment in which they are to be installed, in accordance with 300.6.

Author's Comment: See Article 100 for the definition of "Raceway."

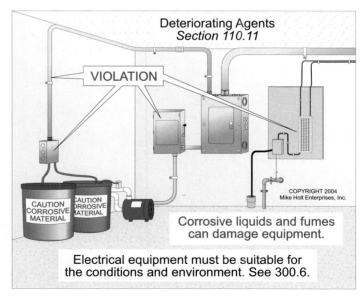

Deteriorating Agents
Section 110.11

VIOLATION

COPYRIGHT 2004
Mike Holt Enterprises, Inc.

CAUTION CORROSIVE MATERIAL
CAUTION CORROSIVE MATERIAL

Corrosive liquids and fumes can damage equipment.

Electrical equipment must be suitable for the conditions and environment. See 300.6.

Figure 110–14

FPN No. 2: Some spray cleaning and lubricating compounds contain chemicals that can deteriorate plastic used for insulating and structural applications in equipment.

110.12 Mechanical Execution of Work. Electrical equipment must be installed in a neat and workmanlike manner.

FPN: Accepted industry practices are described in ANSI/NECA 1-2000, *Standard Practices for Good Workmanship in Electrical Contracting.*

Author's Comment: The National Electrical Contractors Association (NECA) has created a series of National Electrical Installation Standards (NEIS)™ that establish the industry's first quality guidelines for electrical installations. These standards define a benchmark or baseline of quality and workmanship for installing electrical products and systems. They explain what installing electrical products and systems in a "neat and workmanlike manner" means. For more information about these standards, visit http://www.neca-neis.org/.

(A) Unused Openings. Unused cable or raceway openings in electrical equipment must be effectively closed by fittings that provide protection substantially equivalent to the wall of the equipment. **Figure 110–15**

Author's Comments:

- See Article 100 for the definition of "Fitting."
- Unused openings for circuit breakers must be closed using identified closures, or other means approved by the authority having jurisdiction that provide protection substantially equivalent to the wall of the enclosure [408.7]. **Figure 110–16**

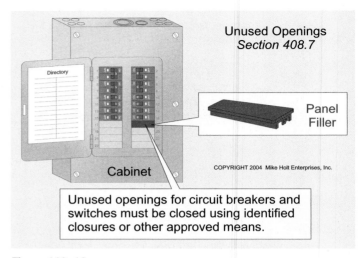

Figure 110–16

(C) Integrity of Electrical Equipment. Internal parts of electrical equipment must not be damaged or contaminated by foreign material, such as paint, plaster, cleaners, etc.

Author's Comment: Precautions must be taken to provide protection from contaminating the internal parts of panelboards and receptacles during the building construction. **Figure 110–17**

Electrical equipment that contains damaged parts that may adversely affect safe operation or mechanical strength of the equipment must not be installed. This would include parts that are broken, bent, cut, or deteriorated by corrosion, chemical action, or overheating.

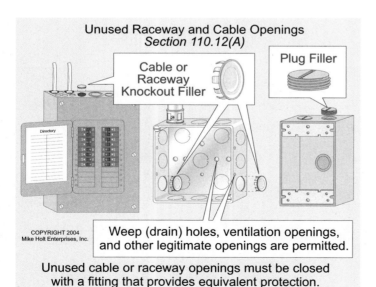

Figure 110–15

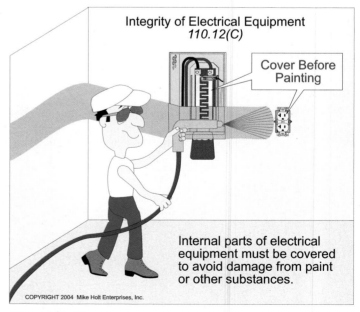

Figure 110–17

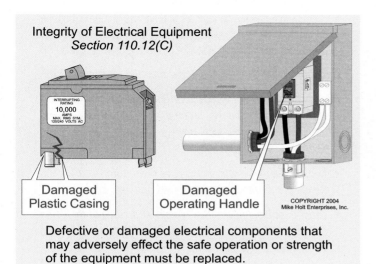

Integrity of Electrical Equipment
Section 110.12(C)

INTERRUPTING
RATING
10,000
AMPS
MAX. RMS SYM.
120/240 VOLTS AC

Damaged
Plastic Casing

Damaged
Operating Handle

COPYRIGHT 2004
Mike Holt Enterprises, Inc.

Defective or damaged electrical components that
may adversely effect the safe operation or strength
of the equipment must be replaced.

Figure 110–18

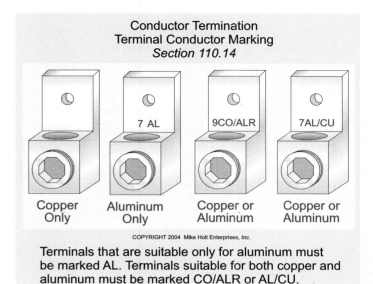

Conductor Termination
Terminal Conductor Marking
Section 110.14

7 AL   9CO/ALR   7AL/CU

Copper   Aluminum   Copper or   Copper or
Only   Only   Aluminum   Aluminum

COPYRIGHT 2004 Mike Holt Enterprises, Inc.

Terminals that are suitable only for aluminum must
be marked AL. Terminals suitable for both copper and
aluminum must be marked CO/ALR or AL/CU.

Figure 110–19

Author's Comment: Damaged parts include anything broken, bent, cut, or deteriorated by corrosion or chemical action, or by overheating, that may adversely affect the safe operation or the mechanical strength of the equipment. This includes cracked insulators, arc shields not in place, overheated fuse clips, and damaged or missing switch or circuit breaker handles. **Figure 110–18**

110.13 Mounting and Cooling of Equipment.

(A) Mounting. Electrical equipment must be firmly secured to the surface on which it's mounted. See 314.23(A)

(B) Cooling. Electrical equipment that depends on natural air circulation must be installed so that walls or equipment do not prevent airflow over the surfaces. The clearances between top surfaces and side surfaces must be maintained to dissipate rising warm air for equipment designed for floor mounting.

Electrical equipment that is constructed with ventilating openings must be installed so that free air circulation isn't inhibited.

Author's Comment: Transformers with ventilating openings must be installed so that the ventilating openings aren't blocked, and the required wall clearances are clearly marked on the transformer case [450.9].

110.14 Conductor Termination. Terminal Conductor
Material. Conductor terminal and splicing devices must be identified for the conductor material and they must be properly installed and used. Devices that are suitable only for aluminum must be marked AL, and devices that are suitable for both copper and aluminum must be marked CO/ALR [404.14(C) and 406.2(C)]. **Figure 110–19**

Author's Comments:

- See Article 100 for the definition of "Identified."

- Existing inventories of equipment or devices might be marked AL/CU to indicate a terminal suitable for both copper and aluminum conductors.

- Conductor terminations must comply with manufacturer's instructions as required by 110.3(B). For example, if the instructions for the device state "Suitable for 18-2 AWG Stranded," then only stranded conductors can be used with the terminating device. If the instructions state "Suitable for 18-2 AWG Solid," then only solid conductors are permitted, and if the instructions state "Suitable for 18-2 AWG," then either solid or stranded conductors can be used with the terminating device.

Aluminum: To reduce the contact resistance between the aluminum conductor and the terminal, terminals listed for aluminum conductors are often filled with an antioxidant gel.

Copper: Some terminal manufacturers sell a compound intended to reduce corrosion and heat at copper conductor terminations that is especially helpful at high-amperage terminals. This compound is messy, but apparently it's effective.

Copper and Aluminum Mixed: Copper and aluminum conductors must not make contact with each other in a device unless the device is listed and identified for this purpose.

Author's Comment: Few terminations are listed for the mixing of aluminum wire and copper, but if they are, they will be marked on the product package or terminal device. The reason copper and aluminum should not be in contact with each other is because corrosion will develop between the two different metals due to galvanic action, resulting in increased contact resistance at the splicing device. This increased resistance can cause overheating of the splice and cause a fire. See http://tis-hq.eh.doe.gov/docs/sn/nsh9001.html for more information on how to properly terminate aluminum and copper conductors together.

FPN: Many terminations and equipment are marked with a tightening torque.

Author's Comment: All conductors must terminate in devices that have been properly tightened in accordance with manufacturer's torque specifications included with equipment instructions. Failure to torque terminals can result in excessive heating of terminals or splicing devices (due to loose connection), which could result in a fire because of a short circuit or ground fault. In addition, this is a violation of 110.3(B), which requires all equipment to be installed in accordance with listed or labeling instructions. **Figure 110–20**

Question: What do you do if the torque value isn't provided with the device?

Answer: Call the manufacturer, visit the manufacturer's website, or have the supplier make a copy of the installation instructions.

Author's Comment: Terminating conductors without a torque tool can result in an improper and unsafe installation. If a torque screwdriver is not used, there's a good chance the conductors are not properly terminated.

(A) Terminations. Conductor terminals must ensure a good connection without damaging the conductors and must be made by pressure connectors (including set-screw type) or splices to flexible leads.

Author's Comments:
- See Article 100 for the definition of "Pressure Connector."

- Grounding (earthing) conductors and bonding jumpers must be connected by exothermic welding, pressure connectors, clamps, or other means listed for grounding (earthing) [250.8].

Question: What if the wire is larger than the terminal device?

Answer: This condition needs to be anticipated in advance, and the equipment should be ordered with terminals that will accommodate the larger wire. However, if you're in the field, you should:

- Contact the manufacturer and have them express deliver you the proper terminals, bolts, washers and nuts, or

- Order a terminal device that crimps on the end of the larger conductor and reduces the termination size, or splice the conductors to a smaller wire.

One Wire Per Terminal: Terminals for more than one wire must be identified for this purpose, either within the equipment instructions or on the terminal itself. **Figure 110–21**

Author's Comment: Split-bolt connectors are commonly listed for only two conductors although some are listed for three conductors. However, it's a common industry practice to terminate as many conductors as possible within a split-bolt connector, even though this violates the *NEC*. **Figure 110–22**
   Split-bolt connectors for aluminum-to-aluminum or aluminum-to-copper conductors must be identified as suitable for the application.

(B) Conductor Splices. Conductors must be spliced by a splicing device identified for the purpose or by exothermic welding.

Author's Comment: Conductors are not required to be twisted together prior to the installation of a twist-on wire connector. **Figure 110–23**

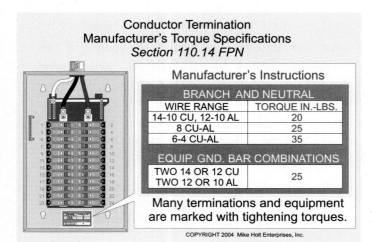

Figure 110–20

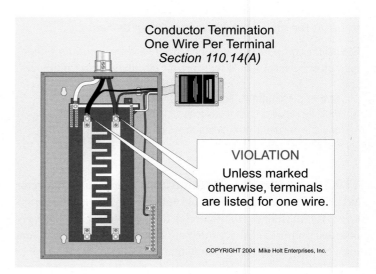

Figure 110–21

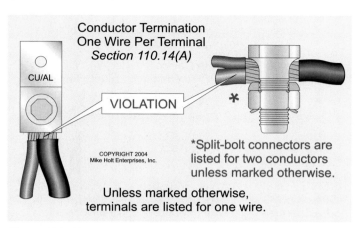

Figure 110–22

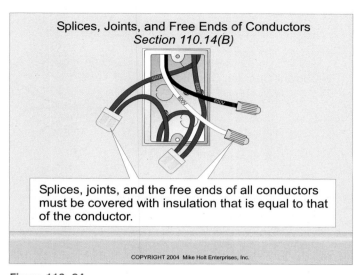

Figure 110–24

Splices, joints, and the free ends of all conductors must be covered with insulation that is equal to that of the conductor.

Author's Comments:

- Circuit conductors not being used are not required to be removed. However, to prevent an electrical hazard, the free ends of the conductors must be insulated to prevent the exposed end of the conductor from touching energized parts. This requirement can be met by the use of an insulated twist-on or push-on wire connector. **Figure 110–24**

- See Article 100 for the definition of "Energized."

Underground Splices: Single Conductors: Single direct burial conductors of Type UF or USE can be spliced underground without a junction box, but the conductors must be spliced with a device that is listed for direct burial. See 300.5(E) and 300.15(G). **Figure 110–25**

Multiconductor Cable: Multiconductor Type UF or Type USE cable can have the individual conductors spliced underground with a listed splice kit that encapsulates the conductors and the cable jacket.

(C) Temperature Limitations (Conductor Size). Conductors are to be sized to the lowest temperature rating of any terminal, device, or conductor of the circuit in accordance with (1) for terminals of equipment, and (2) for independent pressure connectors on a bus.

Conductor Ampacity. Conductors with insulation temperature ratings higher than the termination's temperature rating can be used for conductor ampacity adjustment, correction, or both.

Author's Comments:

- See Article 100 for the definition of "Ampacity."

- This means that conductor ampacity must be based on the conductor's insulation temperature ratings listed in Table 310.16, as adjusted for ambient temperature correction factors, conductor bundling adjustment factors, or both. This

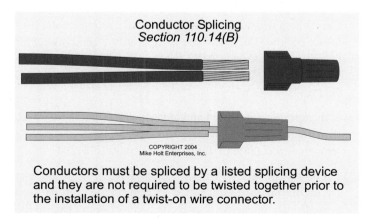

Figure 110–23

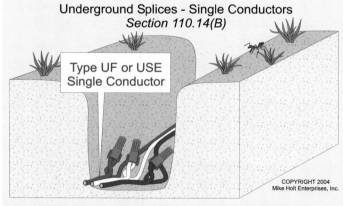

Figure 110–25

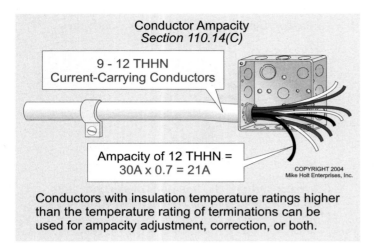

Conductor Ampacity
Section 110.14(C)

9 - 12 THHN
Current-Carrying Conductors

Ampacity of 12 THHN =
30A x 0.7 = 21A

COPYRIGHT 2004
Mike Holt Enterprises, Inc.

Conductors with insulation temperature ratings higher than the temperature rating of terminations can be used for ampacity adjustment, correction, or both.

Figure 110–26

means that conductor ampacity, when required to be adjusted, is based on the conductor insulation temperature rating in accordance with Table 310.16. For example, the ampacity of each 12 THHN conductor is 30A, based on the values listed in the 90°C column of Table 310.16.

If we bundle nine current-carrying 12 THHN conductors in the same raceway or cable, the ampacity for each conductor (30A at 90°C, Table 310.16) needs to be adjusted by a 70 percent adjustment factor [Table 310.15(B)(2)(a)]. **Figure 110–26**

Adjusted Conductor Ampacity = 30A x 0.70
Adjusted Conductor Ampacity = 21A

See *necdigest* magazine, winter 2003 issue, page 32, and the *NEC Handbook*, 310.15(B)(2)(a) Ex. 5, for examples of 90°C ampacity for conductor ampacity adjustment.

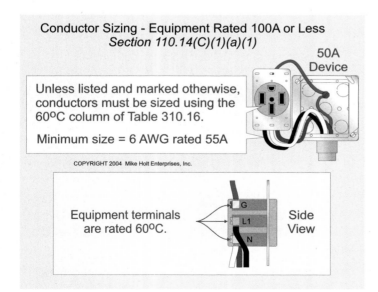

Conductor Sizing - Equipment Rated 100A or Less
Section 110.14(C)(1)(a)(1)

50A
Device

Unless listed and marked otherwise, conductors must be sized using the 60°C column of Table 310.16.

Minimum size = 6 AWG rated 55A

COPYRIGHT 2004 Mike Holt Enterprises, Inc.

Equipment terminals
are rated 60°C.

Side
View

Figure 110–27

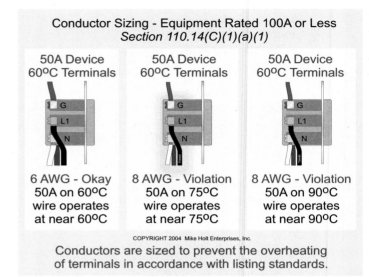

Conductor Sizing - Equipment Rated 100A or Less
Section 110.14(C)(1)(a)(1)

50A Device 60°C Terminals	50A Device 60°C Terminals	50A Device 60°C Terminals
6 AWG - Okay 50A on 60°C wire operates at near 60°C	8 AWG - Violation 50A on 75°C wire operates at near 75°C	8 AWG - Violation 50A on 90°C wire operates at near 90°C

COPYRIGHT 2004 Mike Holt Enterprises, Inc.

Conductors are sized to prevent the overheating of terminals in accordance with listing standards.

Figure 110–28

(1) Equipment Provisions. Unless the equipment is listed and marked otherwise, conductor sizing for equipment termination must be based on Table 310.16 in accordance with (a) or (b):

(a) Equipment Rated 100A and Less.

(1) Conductor sizing for equipment rated 100A or less must be sized using the 60°C temperature column of Table 310.16. **Figure 110–27**

Author's Comment: Conductors are sized to prevent the overheating of terminals, in accordance with listing standards. For example, a 50A circuit with 60°C terminals requires the circuit conductors to be sized not smaller than 6 AWG, in accordance with the 60°C ampacity listed in Table 310.16. However, an 8 THHN insulated conductor has a 90°C ampacity of 50A, but 8 AWG cannot be used for this circuit because the conductor's operating temperature at full-load ampacity (50A) will be near 90°C, which is well in excess of the 60°C terminal rating. **Figure 110–28**

(2) Conductors with an insulation temperature rating greater than 60°C, such as THHN, which is rated 90°C, can be used on terminals that are rated 60°C, but the conductor must be sized based on the 60°C temperature column of Table 310.16. **Figure 110–29A**

(3) If the terminals are listed and identified as suitable for 75°C, then conductors rated at least 75°C can be sized to the 75°C temperature column of Table 310.16. **Figure 110–29B**

(4) For motors marked with design letters B, C, or D, conductors having an insulation rating of 75°C or higher can be used provided the ampacity of such conductors doesn't exceed the 75°C ampacity.

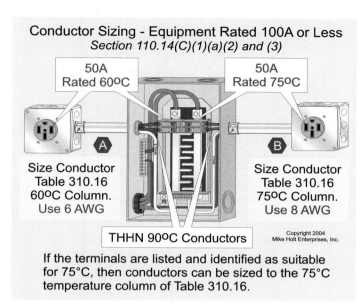

Conductor Sizing - Equipment Rated 100A or Less
Section 110.14(C)(1)(a)(2) and (3)

50A Rated 60ºC

50A Rated 75ºC

A

B

Size Conductor
Table 310.16
60ºC Column.
Use 6 AWG

Size Conductor
Table 310.16
75ºC Column.
Use 8 AWG

THHN 90ºC Conductors

Copyright 2004
Mike Holt Enterprises, Inc.

If the terminals are listed and identified as suitable for 75°C, then conductors can be sized to the 75°C temperature column of Table 310.16.

Figure 110–29

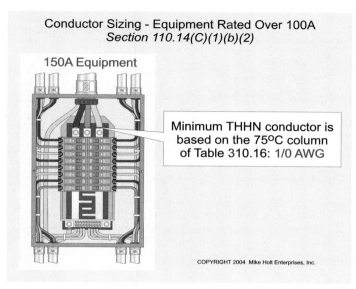

Conductor Sizing - Equipment Rated Over 100A
Section 110.14(C)(1)(b)(2)

150A Equipment

Minimum THHN conductor is based on the 75ºC column of Table 310.16: 1/0 AWG

COPYRIGHT 2004 Mike Holt Enterprises, Inc.

Figure 110–31

(b) Equipment Rated Over 100A.

(1) Conductors for equipment rated over 100A must be sized based on the 75°C temperature column of Table 310.16. Figure 110–30

(2) Conductors with an insulation temperature rating greater than 75°C can be used on terminals that are rated 75°C, but the conductor must be sized based on the 75°C temperature column of Table 310.16. Figure 110–31

(2) Separate Connector Provisions. Conductors can be sized to the 90°C ampacity rating of THHN, if the conductor terminates to a bus connector that is rated 90°C. Figure 110–32

110.15 High-Leg Conductor Identification. Identification. On a 4-wire three-phase delta-connected system, where the midpoint of one phase winding is grounded, the conductor with the higher phase voltage-to-ground (208V) must be durably and permanently marked by an outer finish that is orange in color, or other effective means. Such identification must be placed at each point on the system where a connection is made if the grounded neutral conductor is present [110.15, 215.8, and 230.56]. Figure 110–33

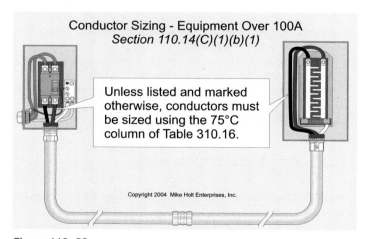

Conductor Sizing - Equipment Over 100A
Section 110.14(C)(1)(b)(1)

Unless listed and marked otherwise, conductors must be sized using the 75°C column of Table 310.16.

Copyright 2004 Mike Holt Enterprises, Inc.

Figure 110–30

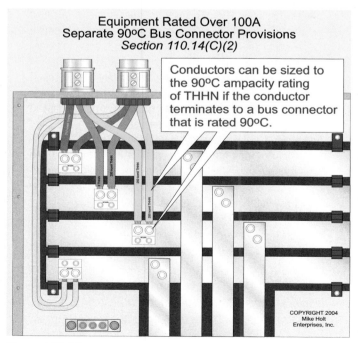

Equipment Rated Over 100A
Separate 90ºC Bus Connector Provisions
Section 110.14(C)(2)

Conductors can be sized to the 90ºC ampacity rating of THHN if the conductor terminates to a bus connector that is rated 90ºC.

COPYRIGHT 2004
Mike Holt
Enterprises, Inc.

Figure 110–32

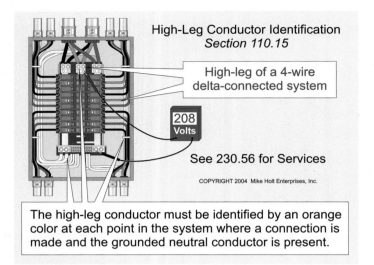

High-Leg Conductor Identification
Section 110.15

High-leg of a 4-wire delta-connected system

208 Volts

See 230.56 for Services

COPYRIGHT 2004 Mike Holt Enterprises, Inc.

The high-leg conductor must be identified by an orange color at each point in the system where a connection is made and the grounded neutral conductor is present.

Figure 110–33

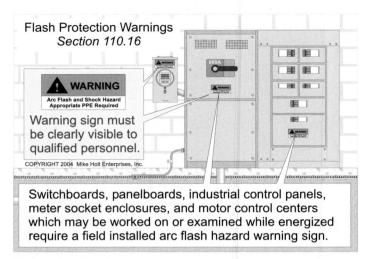

Flash Protection Warnings
Section 110.16

⚠ **WARNING**
Arc Flash and Shock Hazard
Appropriate PPE Required

Warning sign must be clearly visible to qualified personnel.

COPYRIGHT 2004 Mike Holt Enterprises, Inc.

Switchboards, panelboards, industrial control panels, meter socket enclosures, and motor control centers which may be worked on or examined while energized require a field installed arc flash hazard warning sign.

Figure 110–34

Author's Comments:

- The high-leg conductor is also called the "wild-leg," "stinger-leg," or "bastard-leg."
- Other important *NEC* rules relating to the high-leg are as follows:
  - Panelboards. Since 1975, panelboards supplied by a 4-wire three-phase delta-connected system must have the high-leg conductor (208V) terminate to the "B" (center) phase of a panelboard [408.3(E)].

    An exception to 408.3(E) permits the high-leg conductor to terminate to the "C" phase when the meter is located in the same section of a switchboard or panelboard.
  - Disconnects. The *NEC* does not specify the termination location for the high-leg conductor in switch equipment (Switches—Article 404), but the generally accepted practice is to terminate this conductor to the "B" phase.
  - Utility Equipment: It's my understanding that the ANSI standard for meter equipment requires the high-leg conductor (208V-to-neutral) to terminate on the "C" (right) phase of the meter enclosure. This is because the demand meter needs 120V and it gets this from the "B" phase.

    Also hope the utility lineman is not color blind and doesn't inadvertently cross the "orange" high-leg conductor (208V) with the red (120V) service conductor at the weatherhead. It's happened before...

WARNING: *When replacing equipment in existing facilities that contain a high-leg conductor, care must be taken to ensure that the high-leg conductor is replaced in the original location. Prior to 1975, the high-leg conductor was required to terminate on the "C" phase of panelboards and switchboards. Failure to re-terminate the high-leg in accordance with the existing installation can result in 120V circuits inadvertently connected to the 208V high-leg, with disastrous results.*

110.16 Flash Protection Warning.

Switchboards, panelboards, industrial control panels, meter socket enclosures, and motor control centers in commercial and industrial occupancies that are likely to require examination, adjustment, servicing, or maintenance while energized must be <u>field marked</u> to warn qualified persons of the danger associated with an arc flash from line-to-line or ground faults. The field marking must be clearly visible to qualified persons before they examine, adjust, service, or perform maintenance on the equipment. **Figure 110–34**

Author's Comments:

- See Article 100 for the definitions of "Panelboard" and "Qualified Persons."
- This rule is meant to warn qualified persons who work on energized electrical systems that an arc flash hazard exists so they will select proper personal protective equipment (PPE) in accordance with industry accepted safe work practice standards.

FPN No. 1: NFPA 70E, *Standard for Electrical Safety in the Workplace*, provides assistance in determining the severity of potential exposure, planning safe work practices, and selecting personal protective equipment.

Author's Comment: In some installations, the use of current-limiting protection devices may significantly reduce the degree of arc flash hazards. For more information about flash protection, visit http://bussmann.com/safetybasics. **Figure 110–35**

110.21 Manufacturer's Markings.

The manufacturer's name, trademark, or other descriptive marking must be placed on all electrical equipment. Where required by the *Code*, markings such as voltage, current, wattage, or other ratings must be provided with sufficient durability to withstand the environment involved.

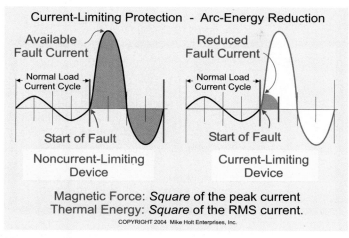

Figure 110–35

110.22 Identification of Disconnecting Means. All

installed disconnecting means must be legibly marked to indicate their purpose unless they are located and arranged so their purpose is evident. In addition, the marking must be of sufficient durability to withstand the environment involved. **Figure 110–36**

> **Author's Comment:** See Article 100 for the definition of "Disconnecting Means."

PART II. 600V, NOMINAL OR LESS

110.26 Spaces About Electrical Equipment. For the

purpose of safe operation and maintenance of equipment, sufficient access and working space must be provided. Enclosures housing electrical apparatus that are controlled by locks are considered accessible to qualified persons who require access. **Figure 110–37**

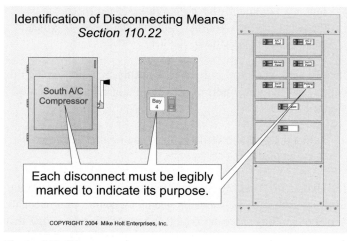

Figure 110–36

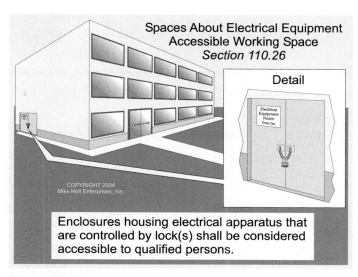

Figure 110–37

Author's Comments:

- See Article 100 for the definition of "Accessible" as it applies to equipment.
- It might be unwise to use an electrically operated lock, if it locks in the de-energized condition!

(A) Working Space. Working space for equipment that may need examination, adjustment, servicing, or maintenance <u>while energized</u> must have sufficient working space in accordance with (1), (2), and (3):

> **Author's Comment:** The phrase "while energized" is the root of many debates. Since electric power to almost all equipment can be turned off, one could argue that working space is never required!

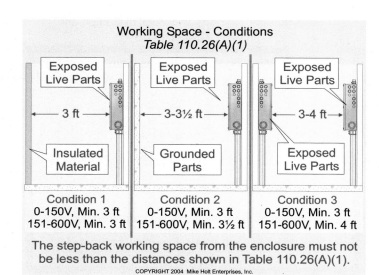

Figure 110–38

(1) Depth of Working Space. The step-back working space, measured from the enclosure front must not be less than the distances contained in Table 110.26(A)(1). **Figure 110–38**

Table 110.26(A)(1) Step-Back Working Space

Voltage-to-Ground	Condition 1	Condition 2	Condition 3
0–150V	3 ft	3 ft	3 ft
151–600V	3 ft	3½ ft	4 ft

- **Condition 1**—Exposed live parts on one side of the working space and no live or grounded parts on the other side of the working space.
- **Condition 2**—Exposed live parts on one side of the working space and grounded parts on the other side of the working space. For this table, concrete, brick, or tile walls are considered grounded.
- **Condition 3**—Exposed live parts on both sides of the working space.

(a) Rear and Sides. Step-back working space isn't required for the back or sides of assemblies where all connections are accessible from the front. **Figure 110–39**

(b) Low Voltage. Where special permission is granted in accordance with 90.4, working space for equipment that operates at not more than 30V ac or 60V dc can be smaller than the distance in Table 110.26(A)(1). **Figure 110–40**

> **Author's Comment:** See Article 100 for the definition of "Special Permission."

Working Space - Sides and Back of Equipment
Section 110.26(A)(1)(a)

Working Space Not Required

Equipment Assembly

Work Space

Step-back working space isn't required for the back or sides of assemblies where all connections are accessible from the front.

COPYRIGHT 2004 Mike Holt Enterprises, Inc.

Figure 110–39

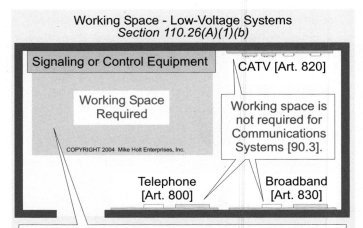

Working Space - Low-Voltage Systems
Section 110.26(A)(1)(b)

Signaling or Control Equipment

CATV [Art. 820]

Working Space Required

Working space is not required for Communications Systems [90.3].

COPYRIGHT 2004 Mike Holt Enterprises, Inc.

Telephone [Art. 800]

Broadband [Art. 830]

Where special permission is granted, working space for equipment operating at not more than 30V ac or 60V dc can be smaller than the distance in Table 110.26(A)(1).

Figure 110–40

(c) Existing Buildings. Where electrical equipment is being replaced, Condition 2 working clearance is permitted between dead-front switchboards, panelboards, or motor control centers located across the aisle from each other where conditions of maintenance and supervision ensure that written procedures have been adopted to prohibit equipment on both sides of the aisle to be open at the same time and only authorized, qualified persons will service the installation.

> **Author's Comment:** The step-back working space requirements of 110.26 do not apply to equipment included in Chapter 8 Communications Circuits [90.3]. **See Figure 110–40.**

(2) Width of Working Space. The width of the working space must be a minimum of 30 in., but in no case less than the width of the equipment. **Figure 110–41**

> **Author's Comment:** The width of the working space can be measured from left-to-right, from the right-to-left, or simply centered on the equipment. **Figure 110–42**

In all cases, the working space must be of sufficient width, depth, and height to permit all equipment doors to open 90°. **Figure 110–43**

> **Author's Comment:** Working space can overlap the working space for other electrical equipment.

(3) Height of Working Space (Headroom). For service equipment, switchboards, panelboards, and motor control equipment, the height of working space in front of equipment must not be less than 6½ ft, measured from the grade, floor, or platform [110.26(E)].

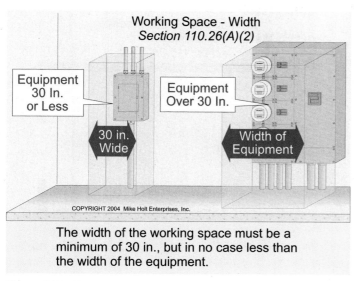

Working Space - Width
Section 110.26(A)(2)

Equipment 30 In. or Less

Equipment Over 30 In.

30 in. Wide

Width of Equipment

COPYRIGHT 2004 Mike Holt Enterprises, Inc.

The width of the working space must be a minimum of 30 in., but in no case less than the width of the equipment.

Figure 110–41

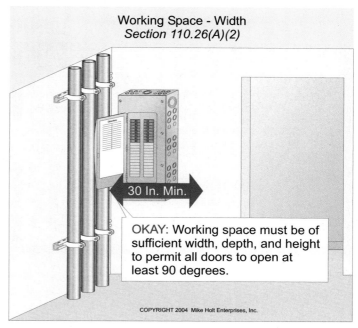

Working Space - Width
Section 110.26(A)(2)

30 In. Min.

OKAY: Working space must be of sufficient width, depth, and height to permit all doors to open at least 90 degrees.

COPYRIGHT 2004 Mike Holt Enterprises, Inc.

Figure 110–43

Equipment such as raceways, cables, wireways, cabinets, panels, etc., can be located above or below electrical equipment, but it must not extend more than 6 in. into the equipment's working space. **Figure 110–44**

(B) Clear Working Space. The working space required by this section must be clear at all times. Therefore, this space is not permitted for storage.

CAUTION: *It's very dangerous to service energized parts in the first place, and it's unacceptable to be subjected to additional dangers by working about, around, over, or under bicycles, boxes, crates, appliances, and other impediments.* Figure 110–45

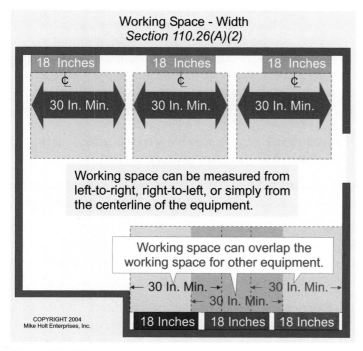

Working Space - Width
Section 110.26(A)(2)

18 Inches | 18 Inches | 18 Inches

30 In. Min. | 30 In. Min. | 30 In. Min.

Working space can be measured from left-to-right, right-to-left, or simply from the centerline of the equipment.

Working space can overlap the working space for other equipment.

30 In. Min. | 30 In. Min.
30 In. Min.

18 Inches | 18 Inches | 18 Inches

COPYRIGHT 2004 Mike Holt Enterprises, Inc.

Figure 110–42

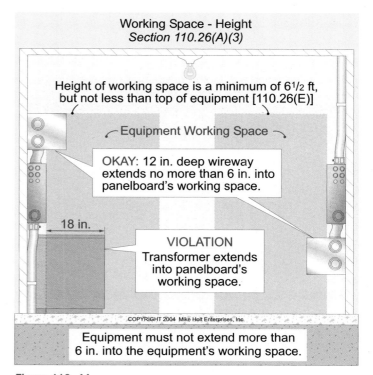

Working Space - Height
Section 110.26(A)(3)

Height of working space is a minimum of 6½ ft, but not less than top of equipment [110.26(E)]

Equipment Working Space

OKAY: 12 in. deep wireway extends no more than 6 in. into panelboard's working space.

18 in.

VIOLATION
Transformer extends into panelboard's working space.

COPYRIGHT 2004 Mike Holt Enterprises, Inc.

Equipment must not extend more than 6 in. into the equipment's working space.

Figure 110–44

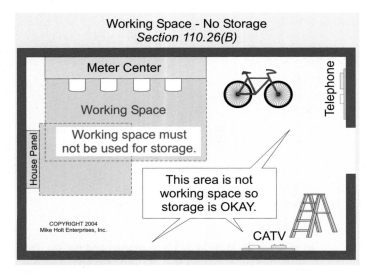

Figure 110–45

Author's Comment: Signaling and communications equipment must not be installed to encroach on the working space of the electrical equipment. **Figure 110–46**

(C) Entrance to Working Space.

(1) Minimum Required. At least one entrance of <u>sufficient area</u> must provide access to the working space.

Author's Comment: Check to see what the authority having jurisdiction considers "sufficient area."

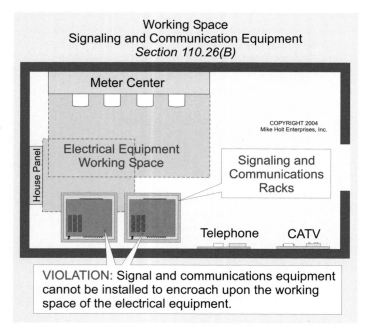

Figure 110–46

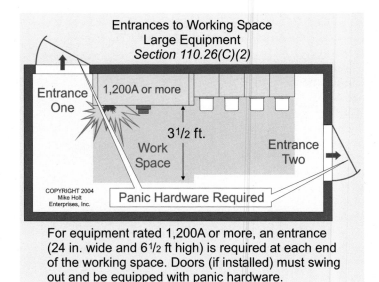

Figure 110–47

(2) Large Equipment. For equipment rated 1,200A or more, an entrance measuring not less than 24 in. wide and 6½ ft high is required at each end of the working space. Where the entrance to the working space has a door, the door must open out and be equipped with panic hardware or other devices that open under simple pressure. **Figure 110–47**

Author's Comment: Since this requirement is in the *NEC*, the electrical contractor is responsible for ensuring that panic hardware is installed where required. Some electrical contractors are offended at being held liable for nonelectrical responsibilities, but this rule should be a little less offensive, given that it's designed to save electricians' lives. For this and other reasons, many construction professionals routinely hold "pre-construction" or "pre-con" meetings to review potential opportunities for miscommunication—before the work begins.

A single entrance to the required working space is permitted where either of the following conditions is met.

(a) Unobstructed Exit. Only one entrance is required where the location permits a continuous and unobstructed way of exit travel.

(b) Double Workspace. Only one entrance is required where the required working space is doubled, and the equipment is located so the edge of the entrance is no closer than the required working space distance. **Figure 110–48**

(D) Illumination. Service equipment, switchboards, panelboards, as well as motor control centers located indoors must have illumination located in or next to the working space. Illumination must not be controlled by automatic means only. **Figure 110–49**

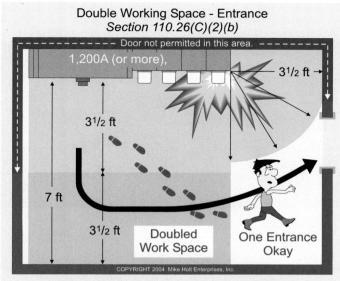

Double Working Space - Entrance
Section 110.26(C)(2)(b)

Only one entrance is required where the required working space is doubled, and the equipment is located so the edge of the entrance is no closer than the required working space distance.

Figure 110–48

(E) Headroom. For service equipment, panelboards, switchboards, or motor control centers, the minimum working space headroom must not be less than 6½ ft. When the height of the equipment exceeds 6½ ft, the minimum headroom must not be less than the height of the equipment.

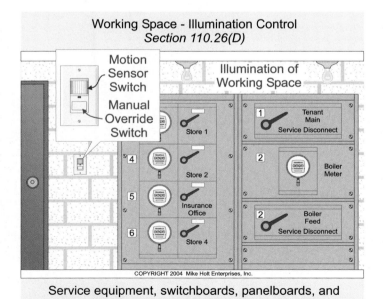

Working Space - Illumination Control
Section 110.26(D)

Service equipment, switchboards, panelboards, and motor control centers located indoors must not have illumination controlled by automatic means only.

Figure 110–49

Exception: The minimum headroom requirement doesn't apply to service equipment or panelboards rated 200A or less located in an existing dwelling unit.

> **Author's Comment:** See Article 100 for the definition of "Dwelling Unit."

(F) Dedicated Equipment Space. Switchboards, panelboards, distribution boards, and motor control centers must comply with the following:

(1) Indoors.

(a) Dedicated Electrical Space. The footprint space (width and depth of the equipment) extending from the floor to a height of 6 ft above the equipment or to the structural ceiling, whichever is lower, must be dedicated for the electrical installation. No piping, duct, or other equipment foreign to the electrical installation can be installed in this dedicated footprint space. **Figure 110–50**

Exception: Suspended ceilings with removable panels can be within the dedicated footprint space.

> **Author's Comment:** Electrical raceways and cables not associated with the dedicated space can be within the dedicated space. It isn't considered "equipment foreign to the electrical installation." See *necdigest* magazine, winter 2003, page 30. **Figure 110–51**

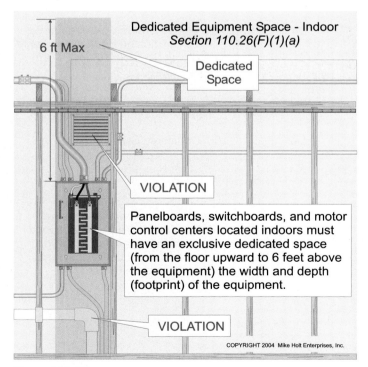

Dedicated Equipment Space - Indoor
Section 110.26(F)(1)(a)

6 ft Max

Dedicated Space

VIOLATION

Panelboards, switchboards, and motor control centers located indoors must have an exclusive dedicated space (from the floor upward to 6 feet above the equipment) the width and depth (footprint) of the equipment.

VIOLATION

Figure 110–50

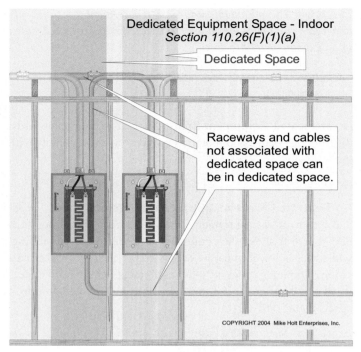

Figure 110-51

(b) Foreign Systems. Foreign systems can be located above the dedicated space if protection is installed to prevent damage to the electrical equipment from condensation, leaks, or breaks in the foreign systems. **Figure 110–52**

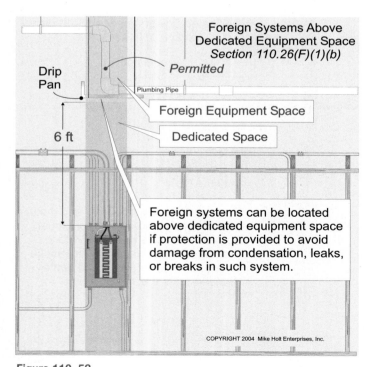

Figure 110-52

Figure 110-53

(c) Sprinkler Protection. Sprinkler protection piping isn't permitted in the dedicated space, but the *NEC* doesn't prohibit sprinklers from spraying water on electrical equipment.

(d) Suspended Ceilings. A dropped, suspended, or similar ceiling isn't considered a structural ceiling.

110.27 Guarding.

(A) Guarding Live Parts. Live parts of electrical equipment operating at 50V or more must be guarded against accidental contact.

(B) Prevent Physical Damage. Electrical equipment must not be installed where it could be subject to physical damage. **Figure 110–53**

Article 110 Questions

(• Indicates that 75% or fewer exam takers get the question correct.)

1. In determining equipment to be installed, considerations such as the following should be evaluated:

   (a) Mechanical strength   (b) Cost   (c) Arcing effects   (d) a and c

2. A wiring method included in the *Code* is recognized as being a(n) _____ wiring method.

   (a) expensive   (b) efficient   (c) suitable   (d) cost-effective

3. •Equipment approved for use in dry locations only must be protected against permanent damage from the weather during _____.

   (a) design   (b) building construction   (c) inspection   (d) none of these

4. The *Code* prohibits damage to the internal parts of electrical equipment by foreign material such as paint, plaster, cleaners, etc. Precautions must be taken to provide protection from the detrimental effects of paint, plaster, cleaners, etc. on internal parts such as _____.

   (a) busbars   (b) wiring terminals   (c) insulators   (d) all of these

5. Soldered splices must first be spliced or joined so as to be mechanically and electrically secure without solder and then be soldered.

   (a) True   (b) False

6. Conductors must have their ampacity determined using the _____ column of Table 310.16 for circuits rated over 100A or marked for conductors larger than 1 AWG, unless the equipment terminals are listed for use with higher temperature rated conductors.

   (a) 60°C   (b) 75°C   (c) 30°C   (d) 90°C

7. Identification of the high leg of a 3Ø, 4-wire delta connected system is required _____.

   (a) at the service disconnect only
   (b) at each point on the system where a connection is made if the equipment grounding (bonding) conductor is also present
   (c) at each point on the system where a connection is made if the grounding electrode conductor is also present
   (d) at each point on the system where a connection is made if the grounded neutral conductor is also present

8. Sufficient access and _____ must be provided and maintained about all electrical equipment to permit ready and safe operation and maintenance of such equipment.

   (a) ventilation   (b) cleanliness   (c) circulation   (d) working space

9. •The required working clearance for access to live parts operating at 300V to ground, where there are exposed live parts on one side and grounded parts on the other side, is _____ according to Table 110.26(A).

   (a) 3 ft   (b) 3½ ft   (c) 4 ft   (d) 4½ ft

10. When normally-enclosed live parts are exposed for inspection or servicing, the working space, if in a passageway or general open space, must be suitably _____.

(a) accessible   (b) guarded   (c) open   (d) enclosed

11. The minimum headroom of working spaces about motor control centers must be _____.

(a) 3 ft   (b) 5 ft   (c) 6 ft   (d) 6½ ft

CHAPTER 2
Wiring and Protection

Introduction

Chapter 2 provides general rules for wiring and protection of conductors. The rules in this chapter apply to all electrical installations covered by the *NEC*—except as modified in Chapters 5, 6, and 7.

Communications systems (Chapter 8 systems) aren't subject to the general requirements of Chapters 1 through 4, or the special requirements of Chapters 5 through 7, unless there's a specific reference in Chapter 8 to a rule in Chapters 1 through 7.

As you go through Chapter 2, remember its purpose. Chapter 2 is primarily concerned with correctly installing and protecting circuits. Every article in Chapter 2 deals with a different aspect of this purpose. This differs from the purpose of Chapter 3, which is to correctly size and install the conductors that make up those circuits.

Chapter 1 introduced you to the *NEC* and provided a solid foundation for *understanding* the *Code*. Chapters 2 and 3 (Wiring Methods and Materials) together form the heart of the *NEC*—they form a solid foundation for *applying* the *NEC*. Chapter 4 applies the preceding chapters to general equipment. So, once again, you find yourself needing to learn the *NEC* in a sequential manner because each chapter builds on the one before it. Once you've mastered the first four chapters, you can learn the next four in any order you wish.

Article 200. Use and Identification of Grounded Neutral Conductors. This article contains the requirements for the use and identification of the grounded neutral conductor and its terminals.

Article 210. Branch Circuits. This article contains the requirements for branch circuits, such as conductor sizing, identification, GFCI protection of receptacles, and receptacle and lighting outlet requirements.

- PART I. GENERAL PROVISIONS
- PART II. BRANCH-CIRCUIT RATINGS
- PART III. REQUIRED OUTLETS

Article 215. Feeders. This article covers the requirements for installation, minimum size, and ampacity of feeders.

Article 220. Branch-Circuit, Feeder, and Service Calculations. This article provides the requirements for sizing branch circuits, feeders, and services, and for determining the number of receptacles on a circuit and the number of branch circuits required.

- PART I. GENERAL
- PART II. BRANCH-CIRCUIT LOAD CALCULATIONS
- PART III. FEEDERS AND SERVICE CALCULATIONS
- PART IV. OPTIONAL CALCULATIONS

Article 225. Outside Branch Circuits and Feeders. This article covers installation requirements for equipment, including conductors located outside, on, or between buildings, poles, and other structures on the premises.

- PART I. GENERAL
- PART II. MORE THAN ONE BUILDING OR STRUCTURE

Article 230. Services. This article covers the installation requirements for service conductors and equipment. It's very important to know where the service begins and ends when applying Articles 230 and 250.

Conductors supplied from a battery, uninterruptible power supply, solar photovoltaic system, generator, or transformer are not considered service conductors; they are feeder conductors.

- PART I. GENERAL
- PART II. OVERHEAD SERVICE-DROP CONDUCTORS
- PART III. UNDERGROUND SERVICE-LATERAL CONDUCTORS
- PART IV. SERVICE-ENTRANCE CONDUCTORS
- PART V. SERVICE EQUIPMENT—GENERAL
- PART VI. SERVICE EQUIPMENT—DISCONNECTING MEANS
- PART VII. SERVICE EQUIPMENT—OVERCURRENT PROTECTION

Article 240. Overcurrent Protection. This article provides the general requirements for overcurrent protection and overcurrent protective devices.

Overcurrent protection for conductors and equipment is provided to open the circuit if the current reaches a value that will cause an excessive or dangerous temperature on the conductors or conductor insulation.

- PART I. GENERAL
- PART II. LOCATION
- PART III. ENCLOSURES
- PART IV. DISCONNECTING AND GUARDING
- PART V. PLUG FUSES, FUSEHOLDERS AND ADAPTERS
- PART VI. CARTRIDGE FUSES AND FUSEHOLDERS
- PART VII. CIRCUIT BREAKERS

Article 250. Grounding (Earthing) and Bonding. Article 250 covers the requirements for providing a low-impedance path to conduct undesired high voltage to the earth, and requirements for the low-impedance fault-current path necessary to facilitate the operation of overcurrent protection devices.

- PART I. GENERAL
- PART II. SYSTEM GROUNDING (BONDING)
- PART III. GROUNDING (EARTHING) ELECTRODE SYSTEM AND GROUNDING ELECTRODE CONDUCTOR
- PART IV. ENCLOSURE, RACEWAY, AND SERVICE CABLE GROUNDING (BONDING)
- PART V. BONDING
- PART VI. EQUIPMENT GROUNDING (BONDING) AND EQUIPMENT GROUNDING (BONDING) CONDUCTORS
- PART VII. METHODS OF EQUIPMENT GROUNDING (BONDING)

Article 280. Surge Arresters. This article covers general requirements, installation requirements, and connection requirements for surge arresters installed on the line side of service equipment.

Article 285. Transient Voltage Surge Suppressors (TVSSs). This article covers general requirements, installation requirements, and connection requirements for transient voltage surge suppressors (TVSSs) permanently installed on the load side of service equipment. It doesn't apply to cord-and-plug connected units, such as "computer power strips."

Use and Identification of Grounded Netural Conductors

Introduction

This article contains the requirements for identification of the grounded neutral conductor and its terminals. If you go back to Article 100, you will see that the *grounded conductor* isn't the same as the *grounding conductor*. Make sure you clearly understand the difference between the two before you begin your study of Article 200.

This article isn't very long, and it's not very complicated, and following these requirements can mean the difference between a safe installation and an electrocution hazard. The illustrations will help you form mental pictures of the key points.

Grounded Conductor. Most electrical power supplies have one output terminal of the power supply bonded to the case of the power supply (system bonding jumper). The conductor that is connected to this grounded terminal is called a "grounded conductor." Figure 200–1

Neutral Conductor. The IEEE dictionary defines a neutral conductor as the conductor with an equal potential difference between it and the other output conductors of a 3- or 4-wire system. Therefore, a neutral conductor is the white/gray wire of a 3-wire single-phase 120/240V system, or of a 4-wire three-phase 120/208V or 277/480V system. Figure 200–2

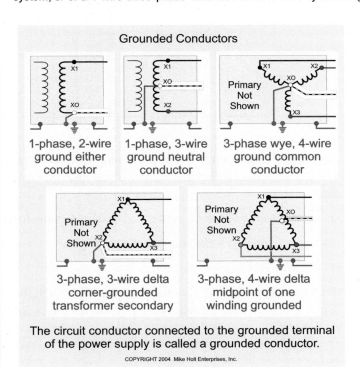

Grounded Conductors

1-phase, 2-wire ground either conductor

1-phase, 3-wire ground neutral conductor

3-phase wye, 4-wire ground common conductor

3-phase, 3-wire delta corner-grounded transformer secondary

3-phase, 4-wire delta midpoint of one winding grounded

The circuit conductor connected to the grounded terminal of the power supply is called a grounded conductor.

COPYRIGHT 2004 Mike Holt Enterprises, Inc.

Figure 200–1

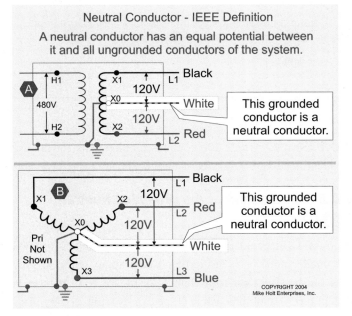

Neutral Conductor - IEEE Definition

A neutral conductor has an equal potential between it and all ungrounded conductors of the system.

This grounded conductor is a neutral conductor.

This grounded conductor is a neutral conductor.

COPYRIGHT 2004
Mike Holt Enterprises, Inc.

Figure 200–2

Since a neutral conductor must have equal potential between it and all ungrounded conductors in a 3- or 4-wire system, the white wire of a 2-wire circuit, and the white wire from a 4-wire three-phase 120/240V delta-connected system are not neutral conductors—they're grounded conductors. Figure 200–3

> **Author's Comment:** The electrical trade industry typically uses the term "neutral," when referring to the white/gray wire. However, the proper term for this conductor is "grounded conductor." Figure 200–4

Technically, it's improper to call a "grounded conductor" a "neutral conductor" or "neutral wire" when it's not truly a neutral conductor, but this is a long-standing industry practice. For the purpose of this textbook this conductor will be called a grounded neutral conductor. That should keep *most people happy.*

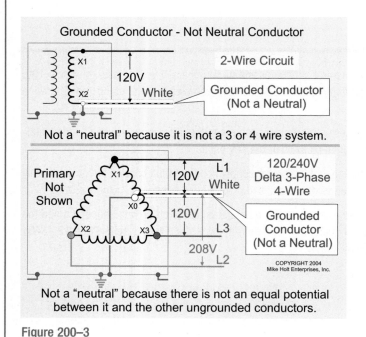

Figure 200–3

Figure 200–4

200.1 Scope.
Article 200 contains requirements for the use and identification of grounded neutral conductors and terminals.

200.6 Identification of the Grounded Neutral Conductor.

(A) 6 AWG or Smaller. grounded neutral conductors 6 AWG and smaller must be identified by a continuous white or gray outer finish along their entire length, or by any color insulation (except green) with three white stripes, or by white or gray insulation with any color stripes (except green). Figure 200–5

> **Author's Comment:** The use of white tape, paint, or other methods of identification isn't permitted for conductors 6 AWG or smaller. Figure 200–6

(B) Larger than 6 AWG. grounded neutral conductors larger than 6 AWG must be identified by one of the following means: Figure 200–7

(1) A continuous white or gray outer finish along its entire length

(2) Three continuous white stripes along its length

(3) White or gray tape or paint at terminations.

(D) Grounded Neutral Conductors of Different Systems. Where grounded neutral conductors of different wiring systems are installed in the same raceway, cable, or enclosure, each grounded neutral conductor must be identified to distinguish the systems by one of the following means:

(1) One system grounded neutral conductor must have an outer covering conforming to 200.6(A) or 200.6(B). Figure 200–8

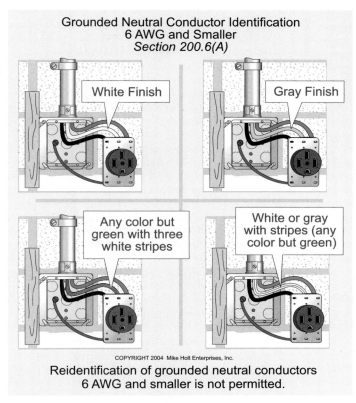

Grounded Neutral Conductor Identification
6 AWG and Smaller
Section 200.6(A)

White Finish

Gray Finish

Any color but green with three white stripes

White or gray with stripes (any color but green)

COPYRIGHT 2004 Mike Holt Enterprises, Inc.

Reidentification of grounded neutral conductors
6 AWG and smaller is not permitted.

Figure 200–5

Author's Comment: This means that you can use either white or gray to identify the grounded neutral conductor of a single wiring system.

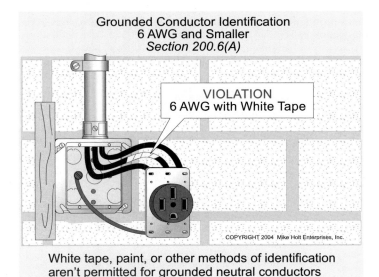

Grounded Conductor Identification
6 AWG and Smaller
Section 200.6(A)

VIOLATION
6 AWG with White Tape

COPYRIGHT 2004 Mike Holt Enterprises, Inc.

White tape, paint, or other methods of identification aren't permitted for grounded neutral conductors 6 AWG or smaller.

Figure 200–6

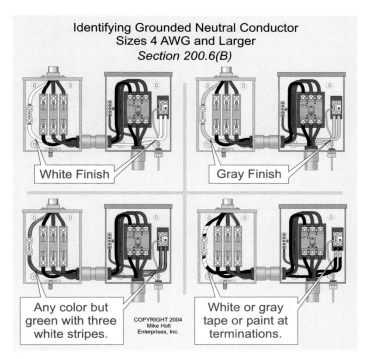

Identifying Grounded Neutral Conductor
Sizes 4 AWG and Larger
Section 200.6(B)

White Finish

Gray Finish

Any color but green with three white stripes.

COPYRIGHT 2004 Mike Holt Enterprises, Inc.

White or gray tape or paint at terminations.

Figure 200–7

(2) The grounded neutral conductor of the other system must have a different outer covering conforming to either 200.6(A) or 200.6(B). Figure 200–9

Or the grounded neutral conductor of the other system must have an outer covering of white or gray with a readily distinguishable, different colored stripe other than green running along the insulation.

(3) Other and different means of identification, as permitted by 200.6(A) or (B), that will distinguish each system grounded neutral conductor.

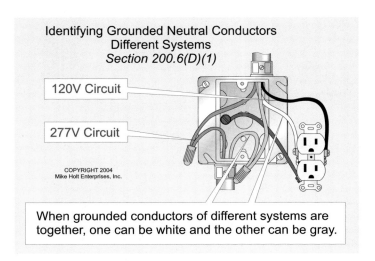

Identifying Grounded Neutral Conductors
Different Systems
Section 200.6(D)(1)

120V Circuit

277V Circuit

COPYRIGHT 2004 Mike Holt Enterprises, Inc.

When grounded conductors of different systems are together, one can be white and the other can be gray.

Figure 200–8

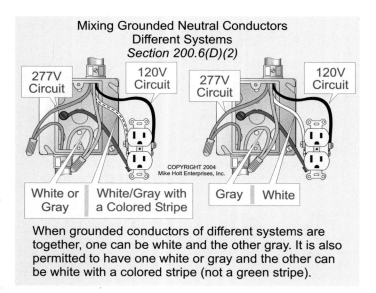

Mixing Grounded Neutral Conductors Different Systems
Section 200.6(D)(2)

When grounded conductors of different systems are together, one can be white and the other gray. It is also permitted to have one white or gray and the other can be white with a colored stripe (not a green stripe).

Figure 200–9

Author's Comment: I guess this means get creative! But remember, you have to have the identification means approved by the authority having jurisdiction.

This means of identification must be permanently posted at each branch-circuit panelboard. Figure 200–10

Author's Comment: Where a premise has branch circuits supplied from more than one voltage system, each ungrounded conductor must be identified by system [210.5(C)].

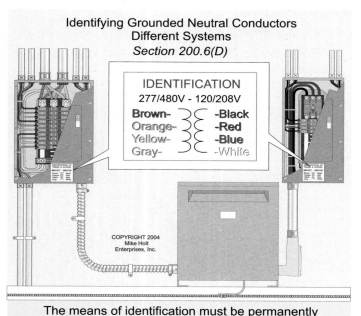

Identifying Grounded Neutral Conductors Different Systems
Section 200.6(D)

IDENTIFICATION
277/480V - 120/208V
Brown-   -Black
Orange-   -Red
Yellow-   -Blue
Gray-   -White

The means of identification must be permanently posted at each branch-circuit panelboard.

Figure 200–10

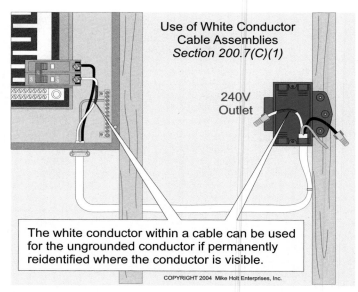

Use of White Conductor Cable Assemblies
Section 200.7(C)(1)

240V Outlet

The white conductor within a cable can be used for the ungrounded conductor if permanently reidentified where the conductor is visible.

Figure 200–11

FPN: Care should be taken when working on existing systems because a gray insulated conductor may have been used in the past as an ungrounded (hot) conductor.

200.7 Use of White or Gray Color.

(C) Circuits Over 50V. A conductor with white insulation can only be used for the ungrounded conductor as permitted in (1), (2), and (3) below.

(1) Cable Assembly. The white conductor within a cable can be used for the ungrounded conductor if permanently reidentified at each location where the conductor is visible to indicate its use as an ungrounded conductor. Identification must encircle the insulation and must be a color other than white, gray, or green. Figure 200–11

(2) Switches. The white conductor within a cable can be used for single-pole, 3-way or 4-way switch loops if permanently identified at each location where the conductor is visible to indicate its use as an ungrounded conductor. Figure 200–12

(3) Flexible Cord. The white conductor within a flexible cord can be used for the ungrounded conductor for connecting an appliance or equipment permitted by 400.7.

FPN: Care should be taken when working on existing systems because a gray insulated conductor may have been used in the past as an ungrounded (hot) conductor.

Author's Comment: The *NEC* doesn't permit the use of white or gray conductor insulation for ungrounded conductors in a raceway, even if the conductors are permanently reidentified. Figure 200–13

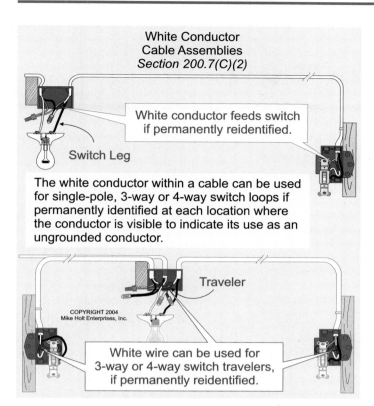

White Conductor
Cable Assemblies
Section 200.7(C)(2)

White conductor feeds switch
if permanently reidentified.

Switch Leg

The white conductor within a cable can be used for single-pole, 3-way or 4-way switch loops if permanently identified at each location where the conductor is visible to indicate its use as an ungrounded conductor.

Traveler

COPYRIGHT 2004
Mike Holt Enterprises, Inc.

White wire can be used for
3-way or 4-way switch travelers,
if permanently reidentified.

Figure 200–12

Screw Shell Terminal Identification
Section 200.10(C)

The grounded conductor must
be connected to the screw shell.

VIOLATION
Reverse Polarity

Correct polarity of a screw shell keeps the screw shell threads from being energized. This reduces the chance of getting a shock when replacing a lamp. See 200.11.

COPYRIGHT 2004 Mike Holt Enterprises, Inc.

Figure 200–14

200.9 Terminal Identification.

The terminal for the grounded neutral conductor must be colored white (actually silver). The terminal for the ungrounded conductor must be a color that is readily distinguishable from white (brass or copper).

Author's Comment: Terminals for the equipment grounding (bonding) conductor must be green [250.126 and 406.9(B)].

200.10 Identification of Terminals.

(C) Screw Shell. To prevent electric shock, the screw shell of a luminaire or lampholder must be connected to the grounded neutral conductor [410.47]. Figure 200–14.

Author's Comment: See Article 100 for the definition of "Luminaire."

200.11 Polarity.

A grounded neutral conductor must not be connected to terminals or leads that will cause reversed polarity [410.23]. See Figure 200–14.

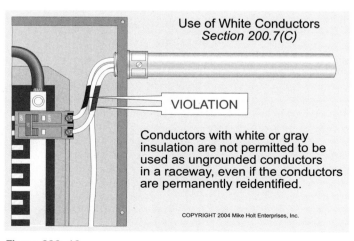

Use of White Conductors
Section 200.7(C)

VIOLATION

Conductors with white or gray insulation are not permitted to be used as ungrounded conductors in a raceway, even if the conductors are permanently reidentified.

COPYRIGHT 2004 Mike Holt Enterprises, Inc.

Figure 200–13

1. Premises wiring must not be electrically connected to a supply system unless the supply system contains, for any grounded neutral conductor of the interior system, a corresponding conductor that is ungrounded.

   (a) True   (b) False

2. Grounded neutral conductors _____ and larger must be identified by a continuous white or gray outer finish along their entire length, by three continuous white stripes along their entire length, or by distinctive white or gray markings such as tape, paint, or other effective means at their terminations.

   (a) 10 AWG   (b) 8 AWG   (c) 6 AWG   (d) 4 AWG

3. Where grounded neutral conductors of different wiring systems are installed in the same raceway, cable, or enclosure, each grounded neutral conductor must be identified by a different one of the acceptable methods in order to distinguish the grounded neutral conductors of each system from the other.

   (a) True   (b) False

4. The white conductor within a cable can be used for the ungrounded (hot) conductor, but the white conductor must be permanently reidentified to indicate its use as an ungrounded (hot) conductor at each location where the conductor is visible and accessible. Identification must _____.

   (a) be by painting or other effective means   (b) be a color other than white, gray, or green
   (c) both a and b   (d) none of these

5. The identification of _____ to which a grounded neutral conductor is to be connected must be substantially white in color.

   (a) wire connectors   (b) circuit breakers   (c) terminals   (d) ground rods

210 Branch Circuits

Introduction

This article contains the requirements for branch circuits, such as conductor sizing and identification, GFCI receptacle protection, and receptacle and lighting outlet requirements. It consists of three parts:

- PART I. GENERAL PROVISIONS
- PART II. BRANCH-CIRCUIT RATINGS
- PART III. REQUIRED OUTLETS

Table 210.2 of this article identifies specific purpose branch circuits. When people complain that the *Code* "buries stuff in the last few chapters and doesn't provide you with any way of knowing where to find things," that is because they didn't pay attention to this table.

The following Sections and Tables contain a few key items to spend extra time on as you study Article 210:

- *210.4. Multiwire Branch Circuits.* The conductors of these circuits must originate from the same panel. These circuits can supply only line-to-neutral loads.
- *210.8. GFCI Protected Receptacles.* Crawl spaces, unfinished basements, and boathouses are just some of the eight locations that require GFCI protection.
- *210.11. Branch Circuits Required.* With three subheadings, 210.11 gives summarized requirements for the number of branch circuits in a given system, states that a load computed on a VA/area basis must be evenly proportioned, and covers rules for dwelling units.
- *210.12. Arc-Fault Circuit-Interrupter Protection.* An AFCI isn't a GFCI, though combination units do exist. The purpose of an AFCI (trips at 30 mA) is to protect equipment. The purpose of a GFCI (trips at 4 to 6 mA) is to protect people.
- *210.19. Conductors—Minimum Ampacity and Size.* This gets complicated in a hurry, but we'll guide you through it.
- *Table 210.21(B)(2)* shows that the maximum load on a given circuit is 80 percent of the receptacle rating and circuit rating. We'll explain more about the implications of this later.
- *210.23. Permissible Loads.* This is intended to prevent a circuit overload from occurring just because someone plugs in a lamp or vacuum cleaner. We'll show you how to conform.
- *210.52. Dwelling Unit Receptacle Outlets.* An area rife with confusion is receptacle spacing. We cut through the confusion, and you'll understand the meaning of 210.52 and how to apply it correctly.

The rest of the material is also important. But mastering these key items will give you a decided edge in your ability to do work free of *Code* violations.

PART I. GENERAL PROVISIONS

210.1 Scope. Article 210 contains the requirements for conductor sizing, overcurrent protection, identification, and GFCI protection of branch circuits, as well as receptacle outlets and lighting outlet requirements.

Author's Comments:

- See Article 100 for the definition of "Branch Circuit."
- Article 100 defines a "branch circuit" as the conductors between the final overcurrent device and the receptacle outlets, lighting outlets, or other outlets. **Figure 210–1**

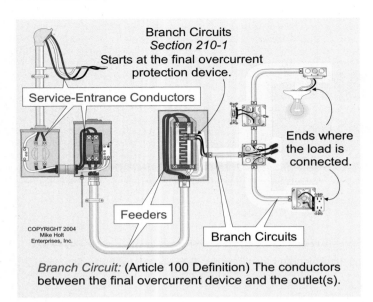

Branch Circuit: (Article 100 Definition) The conductors between the final overcurrent device and the outlet(s).

Figure 210–1

210.2 Other Articles.
Other *NEC* sections that have specific requirements for branch circuits include:

- Air Conditioning and Refrigeration, 440.6, 440.31, and 440.32
- Appliances, 422.10
- Data-Processing (Information Technology) Equipment, 645.5
- Electric Space Heating, 424.3(B)
- Motors, 430.22
- Signs, 600.5

210.3 Branch-Circuit Rating.
The rating of a branch circuit is determined by the rating of the branch-circuit overcurrent protection device, not the conductor size.

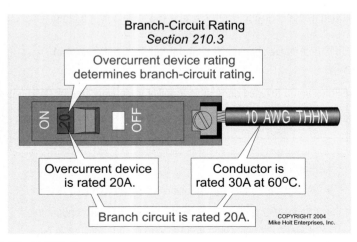

Figure 210–2

Author's Comment: For example, the branch-circuit ampere rating of 10 THHN (rated 30A at 60°C in Table 310.16) on a 20A circuit breaker is 20A. **Figure 210–2**

210.4 Multiwire Branch Circuits.

(A) General. A multiwire branch circuit can be considered a single circuit or a multiple circuit.

Author's Comments:

- See Article 100 for the definition of "Multiwire Branch Circuit."

- Two small-appliance circuits are required for receptacles that serve countertops in dwelling unit kitchens [210.11(C)(1) and 210.52(B)]. One 3-wire, single-phase, 120/240V branch circuit could be used for this purpose. In such a case it's considered a multiwire branch circuit.

To prevent inductive heating and to reduce conductor impedance for fault currents, all multiwire branch-circuit conductors must originate from the same panelboard or distribution equipment.

Author's Comment: For more information on inductive heating of metal parts, see 300.3(B), 300.5(I), and 300.20.

FPN: Unwanted and potentially hazardous harmonic currents can cause additional heating of the neutral conductor of a 4-wire three-phase 120/208V or 277/480V wye-connected system, which supplies nonlinear loads. To prevent fire or equipment damage from excessive harmonic neutral current, the designer should consider: (1) increasing the size of the neutral conductor, or (2) installing a separate neutral for each phase. Also see 220.61(C)(2) FPN 2, and 310.15(B)(4)(c). **Figure 210–3**

Author's Comments:

- See Article 100 for the definition of "Nonlinear Load."

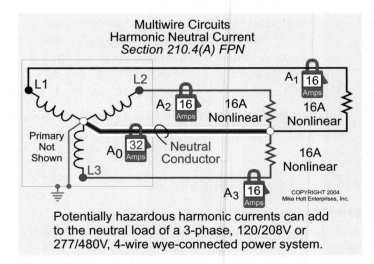

Potentially hazardous harmonic currents can add to the neutral load of a 3-phase, 120/208V or 277/480V, 4-wire wye-connected power system.

Figure 210–3

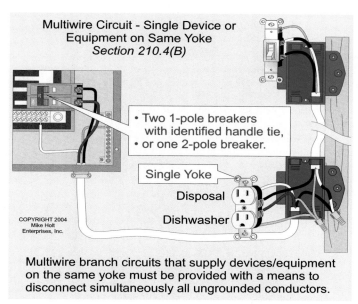

Multiwire Circuit - Single Device or
Equipment on Same Yoke
Section 210.4(B)

• Two 1-pole breakers
  with identified handle tie,
• or one 2-pole breaker.

Single Yoke

Disposal

Dishwasher

COPYRIGHT 2004
Mike Holt
Enterprises, Inc.

Multiwire branch circuits that supply devices/equipment
on the same yoke must be provided with a means to
disconnect simultaneously all ungrounded conductors.

Figure 210–4

• For more information, please visit www.MikeHolt.com. Click
  on "Technical Information" on the left side of the page, and
  then select "Power Quality."

(B) Devices or Equipment. Multiwire branch circuits that
supply devices or equipment on the same yoke (also called a
strap) must be provided with a means to disconnect simultaneously all ungrounded conductors that supply those devices or
equipment at the point where the branch circuit originates.
Figure 210–4

Author's Comments:

• See 210.7(B) for similar requirements for devices or equipment supplied by multiple branch circuits.

• A "yoke" is the metal mount structure for a switch, receptacle, switch and receptacle, switch and pilot light, etc. It's
  also known as a "strap." **Figure 210–5**

• Individual single-pole circuit breakers with handle ties identified for the purpose, or a breaker with common internal trip,
  can be used for this application [240.20(B)(1)].

CAUTION: *This rule is intended to prevent people from
working on energized circuits that they thought were disconnected.*

Author's Comment: Two or more branch circuits that supply
devices or equipment on the same yoke must be provided with
a means to disconnect simultaneously all ungrounded conductors that supply those devices or equipment [210.7(B)].

(C) Line-to-Neutral Loads. Multiwire branch circuits must
supply only line-to-neutral loads.

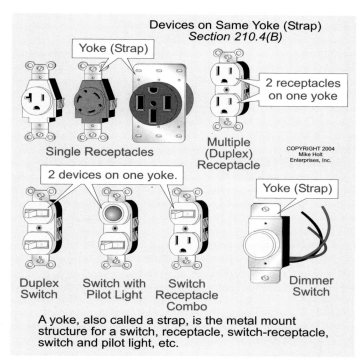

Devices on Same Yoke (Strap)
Section 210.4(B)

Yoke (Strap)

2 receptacles
on one yoke

Single Receptacles

Multiple
(Duplex)
Receptacle

COPYRIGHT 2004
Mike Holt
Enterprises, Inc.

2 devices on one yoke.

Yoke (Strap)

Duplex
Switch

Switch with
Pilot Light

Switch
Receptacle
Combo

Dimmer
Switch

A yoke, also called a strap, is the metal mount
structure for a switch, receptacle, switch-receptacle,
switch and pilot light, etc.

Figure 210–5

*Exception 1: A multiwire branch circuit is permitted to supply
line-to-line utilization equipment, such as a range or dryer.*

*Exception 2: A multiwire branch circuit is permitted to supply
both line-to-line and line-to-neutral loads if the circuit is protected by a device (multipole circuit breaker) that opens all
ungrounded conductors of the multiwire branch circuit simultaneously (common internal trip) under a fault condition.*
Figure 210–6

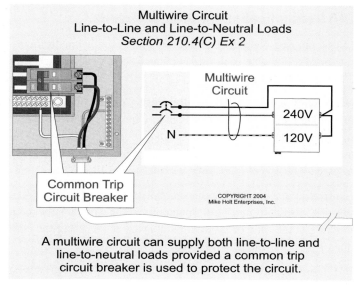

Multiwire Circuit
Line-to-Line and Line-to-Neutral Loads
Section 210.4(C) Ex 2

Multiwire
Circuit

240V

N

120V

Common Trip
Circuit Breaker

COPYRIGHT 2004
Mike Holt Enterprises, Inc.

A multiwire circuit can supply both line-to-line and
line-to-neutral loads provided a common trip
circuit breaker is used to protect the circuit.

Figure 210–6

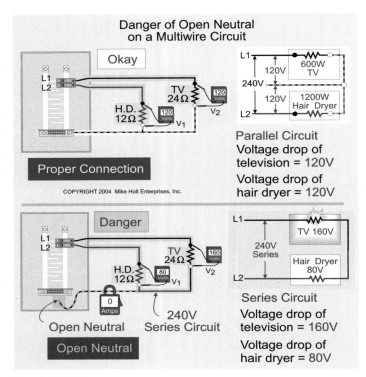

Danger of Open Neutral
on a Multiwire Circuit

Okay

Proper Connection

COPYRIGHT 2004 Mike Holt Enterprises, Inc.

Danger

Open Neutral

Open Neutral

Parallel Circuit
Voltage drop of
television = 120V
Voltage drop of
hair dryer = 120V

Series Circuit
Voltage drop of
television = 160V
Voltage drop of
hair dryer = 80V

Figure 210–7

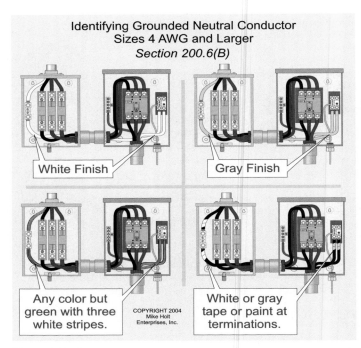

Identifying Grounded Neutral Conductor
Sizes 4 AWG and Larger
Section 200.6(B)

White Finish

Gray Finish

Any color but
green with three
white stripes.

COPYRIGHT 2004
Mike Holt
Enterprises, Inc.

White or gray
tape or paint at
terminations.

Figure 210–8

FPN: See 300.13(B) for the requirements relating to the continuity of the grounded neutral conductor on multiwire circuits.

CAUTION: *If the continuity of the grounded neutral conductor of a multiwire circuit is interrupted (open), the resultant over- or undervoltage could cause a fire and/or destruction of electrical equipment. For details on how this occurs, see 300.13(B) in this textbook.* **Figure 210–7**

210.5 Identification for Branch Circuits.

(A) Grounded Neutral Conductor. The grounded neutral conductor of a branch circuit must be identified in accordance with 200.6. **Figure 210–8**

(B) Equipment Grounding (Bonding) Conductor. Equipment grounding (bonding) conductors can be bare, covered, or insulated. Insulated equipment grounding (bonding) conductors sized 6 AWG and smaller must have a continuous outer finish that is either green or green with one or more yellow stripes [250.119]. **Figure 210–9**

Equipment grounding (bonding) conductors larger than 6 AWG, that are insulated can be permanently reidentified with green marking at the time of installation at every point where the conductor is accessible [250.119(A)].

(C) Ungrounded Conductors. Where the premises wiring system contains branch circuits supplied from more than one

voltage system, each ungrounded conductor, where accessible, must be identified by system. Identification can be by color-coding, marking tape, tagging, or other means approved by the authority having jurisdiction. Such identification must be permanently posted at each branch-circuit panelboard or branch-circuit distribution equipment. **Figure 210–10**

Author's Comments:

• Electricians often use the following color system for power and lighting conductor identification:
  • 120/240V single-phase—black, red, and white

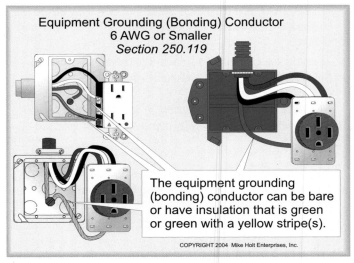

Equipment Grounding (Bonding) Conductor
6 AWG or Smaller
Section 250.119

The equipment grounding
(bonding) conductor can be bare
or have insulation that is green
or green with a yellow stripe(s).

COPYRIGHT 2004 Mike Holt Enterprises, Inc.

Figure 210–9

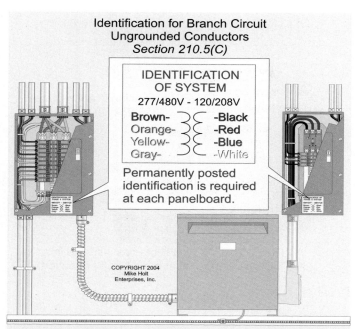

Figure 210–10

- 120/208V three-phase—black, red, blue, and white
- 120/240V three-phase—black, orange, blue, and white
- 277/480V three-phase—brown, orange, yellow, and gray; or, brown, purple, yellow, and gray
- Conductors with insulation that is green or green with one or more yellow stripes cannot be used for an ungrounded or grounded neutral conductor [250.119].

210.6 Branch-Circuit Voltage Limitation.

(A) Occupancy Limitation. In dwelling units and in guest rooms or guest suites of hotels, motels, and similar occupancies, the voltage between conductors that supply the terminals of the following must not exceed 120V, nominal:

(1) Luminaires

(2) Cord-and-plug connected loads rated not more than 1,440 volt-amperes (VA), or less than ¼ horsepower

(C) 277/480V Circuits. Circuits not exceeding 277V-to-ground are permitted to supply any of the following luminaires:

(1) Listed electric-discharge luminaires

(3) Luminaires with mogul base screw shells

> **CAUTION:** *An Edison base lampholder rated for 120V isn't permitted on a 277V circuit.* **Figure 210–11**

(4) Lampholders other than the screw-shell type.

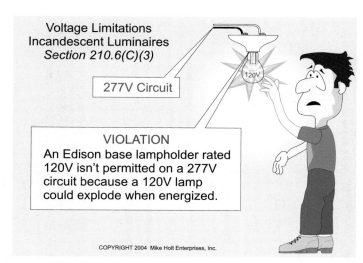

Figure 210–11

210.7 Branch-Circuit Receptacle Requirements.

(B) Multiple Branch Circuits. Where two or more branch circuits supply devices or equipment on the same yoke, a means to disconnect simultaneously all ungrounded conductors that supply those devices or equipment is required at the point where the branch circuit originates. **Figure 210–12**

Author's Comments:

- See 210.4(B) for similar requirements for devices or equipment supplied by multiwire branch circuits.

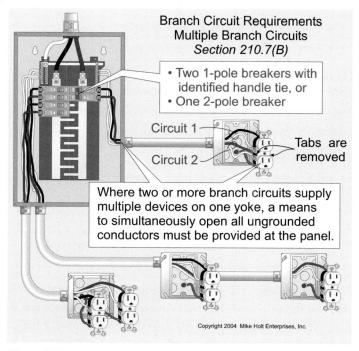

Figure 210–12

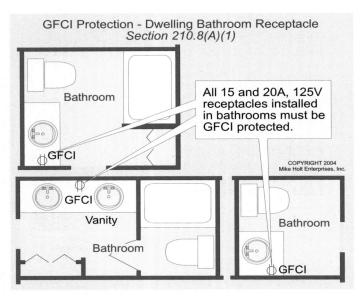

Figure 210–13

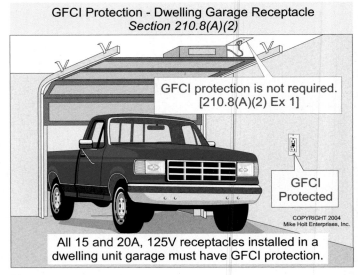

Figure 210–14

- Individual single-pole circuit breakers with handle ties identified for the purpose, or a breaker with common internal trip, can be used for this application [240.20(B)(1)].

210.8 GFCI-Protected Receptacles.

(A) Dwelling Units. GFCI protection is required for all 15 and 20A, 125V receptacles located in the following areas of a dwelling unit:

Author's Comments:
- See Article 100 for the definition of "GFCI."
- Circuits are rated 120V and receptacles are rated 125V.

(1) Bathroom Area. GFCI protection is required for all 15 and 20A, 125V receptacles in the bathroom area of a dwelling unit. See 210.52(D) for acceptable locations for the required bathroom receptacle. **Figure 210–13**

Author's Comments:
- See Article 100 for the definition of "Bathroom."
- *Code*-change proposals that would allow receptacles for dedicated equipment in the bathroom area to be exempted from the GFCI protection requirements have all been rejected because it was not in the interest of safety to allow appliances without GFCI protection in this area.

(2) Garage and Accessory Buildings. GFCI protection is required for all 15 and 20A, 125V receptacles in garages, and in grade-level portions of unfinished or finished accessory buildings used for storage or work areas of a dwelling unit. **Figure 210–14**

Author's Comments:
- See Article 100 for the definition of "Garage."
- A receptacle outlet is required in a dwelling unit attached garage [210.52(G)], but a receptacle outlet isn't required in an accessory building or a detached garage without power. If a 15 or 20A, 125V receptacle is installed in an accessory building, it must be GFCI protected. **Figure 210–15**

Exception 1: GFCI protection isn't required for receptacles that aren't readily accessible, such as those located in the ceiling for the garage door opener motor. **See Figure 210–14.**

   Author's Comment: See Article 100 for the definition of "Readily Accessible."

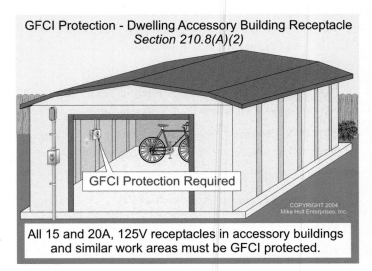

Figure 210–15

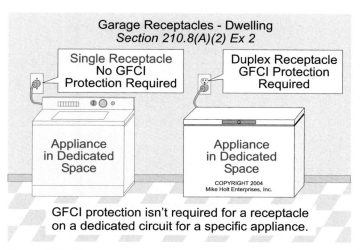

Garage Receptacles - Dwelling
Section 210.8(A)(2) Ex 2

Single Receptacle
No GFCI
Protection Required

Duplex Receptacle
GFCI Protection
Required

Appliance
in Dedicated
Space

Appliance
in Dedicated
Space

COPYRIGHT 2004
Mike Holt Enterprises, Inc.

GFCI protection isn't required for a receptacle
on a dedicated circuit for a specific appliance.

Figure 210–16

Exception 2: GFCI protection isn't required for a receptacle on a dedicated branch circuit located and identified for a specific cord-and-plug connected appliance, such as a refrigerator or freezer. Figure 210–16

Receptacles that aren't readily accessible, or those for a dedicated branch circuit for a specific cord-and-plug connected appliance, as permitted in the two exceptions above, aren't considered as meeting the requirement for a garage receptacle contained in 210.52(G).

(3) Outdoors. All 15 and 20A, 125V receptacles located outdoors of dwelling units, including receptacles installed under the eaves of roofs must be GFCI protected. Figure 210–17

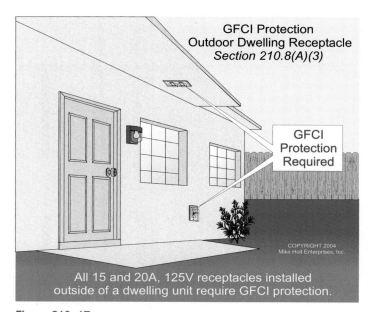

GFCI Protection
Outdoor Dwelling Receptacle
Section 210.8(A)(3)

GFCI
Protection
Required

COPYRIGHT 2004
Mike Holt Enterprises, Inc.

All 15 and 20A, 125V receptacles installed
outside of a dwelling unit require GFCI protection.

Figure 210–17

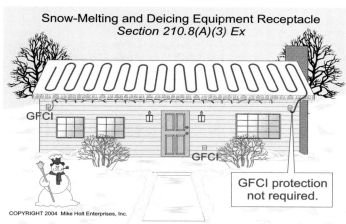

Snow-Melting and Deicing Equipment Receptacle
Section 210.8(A)(3) Ex

GFCI

GFCI

GFCI protection
not required.

COPYRIGHT 2004 Mike Holt Enterprises, Inc.

GFCI protection is not required for receptacles
that supply snow-melting or deicing equipment
if they are not readily accessible.

Figure 210–18

Author's Comment: The ground floor units of a multifamily dwelling (one that contains three or more dwelling units) with individual exterior entrances require at least one outdoor receptacle outlet with GFCI protection. Dwelling units above ground level of a multifamily dwelling unit do not require an outdoor receptacle outlet, but if one is installed, then it must be GFCI protected [210.52(E)].

Exception: GFCI protection isn't required for a fixed electric snow-melting or deicing equipment receptacle supplied by a dedicated branch circuit, if the receptacle isn't readily accessible. See 426.28. Figure 210–18

(4) Crawl Space. All 15 and 20A, 125V receptacles installed in crawl spaces at or below grade of a dwelling unit must be GFCI protected.

Author's Comment: The *Code* doesn't require a receptacle to be installed in the crawl space, except when heating, air-conditioning, and refrigeration equipment is installed there [210.63].

(5) Unfinished Basement. GFCI protection is required for all 15 and 20A, 125V receptacles located in the unfinished portion of a basement not intended as a habitable room and limited to storage and work areas. Figure 210–19

Author's Comment: A receptacle outlet is required in each unfinished portion of a dwelling unit basement [210.52(G)].

Exception 1: GFCI protection isn't required for receptacles that aren't readily accessible.

Exception 2: GFCI protection isn't required for a receptacle on a dedicated branch circuit located and identified for a specific cord-and-plug connected appliance.

GFCI Protection - Dwelling Basement Receptacle
Section 210.8(A)(5)

Finished Basement Area:
GFCI protection not required

COPYRIGHT 2004
Mike Holt Enterprises, Inc.

All 15 and 20A, 125V receptacles in unfinished areas of basements must be GFCI protected.

Figure 210–19

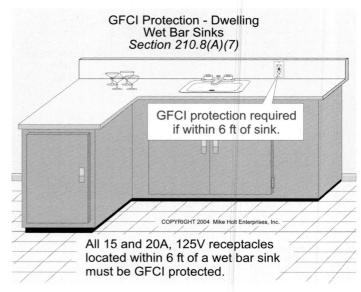

GFCI Protection - Dwelling
Wet Bar Sinks
Section 210.8(A)(7)

GFCI protection required
if within 6 ft of sink.

COPYRIGHT 2004 Mike Holt Enterprises, Inc.

All 15 and 20A, 125V receptacles located within 6 ft of a wet bar sink must be GFCI protected.

Figure 210–21

(6) Kitchen Countertop Surface. GFCI protection is required for all 15 and 20A, 125V receptacles that serve countertop surfaces in a dwelling unit. See 210.52(C) for the location requirements of countertop receptacles. **Figure 210–20**

> **Author's Comment:** GFCI protection is required for all receptacles that serve the countertop surfaces, but GFCI protection isn't required for receptacles that serve built-in appliances, such as dishwashers or kitchen waste disposals.

(7) Laundry, Utility, and Wet Bar Sinks. GFCI protection is required for all 15 and 20A, 125V receptacles located within an arc measurement of 6 ft from the dwelling unit laundry, utility, and wet bar sink. **Figures 210–21 and 210–22**

(8) Boathouse. GFCI protection is required for all 15 and 20A, 125V receptacles located in a dwelling unit boathouse. **Figure 210–23**

> **Author's Comment:** The Code doesn't require a 15 or 20A, 125V receptacle to be installed in a boathouse, but if one is installed, then it must be GFCI protected.

(B) Other than Dwelling Units. GFCI protection is required for all 15 and 20A, 125V receptacles installed in the following commercial/industrial locations:

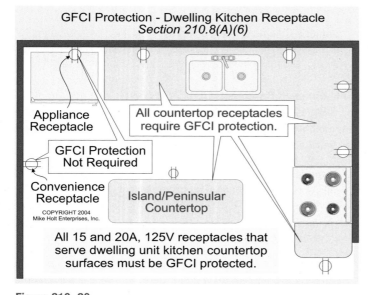

GFCI Protection - Dwelling Kitchen Receptacle
Section 210.8(A)(6)

Appliance
Receptacle

All countertop receptacles require GFCI protection.

GFCI Protection
Not Required

Convenience
Receptacle

Island/Peninsular
Countertop

COPYRIGHT 2004
Mike Holt Enterprises, Inc.

All 15 and 20A, 125V receptacles that serve dwelling unit kitchen countertop surfaces must be GFCI protected.

Figure 210–20

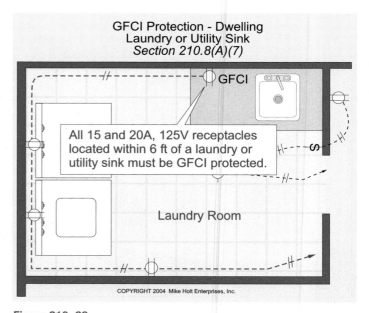

GFCI Protection - Dwelling
Laundry or Utility Sink
Section 210.8(A)(7)

GFCI

All 15 and 20A, 125V receptacles located within 6 ft of a laundry or utility sink must be GFCI protected.

Laundry Room

COPYRIGHT 2004 Mike Holt Enterprises, Inc.

Figure 210–22

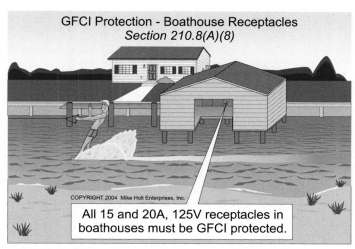

All 15 and 20A, 125V receptacles in boathouses must be GFCI protected.

Figure 210–23

All 15 and 20A, 125V receptacles located in an area with a sink and permanent facilities for food preparation and cooking must be GFCI protected.

Figure 210–25

(1) Bathroom. All 15 and 20A, 125V receptacles installed in commercial or industrial bathrooms must be GFCI protected. See Article 100 for the definition of a bathroom. **Figure 210–24**

> **Author's Comment:** A 15 or 20A, 125V receptacle isn't required in a commercial or industrial bathroom, but if one is installed, then it must be GFCI protected.

(2) Commercial and Institutional Kitchens. All 15 and 20A, 125V receptacles installed in an area with a sink and permanent facilities for food preparation and cooking (kitchens), even those that do not supply the countertop surface, must be GFCI protected. **Figure 210–25**

For the purposes of this section, a kitchen is defined as an area with a sink and permanent facilities for food preparation and cooking.

> **Author's Comment:** GFCI protection is not required for 15 and 20A, 125V receptacles in employee break rooms containing portable cooking appliances.

(3) Rooftops. All 15 and 20A, 125V receptacles installed on rooftops must be GFCI protected. **Figure 210–26**

> **Author's Comment:** A 15 or 20A, 125V receptacle outlet must be installed within 25 ft of heating, air-conditioning, and refrigeration equipment [210.63].

All 15 and 20A, 125V receptacles in nondwelling unit bathrooms must be GFCI protected.

Figure 210–24

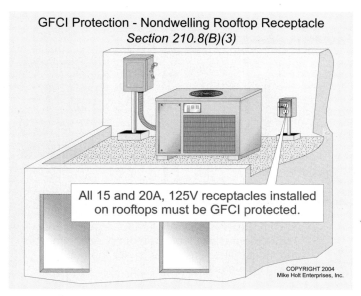

All 15 and 20A, 125V receptacles installed on rooftops must be GFCI protected.

Figure 210–26

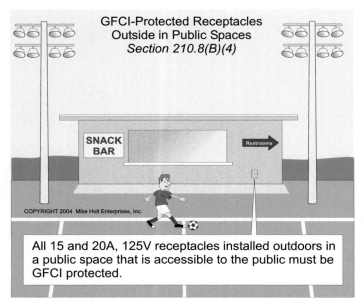

Figure 210–27

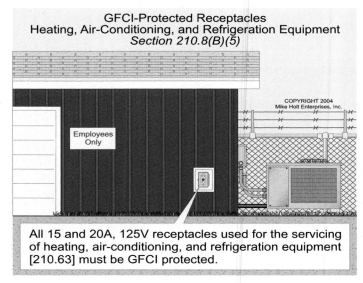

Figure 210–29

Exception: GFCI protection isn't required for a fixed electric snow-melting or deicing equipment receptacle that isn't readily accessible. See 426.28.

(4) Outdoor Public Spaces. All 15 and 20A, 125V receptacles installed outdoors in public spaces used by, or accessible to, the public must be GFCI protected. **Figure 210–27**

> **Author's Comment:** GFCI protection isn't required for receptacles located outdoors of commercial and industrial occupancies where the general public doesn't have access. **Figure 210–28**

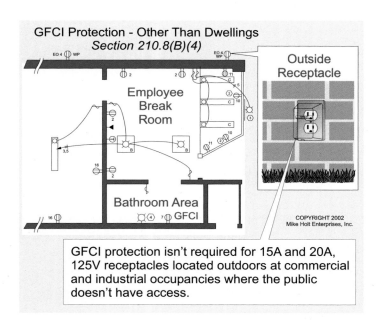

Figure 210–28

(5) Heating, Air-Conditioning, and Refrigeration Equipment. All outdoor 15 and 20A, 125V receptacles installed at an accessible location for the servicing of heating, air-conditioning, and refrigeration equipment in accordance with 210.63 must be GFCI protected. **Figure 210–29**

(C) Boat Hoists. GFCI protection is required for all 15 or 20A, 125V outlets that supply dwelling unit boat hoists.

> **Author's Comments:**
> * See Article 100 for the definition of "Outlet."
> * This ensures GFCI protection regardless of whether the boat hoist is cord-and-plug connected or hard-wired.
> * Since 1971, the *NEC* has been expanding GFCI protection requirements to include the following locations:
>   * Aircraft Hangars, 513.12
>   * Agricultural Buildings, 547.5(G)
>   * Carnivals, Circuses, and Fairs, 525.23
>   * Commercial Garages, 511.12
>   * Elevator Pits, 620.85
>   * Health Care Facilities, 517.20(A)
>   * Marinas and Boatyards, 555.19(B)(1)
>   * Portable or Mobile Signs, 600.10(C)(2)
>   * Swimming Pools, 680.22(A)(5)
>   * Temporary Installations, 590.6

210.11 Branch-Circuits Required.

(A) Number of Branch Circuits. The minimum number of general lighting and general-use receptacle branch circuits must be determined by dividing the total calculated load in amperes by the ampere rating of the circuits used. See Example D1(a) in Annex D.

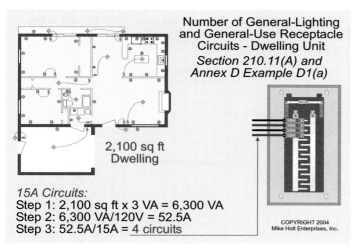

Figure 210–30

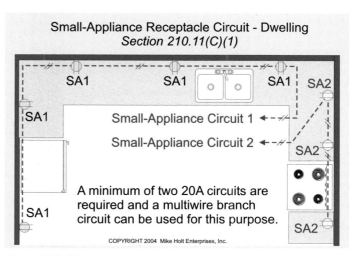

Figure 210–31

Question: How many 15A, 120V circuits are required for the general lighting and general-use receptacles of a 2,100 sq ft dwelling unit? **Figure 210–30**

(a) 1   (b) 2   (c) 3   (d) 4

Answer: (d) 4

Step 1. Determine the total VA load.
  VA = 2,100 sq ft x 3 VA per sq ft [Table 220.12]
  VA = 6,300 VA

Step 2. Determine the total in amperes.
  I = VA/E
  I = 6,300VA/ 120V
  I = 52.5A

Step 3. Determine the number of circuits.
  Number of Circuits = 52.5A/15A
  Number of Circuits = 4

(C) Dwelling Unit.

(1) Small-Appliance Branch Circuits. Two or more 20A, 120V small-appliance receptacle branch circuits are required for the 15 or 20A receptacle outlets in a dwelling unit kitchen, dining room, breakfast room, pantry, or in similar dining areas as required by 210.52(B). The 20A small-appliance receptacle circuits must supply no other outlets [210.52(B)(2)]. **Figure 210–31**

Author's Comments:

- See Article 100 for the definition of "Receptacle Outlets."
- A 15A, 125V receptacle is rated for 20A feed-through, so it can be used for this purpose [210.21(B)(3)].
- Lighting outlets or receptacles located in other areas of a dwelling unit cannot be connected to the small-appliance branch circuit [210.52(B)(2)].

- The two 20A small-appliance branch circuits can be supplied by one 3-wire multiwire circuit or by two separate 120V circuits [210.4(A)].
- The NEC doesn't require each separate countertop to be supplied with two small-appliance circuits [210.52(B)(3)].

(2) Laundry Branch Circuit. One 20A, 120V branch circuit must be provided for the receptacle outlets required by 210.52(F) for a dwelling unit laundry room. The 20A laundry room receptacle circuit is permitted to supply more than one receptacle in the laundry room. The 20A laundry receptacle must not serve any other outlets, such as the laundry room lighting or receptacles in other rooms. **Figure 210–32**

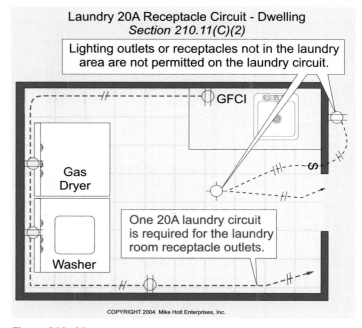

Figure 210–32

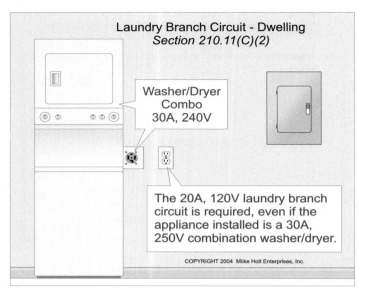

Figure 210–33

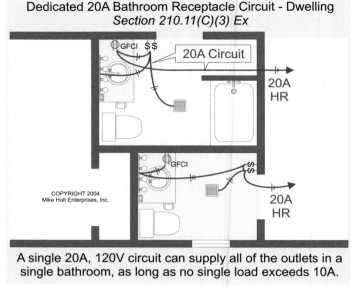

A single 20A, 120V circuit can supply all of the outlets in a single bathroom, as long as no single load exceeds 10A.

Figure 210–35

Author's Comments:

- See Article 100 for the definition of "Continuous Load."

- The 20A, 120V laundry branch circuit is required, even if the laundry appliance installed is a 30A, 250V combination washer/dryer. **Figure 210–33**

- A 15A receptacle is rated for 20A feed-through, so it can be used for this purpose [210.21(B)(3)].

- GFCI protection isn't required for 15 and 20A, 125V receptacles located in a laundry room, unless they are within 6 ft of a sink [210.8(A)(7)].

(3) Bathroom Branch Circuit. One 20A, 120V branch circuit must be provided for the receptacle outlets required by 210.52(D) for a dwelling unit bathroom. This 20A bathroom receptacle circuit must not serve any other outlet, such as bathroom lighting outlets or receptacles in other rooms. **Figure 210–34**

Author's Comments:

- A 15A, 125V receptacle is rated for 20A feed-through, so it can be used for this purpose [210.21(B)(3)].

- A single 20A, 120V receptacle branch circuit can be used to supply multiple bathroom receptacles, including receptacles in different bathrooms.

Exception: A single 20A, 120V branch circuit is permitted to supply all of the outlets in a single bathroom, as long as no single load that is fastened in place is rated more than 10A [210.23(A)]. **Figure 210–35**

> **Question:** *Can a luminaire, ceiling fan, or bath fan be connected to the 20A, 120V branch circuit that supplies one bathroom?*
>
> **Answer:** *Yes.*

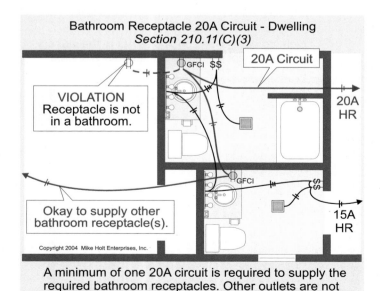

A minimum of one 20A circuit is required to supply the required bathroom receptacles. Other outlets are not permitted on the bathroom receptacle circuit.

Figure 210–34

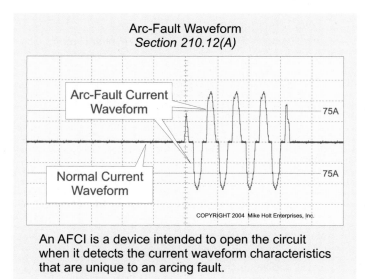

Arc-Fault Waveform
Section 210.12(A)

Arc-Fault Current Waveform

75A

Normal Current Waveform

75A

COPYRIGHT 2004 Mike Holt Enterprises, Inc.

An AFCI is a device intended to open the circuit when it detects the current waveform characteristics that are unique to an arcing fault.

Figure 210–36

210.12 Arc-Fault Circuit-Interrupter (AFCI) Protection.

(A) AFCI Definition. An arc-fault circuit interrupter is a device intended to de-energize the circuit when it detects the current waveform characteristics that are unique to an arcing fault. **Figure 210–36**

(B) Dwelling Unit Bedroom Circuits. All 15 or 20A, 120V branch circuits that supply outlets in dwelling unit bedrooms must be protected by a listed AFCI device. **Figure 210–37**

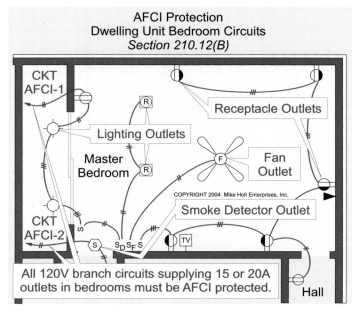

AFCI Protection
Dwelling Unit Bedroom Circuits
Section 210.12(B)

CKT AFCI-1

R

Receptacle Outlets

Lighting Outlets

Master Bedroom

R

F

Fan Outlet

COPYRIGHT 2004 Mike Holt Enterprises, Inc.

Smoke Detector Outlet

CKT AFCI-2

TV

All 120V branch circuits supplying 15 or 20A outlets in bedrooms must be AFCI protected.

Hall

Figure 210–37

Author's Comments:

- The 120V circuit limitation means that AFCI protection isn't required for equipment rated 230V, such as a baseboard heater or room air conditioner. For more information, visit www.MikeHolt.com, click on the "Search" link and search for "AFCI."

- Smoke detectors connected to a 15 or 20A circuit must be AFCI protected if the smoke detector is located in the bedroom of a dwelling unit. The exemption from AFCI protection for the fire alarm circuit [760.21 and 760.41] doesn't apply to the smoke detector's circuit, because a smoke detector circuit isn't defined as a fire alarm circuit; it's an "alarm circuit" [See NFPA 72, *National Fire Alarm Code*].

After January 1, 2008 (basically a 2008 *NEC* requirement), AFCI protection must be provided by a combination-type AFCI protection device.

Author's Comment: Combination type AFCI protection devices provide improved safety performance over existing AFCI protection devices, because the combination type is designed to detect arcs as low as 5A peak. Existing AFCI circuit breakers are designed to operate when the arcs exceed 75A peak. See UL 1699, *Standard for Arc-Fault Circuit Interrupters* (www.UL.com) for information on differences between a branch-circuit type AFCI (circuit breaker) and a combination type AFCI (receptacle).

Exception: The AFCI can be located within 6 ft of the branch-circuit overcurrent device, as measured along the branch-circuit conductors, if the circuit conductors are installed in a metal raceway or a cable with a metallic sheath.

210.18 Guest Rooms and Guest Suites.
Guest rooms and guest suites that are provided with permanent provisions for cooking must have branch circuits and outlets installed to meet the rules for dwelling units.

Author's Comments:

- See Article 100 for the definition of "Guest Room" and "Guest Suite."

- See 210.60 for the requirements for the placement of receptacle outlets in guest rooms and guest suites with permanent provisions for cooking. Also see 210.70(B) for the requirements for the placement of lighting outlets in guest rooms and guest suites with or without provisions for cooking.

PART II. BRANCH-CIRCUIT RATINGS

210.19 Conductor Sizing.

(A) Branch Circuits.

(1) Continuous and Noncontinuous Loads. Conductors must be sized no less than 125 percent of the continuous loads, plus 100 percent of the noncontinuous loads, based on the terminal temperature rating ampacities as listed in Table 310.16, before any ampacity adjustment [110.14(C)].

Author's Comments:

- See Article 100 for the definition of "Continuous Load."

- See 210.20 for the sizing requirements for the branch-circuit overcurrent protection device for continuous and noncontinuous loads.

- Circuit conductors must have sufficient ampacity, after applying adjustment factors, to carry the load, and they must be protected against overcurrent in accordance with their ampacity [210.20(A) and 240.4].

Question: *What size branch-circuit conductor is required for a 44A continuous load if the equipment terminals are rated 75°C as permitted by 110.14(C)(1)(a)?* **Figure 210–38**

(a) 10 AWG   (b) 8 AWG   (c) 6 AWG   (d) 4 AWG

Answer: *(c) 6 AWG*

Since the load is 44A continuous, the conductor must be sized to have an ampacity not less than 55A (44A x 1.25). According to Table 310.16, 75°C column, a 6 AWG conductor is suitable, because it has an ampere rating of 65A at 75°C, before any conductor ampacity adjustment and/or correction.

Exception: Where the assembly and the overcurrent protection device are both listed for 100 percent continuous load operation, the branch-circuit conductors can be sized at 100 percent of the continuous load.

Author's Comment: Equipment suitable for 100 percent continuous loading is rarely available in ratings under 400A.

FPN No. 4: To provide reasonable efficiency of operation of electrical equipment, branch-circuit conductors should be sized to prevent a voltage drop not to exceed three percent. In addition, the maximum total voltage drop on both feeders and branch circuits should not exceed five percent. **Figures 210–39** and **210–40**

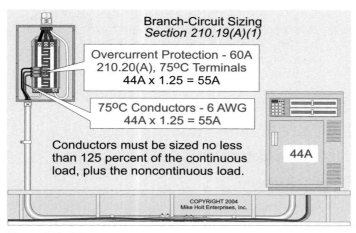

Figure 210–38

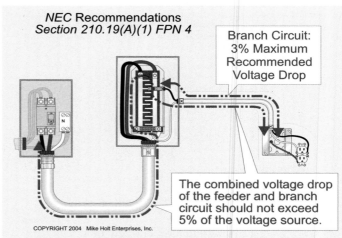

To provide reasonable efficiency of operation of electrical equipment, branch circuit conductors should be sized to prevent a voltage drop not to exceed three percent.

Figure 210–39

Author's Comment: The purpose of the *National Electrical Code* is the practical safeguarding of persons and property from hazards caused by the use of electricity. The *NEC* doesn't consider voltage drop to be a safety issue, except for fire pumps [695.7].

Calculating Conductor Voltage Drop. Conductor voltage drop can be determined by multiplying the current flowing through the circuit by the resistance of the circuit conductors: Voltage Drop = I x R. Where "I" is equal to the load in amperes and "R" is the resistance of the conductor [Chapter 9, Table 8 for direct-current circuits, or Chapter 9, Table 9 for alternating-current circuits]. For three-phase circuits, simply adjust the single-phase voltage-drop value by a multiplier of 0.866

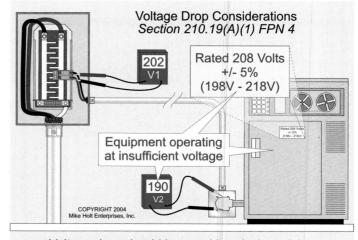

Voltage drop should be considered when sizing branch-circuit conductors, but this is not a requirement.

Figure 210–40

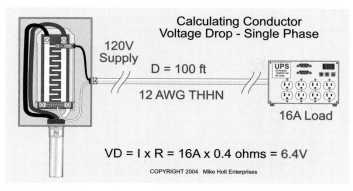

Calculating Conductor Voltage Drop - Single Phase

120V Supply

D = 100 ft

12 AWG THHN

16A Load

VD = I x R = 16A x 0.4 ohms = 6.4V

COPYRIGHT 2004 Mike Holt Enterprises

Figure 210–41

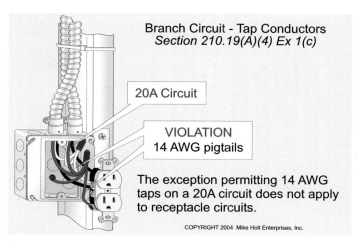

Branch Circuit - Tap Conductors
Section 210.19(A)(4) Ex 1(c)

20A Circuit

VIOLATION
14 AWG pigtails

The exception permitting 14 AWG taps on a 20A circuit does not apply to receptacle circuits.

COPYRIGHT 2004 Mike Holt Enterprises, Inc.

Figure 210–42

Question: *What is the conductor voltage drop for two 12 AWG conductors that supply a single-phase 16A, 120V load located 100 ft from the power supply (200 ft of wire)?* **Figure 210–41**

(a) 3.2V   (b) 6.4V   (c) 9.6V   (d) 12.8V

Answer: *(b) 6.4V*

Voltage Drop = I x R
I = 16A
R = (2 ohms per 1,000 ft/1,000 ft) x 200 ft
R = 0.4 ohms, Chapter 9, Table 9
Voltage Drop = 16A x 0.4 ohms
Voltage Drop = 6.4V
Voltage Drop % = 6.4V/120V
Voltage Drop % = 5.3%

The 5.3 percent voltage drop exceeds the NEC's recommendations of three percent, but this isn't a violation of the Code.

(2) Multioutlet Branch Circuits. Branch-circuit conductors that supply more than one receptacle for cord-and-plug connected portable loads must have an ampacity not less than the rating of the circuit [210.3 and 240.4(D)].

(3) Household Ranges and Cooking Appliances. Branch-circuit conductors that supply household ranges, wall-mounted ovens or counter-mounted cooking units must have an ampacity not less than the rating of the branch circuit and not less than the maximum load to be served. For ranges of 8¾ kW or more rating, the minimum branch-circuit ampere rating must be 40A.

Exception 1: Tap conductors for electric ranges, wall-mounted electric ovens and counter-mounted electric cooking units from a 50A branch circuit must have an ampacity not less than 20A and must have sufficient ampacity for the load to be served. These tap conductors include any conductors that are a part of the lead (pigtail) supplied with the appliance that are smaller than the branch-circuit conductors. The taps must not be longer than necessary for servicing the appliance.

(4) Other Loads. Branch-circuit conductors must have an ampacity sufficient for the loads served and must not be smaller than 14 AWG.

Exception 1: Tap conductors must have an ampacity not less than 15A for circuits rated less than 40A and not less than 20A for circuits rated at 40 or 50A for the following loads:

(c) Individual outlets, other than receptacle outlets, with taps not over 18 in. long.

Author's Comment: Branch-circuit tap conductors aren't permitted for receptacle outlets. **Figure 210–42**

210.20 Overcurrent Protection.

(A) Continuous and Noncontinuous Loads. Branch-circuit overcurrent protection devices must have an ampacity not less than 125 percent of the continuous loads, plus 100 percent of the noncontinuous loads.

Author's Comment: See 210.19(A)(1) for branch-circuit conductor sizing requirements.

Exception; Where the assembly and the overcurrent protection devices are both listed for 100 percent continuous load operation, the branch-circuit protection device can be sized at 100 percent of the continuous load.

Author's Comment: Equipment suitable for 100 percent continuous loading is rarely available in ratings under 400A.

(B) Conductor Protection. Branch-circuit conductors must be protected against overcurrent in accordance with 240.4.

(C) Equipment Protection. Branch-circuit equipment must be protected in accordance with 240.3.

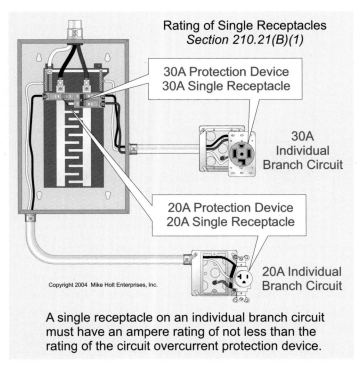

Rating of Single Receptacles
Section 210.21(B)(1)

30A Protection Device
30A Single Receptacle

30A
Individual
Branch Circuit

20A Protection Device
20A Single Receptacle

20A Individual
Branch Circuit

Copyright 2004 Mike Holt Enterprises, Inc.

A single receptacle on an individual branch circuit must have an ampere rating of not less than the rating of the circuit overcurrent protection device.

Figure 210–43

210.21 Outlet Device Rating.

(A) Lampholder Ratings. Lampholders connected to a branch circuit rated over 20A must be of the heavy-duty type.

> **WARNING:** *Fluorescent lampholders aren't rated heavy duty, so fluorescent luminaires cannot be installed on circuits rated over 20A.*

(B) Receptacle Ratings and Loadings.

(1) Single Receptacle. A single receptacle on an individual branch circuit must have an ampacity not less than the rating of the overcurrent protection device. **Figure 210–43**

> **Author's Comment:** A single receptacle has only one contact device on its yoke [Article 100]. This means a duplex receptacle is considered two receptacles.

(2) Multiple Receptacle Loading. Where connected to a branch circuit that supplies two or more receptacles, the total cord-and-plug connected load must not exceed 80 percent of the receptacle rating.

> **Author's Comment:** A duplex receptacle has two contact devices on the same yoke [Article 100]. This means that even one duplex receptacle on a circuit makes that circuit a multi-outlet branch circuit.

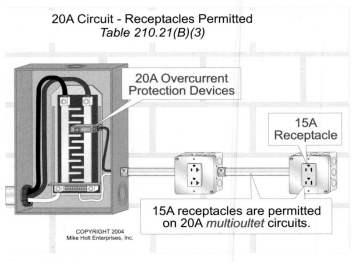

20A Circuit - Receptacles Permitted
Table 210.21(B)(3)

20A Overcurrent
Protection Devices

15A
Receptacle

COPYRIGHT 2004
Mike Holt Enterprises, Inc.

15A receptacles are permitted
on 20A *multioultet* circuits.

Figure 210–44

(3) Multiple Receptacle Rating. Where connected to a branch circuit that supplies two or more receptacles, receptacles must have an ampere rating in accordance with the values listed in Table 210.21(B)(3). **Figure 210–44**

Table 210.21(B)(3) Receptacle Ratings

Circuit Rating	Receptacle Rating
15A	15A
20A	15 or 20A
30A	30A
40A	40 or 50A
50A	50A

210.23 Permissible Loads.
An individual branch circuit is permitted to supply any load for which it's rated. A multioutlet branch circuit must supply loads only in accordance with 210.23(A).

(A) 15 and 20A Circuit. A 15 or 20A branch circuit is permitted to supply lighting, equipment, or any combination of both.

> **Author's Comment:** Except for temporary installations [590.4(D)], 15 or 20A circuits can be used to supply both lighting and receptacles on the same circuit. **Figure 210–45**

(1) Cord-and-Plug Connected Equipment Not Fastened in Place. Cord-and-plug connected equipment not fastened in place, such as a drill press or table saw, must not have an ampere rating more than 80 percent of the branch-circuit rating. **Figure 210–46**

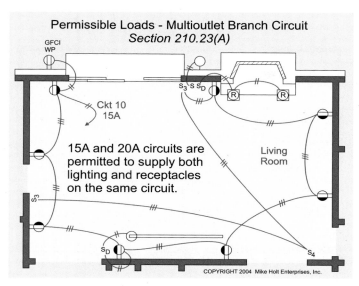

Figure 210–45

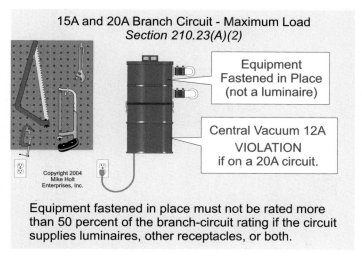

Figure 210–47

Author's Comment: UL and other testing laboratories list portable equipment (such as hair dryers) up to 100 percent of the circuit rating. The *NEC* is an installation standard, not a product standard, so it cannot prohibit this practice. There really is no way to limit the load to 80 percent of the branch-circuit rating if testing laboratories permit equipment to be listed for 100 percent of the circuit rating.

(2) Fixed Equipment. Equipment fastened in place (not a luminaire) must not be rated more than 50 percent of the branch-circuit ampere rating if this circuit supplies luminaires or other receptacles, or both. Figure 210–47

210.25 Common Area Branch Circuits. Dwelling unit branch circuits are only permitted to supply loads within or associated with the dwelling unit.

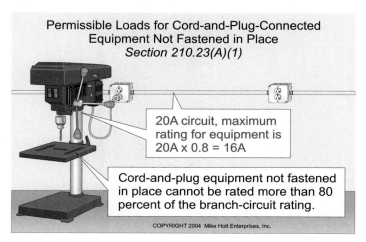

Figure 210–46

Branch circuits for house lighting, central alarm, signal, fire alarm, communications, or other public safety needs must not originate from a dwelling unit.

Author's Comment: This rule prohibits common area branch circuits in two-family or multifamily dwellings from being supplied from an individual dwelling unit. In addition, this prevents common area circuits from being turned off by tenants or by the utility because of nonpayment of electric bills.

PART III. REQUIRED OUTLETS

210.50 General. Receptacle outlets must be installed in accordance with 210.52 through 210.63.

(A) Cord Pendant Receptacle Outlet. A permanently installed flexible cord pendant receptacle is considered a receptacle outlet. Figure 210–48

Author's Comment: See Article 400 for the installation requirements for flexible cords.

(C) Appliance Receptacle Outlet Location. Receptacle outlets installed for a specific appliance, such as a clothes washer, dryer, range, or refrigerator, must be within 6 ft of the intended location of the appliance.

210.52 Dwelling Unit Receptacle Outlet Requirements. This section contains the requirements for 15 and 20A, 125V receptacle outlets for dwelling units.

Author's Comment: Circuits are rated 120V and receptacles are rated 125V.

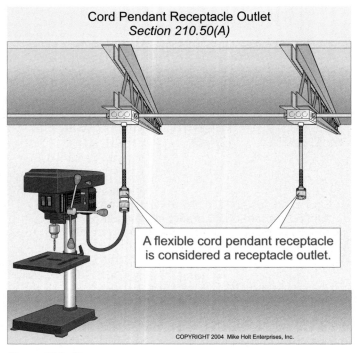

Cord Pendant Receptacle Outlet
Section 210.50(A)

A flexible cord pendant receptacle is considered a receptacle outlet.

COPYRIGHT 2004 Mike Holt Enterprises, Inc.

Figure 210–48

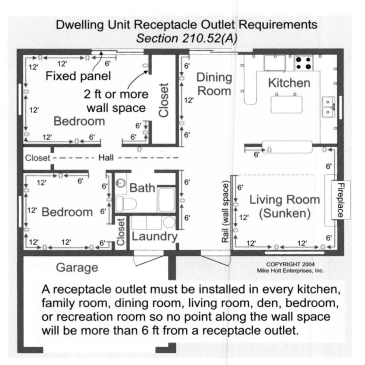

Dwelling Unit Receptacle Outlet Requirements
Section 210.52(A)

A receptacle outlet must be installed in every kitchen, family room, dining room, living room, den, bedroom, or recreation room so no point along the wall space will be more than 6 ft from a receptacle outlet.

Figure 210–49

Receptacle outlets that are part of a luminaire or appliance, or are located within cabinets or cupboards, or are located more than 5½ ft above the floor are not permitted to meet the requirements of this section.

(A) General Requirements—Dwelling Unit. A receptacle outlet must be installed in every kitchen, family room, dining room, living room, parlor, library, den, bedroom, recreation room, and similar room or area in accordance with (1), (2), or (3): Figure 210–49

(1) Receptacle Placement. A receptacle outlet must be installed so no point along the wall space will be more than 6 ft, measured horizontally, from a receptacle outlet.

> **Author's Comment:** The purpose of this rule is to ensure that a general-purpose receptacle is conveniently located to reduce the chance that an extension cord will travel across openings, such as doorways or fireplaces.

(2) Definition of Wall Space.

(1) Any space 2 ft or more in width, unbroken along the floor line by doorways, fireplaces, and similar openings.

(2) The space occupied by fixed panels in exterior walls.

(3) The space occupied by fixed room dividers, such as free-standing bar-type counters or railings.

(3) Floor Receptacle Outlets. Floor receptacle outlets cannot be counted as the required receptacle wall outlet if located more than 18 in. from the wall. Figure 210–50

(B) Small-Appliance Circuit—Dwelling Unit.

(1) Receptacle Outlets. The 20A, 120V small-appliance branch circuits serving the kitchen, pantry, breakfast room, and dining room area of a dwelling unit [210.11(C)(1)] must serve all wall, floor, countertop receptacle outlets [210.52(C)], and the receptacle outlet for refrigeration equipment. Figure 210–51

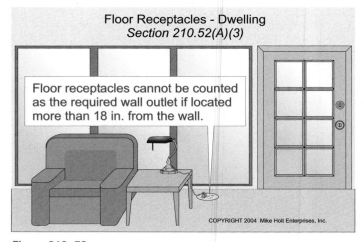

Floor Receptacles - Dwelling
Section 210.52(A)(3)

Floor receptacles cannot be counted as the required wall outlet if located more than 18 in. from the wall.

COPYRIGHT 2004 Mike Holt Enterprises, Inc.

Figure 210–50

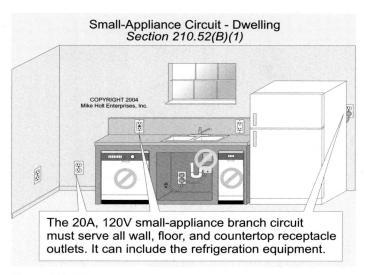

Small-Appliance Circuit - Dwelling
Section 210.52(B)(1)

The 20A, 120V small-appliance branch circuit must serve all wall, floor, and countertop receptacle outlets. It can include the refrigeration equipment.

Figure 210–51

Exception 2: The receptacle outlet for refrigeration equipment can be supplied from an individual branch circuit rated 15A or greater.

(2) Not Supply Other Outlets. The 20A, 120V small-appliance circuits [210.11(C)(1)] cannot be used to supply any other outlet, including outlets for luminaires or appliances.

Exception 1: The 20A, 120V small-appliance branch circuit can be used to supply a receptacle for an electric clock.

Exception 2: A receptacle can be connected to the small-appliance circuit to supply a gas-fired range, oven, or counter-mounted cooking unit. **Figure 210–52**

(3) Kitchen Countertop Receptacles. Kitchen countertop receptacles, as required by 210.52(C) for dwelling units, must be supplied by not less than two 20A, 120V small-appliance branch circuits [210.11(C)(1)]. Either or both of these circuits are per-

mitted to supply receptacle outlets in the same kitchen, pantry, breakfast room, or dining room of the dwelling unit [210.52(B)(1)]. See 210.11(C)(1).

(C) Countertop Receptacle—Dwelling Unit. In kitchens and dining rooms of dwelling units, receptacle outlets for countertop spaces must be installed according to (1) through (5) below.

> **Author's Comment:** GFCI protection is required for all 15 and 20A, 125V receptacles that supply kitchen countertop appliances [210.8(A)(6)].

(1) Wall Counter Space. A receptacle outlet must be installed for each kitchen and dining area countertop wall space that is 1 ft or wider, and receptacles must be placed so no point along the countertop wall space is more than 2 ft, measured horizontally from a receptacle outlet. **Figure 210–53**

Exception: A receptacle outlet isn't required on a wall directly behind a range or sink as shown in **Figure 210–54**

> **Author's Comment:** If the countertop space behind a range or sink is larger that the dimensions noted in Figure 210.52 of the *NEC*, then a GFCI-protected receptacle must be installed in that space. This is because, for all practical purposes, if there's sufficient space for an appliance, an appliance will be placed there.

(2) Island Countertop Space. One receptacle outlet must be installed at each island countertop space with a long dimension of 2 ft or greater, and a short dimension of 1 ft or greater. When breaks occur in countertop spaces for appliances, sinks, etc., and the width of the counter space behind the appliance or sink is less than 1 ft, each countertop space is considered as a separate island for determining receptacle placement [210.52(C)(4)]. **Figure 210–55**

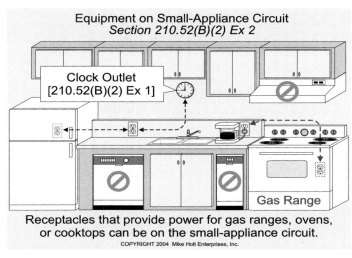

Equipment on Small-Appliance Circuit
Section 210.52(B)(2) Ex 2

Clock Outlet
[210.52(B)(2) Ex 1]

Gas Range

Receptacles that provide power for gas ranges, ovens, or cooktops can be on the small-appliance circuit.
COPYRIGHT 2004 Mike Holt Enterprises, Inc.

Figure 210–52

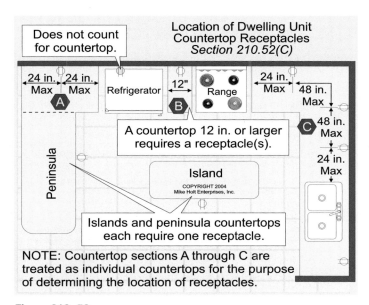

Location of Dwelling Unit
Countertop Receptacles
Section 210.52(C)

Does not count for countertop.

24 in. Max | 24 in. Max — Refrigerator — **A** — 12" — **B** — Range — 24 in. Max | 48 in. Max

48 in. Max
24 in. Max
C

Peninsula

A countertop 12 in. or larger requires a receptacle(s).

Island
COPYRIGHT 2004
Mike Holt Enterprises, Inc.

Islands and peninsula countertops each require one receptacle.

NOTE: Countertop sections A through C are treated as individual countertops for the purpose of determining the location of receptacles.

Figure 210–53

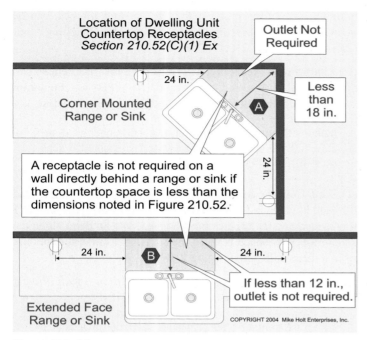

Figure 210–54

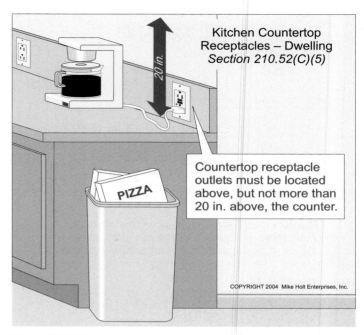

Figure 210–56

(3) Peninsular Countertop Space. One receptacle outlet must be installed at each peninsular countertop with a long dimension of 2 ft or greater, and a short dimension of 1 ft or greater, measured from the connecting edge.

(4) Separate Countertop Spaces. When breaks occur in countertop spaces for appliances, sinks, etc., each countertop space is considered as a separate countertop for determining receptacle placement.

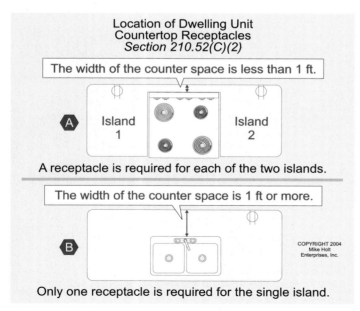

Figure 210–55

(5) Receptacle Location. Receptacle outlets required by 210.52(C)(1) for the countertop space must be located above, but not more than 20 in. above, the counter surface. **Figure 210–56**

Receptacle outlets rendered not readily accessible by appliances fastened in place, located in an appliance garage, behind sinks, or rangetops [210.52(C)(1) Ex.], or supplying appliances that occupy dedicated space are not permitted to be used as the required counter surface receptacles.

> **Author's Comment:** An "appliance garage" is an enclosed area on the counter surface where an appliance can be stored and hidden from view when not in use. If a receptacle is installed inside an appliance garage, it cannot count as a required countertop surface receptacle outlet.

> **Question:** Can a receptacle installed inside an appliance garage be connected to the small-appliance circuit?

> **Answer:** This is a judgment call by the authority having jurisdiction, but receptacles for garbage disposals, dishwashers, compactors, etc., cannot be on the 20A, 120V small-appliance circuits [210.52(B)(2)].

Exception: The receptacle outlet for the countertop space can be installed below the countertop where no wall space is available, such as an island or peninsular counter. Under these conditions, the required receptacle(s) must be located no more than 1 ft below the countertop surface and no more than 6 in. from the counter's edge, measured horizontally. **Figure 210–57**

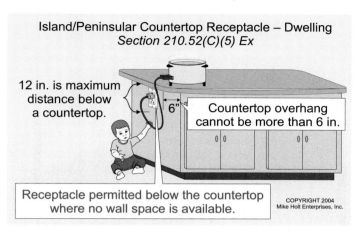

Island/Peninsular Countertop Receptacle – Dwelling
Section 210.52(C)(5) Ex

12 in. is maximum distance below a countertop.

6"

Countertop overhang cannot be more than 6 in.

Receptacle permitted below the countertop where no wall space is available.

COPYRIGHT 2004 Mike Holt Enterprises, Inc.

Figure 210–57

Dwelling Unit Bathroom Receptacle
Section 210.52(D)

Okay if within 3 ft of each.

Basin   Basin

COPYRIGHT 2004 Mike Holt Enterprises, Inc.

One receptacle outlet can meet the requirement if located within 3 ft of the outside edge of each basin.

Figure 210–59

(D) Bathrooms—Dwelling Unit. In dwelling units, not less than one 15 or 20A, 125V receptacle outlet must be installed within an arc measurement of 3 ft from the outside edge of each bathroom basin. The receptacle outlet must be located on a wall or partition that is adjacent to the basin counter surface. See 210.11(C)(3). **Figure 210–58**

Author's Comments:

- One receptacle outlet could be located between two basins to meet the requirement, but only if the receptacle outlet is located within 3 ft of the outside edge of each basin. **Figure 210–59**

- The bathroom receptacles must be GFCI protected [210.8(A)(1)].

Exception: The required bathroom receptacle can be installed on the face or side of the basin cabinet not more than 12 in. below the countertop surface. **Figure 210–60**

(E) Outdoor Receptacle—Dwelling Units.

One-Family Dwelling Unit. Two GFCI-protected receptacle outlets accessible at grade level must be installed outdoors for each one-family dwelling unit, one at the front and one at the back of the dwelling unit, no more than 6½ ft above grade. **Figure 210–61**

> **Author's Comment:** These receptacles must be GFCI protected [210.8(A)(3)].

Two-Family Dwelling Unit. Each dwelling unit of a two-family dwelling that is at grade level must have two GFCI-protected [210.8(A)(3)] receptacle outlets accessible at grade level installed outdoors for each dwelling unit, one at the front and one at the back of each dwelling, no more than 6½ ft above grade.

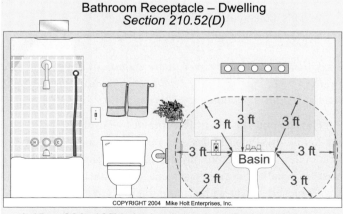

Bathroom Receptacle – Dwelling
Section 210.52(D)

3 ft   3 ft   3 ft
3 ft   3 ft
Basin
3 ft   3 ft

COPYRIGHT 2004 Mike Holt Enterprises, Inc.

A 15 or 20A, 125V single-phase receptacle outlet must be installed within 3 ft of the outside edge of each basin.

Figure 210–58

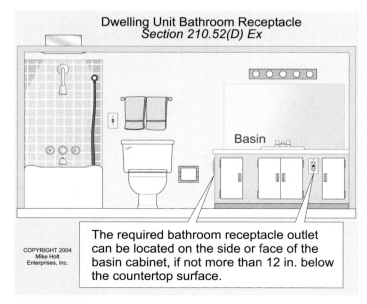

Dwelling Unit Bathroom Receptacle
Section 210.52(D) Ex

Basin

COPYRIGHT 2004 Mike Holt Enterprises, Inc.

The required bathroom receptacle outlet can be located on the side or face of the basin cabinet, if not more than 12 in. below the countertop surface.

Figure 210–60

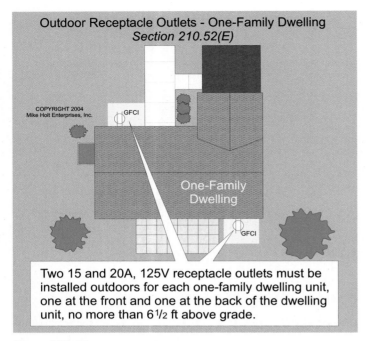

Outdoor Receptacle Outlets - One-Family Dwelling
Section 210.52(E)

COPYRIGHT 2004
Mike Holt Enterprises, Inc.

One-Family Dwelling

GFCI

GFCI

Two 15 and 20A, 125V receptacle outlets must be installed outdoors for each one-family dwelling unit, one at the front and one at the back of the dwelling unit, no more than 6¹/₂ ft above grade.

Figure 210–61

Author's Comment: A receptacle is not required to be located outdoors for dwelling units above grade level, but if installed outdoors, they must be GFCI protected [210.8(A)(3)].

Multifamily Dwelling Unit Building. Each dwelling unit of a multifamily dwelling that has an individual entrance at grade level must have at least one GFCI-protected receptacle outlet accessible from grade level located not more than 6½ ft above grade. **Figure 210-62**

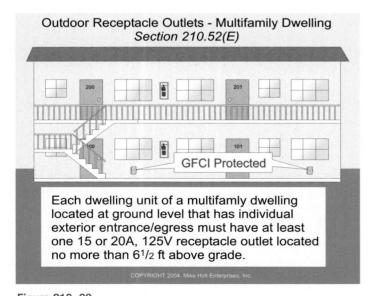

Outdoor Receptacle Outlets - Multifamily Dwelling
Section 210.52(E)

200   201

GFCI Protected

100   101

Each dwelling unit of a multifamily dwelling located at ground level that has individual exterior entrance/egress must have at least one 15 or 20A, 125V receptacle outlet located no more than 6¹/₂ ft above grade.

COPYRIGHT 2004 Mike Holt Enterprises, Inc.

Figure 210–62

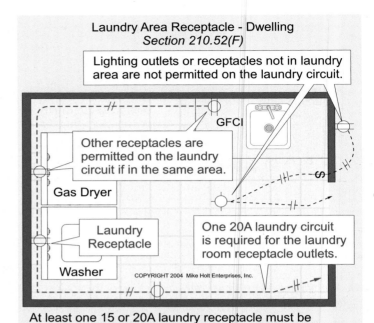

Laundry Area Receptacle - Dwelling
Section 210.52(F)

Lighting outlets or receptacles not in laundry area are not permitted on the laundry circuit.

GFCI

Other receptacles are permitted on the laundry circuit if in the same area.

Gas Dryer

Laundry Receptacle

One 20A laundry circuit is required for the laundry room receptacle outlets.

Washer   COPYRIGHT 2004 Mike Holt Enterprises, Inc.

At least one 15 or 20A laundry receptacle must be installed on the 20A, 120V laundry branch circuit required by 210.11(C)(2). GFCI protection is not required unless receptacle(s) is located within 6 ft of a sink [210.8(A)(7)].

Figure 210–63

Author's Comments:

* See Article 100 for the definition of "Multifamily Dwelling."

* The 2005 *NEC* doesn't specify the rating of the required receptacle, but all 15 or 20A, 125V receptacles located outdoors of a dwelling unit must be GFCI protected [210.8(A)(3)].

(F) Laundry Area Receptacle—Dwelling Unit. Each dwelling unit must have not less than one 15 or 20A, 125V receptacle installed in the laundry area. This receptacle(s) must be supplied by the 20A, 120V laundry branch circuit, which must not supply any other outlets [210.11(C)(2)]. **Figure 210–63**

Author's Comment: All receptacles located within 6 ft of a laundry room sink require GFCI protection [210.8(A)(7)].

Exception 1: A laundry receptacle outlet isn't required in a dwelling unit that is located in a multifamily building with laundry facilities available to all occupants.

(G) Garage and Basement Receptacles—Dwelling Unit. For a one-family dwelling, not less than one 15 or 20A, 125V receptacle outlet, in addition to any provided for laundry equipment, must be installed in each basement, each attached garage, and each detached garage with electric power. **Figure 210–64**

Author's Comment: The required garage and basement receptacles must be GFCI protected in accordance with 210.8(A)(2) for garages and 210.8(A)(5) for unfinished basements.

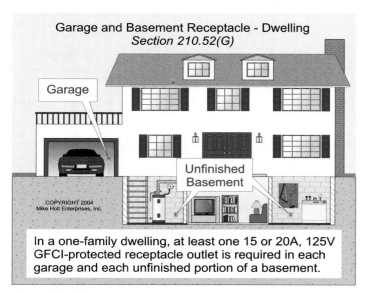

Garage and Basement Receptacle - Dwelling
Section 210.52(G)

In a one-family dwelling, at least one 15 or 20A, 125V GFCI-protected receptacle outlet is required in each garage and each unfinished portion of a basement.

Figure 210–64

Where a portion of the basement is finished into one or more habitable rooms, each separate unfinished portion must have a 15 or 20A, 125V receptacle outlet installed.

> **Author's Comment:** The purpose is to prevent an extension cord from a non-GFCI-protected receptacle to be used to supply power to loads in the unfinished portion of the basement. **See Figure 210–64.**

(H) Hallway Receptacle—Dwelling Unit. One 15 or 20A, 125V receptacle outlet must be installed in each hallway that is at least 10 ft long, measured along the centerline of the hall without passing through a doorway. **Figure 210–65**

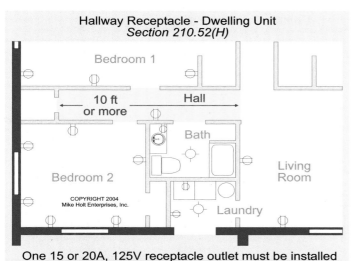

Hallway Receptacle - Dwelling Unit
Section 210.52(H)

One 15 or 20A, 125V receptacle outlet must be installed in each hallway that is at least 10 ft long (measured along the centerline without passing through a doorway).

Figure 210–65

210.60 Receptacles in Guest Rooms or Guest Suites for Hotels and Motels.

(A) General Requirements. Guest rooms or guest suites provided with permanent provisions for cooking must have receptacle outlets installed in accordance with all of the applicable requirements for a dwelling unit as described in 210.52.

210.52(A). Receptacle outlets must be installed so no point along the floor line in any wall space is more than 6 ft, measured horizontally from an outlet in that space, including any wall space 2 ft or more in width.

210.52(D). At least one 15 or 20A, 125V GFCI-protected receptacle outlet must be installed within 3 ft of the outside edge of each bathroom basin [210.8(B)(1)].

(B) Receptacle Placement. The number of receptacle outlets required for guest rooms must not be less than that required for a dwelling unit, in accordance with 210.52(A). To eliminate the need for extension cords by guests for ironing, computers, refrigerators, etc., receptacles can be located to be convenient for permanent furniture layout, but not less than two receptacle outlets must be readily accessible.

Receptacle outlets behind a bed must be located so the bed will not make contact with an attachment plug, or the receptacle must be provided with a suitable guard. **Figure 210–66**

> **Author's Comment:** See Article 100 for the definition of "Attachment Plug."

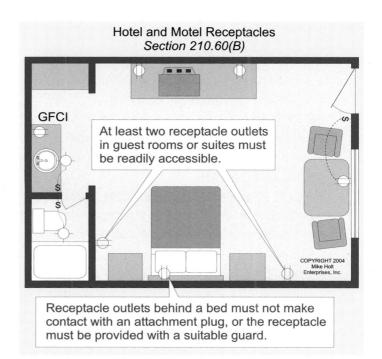

Hotel and Motel Receptacles
Section 210.60(B)

At least two receptacle outlets in guest rooms or suites must be readily accessible.

Receptacle outlets behind a bed must not make contact with an attachment plug, or the receptacle must be provided with a suitable guard.

Figure 210–66

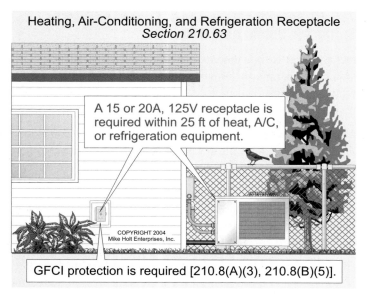

Heating, Air-Conditioning, and Refrigeration Receptacle
Section 210.63

A 15 or 20A, 125V receptacle is required within 25 ft of heat, A/C, or refrigeration equipment.

COPYRIGHT 2004
Mike Holt Enterprises, Inc.

GFCI protection is required [210.8(A)(3), 210.8(B)(5)].

Figure 210–67

210.63 Heating, Air-Conditioning, and Refrigeration Equipment (HACR).

A 15 or 20A, 125V receptacle outlet must be installed at an accessible location for the servicing of heating, air-conditioning, and refrigeration equipment. The receptacle must be located within 25 ft of, and on the same level as, the heating, air-conditioning, and refrigeration equipment. **Figure 210–67**

This receptacle must not be connected to the load side of the equipment disconnecting means.

Author's Comments:

- A receptacle outlet isn't required for ventilation equipment (fans), because it's not HACR equipment (it's HVACR equipment).

- The HACR receptacle must be GFCI protected if located outdoors [210.8(A)(3) and 210.8(B)(5)] or in the crawl space of a dwelling unit [210.8(A)(4)].

- The outdoor 15 or 20A, 125V receptacle outlet required for dwelling units [210.52(E)] can be used to satisfy this requirement, if located within 25 ft of the HACR equipment.

Exception: A receptacle outlet isn't required at one- and two-family dwellings for the service of evaporative coolers.

Author's Comment: This exception doesn't make sense. Since evaporative coolers are listed per UL Standard 507, *Electric Fans*, they aren't heating, air-conditioning, or refrigeration equipment and a receptacle outlet isn't required anyway.

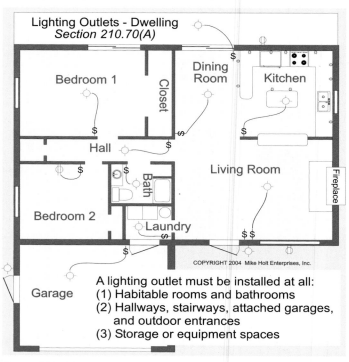

Lighting Outlets - Dwelling
Section 210.70(A)

Bedroom 1   Closet   Dining Room   Kitchen

Hall   Bath   Living Room   Fireplace

Bedroom 2   Laundry

Garage

COPYRIGHT 2004 Mike Holt Enterprises, Inc.

A lighting outlet must be installed at all:
(1) Habitable rooms and bathrooms
(2) Hallways, stairways, attached garages, and outdoor entrances
(3) Storage or equipment spaces

Figure 210–68

210.70 Lighting Outlet Requirements.

(A) Dwelling Unit Lighting Outlet. Lighting outlets must be installed: **Figure 210–68**

(1) Habitable Rooms. At least one wall switch-controlled lighting outlet must be installed in every habitable room and bathroom of a dwelling unit. **Figure 210–69**

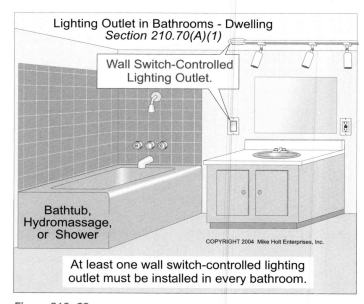

Lighting Outlet in Bathrooms - Dwelling
Section 210.70(A)(1)

Wall Switch-Controlled Lighting Outlet.

Bathtub, Hydromassage, or Shower

COPYRIGHT 2004 Mike Holt Enterprises, Inc.

At least one wall switch-controlled lighting outlet must be installed in every bathroom.

Figure 210–69

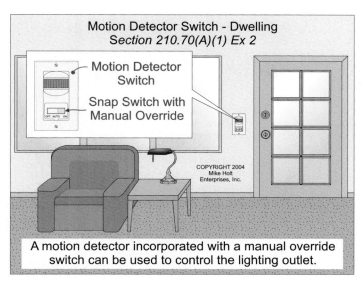

Figure 210–70

Author's Comment: See Article 100 for the definition of "Lighting Outlet."

Exception 1: In other than kitchens and bathrooms, a receptacle controlled by a wall switch can be used instead of a lighting outlet.

Exception 2: Lighting outlets can be controlled by occupancy sensors equipped with a manual override that permits the sensor to function as a wall switch. **Figure 210–70**

(2) Other Areas.

(a) Hallways, Stairways, Garages. In dwelling units, not less than one wall switch-controlled lighting outlet must be installed in hallways, stairways, attached garages, and detached garages with electric power.

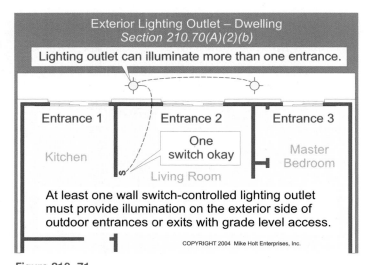

Figure 210–71

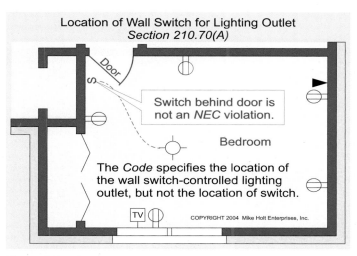

Figure 210–72

(b) Exterior Entrances. At least one wall switch-controlled lighting outlet must provide illumination on the exterior side of outdoor entrances or exits with grade level access. **Figure 210–71**

Author's Comments:

- The *NEC* doesn't require a switch adjacent to each outdoor entrance or exit.

- The *Code* contains the location requirement for the wall switched-controlled lighting outlet, but it doesn't specify the location for the switch. Naturally, you wouldn't want to install a switch behind a door or other inconvenient location, but the *NEC* doesn't require you to relocate the switch to suit the swing of the door. When in doubt as to the best location to place a light switch, consult the job plans or ask the customer. If you're the boss and you don't know, check with the authority having jurisdiction. **Figure 210–72**

- A lighting outlet isn't required to provide illumination on the exterior side of outdoor entrances or exits for a commercial or industrial occupancy.

(c) Stairway. Where the stairway between floor levels has six risers or more, a wall switch must be located at each floor level and landing level that includes an entryway to control the illumination for the stairway.

Exception to (a), (b), and (c): Lighting outlets for hallways, stairways, and outdoor entrances can be switched by a remote, central, or automatic control device. **Figure 210–73**

(3) Storage and Equipment Rooms. At least one lighting outlet that contains a switch or is controlled by a wall switch must be installed in attics, underfloor spaces, utility rooms, and basements used for storage or containing equipment that requires servicing. The switch must be located at the usual point of entry to these spaces, and the lighting outlet must be located at or near the equipment that requires servicing. **Figure 210–74**

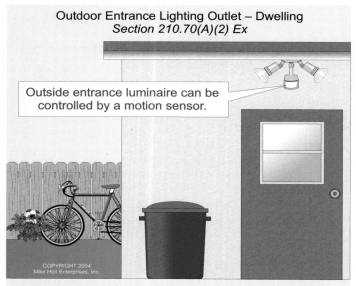

Outdoor Entrance Lighting Outlet – Dwelling
Section 210.70(A)(2) Ex

Outside entrance luminaire can be controlled by a motion sensor.

At outdoor entrances of a dwelling unit, remote, central, or automatic control of lighting is permitted in lieu of a switch.

Figure 210–73

(B) Guest Rooms or Guest Suites. At least one wall switch-controlled lighting outlet must be installed in every habitable room and bathroom of a guest room or guest suite of hotels, motels, and similar occupancies.

Exception 1: In other than bathrooms and kitchens, a receptacle controlled by a wall switch is permitted in lieu of lighting outlets. Figure 210–75

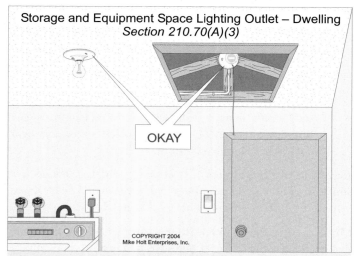

Storage and Equipment Space Lighting Outlet – Dwelling
Section 210.70(A)(3)

OKAY

For attics, underfloor spaces, utility rooms, and basements, at least one lighting outlet containing a switch, or controlled by a wall switch, must be installed where these spaces are used for storage or contain equipment for servicing.

Figure 210–74

Lighting Outlets - Guest Rooms or Guest Suites
Section 210.70(B) Ex

Bathroom and kitchen areas must have a wall-switch controlled lighting outlet [210.70(B)].

In other than bathrooms and kitchens, one or more wall-switch controlled receptacles can be used instead of lighting outlets.

GFCI

Figure 210–75

Exception 2: Lighting outlets can be controlled by occupancy sensors equipped with a manual override that permits the sensor to function as a wall switch.

(C) Commercial Spaces. At least one lighting outlet that contains a switch or is controlled by a wall switch must be installed in attics and underfloor spaces containing equipment that requires servicing. The switch must be located at the usual point of entry to these spaces, and the lighting outlet must be located at or near the equipment requiring servicing.

Author's Comment: A 15 or 20A, 125V receptacle must be installed within 25 ft of HACR equipment [210.63].

1. The rating of a branch circuit is determined by the rating of the _____.

   (a) ampacity of the largest device connected to the circuit
   (b) average of the ampacity of all devices
   (c) branch-circuit overcurrent protection
   (d) ampacity of the branch circuit conductors according to Table 310.16

2. In dwelling units, the voltage between conductors that supply the terminals of _____ must not exceed 120V, nominal.

   (a) luminaires
   (b) cord-and-plug connected loads of 1,440 VA, nominal, or less
   (c) cord-and-plug connected loads of more than ¼ hp
   (d) a and b

3. GFCI protection for personnel is required for all 15 and 20A, 125V single-phase receptacles installed in a dwelling unit _____.

   (a) attic   (b) garage   (c) laundry room   (d) all of these

4. GFCI protection for personnel is required for all 15 and 20A, 125V single-phase receptacles installed to serve the countertop surfaces in dwelling unit kitchens.

   (a) True   (b) False

5. All 15 and 20A, 125V single-phase receptacles _____ of commercial occupancies must have GFCI protection for personnel.

   (a) in bathrooms   (b) on rooftops   (c) in kitchens   (d) all of these

6. Two or more _____, 120V small-appliance branch circuits must be provided to supply power for the receptacle outlets in the dwelling unit kitchen, dining room, breakfast room, pantry, or similar dining areas.

   (a) 15A   (b) 20A   (c) 30A   (d) either 20A or 30A

7. The location of the arc-fault circuit interrupter can be at other than the origination of the branch circuit if _____.

   (a) the arc-fault circuit interrupter is installed within 6 ft of the branch-circuit overcurrent device
   (b) the circuit conductors up to the arc-fault circuit interrupter are in a metal raceway or a cable with a metallic sheath
   (c) both a and b
   (d) none of these

8. Where a branch circuit supplies continuous loads, or any combination of continuous and noncontinuous loads, the rating of the overcurrent device must not be less than the noncontinuous load plus 125 percent of the continuous load.

   (a) True   (b) False

9. It is permitted to base the _____ rating of a range receptacle on a single range demand load specified in Table 220.19.

   (a) circuit   (b) voltage   (c) ampere   (d) resistance

10. Receptacle outlets installed for a specific appliance in a dwelling unit, such as a clothes washer, dryer, range, or refrigerator, must be within _____ of the intended location of the appliance.

    (a) sight   (b) 6 ft
    (c) 3 ft   (d) readily accessible, no maximum distance

11. Receptacle outlets in floors are not counted as part of the required number of receptacle outlets to service dwelling unit wall spaces unless they are located _____ the wall.

    (a) within 6 in. of   (b) within 12 in. of   (c) within 18 in. of   (d) close to

12. Receptacles installed in a kitchen to serve countertop surfaces must be supplied by not fewer than _____ small-appliance branch circuits.

    (a) one   (b) two   (c) three   (d) no minimum

13. One receptacle outlet must be installed at each island or peninsular countertop space with a long dimension of 2 ft or greater, and a short dimension of 12 in. or greater. When breaks occur in countertop spaces for appliances, sinks, etc., there is never a need for more than one receptacle outlet.

    (a) True   (b) False

14. The required receptacle for a dwelling unit countertop surface can be mounted a maximum height of _____ above a dwelling unit kitchen counter surface.

    (a) 10 in.   (b) 12 in.   (c) 18 in.   (d) 20 in.

15. A receptacle outlet for the laundry is not required in a dwelling unit in a multifamily building when laundry facilities available to all building occupants are provided on the premises.

    (a) True   (b) False

16. Guest rooms or guest suites provided with permanent provisions for _____ must have receptacle outlets installed in accordance with all of the applicable requirements for a dwelling unit in accordance with 210.52.

    (a) whirlpool tubs   (b) bathing   (c) cooking   (d) internet access

17. In a dwelling unit, at least _____ wall switch-controlled lighting outlet(s) must be installed in every dwelling unit habitable room and bathroom.

    (a) one   (b) three   (c) six   (d) none of these

18. When considering lighting outlets in dwelling units, a vehicle door in a garage is considered an outdoor entrance.

    (a) True   (b) False

215 Feeders

Introduction

The next logical step up from the branch circuit is the feeder circuit. Consequently, Article 215 follows Article 210. This article covers the rules for installation, minimum size, and ampacity of feeders.

This is a very short article, and that's puzzling at first glance. It might seem feeders would just be "heavier" branch circuits, and that Article 215 should just be another Article 210 but with more stringent requirements. But this isn't the case at all.

If you go back and look at Article 210 again, you'll see it covers many permutations of branch circuits. It also devotes extensive space to dwelling-area branch circuits. Dwelling units don't have many feeders. A multifamily dwelling building will have at least one feeder for each occupancy.

Here's an object lesson in the value of Article 100. Go there now and review the definitions of branch circuit and feeder. Once you've done that, you will understand why Article 215 is so much shorter than Article 210.

215.1 Scope. Article 215 covers the installation, conductor sizing, and protection requirements for feeders. Figure 215–1

Author's Comment: See Article 100 for the definition of "Feeder."

215.2 Minimum Rating.

(A) Feeder Conductor Size.

(1) Continuous and Noncontinuous Loads. The minimum feeder-circuit conductor ampacity, before the application of any adjustment and/or correction factors, must be no less than 125 percent of the continuous load, plus 100 percent of the noncontinuous load, based on the terminal temperature rating ampacities as listed in Table 310.16 [110.14(C)].

Author's Comments:

- See 215.3 for the sizing requirements of the feeder overcurrent protection device for continuous and noncontinuous loads.
- Circuit conductors must have sufficient ampacity, after adjustment, to carry the load, and the conductors must be protected against overcurrent in accordance with their ampacity [215.3 and 240.4].

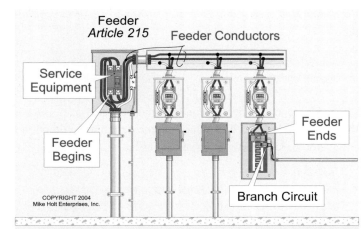

Feeder: [Article 100 Definition] The circuit conductors between the service equipment or the source of a separately derived system and the final branch-circuit overcurrent device.

Figure 215–1

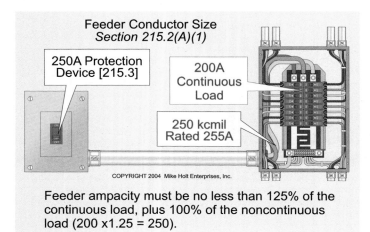

Feeder ampacity must be no less than 125% of the continuous load, plus 100% of the noncontinuous load (200 x1.25 = 250).

Figure 215–2

Question: *What size feeder conductor is required for a 200A continuous load if the terminals are rated 75°C?* **Figure 215–2**

(a) 2/0 AWG   (b) 3/0 AWG   (c) 4/0 AWG   (d) 250 kcmil

Answer: *(d) 250 kcmil*

Since the load is 200A continuous, the feeder conductors must have an ampacity not less than 250A (200A x 1.25). According to Table 310.16, 75°C column, 250 kcmil conductors are suitable, because they have an ampere rating of 255A at 75°C.

Exception: Where the assembly and the overcurrent protection device are both listed for 100 percent continuous load operation, the feeder conductors can be sized at 100 percent of the continuous load.

Author's Comment: Equipment suitable for 100 percent continuous loading is rarely available in ratings under 400A.

The feeder grounded neutral conductor must not be sized smaller than specified in 250.122 for the equipment grounding (bonding) conductor, based on the rating of the feeder protection device.

Question: *What size grounded neutral conductor is required for a feeder consisting of 250 kcmil ungrounded conductors and one grounded neutral conductor protected by a 250A protection device, where the unbalanced load is only 50A, with 75°C terminals?* **Figure 215–3**

(a) 6 AWG   (b) 4 AWG   (c) 1/0 AWG   (d) 3/0 AWG

Answer: *(b) 4 AWG, based on Table 250.122*

Table 310.16 and 220.61 would permit an 8 AWG grounded neutral conductor, rated 50A at 75°C to carry the 50A unbalanced load, but the grounded neutral conductor is not permitted to be smaller than 4 AWG, as required by Table 250.122.

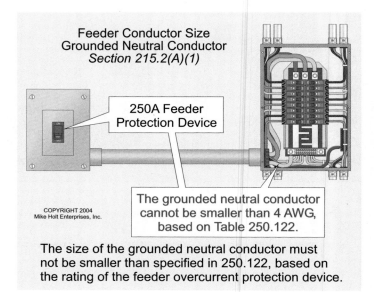

The size of the grounded neutral conductor must not be smaller than specified in 250.122, based on the rating of the feeder overcurrent protection device.

Figure 215–3

(2) Ampacity Relative to Service Conductors. The feeder conductor ampacity must not be less than that of the service conductors where the feeder conductors carry the total load supplied by service conductors with an ampacity of 55A or less.

(3) Dwelling Unit and Mobile Home Feeder Sizing. Feeder conductors for individual dwelling units or mobile homes need not be larger than service conductors sized in accordance with 310.15(B)(6).

> **FPN No. 2:** Voltage drop should be considered when sizing feeder conductors, but this isn't a *Code* requirement. For examples, see 210.19(A)(1) FPN 4 in this textbook.

> **Author's Comment:** For more information on this topic, visit www.MikeHolt.com and search for the phrase "Voltage Drop."

215.3 Overcurrent Protection. Feeder overcurrent protection devices must have an ampacity not less than 125 percent of the continuous loads, plus 100 percent of the noncontinuous loads.

> **Author's Comment:** See 215.2(A)(1) for feeder conductor sizing requirements.

Exception: Where the assembly and the overcurrent protection device are both listed for 100 percent continuous load operation, the protection device can be sized at 100 percent of the continuous load.

> **Author's Comment:** Equipment suitable for 100 percent continuous loading is rarely available in ratings under 400A.

215.8 High-Leg Conductor Identification.

On a 4-wire three-phase delta-connected system, where the midpoint of one phase winding is grounded, the conductor with the higher phase voltage-to-ground (208V) must be durably and permanently marked by an outer finish that is orange in color, or other effective means. Such identification must be placed at each point on the system where a connection is made if the grounded neutral conductor is present [110.15, 215.8, and 230.56].

The high-leg conductor is also called the "wild-leg," "stinger-leg," or "bastard-leg."

Author's Comments:

- See 110.15 in this textbook for additional details.
- Panelboards. Since 1975, panelboards supplied by a 4-wire three-phase delta-connected system must have the high-leg conductor (208V) terminate to the "B" (center) phase of a panelboard [408.3(E)].

An exception to 408.3(E) permits the high-leg conductor to terminate to the "C" phase when the meter is located in the same section of a switchboard or panelboard.

215.10 Ground-Fault Protection of Equipment.

Each solidly grounded wye electrical 277/480V feeder disconnecting means rated 1,000A or more must be provided with ground-fault protection of equipment in accordance with 230.95 and 240.13.

Author's Comment:
See Article 100 for the definition of "Ground-Fault Protection of Equipment."

Exception 2: The provisions of this section do not apply to fire pumps [695.6(H)].

Exception 3. Equipment ground-fault protection isn't required if ground-fault protection is provided on the supply side of the feeder.

Author's Comment:
Ground-fault protection of equipment isn't required for emergency systems [700.26] or legally required standby systems [701.17].

215.12 Identification for Feeders.

(A) Grounded Neutral Conductor. The grounded neutral conductor of a feeder must be identified in accordance with 200.6. **Figure 215–4**

(B) Equipment Grounding (Bonding) Conductor. Equipment grounding (bonding) conductors can be bare, and individually covered or insulated equipment grounding (bonding) conductors sized 6 AWG and smaller must have a continuous outer finish that is either green or green with one or more yellow stripes [250.119].

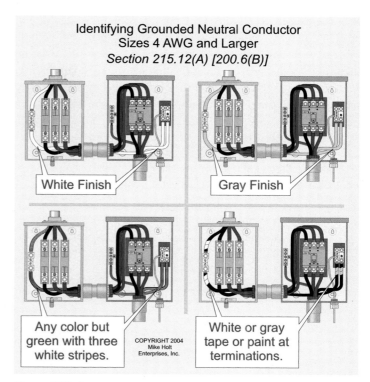

Identifying Grounded Neutral Conductor
Sizes 4 AWG and Larger
Section 215.12(A) [200.6(B)]

White Finish

Gray Finish

Any color but green with three white stripes.

White or gray tape or paint at terminations.

COPYRIGHT 2004 Mike Holt Enterprises, Inc.

Figure 215–4

Equipment grounding (bonding) conductors larger than 6 AWG that are insulated, can be permanently reidentified with green marking at the time of installation at every point where the conductor is accessible [250.119(A)]. **Figure 215–5**

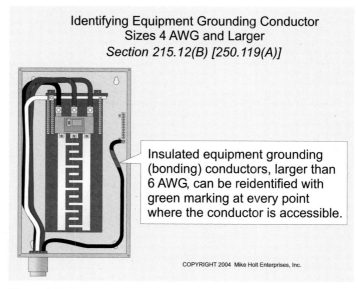

Identifying Equipment Grounding Conductor
Sizes 4 AWG and Larger
Section 215.12(B) [250.119(A)]

Insulated equipment grounding (bonding) conductors, larger than 6 AWG, can be reidentified with green marking at every point where the conductor is accessible.

COPYRIGHT 2004 Mike Holt Enterprises, Inc.

Figure 215–5

(C) Ungrounded Conductors. Where the premises wiring system contains feeders supplied from more than one voltage system, each ungrounded conductor, where accessible, must be identified by the system. Identification can be by color-coding, marking tape, tagging, or other means approved by the authority having jurisdiction. Such identification must be permanently posted at each feeder panelboard or similar feeder distribution equipment. Figure 215–6

Author's Comment: Electricians often use the following color system for power and lighting conductor identification:

- 120/240V single-phase—black, red, and white
- 120/208V three-phase—black, red, blue, and white
- 120/240V three-phase—black, orange, blue, and white
- 277/480V three-phase—brown, orange, yellow, and gray; or, brown, purple, yellow, and gray

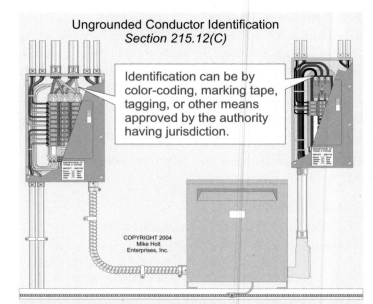

Ungrounded Conductor Identification
Section 215.12(C)

Identification can be by color-coding, marking tape, tagging, or other means approved by the authority having jurisdiction.

COPYRIGHT 2004
Mike Holt
Enterprises, Inc.

Where the premises wiring system contains feeders supplied from more than one voltage system, each ungrounded conductor, where accessible, must be identified by the system.

Figure 215–6

Article 215 Questions

1. The feeder conductor ampacity must not be less than that of the service-entrance conductors where the feeder conductors carry the total load supplied by service-entrance conductors with an ampacity of _____ or less.

    (a) 100A   (b) 60A   (c) 55A   (d) 30A

2. If required by the authority having jurisdiction, a diagram showing feeder details must be provided _____ of the feeders.

    (a) after the installation   (b) prior to the installation   (c) before the final inspection   (d) diagrams are not required

3. Ground-fault protection of equipment is required for the feeder disconnect if _____.

    (a) the feeder is rated 1,000A or more
    (b) it is a solidly-grounded wye system
    (c) it is more than 150 volts-to-ground, but not exceeding 600V phase-to-phase
    (d) all of these

4. Ground-fault protection of equipment is not required at the feeder disconnect if ground-fault protection of equipment is provided on the _____ side of the feeder.

    (a) load   (b) supply   (c) service   (d) none of these

5. Where the premises wiring system contains feeders supplied from more than one voltage system, each ungrounded (hot) conductor, where accessible, must be identified by the system. Identification can be by _____ or other approved means. Such identification must be permanently posted at each feeder panelboard or similar feeder distribution equipment.

    (a) color-coding   (b) marking tape   (c) tagging   (d) a, b, or c

220 Branch-Circuit, Feeder, and Service Calculations

Introduction

This article provides the requirements for sizing branch circuits, feeders, and services, and for determining the number of receptacles on a circuit and the number of branch circuits required. It consists of five parts:

- PART I. GENERAL
- PART II. BRANCH-CIRCUIT LOAD CALCULATIONS
- PART III. FEEDER AND SERVICE CALCULATIONS
- PART IV. OPTIONAL CALCULATIONS
- PART V. FARM LOAD CALCULATIONS

Part I describes the layout of Article 220 and provides a table of where other types of load calculations can be found in the *NEC*. Part II provides requirements for branch-circuit calculations and for specific types of branch circuits. Part III provides requirements for feeder and service calculations, just as the title says. Part IV provides some shortcut calculations you can use in place of the more complicated calculations provided in Parts II and III—if your installation meets certain requirements. Part IV covers just what it says, Farm Load Calculations.

The typical electrician is wise to focus on Parts I, II, and III. Whether to do the optional calculations is typically a decision made by the project manager or design engineer. You need to be aware that there can be two right answers when doing the calculations because the *NEC* allows two different methods.

The cost of improperly applying Article 220 can be staggering. In the best of all possible worlds, the price of misapplication is just an expensive call-back and some rework. In reality, the costs can easily involve catastrophic destruction and the loss of human life.

So study Article 220 carefully. If something doesn't make sense at first, make a note of it and take a short break from your studies. Then go back to that item and read through the explanation, using the illustrations to help you understand. Your learning will really stick if you also consider the why, not just the how.

PART I. GENERAL

220.1 Scope. This article contains the requirements necessary for sizing branch circuits, feeders, and services. In addition, this article can be used to determine the number of receptacles on a circuit and the number of general-purpose branch circuits required.

220.3 Application of Other Articles. Other articles contain calculations that are in addition to, or modify those, contained within Article 220. Take a moment to review the following additional calculation requirements:

- Air Conditioning and Refrigeration, 440.6, 440.21, 440.22, 440.31, 440.32, and 440.62
- Appliances, 422.10 and 422.11
- Branch Circuits, 210.19 and 210.20(A)
- Computers (Data Processing Equipment), 645.4 and 645.5(A)
- Conductors, 310.15

- Feeders, 215.2(A) and 215.3
- Fire Pumps, 695.7
- Fixed Electric Heat, 424.3(B)
- Marinas, 555.12, 555.19(A)(4), and 555.19(B)
- Mobile Homes and Manufactured Homes, 550.12 and 550.18
- Motors, 430.6(A), 430.22(A), 430.24, 430.52, and 430.62
- Overcurrent Protection, 240.4 and 240.20
- Refrigeration (Hermetic), 440.6 and Part IV
- Recreational Vehicle Parks, 551.73(A)
- Sensitive Electronic Equipment, 647.4(D)
- Services, 230.42(A) and 230.79
- Signs, 600.5
- Transformers, 450.3

220.5 Calculations.

(A) Voltage Used for Calculations. Unless other voltages are specified, branch-circuit, feeder, and service loads must be calculated on nominal system voltage, such as 120V, 120/240V, 120/208V, 240V, 277/480V, or 480V. **Figure 220–1**

> **Author's Comment:** A nominal value is assigned to a circuit for the purpose of convenient circuit identification. The actual voltage at which a circuit operates can vary from the nominal within a range that permits satisfactory operation of equipment [Article 100].

(B) Fractions of an Ampere (Rounding Amperes). Calculations that result in a fraction of less than one-half of an ampere can be dropped.

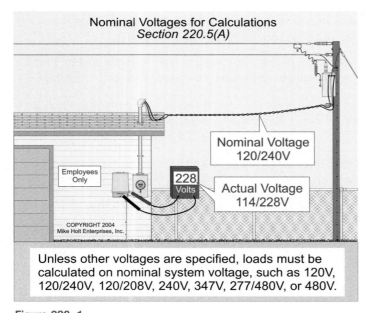

Nominal Voltages for Calculations
Section 220.5(A)

Nominal Voltage
120/240V

228
Volts

Actual Voltage
114/228V

Employees Only

COPYRIGHT 2004
Mike Holt Enterprises, Inc.

Unless other voltages are specified, loads must be calculated on nominal system voltage, such as 120V, 120/240V, 120/208V, 240V, 347V, 277/480V, or 480V.

Figure 220–1

> **Author's Comment:** When do you round—after each calculation, or at the final calculation? The *NEC* isn't specific on this issue, but I guess it all depends on the answer you want to see!

Question: *According to 424.3(B), the branch-circuit conductors and overcurrent protection device for electric space-heating equipment must be sized no less than 125 percent of the total load. What size conductor is required to supply a 9 kW (37.5A), 240V single-phase fixed space heater with a 3A blower motor, if equipment terminals are rated 75°C?* **Figure 220–2**

(a) 10 AWG   (b) 8 AWG   (c) 6 AWG   (d) 4 AWG

Answer: (c) 6 AWG

Step 1: Determine the total load
$$I = VA/E$$
$$I = 9,000\ VA/240V$$
$$I = 37.5A$$

Step 2: Conductor size at 125% of the load.
$$Conductor\ Size = (37.5A + 3A) \times 1.25$$
$$Conductor\ Size = 50.63A,\ round\ up\ to\ 51A$$

If we rounded down, then 8 AWG rated 50A at 75°C could be used, but since we have to round up, 6 AWG rated 65A at 75°C is required.

PART II. BRANCH-CIRCUIT LOAD CALCULATIONS

220.12 General Lighting. The general lighting load specified in Table 220.12 must be calculated from the outside dimensions of the building or area involved. For dwelling units, the calculated floor area must not include open porches, garages, or unfinished spaces not adaptable for future use.

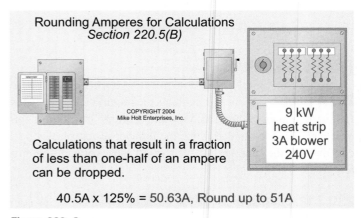

Rounding Amperes for Calculations
Section 220.5(B)

COPYRIGHT 2004
Mike Holt Enterprises, Inc.

9 kW
heat strip
3A blower
240V

Calculations that result in a fraction of less than one-half of an ampere can be dropped.

40.5A x 125% = 50.63A, Round up to 51A

Figure 220–2

Table 220.12 General Lighting Loads by Occupancy

Occupancy	VA per Square Ft
Armories and auditoriums	1
Assembly halls and auditoriums	1
Banks	3 ½[b]
Barber shops and beauty parlors	3
Churches	1
Clubs	2
Courtrooms	2
Dwelling units	3[a]
Garages — commercial (storage)	½
Halls, corridors, closets, stairways	½
Hospitals	2
Hotels and motels without cooking facilities	2
Industrial commercial (loft buildings)	2
Lodge rooms	1 ½
Office buildings	3 ½[b]
Restaurants	2
Schools	3
Storage spaces	¼
Stores	3
Warehouses (storage)	¼

Table 220.12 Note a: *The VA load for general-use receptacles, bathroom receptacles [220.14(J)(1) and 210.11(C)(3)], outside receptacles, as well as garage, basement receptacles [220.14(J)(2), 210.52(E) and (G)], and lighting outlets [220.14(J)(3), 210.70(A) and (B)] in a dwelling unit are included in the 3VA per sq ft general lighting [220.14(J)].*

Table 220.12 Note b: *The receptacle calculated load for banks and office buildings is the largest calculation of either (1) or (2) [220.14(K)].*

(1) Determine the receptacle calculated load at 180 VA per receptacle yoke [220.14(I)], then apply the demand factor from Table 220.44, or
(2) Determine the receptacle calculated load at 1 VA per sq ft.

Question: *What is the general lighting and general-use receptacle load for a 40 ft x 50 ft (2,000 sq ft) dwelling unit?* Figure 220–3

(a) 2,000 VA   (b) 3,000 VA   (c) 5,000 VA   (d) 6,000 VA

Answer: *(d) 6,000 VA*

General lighting/receptacle load: 40 ft x 50 ft = 2,000 sq ft x 3 VA per sq ft = 6,000 VA

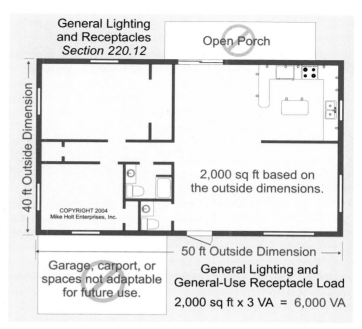

General Lighting and Receptacles *Section 220.12*

Open Porch

40 ft Outside Dimension

2,000 sq ft based on the outside dimensions.

COPYRIGHT 2004 Mike Holt Enterprises, Inc.

Garage, carport, or spaces not adaptable for future use.

50 ft Outside Dimension

General Lighting and General-Use Receptacle Load
2,000 sq ft x 3 VA = 6,000 VA

Figure 220–3

220.14 Other Loads—All Occupancies.
The minimum VA load for each outlet must comply with (A) through (L).

(A) Specific Equipment. The branch-circuit VA load for equipment and appliance outlets must be calculated on the VA rating of the equipment or appliance.

(B) Electric Dryers and Household Electric Cooking Appliances. The branch-circuit VA load for household electric dryers must comply with 220.54, and household electric ranges and other cooking appliances must comply with 220.55.

(C) Motor Loads. The motor branch-circuit VA load must be determined by multiplying the motor full-load current (FLC) listed in Table 430.248 or 430.250 by the motor table voltage, in accordance with 430.22 [430.6(A)(1)].

(D) Recessed Luminaires. The branch-circuit VA load for recessed luminaires must be calculated based on the maximum VA rating for which the luminaires are rated.

(E) Heavy-Duty Lampholders. The branch-circuit VA load for heavy-duty lampholders must be calculated at a minimum of 600 VA.

(F) Sign Outlet. Each commercial occupancy accessible to pedestrians must have at least one 20A sign outlet [600.5(A)], which must have a minimum branch-circuit load of 1,200 VA. Figure 220–4

(G) Show Windows. The branch-circuit VA load for show-window lighting must be calculated in accordance with (1) or (2):

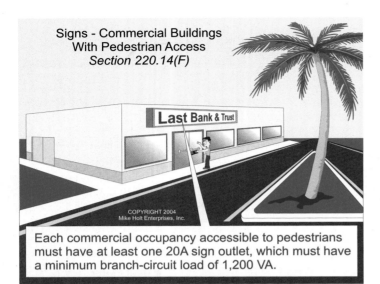

Signs - Commercial Buildings With Pedestrian Access
Section 220.14(F)

Each commercial occupancy accessible to pedestrians must have at least one 20A sign outlet, which must have a minimum branch-circuit load of 1,200 VA.

Figure 220–4

(1) 180 VA per outlet in accordance with 220.14(L), or

(2) 200 VA per linear foot of show-window lighting. See 220.43. **Figure 220–5**

(H) Fixed Multioutlet Assemblies. Fixed multioutlet assemblies in commercial occupancies used in other than dwelling units or in the guest rooms of hotels or motels must be calculated in accordance with (1) or (2). **Figure 220–6**

(1) Where appliances are unlikely to be used simultaneously, each 5 ft or fraction of 5 ft of multioutlet assembly is considered as one outlet of 180 VA.

(2) Where appliances are likely to be used simultaneously, each 1 ft or fraction of a foot of multioutlet assembly is considered as one outlet of 180 VA.

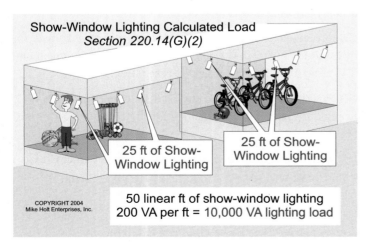

Show-Window Lighting Calculated Load
Section 220.14(G)(2)

25 ft of Show-Window Lighting

25 ft of Show-Window Lighting

50 linear ft of show-window lighting
200 VA per ft = 10,000 VA lighting load

Figure 220–5

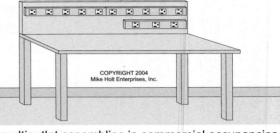

Multioutlet Assembly Calculated Load
Section 220.14(H)

Fixed multioutlet assemblies in commercial occupancies must be calculated in accordance with (1) or (2).
(1) Each 5 ft or fraction is equal to 180 VA.
(2) Simultaneous use; each 1 ft or fraction equals 180 VA.

Figure 220–6

Author's Comments:

- See Article 100 for the definition of "Multioutlet Assembly."
- The feeder or service calculated load for fixed multioutlet assemblies can be calculated in accordance with the demand factors contained in 220.44.

(I) Commercial Receptacle Load. Except as covered in 200.14(J) and (K), each 15 or 20A, 125V general-use receptacle outlet is to be considered as 180 VA per mounting strap. **Figure 220–7**

A single device that consists of four or more receptacles must be considered as 90 VA per receptacle (360 VA for a quad receptacle). **Figure 220–8**

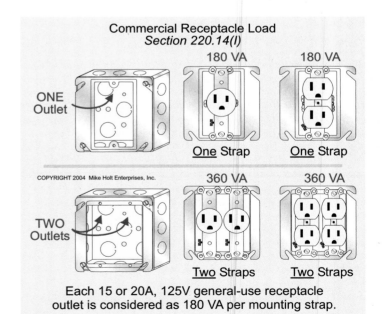

Commercial Receptacle Load
Section 220.14(I)

ONE Outlet   180 VA — One Strap   180 VA — One Strap

TWO Outlets   360 VA — Two Straps   360 VA — Two Straps

Each 15 or 20A, 125V general-use receptacle outlet is considered as 180 VA per mounting strap.

Figure 220–7

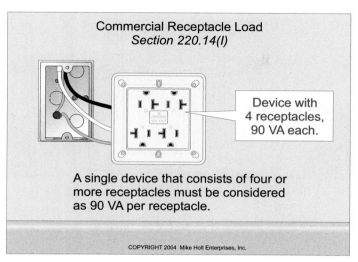

Commercial Receptacle Load
Section 220.14(I)

Device with 4 receptacles, 90 VA each.

A single device that consists of four or more receptacles must be considered as 90 VA per receptacle.

COPYRIGHT 2004 Mike Holt Enterprises, Inc.

Figure 220–8

Question: *What is the maximum number of 15A receptacle outlets permitted on a 20A, 120V circuit in a commercial occupancy?* **Figure 220–9**

(a) 4   (b) 6   (c) 10   (d) 13

Answer: (d) 13

Circuit VA = Volts x Amperes
Circuit VA = 120V x 20A
Circuit VA = 2,400 VA
Number of Receptacles = 2,400 VA/180 VA = 13

Author's Comment: According to the *NEC Handbook*, published by the NFPA, general-purpose receptacles aren't considered a continuous load.

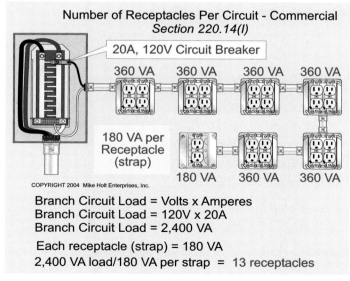

Number of Receptacles Per Circuit - Commercial
Section 220.14(I)

20A, 120V Circuit Breaker

360 VA   360 VA   360 VA   360 VA

180 VA per Receptacle (strap)

180 VA   360 VA   360 VA

COPYRIGHT 2004 Mike Holt Enterprises, Inc.

Branch Circuit Load = Volts x Amperes
Branch Circuit Load = 120V x 20A
Branch Circuit Load = 2,400 VA

Each receptacle (strap) = 180 VA
2,400 VA load/180 VA per strap = 13 receptacles

Figure 220–9

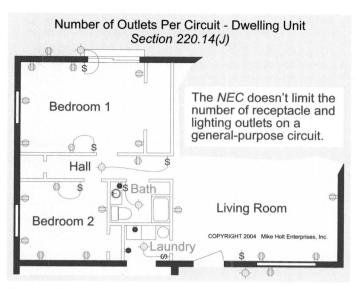

Number of Outlets Per Circuit - Dwelling Unit
Section 220.14(J)

Bedroom 1

Hall

Bath

Bedroom 2

Living Room

Laundry

The *NEC* doesn't limit the number of receptacle and lighting outlets on a general-purpose circuit.

COPYRIGHT 2004 Mike Holt Enterprises, Inc.

Figure 220–10

(J) Residential Receptacle Load. In one-family, two-family, and multifamily dwellings, and in guest rooms of hotels and motels, the outlets specified in (1), (2), and (3) are included in the general lighting load calculations of 220.12.

(1) General-use receptacle outlets, including the receptacles connected to the 20A bathroom circuit [210.11(C)(3)].

(2) Outdoor, garage, and basement receptacle outlets [210.52(E) and (G)].

(3) Lighting outlets [210.70(A) and (B)].

There's no VA load for 15 and 20A, 125V general lighting and general-use receptacle outlets, because the loads for these devices are part of the 3 VA per sq ft for general lighting as listed in Table 220.12 for dwelling units.

Question: What is the maximum number of 15 or 20A, 125V receptacle and lighting outlets permitted on a 15A, 120V general-purpose branch circuit in a dwelling unit? **Figure 220–10**

(a) 4   (b) 6   (c) 8   (d) No limit

Answer: (d) No limit

Author's Comment: The *NEC* doesn't limit the number of receptacle and lighting outlets on a general-purpose branch circuit in a dwelling unit. See the NFPA's *NEC Handbook* for more information.

CAUTION: *There might be a local Code requirement that limits the number of receptacles and lighting outlets on a general-purpose branch circuit.*

Author's Comment: Although there's no limit on the number of lighting and/or receptacle outlets on dwelling general-purpose branch circuits, the *NEC* does require a minimum number of circuits to be installed for general-purpose receptacles and lighting outlets [210.11(A)]. In addition, the receptacle and lighting loads must be evenly distributed among the required circuits [210.11(B)].

(K) Banks and Office Buildings. The receptacle calculated load for banks and office buildings is the largest calculation of either (1) or (2).

(1) Determine the receptacle calculated load at 180 VA per receptacle yoke [220.14(I)], then apply the demand factor from Table 220.44, or

(2) Determine the receptacle load at 1 VA per sq ft.

Author's Comment: Often, you don't know in advance the exact number of receptacle outlets that will be installed in an office building or bank. First, the main structure is built, and then individual office spaces are rented out to tenants who do their own custom installations. The 1 VA per sq ft allows a generic feeder/service demand for general-purpose receptacles.

Bank or Office General Lighting and Receptacle— Example 1:

What is the calculated receptacle load for an 18,000 sq ft bank with 160, 15A, 125V receptacles? **Figure 220–11**

(a) 15,400 VA	(b) 19,400 VA
(c) 28,800 VA	(d) 142 kVA

Answer: (b) 19,400 VA

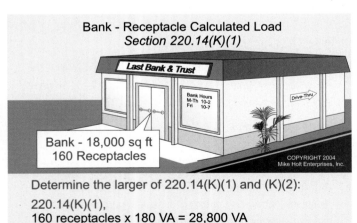

Determine the larger of 220.14(K)(1) and (K)(2):

220.14(K)(1),
160 receptacles x 180 VA = 28,800 VA
First 10,000 VA at 100% - 10,000 VA = 10,000 VA
Remainder at 50%   18,800 VA = + 9,400 VA
Receptacle Demand Load   19,400 VA

220.14(K)(2), 8,000 sq ft x 1 VA per ft = 18,000 VA

Figure 220–11

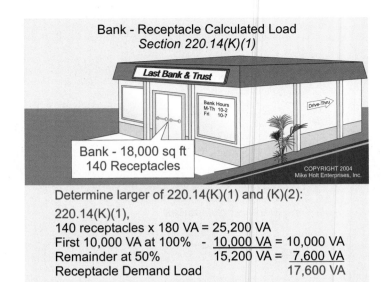

Bank - Receptacle Calculated Load
Section 220.14(K)(1)

Bank - 18,000 sq ft
140 Receptacles

Determine larger of 220.14(K)(1) and (K)(2):

220.14(K)(1),
140 receptacles x 180 VA = 25,200 VA
First 10,000 VA at 100% - 10,000 VA = 10,000 VA
Remainder at 50%   15,200 VA = 7,600 VA
Receptacle Demand Load   17,600 VA

220.14(K)(2), 18,000 sq ft x 1 VA per ft = 18,000 VA

Figure 220–12

220.14(K)(1) and 220.14(I)
   160 receptacles x 180 VA = 28,800 VA
   First 10,000 at 100%   10,000 VA x 1.0 = 10,000 VA
   Remainder at 50%   18,800 VA x 0.5 = 9,400 VA
   Receptacle Calculated load   19,400 VA

220.14(K)(2)
   18,000 x 1 VA per sq ft = 18,000 VA (smaller, omit)

Bank or Office General Lighting and Receptacle— Example 2:

What is the receptacle calculated load for an 18,000 sq ft bank with 140 receptacles? **Figure 220–12**

(a) 15,000 VA	(b) 18,000 VA
(c) 23,000 VA	(d) 31,000 VA

Answer: (b) 18,000 VA

220.14(K)(1) and 220.14(I)
   140 receptacles x 180 VA = 25,200 VA
   First 10,000 at 100%   10,000 VA x 1.0 = 10,000 VA
   Remainder at 50%   15,200 VA x 0.5 = 7,600 VA
   Receptacle Calculated load   17,600 VA

220.14(K)(2)
   18,000 x 1 VA per sq ft = 18,000 VA (smaller, omit)

(L) Other Outlets. Receptacle and lighting outlets not covered in (A) through (K) must have the branch-circuit VA load calculated at 180 VA per outlet.

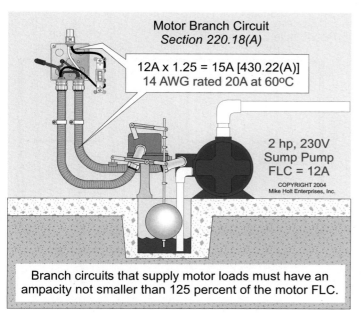

Motor Branch Circuit
Section 220.18(A)

12A x 1.25 = 15A [430.22(A)]
14 AWG rated 20A at 60ºC

2 hp, 230V
Sump Pump
FLC = 12A

COPYRIGHT 2004
Mike Holt Enterprises, Inc.

Branch circuits that supply motor loads must have an
ampacity not smaller than 125 percent of the motor FLC.

Figure 220–13

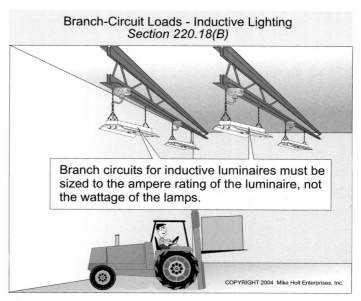

Branch-Circuit Loads - Inductive Lighting
Section 220.18(B)

Branch circuits for inductive luminaires must be
sized to the ampere rating of the luminaire, not
the wattage of the lamps.

COPYRIGHT 2004 Mike Holt Enterprises, Inc.

Figure 220–14

Author's Comment: Where a branch circuit supplies continuous loads, the minimum branch-circuit conductor size, before the application of any adjustment and/or correction factors, must have an allowable ampacity of not less than 125 percent of the continuous load [210.19(A)(1)].

220.18 Maximum Load on a Branch Circuit.

(A) Motor Operated Loads. Branch circuits that supply motor loads must be sized not smaller than 125 percent of the motor FLC, in accordance with 430.6(A) and 430.22(A).

Question: What is the minimum size branch-circuit conductor for a 2 hp, 230V motor, where the conductors' terminals are rated 60°C? **Figure 220–13**

(a) 14 AWG   (b) 12 AWG   (c) 10 AWG   (d) 8 AWG

Answer: (a) 14 AWG

Step 1. Determine the motor full-load current [Table 430.248].
2 hp FLC = 12A

Step 2. Size the branch-circuit conductors at 125 percent in accordance with Table 310.16 [430.22(A)].
Branch-Circuit Conductors = 12A x 1.25 = 15A,
14 AWG rated 20A at 60°C.

(B) Inductive Lighting Loads. Branch circuits that supply inductive luminaires, such as fluorescent and HID fixtures, must have the branch-circuit conductors sized to the ampere rating of the luminaire, not to the wattage of the lamps. **Figure 220–14**

Question: What is the maximum number of 1.34A fluorescent luminaires permitted on a 20A circuit if the luminaires operate for more than three hours?

(a) 8   (b) 11   (c) 13   (d) 15

Answer: (b) 11

The maximum continuous load must not exceed 80 percent of the circuit rating [210.19(A)(1)].

Maximum load = 20A x 0.80
Maximum load = 16A
Luminaires on Circuit = 16A/1.34A
Luminaires on Circuit = 11.94 or 11 luminaires

Author's Comment: Because of power factor (inductive luminaires), the input VA of each luminaire is 162 VA (120V x 1.34A), which is greater than the 136W (34W x 4 lamps) of the lamps. I know this is getting complicated, but just remember—size all circuits that supply inductive loads to the ampere rating of the luminaire, not to the wattage of the lamps.

(C) Household Cooking Appliances. Branch-circuit conductors for household cooking appliances can be sized in accordance with Table 220.55; specifically, Note 4 for branch circuits.

Author's Comment: For ranges rated 8.75 kW or more, the minimum branch-circuit rating is 40A. See 210.19(A)(3).

PART III. FEEDER AND SERVICE CALCULATIONS

220.40 General. The calculated load for a feeder or service must not be less than the sum of the branch-circuit loads, as determined by Part II of this article as adjusted for the demand factors contained in Parts III, IV, or V.

> **FPN:** See Examples D1(A) through D10 in Annex D.

220.42 General Lighting Demand Factors. The *Code* recognizes that not all luminaires will be on at the same time, and it permits the following demand factors to be applied to the general lighting load as determined in Table 220.42.

Table 220–42
Lighting Load Demand Factors

Type of Occupancy	Lighting VA Load	Demand Factor
Dwelling Units	First 3,000 VA	100%
	Next 117,000 VA	35%
	Remainder at	25%
Hotels/motels without provision for cooking	First 20,000 VA	50%
	Next 80,000 VA	40%
	Remainder at	30%
Warehouses (storage)	First 12,500 VA	100%
	Remainder	50%
All others	Total VA	100%

Question: What is the general lighting and receptacle calculated load, after demand factors, for a 40 ft x 50 ft (2,000 sq ft) dwelling unit? **Figure 220–15**

(a) 2,050 VA   (b) 3,050 VA   (c) 4,050 VA   (d) 5,050 VA

Answer: (c) 4,050 VA

General lighting = 40 ft x 50 ft
General lighting = 2,000 sq ft x 3 VA per sq ft
General lighting = 6,000 VA

First 3,000 VA at 100%	3,000 VA x 1.00 =	3,000 VA
Next 117,000 VA at 35%	3,000 VA x 0.35 =	*1,050 VA*

General lighting and general-
use receptacles calculated load   4,050 VA

Author's Comment: For commercial occupancies, the VA load for receptacles [220.14(I)] and fixed multioutlet assemblies [220.14(H)] can be added to the general lighting load and subjected to the demand factors of Table 220.42 [220.44].

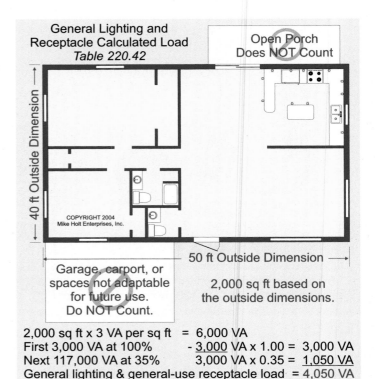

General Lighting and Receptacle Calculated Load
Table 220.42

Open Porch Does NOT Count

40 ft Outside Dimension

50 ft Outside Dimension

Garage, carport, or spaces not adaptable for future use. Do NOT Count.

2,000 sq ft based on the outside dimensions.

2,000 sq ft x 3 VA per sq ft   = 6,000 VA
First 3,000 VA at 100%   - 3,000 VA x 1.00 = 3,000 VA
Next 117,000 VA at 35%   3,000 VA x 0.35 = 1,050 VA
General lighting & general-use receptacle load = 4,050 VA

Figure 220–15

220.43 Commercial—Show Window and Track Lighting Load.

(A) Show Windows. The feeder/service VA load must not be less than 200 VA per linear foot.

(B) Track Lighting. The feeder/service VA load must not be less than 150 VA for every 2 ft of track lighting or fraction thereof. **Figure 220–16**

Author's Comments:

- There is no limit on the length of track that can be supplied by a single branch circuit.
- Where a feeder or service supplies continuous loads, the minimum feeder or service conductor size, before the application of any adjustment and/or correction factors, must have an allowable ampacity of not less than 125 percent of the continuous load [215.2(A)(1) for feeders and 230.42(A) for services].

Question: What is the approximate feeder/service calculated load for conductor sizing for 150 ft of track lighting in a commercial occupancy?

(a) 10,000 VA   (b) 12,000 VA   (c) 14,000 VA   (d) 16,000 VA

Answer: (c) 14,000 VA

150 ft/2 ft = 75 units x 150 VA x 1.25 = 14,063 VA

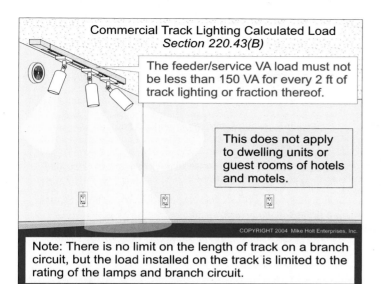

Commercial Track Lighting Calculated Load
Section 220.43(B)

The feeder/service VA load must not be less than 150 VA for every 2 ft of track lighting or fraction thereof.

This does not apply to dwelling units or guest rooms of hotels and motels.

COPYRIGHT 2004 Mike Holt Enterprises, Inc.

Note: There is no limit on the length of track on a branch circuit, but the load installed on the track is limited to the rating of the lamps and branch circuit.

Figure 220–16

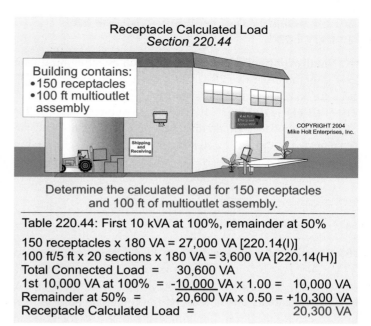

Receptacle Calculated Load
Section 220.44

Building contains:
• 150 receptacles
• 100 ft multioutlet assembly

COPYRIGHT 2004
Mike Holt Enterprises, Inc.

Shipping and Receiving

Determine the calculated load for 150 receptacles and 100 ft of multioutlet assembly.

Table 220.44: First 10 kVA at 100%, remainder at 50%

150 receptacles x 180 VA = 27,000 VA [220.14(I)]
100 ft/5 ft x 20 sections x 180 VA = 3,600 VA [220.14(H)]
Total Connected Load =   30,600 VA
1st 10,000 VA at 100% = -10,000 VA x 1.00 =   10,000 VA
Remainder at 50% =   20,600 VA x 0.50 = +10,300 VA
Receptacle Calculated Load =   20,300 VA

Figure 220–17

Author's Comment: This rule doesn't apply to branch circuits. Therefore, the maximum number of lampholders permitted on a track lighting system is based on the wattage rating of the lamps and the voltage and ampere rating of the circuit. The maximum load on a branch circuit must not exceed 80 percent of the circuit rating [210.19(A)(1)].

Question: How many 75W lampholders can be installed on a 20A, 120V track lighting circuit in a commercial occupancy?

(a) 10   (b) 15   (c) 20   (d) 25

Answer: (d) 25

Maximum load permitted on circuit = 20A x 0.80
Maximum load permitted on circuit = 16A

Maximum load in VA = 120V x 16A
Maximum load in VA = 1,920 VA

Number of Lampholders = 1,920 VA/75 VA
Number of Lampholders = 25.6

Author's Comment: There is no limit on the length of track on a single branch circuit.

220.44 Commercial—Receptacle Load. The feeder/service VA load for general-purpose receptacles [220.14(I)] and fixed multioutlet assemblies [220.14(H)] must be determined by:

• Adding the receptacle and fixed multioutlet assembly VA load with the general lighting load [Table 220.12] and adjusting this VA value by the demand factors contained in Table 220.42, or

• Applying a 50 percent demand factor to that portion of the receptacle and fixed multioutlet receptacle load exceeding 10 kVA.

Question: What is the calculated feeder/service VA load, after demand factors, for 150 general-purpose receptacles and 100 ft of fixed multioutlet assembly in a commercial occupancy?
Figure 220–17

(a) 8,500 VA   (b) 10,000 VA (c) 20,300 VA (d) 27,000 VA

Answer: (c) 20,300 VA

Step 1. Determine the total connected load:

Receptacle Load = 150 receptacles x 180 VA
Receptacle Load = 27,000 VA [220.14(I)].

Multioutlet Load = 100 ft/5 ft = 20 sections x 180 VA

Multioutlet Load = 3,600 VA [220.14(H)]

Step 2. Apply Table 220.44 demand factor:

Total Connected Load   30,600 VA

First 10,000 VA at 100%   10,000 VA x 1.0 = 10,000 VA
Remainder at 50%   20,600 VA x 0.5 = 10,300 VA
Receptacle calculated load   20,300 VA

220.50 Motor Load. The feeder/service load for motors must be sized not smaller than 125 percent of the largest motor load, plus the sum of the other motor loads. See 430.24 for example.

220.51 Fixed Electric Space-Heating Load. The feeder/service load for fixed electric space-heating equipment must be calculated at 100 percent of the total connected load.

220.52 Dwelling Unit—Small-Appliance and Laundry Load.

(A) Small-Appliance Circuit Load. Each dwelling unit must have a minimum of two 20A, 120V small-appliance branch circuits for kitchen and dining room receptacles, as required by 210.52(B)(1) [210.11(C)(1)]. The feeder/service VA load for each small-appliance circuit must be 1,500 VA, and this load can be subjected to the general lighting demand factors contained in Table 220.42.

> **Author's Comment:** Receptacles rated 15 or 20A, 125V, can be installed on the 20A small-appliance branch circuit [Table 210.21(B)(3)].

(B) Laundry Circuit Load. Each dwelling unit must have a 20A, 120V laundry circuit for the laundry room receptacles, as required by 210.52(F) [210.11(C)(2)]. The feeder/service VA load for the laundry circuit must be 1,500 VA, and this load can be subjected to the general lighting demand factors contained in Table 220.42.

> **Author's Comment:** A laundry circuit isn't required in a dwelling unit of a multifamily building if laundry facilities are provided on the premises for all building occupants [210.52(F) Ex. 1].

220.53 Dwelling Unit—Appliance Load.

A demand factor of 75 percent can be applied to the total connected load of four or more appliances on the same feeder/service. This demand factor doesn't apply to electric space-heating equipment [220.51], electric clothes dryers [220.54], electric ranges [220.55], electric air-conditioning equipment [Article 440, Part IV], or motors [220.50].

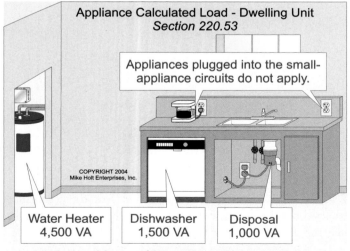

A demand factor of 75 percent can be applied to the total connected load of four or more appliances.

Figure 220–18

Question: What is the feeder/service appliance calculated load for a dwelling unit that contains a 1,000 VA disposal, a 1,500 VA dishwasher, and a 4,500 VA water heater? **Figure 220–18**

(a) 3,000 VA   (b) 4,500 VA   (c) 6,000 VA   (d) 7,000 VA

Answer: (d) 7,000 VA

No demand factor applies for three appliances.

Question: What is the feeder/service appliance calculated load, after demand factors, for a 12-unit multifamily dwelling if each unit contains a 1,000 VA disposal, a 1,500 VA dishwasher, and a 4,500 VA water heater?

(a) 23,000 VA   (b) 43,500 VA   (c) 63,000 VA   (d) 71,000 VA

Answer: (c) 63,000 VA

Calculated load = 7,000 VA x 12 units x 0.75*
Calculated load = 63,000 VA

*Each dwelling unit has only three appliances, but the feeder supplies a total of 36 appliances (12 units x 3 appliances).

220.54 Dwelling Unit—Electric Clothes Dryer Load.

The feeder/service load for electric clothes dryers located in a dwelling unit must not be less than 5,000W, or the nameplate rating of the equipment if greater than 5,000W. When a building contains five or more dryers, it is permissible to apply the demand factors listed in Table 220.54 to the total connected dryer load. **Figure 220–19**

> **Author's Comment:** A clothes dryer load isn't required if the dwelling unit doesn't have an electric clothes dryer circuit receptacle outlet.

Question: What is the feeder/service calculated load for a 10-unit multifamily building that contains a 5 kW dryer in each unit?

(a) 25,000W   (b) 43,500W   (c) 63,000W   (d) 71,000W

Answer: (a) 25,000W

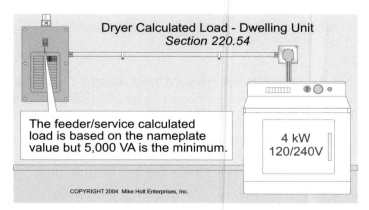

The feeder/service calculated load is based on the nameplate value but 5,000 VA is the minimum.

4 kW
120/240V

Figure 220–19

Table 220.54 demand factor for 10 units is 50%
Calculated load = 10 units x 5,000W x 0.50
Calculated load = 25,000W

220.55 Dwelling Unit—Electric Ranges and Cooking Appliances.

Feeder/Service. Household cooking appliances rated over 1.75 kW can have the feeder/service load calculated according to the demand factors of Table 220.55.

Over 12 kW—Table 220.55, Note 1. For identically sized ranges individually rated more than 12 kW, the maximum demand in Column C must be increased 5 percent for each additional kilowatt of rating, or major fraction thereof, by which the rating of individual ranges exceeds 12 kW.

Table 220.55, Note 2. For ranges individually rated more than 8.75 kW, but none exceeding 27 kW, and of different ratings, an average rating must be calculated by adding together the ratings of all ranges to obtain the total connected load (using 12 kW for any range rated less than 12 kW) and dividing this total by the number of ranges. Then the maximum demand in Column C must be increased 5 percent for each kilowatt, or major fraction thereof, by which this average value exceeds 12 kW.

Branch-Circuit Calculations [220.14B] Table 220.55, Note 4. Branch-Circuit Load for One Range. It is permissible to compute the branch-circuit load for one range in accordance with Table 220.55.

> **Question:** What is the branch-circuit calculated load in amperes for a single 12 kW range connected on a 120/240V circuit? **Figure 220–20**
>
> (a) 20A (b) 33A (c) 41A (d) 50A
>
> **Answer:** (b) 33A

Column C calculated load = 8 kW
Branch-circuit load in amperes, I = P/E
P = 8,000W
E = 240V
I = 8,000W/240V
I = 33.33A

Branch-Circuit Load for One Wall-Mounted Oven or One Counter-Mounted Cooking Unit. The branch-circuit load for one wall-mounted oven or one counter-mounted cooking unit must be the nameplate rating of the appliance.

> **Question:** What size branch-circuit conductors are required for a 6 kW wall-mounted oven connected on a 120/240V circuit? **Figure 220–21**
>
> (a) 14 AWG (b) 12 AWG (c) 10 AWG (d) 8 AWG
>
> **Answer:** (c) 10 AWG

Branch-circuit load in amperes, I = P/E
P = 6,000W
E = 240V
I = 6,000W/240V
I = 25A, 10 AWG rated 30A at 60°C
  [Table 310.16 and 110.14(C)(1)]

Branch-Circuit Load for One Counter-Mounted Cooking Unit and Up to Two Wall-Mounted Ovens. The branch-circuit load for one counter-mounted cooking unit and up to two wall-mounted ovens is determined by adding the nameplate ratings together and treating this value as a single range.

> **Question:** What size branch circuit is required for one 6 kW counter-mounted cooking unit and two 3 kW wall-mounted ovens connected on a 120/240V circuit? **Figure 220–22**
>
> (a) 14 AWG (b) 12 AWG (c) 10 AWG (d) 8 AWG

Household Cooking Appliances - Branch Circuit
Table 220.55 Note 4

12 kW, 120/240V
Household Range

Table 310.16, 8 AWG
branch-circuit conductors.

COPYRIGHT 2004
Mike Holt Enterprises, Inc.

Size the branch circuit for a 12 kW range.

Step 1: Column C, one unit = 8 kW demand

Step 2: Convert the demand load into amperes

I = P/E = 8,000W/240V = 33.33A

Figure 220–20

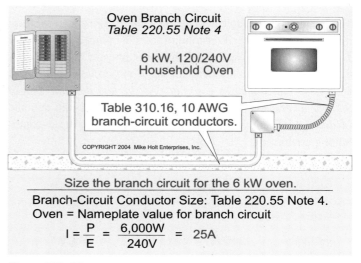

Oven Branch Circuit
Table 220.55 Note 4

6 kW, 120/240V
Household Oven

Table 310.16, 10 AWG
branch-circuit conductors.

COPYRIGHT 2004 Mike Holt Enterprises, Inc.

Size the branch circuit for the 6 kW oven.

Branch-Circuit Conductor Size: Table 220.55 Note 4.
Oven = Nameplate value for branch circuit

$$I = \frac{P}{E} = \frac{6,000W}{240V} = 25A$$

Figure 220–21

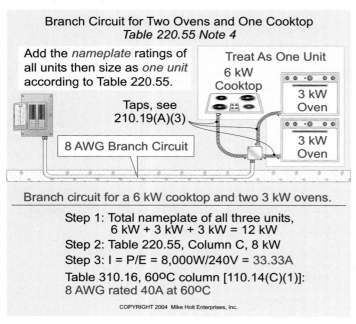

Figure 220–22

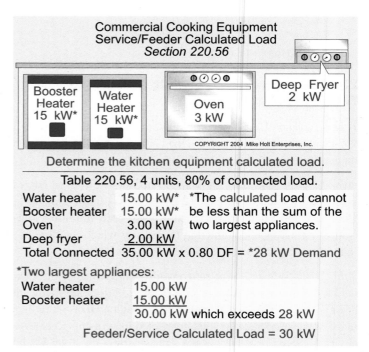

Figure 220–23

Answer: *(d) 8 AWG*

Step 1. Total connected load = 12 kW (6 kW + 3 kW + 3 kW)

Step 2. Determine calculated VA load as a single 12 kW range. Table 220.55 Column C = 8 kW

Step 3. Branch-circuit load in amperes, I = P/E
P = 8,000W
E = 240V
I = 8,000W/240V
I = 33.33A, 8 AWG rated 40A at 60°C [Table 310.16 and 110.14(C)(1)]

220.56 Commercial—Kitchen Equipment Load.

Table 220.56 is used to calculate the feeder/service load for thermostat-controlled or intermittently used commercial electric cooking equipment, such as dishwasher booster heaters, water heaters, and other kitchen loads. The kitchen equipment feeder/service calculated load *must not be less than the sum of the two largest kitchen equipment loads.* Table 220.56 demand factors do not apply to space-heating, ventilating, or air-conditioning equipment.

> **Question:** *What is the commercial kitchen equipment calculated load for one 15 kW booster water heater, one 15 kW water heater, one 3 kW oven, and one 2 kW deep fryer?* **Figure 220–23**
>
> *(a) 15 kW*   *(b) 20 kW*   *(c) 26 kW*   *(d) 30 kW*

Answer: *(d) 30kW*

Step 1. Determine the total connected load:
Total connected load = 15 kW + 15 kW + 3 kW + 2 kW
Total connected load = 35 kW

Step 2. Determine the feeder/service calculated load:
35 kW x 0.8 = 28 kW, but it must not be less than the sum of the two largest appliances, or 30 kW.

220.60 Noncoincident Loads.

Where it's unlikely that two or more loads will be used at the same time, only the largest load(s) must be used to determine the feeder/service VA calculated load. **Figure 220–24**

> **Question:** *What is the feeder/service calculated load for a 5 hp, 230V air conditioner with a current rating of 28A versus 9 kW heating?* **Figure 220–25**
>
> *(a) 5,000W*   *(b) 6,000W*   *(c) 7,500W*   *(d) 9,000W*

Answer: *(d) 9,000W*

Air conditioner load = 230V x 28A
Air conditioner load = 6,440 VA (omit, smaller than 9,000W)
Heat Load = 9,000W

220.61 Feeder/Service Neutral Unbalanced Load.

(A) Basic Calculation. The feeder/service calculated neutral load must be the maximum calculated load between the neutral conductor and any one ungrounded conductor. Line-to-line loads

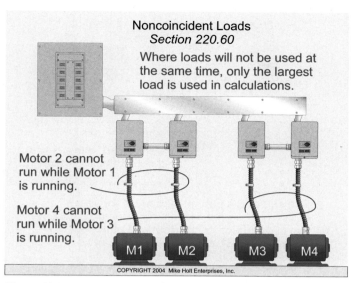

Noncoincident Loads
Section 220.60

Where loads will not be used at the same time, only the largest load is used in calculations.

Motor 2 cannot run while Motor 1 is running.

Motor 4 cannot run while Motor 3 is running.

M1 M2 M3 M4

COPYRIGHT 2004 Mike Holt Enterprises, Inc.

Figure 220–24

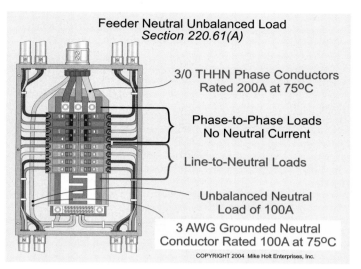

Feeder Neutral Unbalanced Load
Section 220.61(A)

3/0 THHN Phase Conductors Rated 200A at 75°C

Phase-to-Phase Loads No Neutral Current

Line-to-Neutral Loads

Unbalanced Neutral Load of 100A

3 AWG Grounded Neutral Conductor Rated 100A at 75°C

COPYRIGHT 2004 Mike Holt Enterprises, Inc.

Figure 220–26

do not place any load on the neutral, therefore they aren't considered when sizing the feeder/service neutral conductor calculated load. **Figure 220–26**

Question: What is the feeder/service grounded neutral conductor size for a 200A service, of which 100A is line-to-line loads with an unbalanced neutral load of 100A? See Figure 220–26.

(a) 3/0 AWG   (b) 1/0 AWG   (c) 1 AWG   (d) 3 AWG

Answer: (d) 3 AWG

200A total load less 100A line-to-line loads = 100A neutral loads

Table 310.16, 75°C column, 3 AWG rated 100A

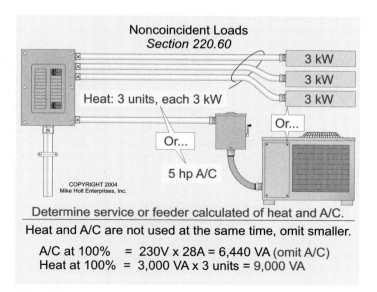

Noncoincident Loads
Section 220.60

Heat: 3 units, each 3 kW

Or...

Or...

5 hp A/C

COPYRIGHT 2004
Mike Holt Enterprises, Inc.

Determine service or feeder calculated of heat and A/C.

Heat and A/C are not used at the same time, omit smaller.

A/C at 100% = 230V x 28A = 6,440 VA (omit A/C)
Heat at 100% = 3,000 VA x 3 units = 9,000 VA

Figure 220–25

(B) Permitted Reductions.

(1) Dwelling Unit Range and Cooking Appliance Load. The feeder/service neutral calculated load for household electric ranges, wall-mounted ovens, or counter-mounted cooking units must be calculated at 70 percent of the cooking equipment calculated load in accordance with Table 220.55.

Question: What is the feeder/service calculated neutral load for nine 12 kW household ranges?

(a) 13 kW   (b) 14.7 kW   (c) 16.8 kW   (d) 24 kW

Answer: (c) 16.8 kW

Step 1. Table 220.55 Column C = 24 kW

Step 2. Neutral Load = 24 kW x 0.70 = 16.8 kW

(1) Dwelling Unit Dryer Load. The feeder/service neutral calculated load for household electric dryers must be calculated at 70 percent of the dryer calculated load in accordance with Table 220.54.

Question: A 10-unit multifamily building has a 5 kW electric clothes dryer in each unit. What is the feeder/service neutral load for these dryers?

(a) 17.5 kW   (b) 23.5 kW   (c) 33 kW   (d) 41 kW

Answer: (a) 17.5 kW

Step 1. Table 220.54 = 10 units x 5 kW x 0.50 = 25 kW

Step 2. Neutral Load = 25 kW x 0.70 = 17.5 kW

(2) Over 200A Neutral Reduction. The feeder/service calculated neutral load for a 3-wire single-phase, or 4-wire, three-phase system, can be reduced for that portion of the unbalanced load over 200A by a multiplier of 70 percent.

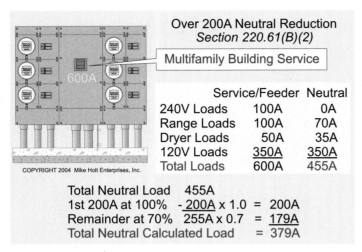

Figure 220–27

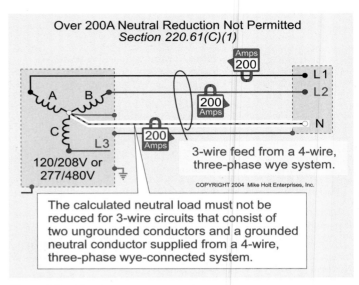

Figure 220–28

Question: *What is the feeder/service neutral calculated load for the following?* **Figure 220–27**
- *100A of line-to-line loads*
- *100A of household ranges*
- *50A of household dryers*
- *350A of line-to-neutral loads*

(a) 200A   *(b) 379A*   *(c) 455A*   *(d) 600A*

Answer: *(b) 379A*

Step 1. Determine the total neutral load:

	Feeder/Service	Neutral Load
Line-to-line	100A	0A
Ranges	100A	70A (100A x 0.7)
Dryers	50A	35A (50A x 0.7)
Line-to-neutral	350A	350A
Total load	600A	455A

Step 2. Determine the demand neutral load:

Total neutral load	455A		
First 200A at 100%	200A x 1.0	=	200A
Remainder at 70%	255A x 0.7	=	179A
Total demand neutral load			379A

(C) Prohibited Reductions.

(1) 3-Wire Circuits from 4-Wire Wye-Connected Systems. The feeder/service neutral calculated load must not be reduced for 3-wire circuits that consist of two ungrounded conductors and a grounded neutral conductor supplied from a 4-wire, three-phase wye-connected system. This is because the neutral load on the 3-wire circuit will carry approximately the same amount of line-to-neutral current as the ungrounded conductors [310.15(B)(4)(c)].

Question: *What is the current on the grounded neutral conductor of a 3-wire feeder supplied from a 4-wire, three-phase*

wye-connected system? The ungrounded conductors carry 200A of line-to-neutral loads. **Figure 220–28**

(a) 200A   *(b) 379A*   *(c) 455A*   *(d) 600A*

Answer: *(a) 200A*

(2) Nonlinear Loads. The feeder/service neutral calculated load must not be reduced for nonlinear loads supplied from a 4-wire, three-phase wye-connected system.

Question: *What is the feeder/service neutral calculated load for the following?*

- *200A of line-to-line loads*
- *200A of line-to-neutral nonlinear loads*
- *200A of line-to-neutral linear loads*

(a) 200A   *(b) 400A*   *(c) 500A*   *(d) 600A*

Answer: *(d) 400A. The current on the neutral conductor for non-linear loads could be as much as twice the maximum neutral load.*

	Feeder/Service	Neutral Load
Line-to-line loads	200A	0A
Nonlinear line-to-neutral loads	200A	200A
Linear line-to-neutral loads	200A	200A
Total calculated load	600A	400A

Author's Comment: Nonlinear loads of the line-to-neutral type from a wye-connected system generate triplen harmonic neutral currents (3rd, 9th, 15th harmonic) that add, instead of cancel, on the neutral conductor. This can result in excessive heating of the neutral conductor if it isn't increased in size to accommodate the excessive neutral current. See 210.4(A) FPN, 220.61(C)(2) FPN 2, and 310.15(B)(4)(c). **Figure 220–29**

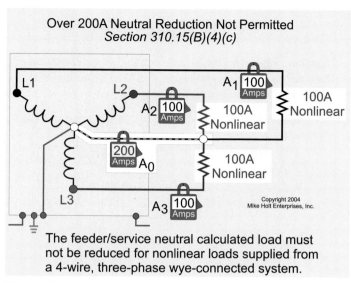

Over 200A Neutral Reduction Not Permitted
Section 310.15(B)(4)(c)

The feeder/service neutral calculated load must not be reduced for nonlinear loads supplied from a 4-wire, three-phase wye-connected system.

Figure 220–29

PART IV. OPTIONAL CALCULATIONS FOR COMPUTING FEEDER AND SERVICE LOADS

220.82 Dwelling Unit—Optional Load Calculation.

(A) Feeder/Service Load. The 3-wire feeder/service load for a dwelling unit can be calculated by adding the calculated loads from 220.82(B) and (C). The feeder/service neutral calculated load must be determined in accordance with 220.61.

(B) General Loads. The feeder/service calculated load must not be less than 100 percent of the first 10 kVA, plus 40 percent of the remainder of the following:

(1) General Lighting. 3 VA per sq ft for general lighting and general-use receptacles. The floor area must be calculated from the outside dimensions of the dwelling unit, not including open porches, garages, or unused or unfinished spaces not adaptable for future use.

(2) Small-Appliance and Laundry Circuits. A load of 1,500 VA for each 20A small-appliance and laundry branch circuit. Since two small-appliance circuits and a laundry circuit are required, the minimum will be 4,500 VA.

(3) Appliances. The nameplate rating of all appliances fastened in place, permanently connected or located to be on a specific circuit must be included.

(4) Motor VA. The VA nameplate rating of all motors.

(C) Heating and Air Conditioning. The largest of:

(1) Air Conditioning. 100 percent of the nameplate rating(s).

(2) Heat Pump without Supplemental Heating. 100 percent of the nameplate rating(s).

(3) Thermal Storage Heating. 100 percent of the nameplate rating(s).

> **Author's Comment:** One form of thermal storage heating involves heating bricks or water at night when the electric rates are lower. Then during the day, the building uses the thermally stored heat.

(4) Heat-Pump Compressor and Supplemental Heating. 100 percent of the nameplate rating(s) of the heat-pump compressor and 65 percent of the supplemental electric heating for central electric space-heating systems. If the heat-pump compressor is prevented from operating at the same time as the supplementary heat, it can be omitted in the calculation.

(5) Space Heating (three or less units). 65 percent of the nameplate rating(s).

(6) Space Heating (four or more units). 40 percent of the nameplate rating(s).

> **Question:** *Using the optional calculation method, what size 3-wire, single-phase, 120/240V feeder/service conductor is required for a 1,500-sq ft dwelling unit that contains the following loads?*
>
> | | |
> |---|---|
> | • Dishwasher | *1,200 VA* |
> | • Water heater | *4,500 VA* |
> | • Disposal | *900 VA* |
> | • Dryer | *4,000 VA* |
> | • Cooktop | *6,000 VA* |
> | • Oven | *3,000 VA* |
> | • Heat pump 5 hp compressor, with a 7 kW supplemental electric heat that operates with the heat pump | |
>
> *(a) 100A   (b) 110A   (c) 125A   (d) 150A*
>
> **Answer:** *(c) 125A*
>
> Step 1. Determine the total feeder/service calculated load.
>   Lighting, receptacles, and appliance calculated load [220.82(B)].
>
> | | | | |
> |---|---|---|---:|
> | Small appliance | 1,500 VA x 2 | = | 3,000 VA |
> | Laundry | 1,500 VA x 1 | = | 1,500 VA |
> | General lighting | 1,500 sq ft x 3 VA/sq ft | = | 4,500 VA |
> | Dishwasher | 1,200 VA x 1 | = | 1,200 VA |
> | Water heater | 4,500 VA x 1 | = | 4,500 VA |
> | Disposal | 900 VA x 1 | = | 900 VA |
> | Dryer | 4,000 VA x 1 | = | 4,000 VA |
> | Cooktop | 6,000 VA x 1 | = | 6,000 VA |
> | Oven | 3,000 VA x 1 | = | 3,000 VA |
> | | | | 28,600 VA |

First 10,000 VA at
   100% -10,000 VA x 1.00   =   10,000 VA
Remainder at 40%
   18,600 VA x 0.40   =   + 7,440 VA
220.82(B) calculated load   =   17,440 VA

Largest of A/C or Heat [220.82(C)(4)]
Heat pump 5 hp compressor at 100%
   230V x 28A   =   6,440 VA

Supplemental heat at 65%
   7,000 VA x 0.65   =   + 4,550 VA

Total calculated load 220.82(B) and (C)
   17,440 VA + 6,440 VA + 4,550 VA   =   28,430 VA

Step 2. Feeder/service calculated load in amperes:
   I = VA/E
   I = 28,430 VA/240V
   I = 119A, 2 AWG [215.2(A)(3) and 310.15(B)(6)]

220.84 Multifamily—Optional Load Calculation.

(A) Feeder or Service Load. The feeder/service calculated load for a building with three or more dwelling units equipped with electric cooking equipment, and either electric space heating or air conditioning, can be in accordance with the demand factors of Table 220.84 based on the number of dwelling units. The feeder/service neutral calculated load must be determined in accordance with 220.61.

(B) House Loads. House loads are calculated in accordance with Part III of Article 220 and then added to the Table 220.84 calculated load.

> **Author's Comment:** House loads are those not directly associated with the individual dwelling units of a multifamily dwelling. Some examples of house loads could be landscape and parking lot lighting, common area lighting, common laundry facilities, common pool and recreation areas, etc.

(C) Connected loads. The connected loads from all of the dwelling units are added together, and then the Table 220.84 demand factors are applied to determine the calculated load.

(1) 3 VA per sq ft for general lighting and general-use receptacles.

(2) 1,500 VA for each small-appliance circuit (minimum of 2 circuits [220.52(A)]), and 1,500 VA for each laundry circuit.

> **Author's Comment:** A laundry circuit isn't required in an individual unit of a multifamily dwelling if common laundry facilities are provided.

(3) The nameplate rating of all appliances.

(4) The nameplate rating of all motors.

(5) The larger of air-conditioning load or space-heating load.

> **Question:** What size 120/208V three-phase service is required for a multifamily building with twenty 1,500 sq ft dwelling units, where each unit contains the following loads?
>
> - Dishwasher   1,200 VA
> - Water heater   4,500 VA
> - Disposal   900 VA
> - Dryer   4,000 VA
> - Cooktop   6,000 VA
> - Oven   3,000 VA
> - Heat   7,000 VA
> - A/C, 5 hp compressor   6,440 VA

(a) 400A   (b) 600A   (c) 800A   (d) 1,200A

Answer: (c) 800A, 240.4 and 240.6(A)

Step 1. Determine the dwelling unit connected load:

General lighting 1,500 sq ft x 3 VA/sq ft	=	4,500 VA
Small appliance 1,500 x 2	=	3,000 VA
Laundry	=	1,500 VA
Dishwasher	=	1,200 VA
Water heater	=	4,500 VA
Disposal	=	900 VA
Dryer	=	4,000 VA
Cooktop	=	6,000 VA
Oven	=	3,000 VA
A/C 5 hp (omit)	=	0 VA
Heat	=	+ 7,000 VA
Total Dwelling Unit Load	=	35,600 VA

Step 2. Determine the calculated load for the
   multifamily building:
     35,600 VA x 20 x 0.38   = 270,560 VA

Step 3. Determine feeder/service conductor size:
   I = VA/(E x √3)
   I = 270,560 VA/(208V x 1.732)
   I = 751A
   I of each conductor parallel set = 751A/2 conductors
   I of each conductor parallel set = 376A
   Conductor = 500 kcmil, rated 380A x 2 = 760A,
      Table 316.16

220.85 Optional Calculation—Two Dwelling Units.

Where two dwelling units are supplied by a single feeder and where the standard calculated load in accordance with Part II of this article exceeds that for three identical units computed in accordance with 220.84, the lesser of the two calculated loads may be used.

Article 220 Questions

1. When computations in Article 220 result in a fraction of an ampere that is less than_____, such fractions can be dropped.

   (a) 0.49   (b) 0.50   (c) 0.51   (d) none of these

2. A single piece of equipment consisting of a multiple receptacle comprised of _____ or more receptacles must be computed at not less than 90 VA per receptacle.

   (a) 1   (b) 2   (c) 3   (d) 4

3. The feeder and service conductors for motors must be computed in accordance with Article _____.

   (a) 450   (b) 240   (c) 430   (d) 100

4. The load for electric clothes dryers in a dwelling unit must be _____ watts or the nameplate rating, whichever is larger, per dryer.

   (a) 1,500   (b) 4,500   (c) 5,000   (d) 8,000

5. The feeder demand load for nine 16 kW ranges is _____.

   (a) 15,000W   (b) 28,800W   (c) 20,000W   (d) 26,000W

6. Where it is unlikely that two or more noncoincident loads will be in use simultaneously, it is permissible to use only the _____ loads on at any given time in computing the total load to a feeder.

   (a) smaller of the   (b) largest of the   (c) difference between the   (d) none of these

7. Under the optional method for calculating a single-family dwelling, general loads beyond the initial 10 kW are assessed at a _____ percent demand factor.

   (a) 40   (b) 50   (c) 60   (d) 75

Notes

ARTICLE 225 Outside Wiring

Introduction

This article covers installation requirements for equipment including conductors located outdoors, on or between buildings, poles, and other structures on the premises. It has two parts:

Part I provides a listing of other articles that may provide additional requirements, addresses some general concerns, and briefly covers conductor sizing. Then it addresses conductor support, attachment, and clearances.

Part II addresses how many supplies you can have to a building and how to disconnect them. This includes such things as where to locate the disconnecting means and how to group them.

PART I GENERAL REQUIREMENTS

225.1 Scope. Article 225 contains the installation requirements for outside branch circuits and feeders run on or between buildings, structures, or poles. **Figure 225–1**

225.2 Other Articles. Other articles containing important requirements include:

- Branch Circuits, Article 210
- Class 1, Class 2, and Class 3 Remote-Control, Signaling and Power-Limited Circuits, Article 725
- Communications Circuits, Article 800
- Community Antenna Television and Radio Distribution Systems, Article 820
- Conductors for General Wiring, Article 310
- Electric Signs and Outline Lighting, Article 600
- Feeders, Article 215
- Floating Buildings, Article 553
- Grounding (Earthing) and Bonding, Article 250
- Marinas and Boatyards, Article 555
- Messenger Supported Wiring, Article 396
- Open Wiring on Insulators, Article 398
- Radio and Television Equipment, Article 810
- Services, Article 230
- Solar Photovoltaic Systems, Article 690

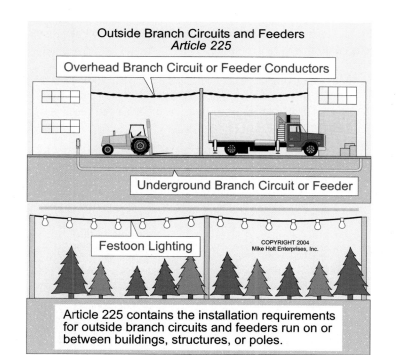

Outside Branch Circuits and Feeders
Article 225

Overhead Branch Circuit or Feeder Conductors

Underground Branch Circuit or Feeder

Festoon Lighting

COPYRIGHT 2004
Mike Holt Enterprises, Inc.

Article 225 contains the installation requirements for outside branch circuits and feeders run on or between buildings, structures, or poles.

Figure 225–1

- Swimming Pools, Fountains, and Similar Installations, Article 680

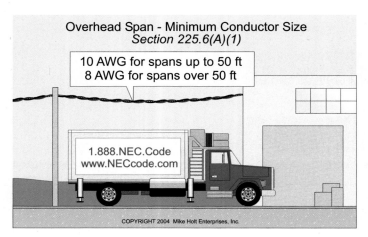

Overhead Span - Minimum Conductor Size
Section 225.6(A)(1)

10 AWG for spans up to 50 ft
8 AWG for spans over 50 ft

1.888.NEC.Code
www.NECcode.com

COPYRIGHT 2004 Mike Holt Enterprises, Inc.

Figure 225–2

225.6 Minimum Size Conductors.

(A) Overhead Conductor.

(1) Spans. Conductors 10 AWG and larger are permitted for overhead spans up to 50 ft. For spans over 50 ft, the minimum size conductor is 8 AWG. **Figure 225–2**

(B) Festoon Lighting. Overhead conductors for festoon lighting must not be smaller than 12 AWG, unless messenger wires support the conductors. The overhead conductors must be supported by messenger wire, with strain insulators, whenever the spans exceed 40 ft. **Figure 225–3**

> **Author's Comment:** Festoon lighting is a string of outdoor lights suspended between two points [Article 100]. It's commonly used at carnivals, circuses, fairs, and Christmas tree lots [525.20(C)].

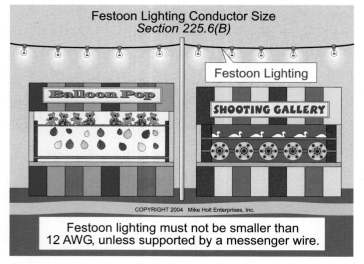

Festoon Lighting Conductor Size
Section 225.6(B)

Festoon Lighting

Balloon Pop

SHOOTING GALLERY

COPYRIGHT 2004 Mike Holt Enterprises, Inc.

Festoon lighting must not be smaller than
12 AWG, unless supported by a messenger wire.

Figure 225–3

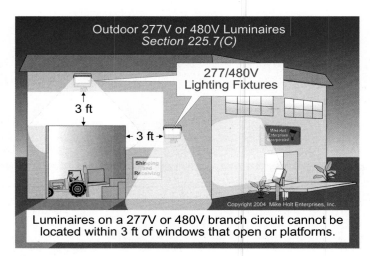

Outdoor 277V or 480V Luminaires
Section 225.7(C)

277/480V
Lighting Fixtures

3 ft

3 ft

Mike Holt
Enterprises
Incorporated

Shipping
and
Receiving

Copyright 2004 Mike Holt Enterprises, Inc.

Luminaires on a 277V or 480V branch circuit cannot be
located within 3 ft of windows that open or platforms.

Figure 225–4

225.7 Luminaires Installed Outdoors.

(C) 277V-to-Ground. Luminaires on a 277V or 480V branch circuit cannot be located within 3 ft of windows that open, platforms, fire escapes, and the like. **Figure 225–4**

> **Author's Comment:** See 210.6(C) for the types of luminaires permitted on 277V or 480V branch circuits.

225.15 Supports Over Buildings.
Conductor spans over a building must be securely supported by substantial structures. Where practicable, such supports must be independent of the building [230.29].

225.16 Attachment.

(A) Point of Attachment. The points of attachment for overhead conductors must not be less than 10 ft above the finished grade, and they must be located so the minimum conductor clearance required by 225.18 can be maintained.

> **CAUTION:** *Conductors might need to have the point of attachment raised so the overhead conductors will comply with the clearances required by 225.19 from building openings and other building areas.* **Figure 225–5**

(B) Means of Attachment to Buildings. Open conductors must be attached to fittings identified for use with conductors, or to noncombustible, nonabsorbent insulators securely attached to the building or other structure.

> **Author's Comment:** The point of attachment of the overhead conductor spans to a building or other structure must provide the minimum clearances as specified in 225.15, 225.18, and 225.19. In no case can this point of attachment be less than 10 ft above the finished grade.

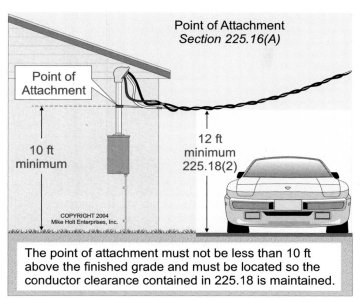

Point of Attachment
Section 225.16(A)

Point of Attachment

10 ft minimum

12 ft minimum 225.18(2)

COPYRIGHT 2004 Mike Holt Enterprises, Inc.

The point of attachment must not be less than 10 ft above the finished grade and must be located so the conductor clearance contained in 225.18 is maintained.

Figure 225–5

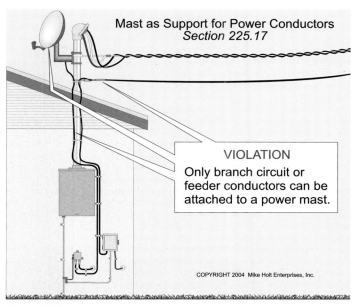

Mast as Support for Power Conductors
Section 225.17

VIOLATION
Only branch circuit or feeder conductors can be attached to a power mast.

COPYRIGHT 2004 Mike Holt Enterprises, Inc.

Figure 225–7

225.17 Masts as Support.
Where a mast is used for overhead conductor support, it must have adequate mechanical strength, braces, or guy wires to withstand the strain caused by the conductors. Only branch-circuit or feeder conductors can be attached to the mast. **Figure 225–6**

Author's Comment: Aerial cables and antennas for radio TV equipment cannot be attached to the electric service mast [810.12]. In addition, 800.133(C) and 830.133(B) prohibit communications cables from being attached to raceways, including a service mast for power conductors. **Figure 225–7**

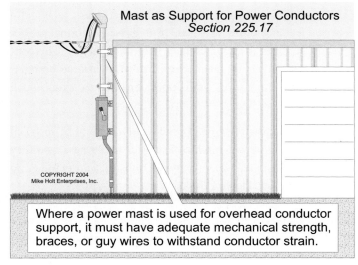

Mast as Support for Power Conductors
Section 225.17

COPYRIGHT 2004 Mike Holt Enterprises, Inc.

Where a power mast is used for overhead conductor support, it must have adequate mechanical strength, braces, or guy wires to withstand conductor strain.

Figure 225–6

225.18 Clearances.
Overhead conductor spans must maintain the following vertical clearances:

(1) 10 ft above finished grade, sidewalks, platforms, or projections from which they might be accessible to pedestrians for 120V, 120/208V, 120/240V, or 240V circuits.

(2) 12 ft above residential property and driveways, and those commercial areas not subject to truck traffic for 120V, 120/208V, 120/240V, 240V, 277V, 277/480V, or 480V circuits.

(4) 18 ft over public streets, alleys, roads, parking areas subject to truck traffic, driveways on other than residential property, and other areas traversed by vehicles (such as those used for cultivation, grazing, forestry, and orchards).

Author's Comment: Overhead conductors not under the exclusive control of the electric utility located above pools, outdoor spas, outdoor hot tubs, diving structures, observation stands, towers, or platforms must be installed in accordance with the clearance requirements in 680.8.

225.19 Clearances from Buildings.

(A) Above Roofs. Overhead conductors must maintain a vertical clearance of 8 ft above the surface of a roof that must be maintained for a distance of not less than 3 ft from the edge of the roof. **Figure 225–8**

Exception 2: The overhead conductor clearances from the roof can be reduced from 8 ft to 3 ft if the slope of the roof meets or exceeds 4 in. for every 12 in.

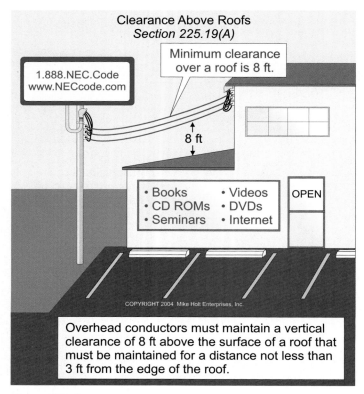

Clearance Above Roofs
Section 225.19(A)

Minimum clearance over a roof is 8 ft.

1.888.NEC.Code
www.NECcode.com

8 ft

• Books   • Videos
• CD ROMs • DVDs
• Seminars • Internet

OPEN

COPYRIGHT 2004 Mike Holt Enterprises, Inc.

Overhead conductors must maintain a vertical clearance of 8 ft above the surface of a roof that must be maintained for a distance not less than 3 ft from the edge of the roof.

Figure 225–8

Exception 3: For 120/208V or 120/240V circuits, the conductor clearance over the roof overhang can be reduced from 8 ft to 18 in., if no more than 6 ft of conductor passes over no more than 4 ft of roof. Figure 225–9

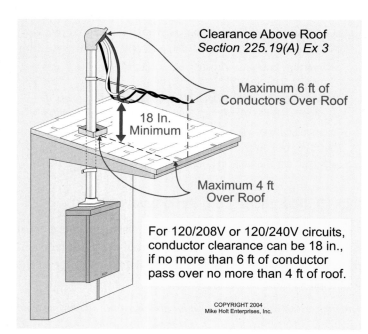

Clearance Above Roof
Section 225.19(A) Ex 3

Maximum 6 ft of Conductors Over Roof

18 In. Minimum

Maximum 4 ft Over Roof

For 120/208V or 120/240V circuits, conductor clearance can be 18 in., if no more than 6 ft of conductor pass over no more than 4 ft of roof.

COPYRIGHT 2004
Mike Holt Enterprises, Inc.

Figure 225–9

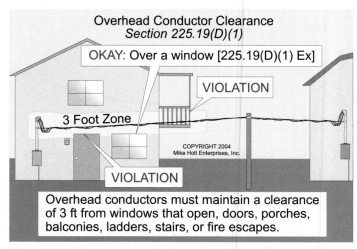

Overhead Conductor Clearance
Section 225.19(D)(1)

OKAY: Over a window [225.19(D)(1) Ex]

VIOLATION

3 Foot Zone

COPYRIGHT 2004
Mike Holt Enterprises, Inc.

VIOLATION

Overhead conductors must maintain a clearance of 3 ft from windows that open, doors, porches, balconies, ladders, stairs, or fire escapes.

Figure 225–10

Exception 4: The 3 ft clearance from the roof edge does not apply when the point of attachment is on the side of the building below the roof.

(B) From Other Structures. Overhead conductors must maintain a vertical, diagonal, and horizontal clearance of not less than 3 ft from signs, chimneys, radio and television antennas, tanks, and other nonbuilding or nonbridge structures.

(D) Final Span Clearance.

(1) Clearance from Windows. Overhead conductors must maintain a clearance of 3 ft from windows that open, doors, porches, balconies, ladders, stairs, fire escapes, or similar locations. **Figure 225–10**

Exception: Overhead conductors that run above a window aren't required to maintain the 3 ft distance.

(2) Vertical Clearance. Overhead conductors must maintain a vertical clearance of not less than 10 ft above platforms, projections, or surfaces from which they might be reached. This vertical clearance must be maintained for 3 ft, measured horizontally from the platforms, projections, or surfaces from which they might be reached.

(3) Below Opening. Overhead conductors must not be installed under an opening through which materials might pass, and they must not be installed where they will obstruct an entrance to building openings. **Figure 225–11**

225.22 Raceways on Exterior Surfaces of Buildings or Other Structures. Raceways on exterior surfaces of buildings or other structures must be arranged to drain, and in wet locations must be raintight.

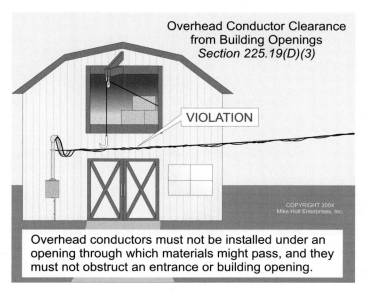

Figure 225–11

Author's Comment: A wet location is an area subject to saturation with water and unprotected locations exposed to weather [Article 100].

225.26 Trees for Conductor Support. Trees or other vegetation cannot be used for the support of overhead conductor spans. Figure 225–12

Author's Comment: Overhead conductor spans for services [230.10] and temporary wiring [590.4(J)] aren't permitted to be supported by vegetation.

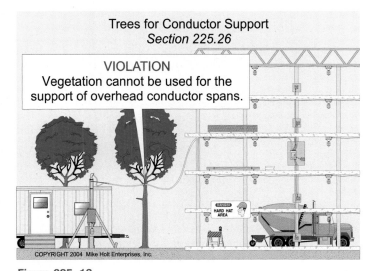

Figure 225–12

PART II. MORE THAN ONE BUILDING OR STRUCTURE

225.30. Number of Supplies. Where more than one building or structure is on the same property, each building or structure must be served by no more than one feeder or branch circuit, except as permitted in (A) through (E).

For the purpose of this section, a multiwire branch circuit is considered a single circuit.

(A) Special Conditions. Additional circuits are permitted for:

(1) Fire pumps

(2) Emergency systems

(3) Legally required standby systems

(4) Optional standby systems

(5) Parallel power production systems

(6) Systems designed for connection to multiple sources of supply for the purpose of enhanced reliability.

> **Author's Comment:** To minimize the possibility of accidental interruption, the disconnecting means for the fire pump or standby power must be located remotely away from the normal power disconnect [225.34(B)].

(B) Special Occupancies. By special permission, additional feeders are permitted for:

(1) Multiple-occupancy buildings where there's no available space for supply equipment accessible to all occupants, or

(2) A building or structure that is so large that two or more feeder supplies are necessary.

(C) Capacity Requirements. Additional feeders are permitted for a building or structure where the capacity requirements exceed 2,000A.

(D) Different Characteristics. Additional feeders or branch circuits are permitted for different voltages, frequencies, or uses, such as control of outside lighting from multiple locations.

(E) Documented Switching Procedures. Additional feeders are permitted where documented safe switching procedures are established and maintained for disconnection.

225.31 Disconnecting Means. A disconnect is required for all conductors that enter or pass through a building or structure.

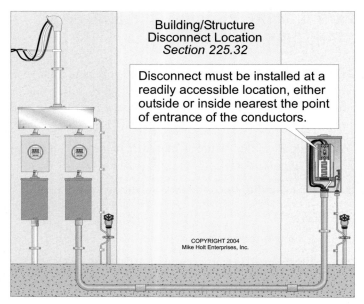

Figure 225-13

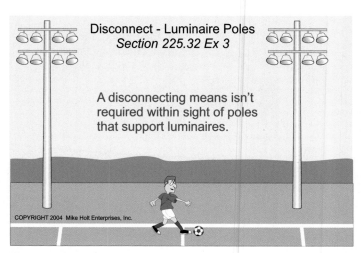

Figure 225-15

225.32 Disconnect Location.

225.32 Disconnect Location. The disconnecting means for a building or structure must be installed at a readily accessible location, either outside the building or structure or inside the building or structure, nearest the point of entrance of the conductors. **Figure 225-13**

Supply conductors are considered outside of a building or other structure where they are encased or installed under not less than 2 in. of concrete or brick [230.6]. **Figure 225-14**

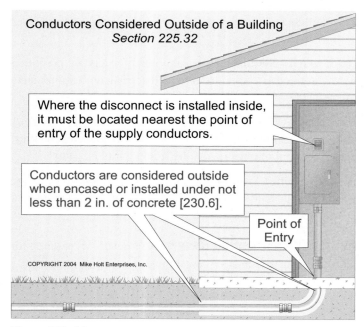

Figure 225-14

Exception 1: Where documented safe switching procedures are established and maintained, the building/structure disconnecting means can be located elsewhere on the premises if monitored by qualified persons.

> **Author's Comment:** A qualified person is one who has skills and knowledge related to the construction and operation of the electrical equipment and installation, and has received safety training on the hazards involved with electrical systems [Article 100].

Exception 3: A disconnecting means isn't required within sight of poles that support luminaires. **Figure 225-15**

Exception 4: The disconnecting means for a sign isn't required to be readily accessible if installed in accordance with the requirements for signs. **Figure 225-16**

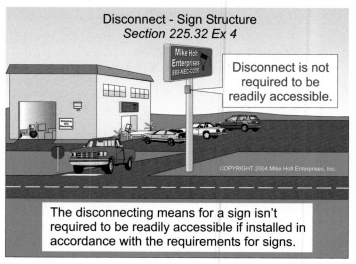

Figure 225-16

Author's Comment: Each sign must be controlled by an externally operable switch or circuit breaker that opens all ungrounded conductors to the sign. The sign disconnecting means must be within sight of the sign, or the disconnecting means must be capable of being locked in the open position [600.6(A)].

225.33 Maximum Number of Disconnects.

(A) General. The building or structure disconnecting means can consist of no more than six switches or six circuit breakers in a single enclosure, or separate enclosures for each supply permitted by 225.30.

225.34 Grouping of Disconnects.

(A) General. The building or structure disconnecting means must be grouped in one location, and they must be marked to indicate the loads they serve [110.22].

(B) Additional Disconnects. To minimize the possibility of accidental interruption of the critical power systems, 225.30(A) requires the disconnecting means for a fire pump or standby power to be located remotely away from the normal power disconnect.

225.35 Access to Occupants. In a multiple-occupancy building, each occupant must have access to the disconnecting means for their occupancy.

Exception: The occupant disconnect can be accessible to building management, if electrical maintenance under continuous supervision is provided by the building management.

225.36 Identified as Suitable for Service Equipment.
The building or structure disconnecting means must be identified as "suitable for use as service equipment."

Author's Comment: "Suitable for use as service equipment" means that the disconnecting means is supplied with a main bonding jumper so that a neutral-to-case connection can be made when required by 250.32(B)(2) and 250.142(A). Figure 225–17

Exception: A snap switch or a set of 3-way or 4-way snap switches can be used as the disconnecting means for garages and outbuildings on residential property, without having a "service equipment" rating.

225.37 Identification of Multiple Supplies. Where more than one feeder supplies a building or structure, a permanent plaque or directory must be installed at each feeder disconnect location denoting all other feeders that supply that building or structure and the area served by each.

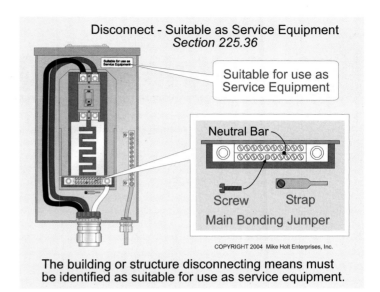

Disconnect - Suitable as Service Equipment
Section 225.36

Suitable for use as Service Equipment

Neutral Bar

Screw   Strap

Main Bonding Jumper

COPYRIGHT 2004 Mike Holt Enterprises, Inc.

The building or structure disconnecting means must be identified as suitable for use as service equipment.

Figure 225–17

225.38 Disconnect Construction.

(A) Manual or Power-Operated Circuit Breakers. The building or structure disconnecting means can consist of either a manually or power-operated switch or circuit breaker that is capable of being operated manually.

Author's Comment: A shunt-trip pushbutton can be used to open a power-operated circuit breaker. The circuit breaker is the disconnecting means, not the pushbutton.

225.39 Rating of Disconnecting Means. The feeder or branch-circuit disconnecting means for a building or structure must have an ampere rating not less than the calculated load determined in accordance with Article 220, and in no case less than:

(A) One-Circuit Installation. For installations consisting of a single branch circuit, the disconnecting means must have a rating not less than 15A.

(B) Two Circuit Installation. For installations consisting of two 2-wire branch circuits, the feeder disconnecting means must have a rating not less than 30A.

(C) One-Family Dwelling. For a one-family dwelling, the feeder disconnecting means must have a rating not less than 100A, 3-wire.

(D) All Others. For all other installations, the feeder or branch-circuit disconnecting means must have a rating not less than 60A.

Article 225 Questions

(• Indicates that 75% or fewer exam takers get the question correct.)

1. Open individual conductors must not be smaller than _____ AWG copper for spans up to 50 ft in length and _____ AWG copper for a longer span, unless supported by a messenger wire.

   (a) 10, 8   (b) 6, 8   (c) 6, 6   (d) 8, 8

2. The minimum clearance for overhead conductors not exceeding 600V that pass over commercial areas subject to truck traffic is _____.

   (a) 10 ft   (b) 12 ft   (c) 15 ft   (d) 18 ft

3. Overhead conductors to a building must maintain a vertical clearance of final spans above, or within _____ measured horizontally from the platforms, projections, or surfaces from which they might be reached.

   (a) 3 ft   (b) 6 ft   (c) 8 ft   (d) 10 ft

4. A building or structure must be supplied by a maximum of _____ feeder(s) or branch circuit(s).

   (a) one   (b) two   (c) three   (d) as many as desired

5. •There must be no more than _____ disconnects installed for each electric supply.

   (a) two   (b) four   (c) six   (d) none of these

230 Services

Introduction

This article covers the installation requirements for service conductors and equipment. The requirements for service conductors differ from those for other conductors. For one thing, service conductors for one structure cannot pass through the interior of another structure [230.3], and you apply different rules depending on whether a service conductor is inside or outside a structure. When are they "outside" as opposed to "inside?" The answer may seem obvious, but isn't.

It's usually good to start a service installation by deciding which conductors actually are parts of the service. What you decide here will determine how you do the rest of the job. To identify a service conductor, you must first determine whether you're dealing with a service (line side) or a premises (load side) distribution point.

Let's review the following definitions in Article 100 to understand when the requirements of Article 230 apply:

- Service Point—The point of connection between the facilities of the serving utility and the premises wiring.

- Service Conductors—The conductors from the service point to the service disconnecting means (service equipment, not meter). Service-entrance conductors may be either overhead (service drop) or underground (service lateral).

- Service Equipment—The necessary equipment, usually consisting of circuit breakers or switches and fuses and their accessories, connected to the load end of service conductors in a building or other structure (or an otherwise designated area), and intended to constitute the main control and cutoff of the electricity supply. Service equipment doesn't include the metering equipment, such as the meter and meter enclosure [230.66].

After reviewing these definitions, you should understand that service conductors originate at the serving utility (service point) and terminate on the line side of the service disconnecting means (service equipment). Conductors and equipment on the load side of service equipment are considered feeder conductors and must be installed in accordance with Articles 215 and 225. These include: Figure 230–1

- Secondary conductors from customer-owned transformers.

- Conductors from generators, UPS systems, or photovoltaic systems.

- Conductors to remote buildings or structures.

Conductors supplied from a battery, uninterruptible power supply system, solar photovoltaic system, generator, transformer, or phase converters aren't considered service conductors; they are feeder conductors [Article 100 Feeder].

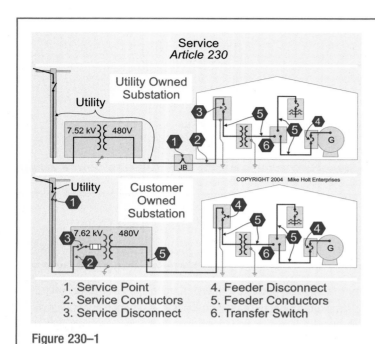

1. Service Point   4. Feeder Disconnect
2. Service Conductors   5. Feeder Conductors
3. Service Disconnect   6. Transfer Switch

Figure 230–1

Article 230 consists of seven parts:

- **PART I. GENERAL**
- **PART II. OVERHEAD SERVICE-DROP CONDUCTORS**
- **PART III. UNDERGROUND SERVICE-LATERAL CONDUCTORS**
- **PART IV. SERVICE-ENTRANCE CONDUCTORS**
- **PART V. SERVICE EQUIPMENT**
- **PART VI. DISCONNECTING MEANS**
- **PART VII. OVERCURRENT PROTECTION**

Before studying the material that comes next, just read through it to get a feel for how the seven parts relate to each other. Then go back and study it.

PART I. GENERAL

230.1 Scope. Article 230 covers the installation requirements for service conductors and service equipment.

> **Author's Comment:** See Article 100 for the definitions of "Service Conductors" and "Service Equipment."

230.2 Number of Services. A building or structure can only be served by one service drop or service lateral, except as permitted by (A) through (D). **Figure 230–2**

> **Author's Comment:** See Article 100 for the definitions of "Service Drop" and "Service Lateral."

Service laterals 1/0 AWG and larger that run to the same location and are connected together at their supply end, but not connected together at their load end, are considered to be a single service.

(A) Special Conditions. Additional services are permitted for the following:

(1) Fire pumps

(2) Emergency power

(3) Legally required standby power

(4) Optional standby power

(5) Parallel power production systems

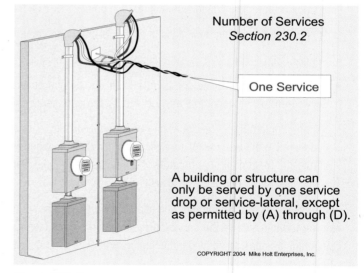

A building or structure can only be served by one service drop or service-lateral, except as permitted by (A) through (D).

Figure 230–2

(6) Systems designed for connection to multiple sources of supply for the purpose of enhanced reliability.

> **Author's Comments:**
> - See Article 100 for the definition of "Service."
> - To minimize the possibility of accidental interruption, the disconnecting means for the fire pump or standby power must be located remotely away from the normal power disconnect [230.72(B)].

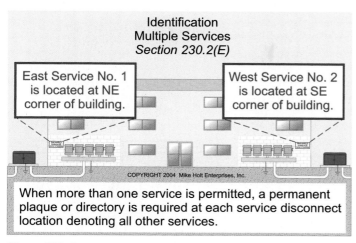

Figure 230–3

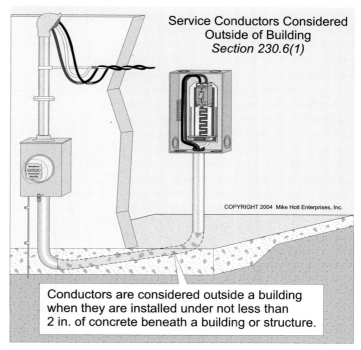

Conductors are considered outside a building when they are installed under not less than 2 in. of concrete beneath a building or structure.

Figure 230–4

(B) Special Occupancies. By special permission, additional services are permitted for:

(1) Multiple-occupancy buildings where there's no available space for supply equipment accessible to all occupants, or

(2) A building or other structure so large that two or more supplies are necessary.

(C) Capacity Requirements. Additional services are permitted:

(1) Where the capacity requirements exceed 2,000A.

(2) Where the load requirements of a single-phase installation exceeds the utility's capacity.

(3) By special permission.

(D) Different Characteristics. Additional services are permitted for different voltages, frequencies, or phases, or for different uses, such as for different electricity rate schedules.

(E) Identification of Multiple Services. Where a building or structure is supplied by more than one service, or a combination of feeders and services, a permanent plaque or directory must be installed at each service and feeder disconnect location to denote all other services and feeders supplying that building or structure, and the area served by each. **Figure 230–3**

230.3 Pass Through a Building or Structure. Service

conductors must not pass through the interior of another building or other structure.

230.6 Conductors Considered Outside a Building.

Conductors are considered outside a building when they are installed:

(1) Under not less than 2 in. of concrete beneath a building or structure. **Figure 230–4**

(2) Within a building or structure in a raceway that is encased in not less than a 2 in. thickness of concrete or brick.

(3) Installed in a vault that meets the construction requirements of Article 450, Part III.

(4) In conduit under not less than 18 in. of earth beneath a building or structure.

230.7 Service Conductors Separate from Other

Conductors. Service conductors cannot be installed in the same raceway or cable with feeder or branch-circuit conductors. **Figure 230–5**

WARNING: *Overcurrent protection for the feeder conductors could be bypassed if service conductors were mixed with nonservice conductors in the same raceway and a fault occurred between the service and nonservice conductors.*

Author's Comments:

• This rule doesn't prohibit the mixing of service, feeder, and branch-circuit conductors in the same "service equipment enclosure." **Figure 230–6**

• This requirement may be the root of the misconception that "line" and "load" conductors cannot be installed in the same raceway. It's true that service conductors cannot be installed in the same raceway with feeder or branch-circuit conductors, but line and load conductors for feeders and branch circuits can be in the same raceway or enclosure. **Figure 230–7**

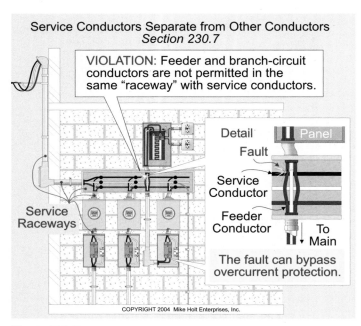

VIOLATION: Feeder and branch-circuit conductors are not permitted in the same "raceway" with service conductors.

Service Conductors Separate from Other Conductors
Section 230.7

Detail

Panel

Fault

Service Conductor

Feeder Conductor

To Main

Service Raceways

The fault can bypass overcurrent protection.

COPYRIGHT 2004 Mike Holt Enterprises, Inc.

Figure 230–5

230.8 Raceway Seals.
Underground raceways (used or unused) must be sealed or plugged to prevent moisture from contacting energized live parts [300.5(G)].

Author's Comment: Sealing can be accomplished with the use of a putty-like material called duct seal or a fitting identified for the purpose. A seal of the type required in Chapter 5 for hazardous (classified) locations isn't required.

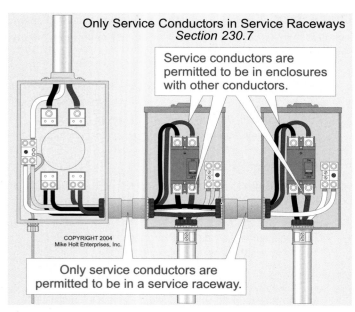

Only Service Conductors in Service Raceways
Section 230.7

Service conductors are permitted to be in enclosures with other conductors.

COPYRIGHT 2004 Mike Holt Enterprises, Inc.

Only service conductors are permitted to be in a service raceway.

Figure 230–6

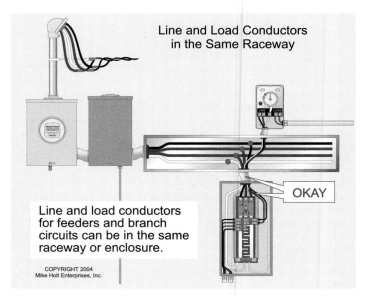

Line and Load Conductors in the Same Raceway

OKAY

Line and load conductors for feeders and branch circuits can be in the same raceway or enclosure.

COPYRIGHT 2004 Mike Holt Enterprises, Inc.

Figure 230–7

230.9 Clearance from Building Openings.

(A) Clearance. Overhead-service conductors must maintain a clearance of 3 ft from windows that open, doors, porches, balconies, ladders, stairs, fire escapes, or similar locations. **Figure 230–8**

Exception: Overhead conductors run above a window aren't required to maintain the 3 ft distance.

(B) Vertical Clearance. Overhead-service conductors must maintain a vertical clearance not less than 10 ft above platforms, projections, or surfaces from which they might be reached [230.24(B)]. This vertical clearance must be maintained for 3 ft, measured horizontally from the platform, projections, or surfaces from which people might reach them.

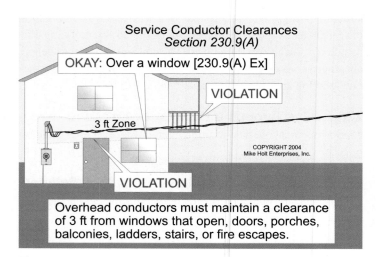

Service Conductor Clearances
Section 230.9(A)

OKAY: Over a window [230.9(A) Ex]

VIOLATION

3 ft Zone

COPYRIGHT 2004 Mike Holt Enterprises, Inc.

VIOLATION

Overhead conductors must maintain a clearance of 3 ft from windows that open, doors, porches, balconies, ladders, stairs, or fire escapes.

Figure 230–8

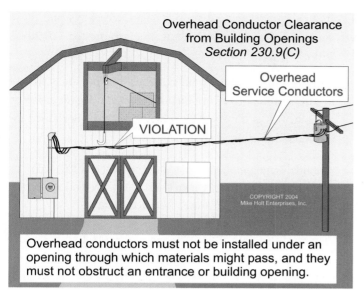

Overhead Conductor Clearance from Building Openings
Section 230.9(C)

Overhead Service Conductors

VIOLATION

Overhead conductors must not be installed under an opening through which materials might pass, and they must not obstruct an entrance or building opening.

Figure 230–9

(C) Below Opening. Service conductors must not be installed under an opening through which materials might pass, and they must not be installed where they will obstruct entrance to building openings. **Figure 230–9**

230.10 Vegetation as Support. Trees or other vegetation cannot be used for the support of overhead service conductor spans. **Figure 230–10**

> **Author's Comment:** Service-drop conductors installed by the electric utility must comply with the *National Electrical Safety Code (NESC)*, not the *NEC* [90.2(B)(5)], but overhead-service conductors not under exclusive control of the electric utility must be installed in accordance with the *NEC*.

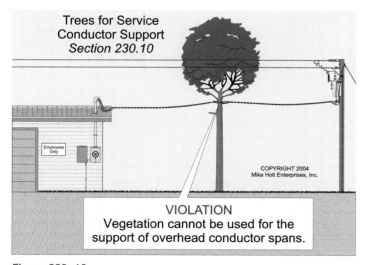

Trees for Service Conductor Support
Section 230.10

Employees Only

VIOLATION
Vegetation cannot be used for the support of overhead conductor spans.

Figure 230–10

PART II. OVERHEAD SERVICE-DROP CONDUCTORS

230.23 Size and Rating.

(A) Ampacity of Service-Drop Conductors. Service-drop conductors must have adequate mechanical strength and sufficient ampacity to carry the load as calculated in accordance with Article 220.

(B) Ungrounded Neutral Conductor Size. Service-drop conductors must not be smaller than 8 AWG copper or 6 AWG aluminum.

Exception: Service-drop conductors can be as small as 12 AWG for limited-load installations.

(C) Grounded Neutral Conductor Size. The grounded neutral service-drop conductor must be sized to carry the maximum unbalanced load, in accordance with 220.61, and must not be sized smaller than required by 250.24(C).

> **WARNING:** *In all cases, the grounded neutral service conductor size must not be smaller than required by 250.24(C) to ensure that it has sufficiently low impedance and current-carrying capacity to safely carry fault current in order to facilitate the operation of the overcurrent protection device.*

> **Question:** *What size grounded neutral conductor is required for a structure with a 400A service supplied with 500 kcmil conductors if the maximum line-to-neutral load is no more than 100A?*
>
> *(a) 3 AWG   (b) 2 AWG   (c) 1 AWG   (d) 1/0 AWG*
>
> **Answer:** *(d) 1/0 AWG*

According to Table 310.16, 3 AWG rated 100A at 75°C [110.14(C)] is sufficient to carry 100A of neutral current. However, the grounded neutral service conductor must be sized not smaller than 1/0 AWG, in accordance with Table 250.66 [250.24(C)].

230.24 Clearances.

Service-drop conductors must be located so they aren't readily accessible, and they must comply with the following clearance requirements:

(A) Above Roofs. Overhead-service conductors must maintain a minimum vertical clearance of 8 ft above the surface of a roof that must be maintained for a minimum distance of 3 ft in all directions from the edge of the roof.

Exception 2: For 120/208V or 120/240V service-drop conductors, overhead conductor clearances from the roof can be reduced from 8 ft to 3 ft if the slope of the roof exceeds 4 in. for every 12 in.

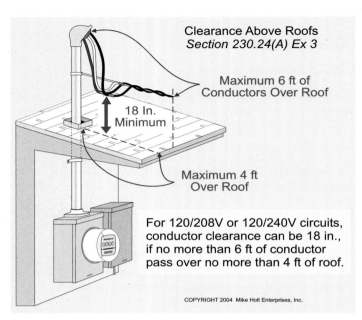

Clearance Above Roofs
Section 230.24(A) Ex 3

Maximum 6 ft of
Conductors Over Roof

18 In.
Minimum

Maximum 4 ft
Over Roof

For 120/208V or 120/240V circuits, conductor clearance can be 18 in., if no more than 6 ft of conductor pass over no more than 4 ft of roof.

COPYRIGHT 2004 Mike Holt Enterprises, Inc.

Figure 230–11

Exception 3: For 120/208V or 120/240V service-drop conductors, conductor clearance over the roof overhang can be reduced from 8 ft to 18 in. if no more than 6 ft of overhead conductors pass over no more than 4 ft of roof. Figure 230–11

Exception 4: The 3 ft vertical clearance that extends from the roof doesn't apply when the point of attachment is on the side of the building below the roof.

(B) Clearances. Overhead conductor spans must maintain the following vertical clearances: Figure 230–12

(1) 10 ft above finished grade, sidewalks, or platforms or projections from which they might be accessible to pedestrians for 120/208V or 120/240V circuits where the voltage doesn't exceed 150V-to-ground.

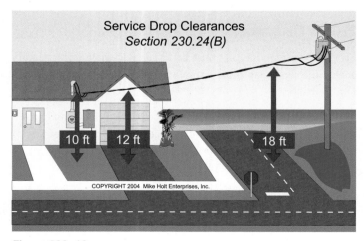

Service Drop Clearances
Section 230.24(B)

10 ft   12 ft   18 ft

COPYRIGHT 2004 Mike Holt Enterprises, Inc.

Figure 230–12

(2) 12 ft above residential property and driveways and those commercial areas not subject to truck traffic for 120/208V, 120/240V, or 277/480V circuits.

(4) 18 ft over public streets, alleys, roads, parking areas subject to truck traffic, driveways on other than residential property, and other areas traversed by vehicles, such as those used for cultivation, grazing, forestry, and orchards.

> **Author's Comment:** Department of Transportation (DOT) type right-of-ways in rural areas are often used by slow-moving and tall farming machinery to avoid impeding road traffic.

(D) Swimming Pools. Overhead-service conductors not under the exclusive control of the electric utility located above pools, outdoor spas, outdoor hot tubs, diving structures, observation stands, towers, or platforms must be installed in accordance with the clearance requirements contained in 680.8.

230.26 Point of Attachment. The point of attachment for service-drop conductors must not be less than 10 ft above the finished grade and must be located so the minimum service conductor clearance required by 230.9 and 230.24 can be maintained.

> **CAUTION:** *The points of attachment for conductors might need to be raised so the overhead conductors will comply with the clearances from building openings required by 230.9 and from other areas by 230.24.* Figure 230–13

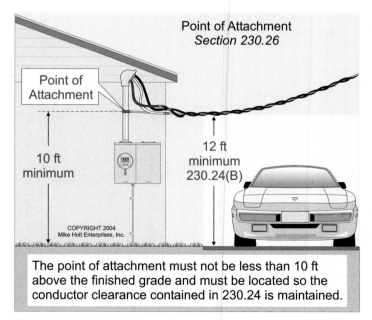

Point of Attachment
Section 230.26

Point of Attachment

10 ft minimum

12 ft minimum
230.24(B)

COPYRIGHT 2004
Mike Holt Enterprises, Inc.

The point of attachment must not be less than 10 ft above the finished grade and must be located so the conductor clearance contained in 230.24 is maintained.

Figure 230–13

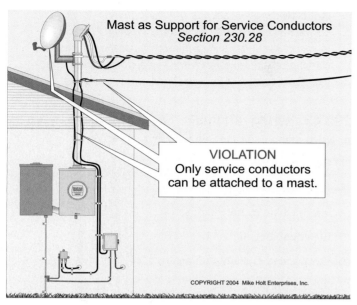

Mast as Support for Service Conductors
Section 230.28

VIOLATION
Only service conductors can be attached to a mast.

COPYRIGHT 2004 Mike Holt Enterprises, Inc.

Figure 230–14

230.27 Means of Attachment.
Multiconductor cables used for service drops must be attached to buildings or other structures by fittings identified for use with service conductors.

Open conductors must be attached to fittings identified for use with service conductors or to noncombustible, nonabsorbent insulators securely attached to the building or other structure.

230.28 Service Masts Used as Supports.

The service mast used as overhead conductor support must have adequate mechanical strength, or braces or guy wires to support it, to withstand the strain caused by the service-drop conductors.

> **Author's Comment:** Some local building codes require a minimum 2 in. rigid metal conduit for the service mast. In addition, many electric utilities contain specific requirements for the service mast.

Only electric utility service-drop conductors can be attached to a service mast.

> **Author's Comment:** 810.12 and 820.44(C) specify that aerial cables for radio, TV, or CATV cannot be attached to the electric service mast, and 810.12 prohibits antennas from being attached to the service mast. In addition, 800.133(C) and 830.133(B) prohibit broadband communications cables from being attached to raceways, including a service mast for power conductors. **Figure 230–14**

PART III. UNDERGROUND SERVICE-LATERAL CONDUCTORS

> **Author's Comment:** Underground service-lateral conductors installed by the electric utility must comply with the *National Electrical Safety Code (NESC)*, not the *NEC* [90.2(B)(5)], but underground conductors not under the exclusive control of the electric utility must be installed in accordance with the *NEC*.

230.31 Size and Rating.

(A) Service-Laterals. Service-lateral conductors must have adequate mechanical strength and sufficient ampacity to carry the load as calculated in accordance with Article 220.

(B) Ungrounded Neutral Conductor Size. Service-lateral conductors must not be smaller than 8 AWG copper or 6 AWG aluminum.

Exception: Service-lateral conductors can be as small as 12 AWG for limited-load installations.

(C) Grounded Neutral Conductor Size. The grounded neutral service-lateral conductor must be sized to carry the maximum unbalanced load in accordance with 220.61 and must not be sized smaller than required by 250.24(C).

> **Author's Comment:** 250.24(C) requires the grounded neutral service conductor to be sized in accordance with Table 250.66.

230.32 Protection Against Damage.
To protect underground service conductors against physical damage, they must be installed in accordance with the minimum burial depths listed in Table 300.5. Service-lateral conductors that enter a building must be installed using a wiring method listed in 230.43.

PART IV. SERVICE–ENTRANCE CONDUCTORS

230.40 Number of Service-Entrance Conductor Sets.
Each service drop or lateral can supply only one set of service-entrance conductors.

Exception 2: Service-entrance conductors are permitted to supply two to six service disconnecting means as permitted in 230.71(A).

Exception 3: A single-family dwelling unit with a separate structure can have one set of service-entrance conductors run to each structure from a single service drop or lateral.

Exception 5: One set of service-entrance conductors connected to the supply side of the normal service disconnecting means are permitted to supply standby power systems, fire pump equipment, and fire and sprinkler alarms [230.82(5) and (6)].

230.42 Size and Rating.

(A) Load Calculations. Service-entrance conductors must have sufficient ampacity for the loads to be served in accordance with Article 220.

(1) Continuous and Noncontinuous Loads. Conductors must be sized no less than 125 percent of the continuous loads, plus 100 percent of the noncontinuous loads based on the terminal temperature rating ampacities as listed in Table 310.16, before any ampacity adjustment [110.14(C)].

> **Author's Comment:** See 215.3 for the sizing requirements of feeder overcurrent protection devices for continuous and non-continuous loads.

> **Question:** *What size service conductors are required for a 200A continuous load if the terminals are rated 75°C?* **Figure 230–15**
>
> (a) 2/0 AWG   (b) 3/0 AWG
> (c) 4/0 AWG   (d) 250 kcmil
>
> **Answer:** *(d) 250 kcmil*

Since the load is 200A continuous, the service conductors must have an ampacity not less than 250A (200A x 1.25). According to Table 310.16, 75°C column, 250 kcmil conductors are suitable because they have an ampere rating of 255A at 75°C, before any conductor ampacity adjustment and/or correction.

(C) Grounded Neutral Conductor Size. The grounded neutral service conductor must be sized to carry the maximum unbalanced load in accordance with 220.61, and must not be sized smaller than required by 250.24(C).

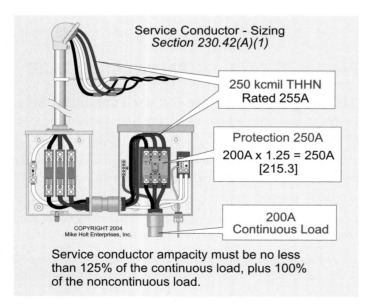

Service conductor ampacity must be no less than 125% of the continuous load, plus 100% of the noncontinuous load.

Figure 230–15

> **WARNING:** *In all cases the grounded neutral service conductor size must not be smaller than required by 250.24(C) to ensure that it has sufficiently low impedance and current-carrying capacity to safely carry fault current in order to facilitate the operation of the overcurrent protection device.*

230.43 Wiring Methods. Service conductors must be installed with one of the following wiring methods:

(1) Open wiring on insulators

(3) Rigid metal conduit

(4) Intermediate metal conduit

(5) Electrical metallic tubing

(6) Electrical nonmetallic tubing

(7) Service-entrance cables

(8) Wireways

(9) Busways

(11) Rigid nonmetallic conduit

(13) Type MC cable

(15) Flexible metal conduit or liquidtight flexible metal conduit not longer than 6 ft

(16) Liquidtight flexible nonmetallic conduit

230.44 Cable Trays. Cable tray systems are permitted to support service-entrance conductors. Cable trays used to support service-entrance conductors can contain only service-entrance conductors.

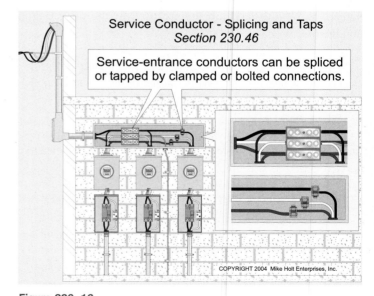

Figure 230–16

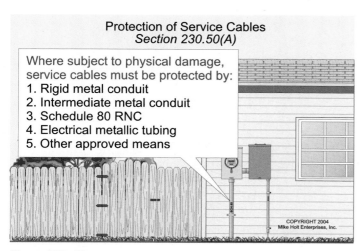

Protection of Service Cables
Section 230.50(A)

Where subject to physical damage, service cables must be protected by:
1. Rigid metal conduit
2. Intermediate metal conduit
3. Schedule 80 RNC
4. Electrical metallic tubing
5. Other approved means

COPYRIGHT 2004
Mike Holt Enterprises, Inc.

Figure 230–17

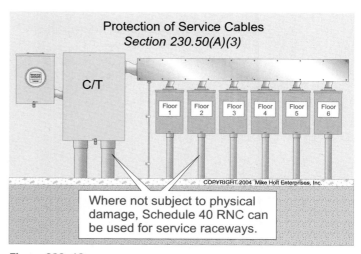

Protection of Service Cables
Section 230.50(A)(3)

C/T

Floor 1 | Floor 2 | Floor 3 | Floor 4 | Floor 5 | Floor 6

COPYRIGHT 2004 Mike Holt Enterprises, Inc.

Where not subject to physical damage, Schedule 40 RNC can be used for service raceways.

Figure 230–19

Exception: Conductors other than service-entrance conductors can be installed in a cable tray with service-entrance conductors provided a solid fixed barrier separates the service-entrance conductors from other conductors.

230.46 Spliced Conductors. Service-entrance conductors can be spliced or tapped in accordance with 110.14, 300.5(E), 300.13, and 300.15. Figure 230–16

230.50 Protection Against Physical Damage—Aboveground.

(A) Service Cables. Service cables subject to physical damage must be protected by one of the following: Figure 230–17

(1) Rigid metal conduit

(2) Intermediate metal conduit

(3) Schedule 80 rigid nonmetallic conduit, Figure 230–18

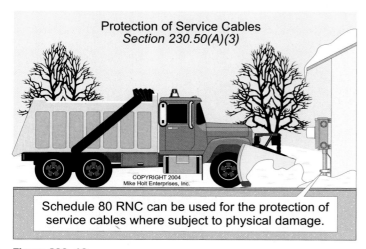

Protection of Service Cables
Section 230.50(A)(3)

COPYRIGHT 2004
Mike Holt Enterprises, Inc.

Schedule 80 RNC can be used for the protection of service cables where subject to physical damage.

Figure 230–18

Author's Comment: If the authority having jurisdiction determines that the raceway isn't subject to physical damage, Schedule 40 rigid nonmetallic conduit can be used. **Figure 230–19**

(4) Electrical metallic tubing

(5) Other means acceptable to the authority having jurisdiction

230.51 Service Cable Supports.

(A) Service Cable Supports. Service-entrance cable must be supported within 1 ft of the weatherhead, raceway connections, or enclosure and at intervals not exceeding 30 in.

230.54 Connections at Weatherheads.

(A) Raceway. Raceways for overhead service drops must have a weatherhead.

(B) Service Cable. Service cables must be equipped with a weatherhead.

Exception: Type SE cable can be formed in a gooseneck and taped with self-sealing weather-resistant thermoplastic.

(C) Above the Point of Attachment. Service heads and goosenecks must be located above the point of attachment. See 230.26.

Exception: Where it's impractical to locate the service head above the point of attachment, it must be located within 2 ft of the point of attachment.

(E) Opposite Polarity Through Separately Bushed Holes. Service heads must provide a bushed opening, and ungrounded conductors must be in separate openings.

(F) Drip Loops. Drip loop conductors must be below the service head or below the termination of the service-entrance cable sheath.

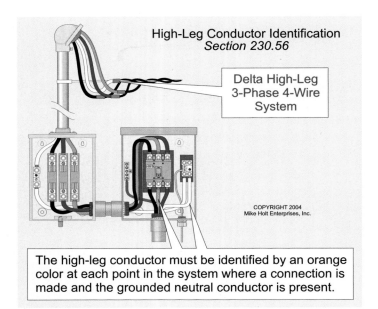

The high-leg conductor must be identified by an orange color at each point in the system where a connection is made and the grounded neutral conductor is present.

Figure 230–20

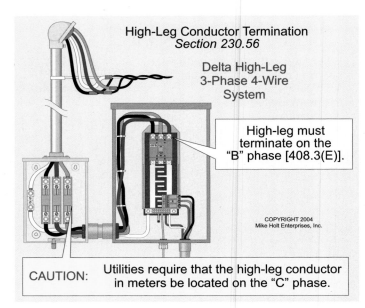

CAUTION: Utilities require that the high-leg conductor in meters be located on the "C" phase.

Figure 230–21

(G) Arranged so Water will not Enter. Service drops and service-entrance conductors must be arranged to prevent water from entering service equipment.

230.56 High-Leg Identification. On a 4-wire three-phase delta-connected system, where the midpoint of one phase winding is grounded, the conductor with the higher phase voltage-to-ground (208V) must be durably and permanently marked by an outer finish that is orange in color, or other effective means. Such identification must be placed at each point on the system where a connection is made if the grounded neutral conductor is present [110.15, 215.8, and 230.56]. **Figure 230-20**

The high-leg conductor is also called the "wild-leg," "stinger-leg," or "bastard-leg."

> **Author's Comment:** Since 1975, panelboards supplied by a 4-wire three-phase delta-connected system must have the high-leg conductor (208V) terminate to the "B" (center) phase of a panelboard [408.3(E)].
>   An exception to 408.3(E) permits the high-leg conductor to terminate to the "C" phase when the meter is located in the same section of a switchboard or panelboard.
>   The ANSI standard for meter equipment requires the high-leg conductor (208V-to-neutral) to terminate on the "C" (right) phase of the meter enclosure. This is because the demand meter needs 120V and it gets this from the "B" phase. Hopefully, the utility lineman is not color blind and doesn't inadvertently cross the "orange" high-leg conductor (208V) with the red (120V) service conductor at the weatherhead. It's happened before... See 110.15 in this textbook for additional details.
> **Figure 230-21**

PART V. SERVICE EQUIPMENT—GENERAL

230.66 Identified as Suitable for Service Equipment.
The service disconnecting means must be identified as suitable for use as service equipment.

> **Author's Comment:** Suitable for use as service equipment means that the service disconnecting means is supplied with a main bonding jumper so that a neutral-to-case connection can be made, as permitted in 250.32(B)(2) and 250.142(A).
> **Figure 230–22**

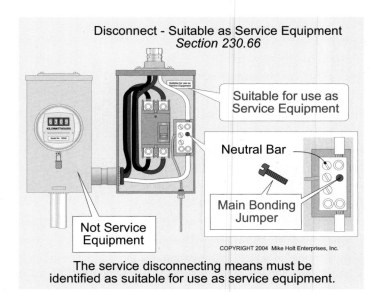

The service disconnecting means must be identified as suitable for use as service equipment.

Figure 230–22

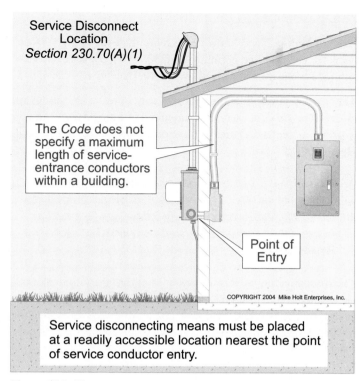

Service Disconnect
Location
Section 230.70(A)(1)

The *Code* does not specify a maximum length of service-entrance conductors within a building.

Point of
Entry

Service disconnecting means must be placed at a readily accessible location nearest the point of service conductor entry.

Figure 230–23

Service Disconnect - In Bathroom
Section 230.70(A)(2)

VIOLATION
Service disconnect is not permitted in a bathroom.

Bathroom

Overcurrent protection devices are not permitted in dwelling unit, guest room, or guest suite bathrooms [240.24(E)].

Figure 230–24

PART VI. SERVICE EQUIPMENT— DISCONNECTING MEANS

230.70 General. The service disconnecting means must open all service-entrance conductors from the building or structure premises wiring.

(A) Location.

(1) Readily Accessible. The service disconnecting means must be placed at a readily accessible location either outside the building or structure, or inside nearest the point of service conductor entry.

> **WARNING:** *Because service-entrance conductors do not have short-circuit or ground-fault protection, they must be limited in length when installed inside a building. Some local jurisdictions have a specific requirement as to the maximum length permitted within a building.* **Figure 230–23**

(2) Bathrooms. The service disconnecting means is not permitted to be installed in a bathroom. **Figure 230–24**

> **Author's Comment:** Overcurrent protection devices must not be located in the bathrooms of dwelling units, or guest rooms or guest suites of hotels or motels [240.24(E)].

(3) Remote Control. Where a remote-control device (shunt-trip) is used to actuate the service disconnecting means, the service disconnecting means must still be at a readily accessible location either outside the building or structure, or nearest the point of entry of the service conductors as required by 230.70(A)(1). **Figure 230–25**

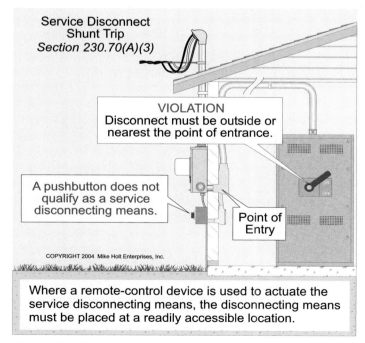

Service Disconnect
Shunt Trip
Section 230.70(A)(3)

VIOLATION
Disconnect must be outside or nearest the point of entrance.

A pushbutton does not qualify as a service disconnecting means.

Point of
Entry

Where a remote-control device is used to actuate the service disconnecting means, the disconnecting means must be placed at a readily accessible location.

Figure 230–25

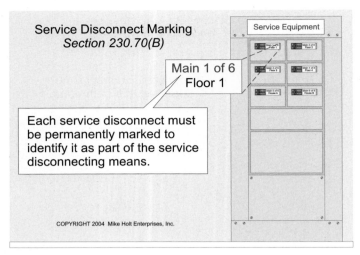

Figure 230–26

Author's Comments:

- See Article 100 for the definition of "Remote Control."

- The service disconnecting means must consist of either a manually operated switch, or a power-operated switch or circuit breaker also capable of being operated manually [230.76].

(B) Disconnect Identification. Each service disconnecting means must be permanently marked to identify it as part of the service disconnecting means. **Figure 230–26**

Author's Comment: When a building or structure has multiple services and/or feeders, a plaque is required at each service or feeder disconnect location to show the location of the other service or feeder disconnect locations. See 230.2(E).

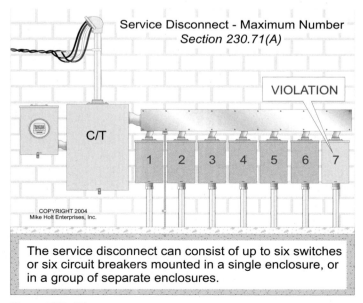

Figure 230–27

230.71 Number of Disconnects.

(A) Maximum. There must be no more than six service disconnects for each service permitted by 230.2, or each set of service-entrance conductors permitted by 230.40, Exceptions 1, 3, 4, or 5.

The service disconnecting means can consist of up to six switches or six circuit breakers mounted in a single enclosure, in a group of separate enclosures, or in or on a switchboard. **Figure 230–27**

> **CAUTION:** *The rule is six disconnecting means for each service, not six service disconnecting means per building. If the building has two services, then there can be a total of twelve service disconnects (six disconnects per service).* **Figure 230–28**

The disconnecting means for power monitoring equipment, transient voltage surge suppressors, the control circuit of the ground-fault protection system, or power-operable service disconnecting means is not considered a service disconnecting means. **Figure 230–29**

230.72 Grouping of Disconnects.

(A) Two to Six Disconnects. The service disconnecting means for each service must be grouped.

(B) Additional Service Disconnecting Means. To minimize the possibility of accidental interruption of power, the disconnecting means for fire pumps [695], and emergency [700], legally required [701], or optional standby [702] systems must be located remote from the one to six service disconnects for normal service.

> **Author's Comment:** Because emergency systems are just as important, if not more so, than fire pumps and standby systems, they should have the same safety precautions to prevent unintended interruption of the supply of electricity.

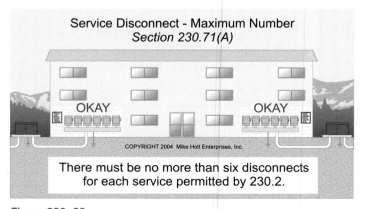

Figure 230–28

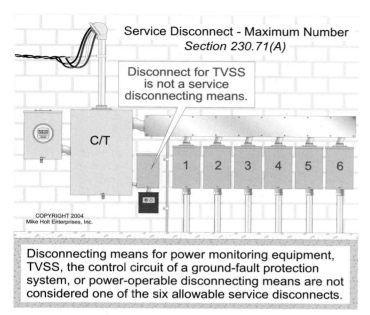

Service Disconnect - Maximum Number
Section 230.71(A)

Disconnect for TVSS
is not a service
disconnecting means.

C/T

1 2 3 4 5 6

COPYRIGHT 2004
Mike Holt Enterprises, Inc.

Disconnecting means for power monitoring equipment, TVSS, the control circuit of a ground-fault protection system, or power-operable disconnecting means are not considered one of the six allowable service disconnects.

Figure 230–29

(C) Access to Occupants. In a multiple-occupancy building, each occupant must have access to his or her service disconnecting means.

Exception: In multiple-occupancy buildings where electrical maintenance is provided by continuous building management, the service disconnecting means can be accessible only to building management personnel.

230.76 Manual or Power Operated. The service disconnecting means can consist of either a manually operated or a power-operated switch, or a circuit breaker that is capable of being operated manually. Figure 230–30

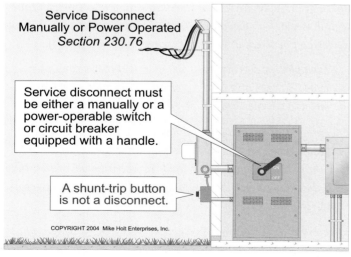

Service Disconnect
Manually or Power Operated
Section 230.76

Service disconnect must be either a manually or a power-operable switch or circuit breaker equipped with a handle.

A shunt-trip button is not a disconnect.

COPYRIGHT 2004 Mike Holt Enterprises, Inc.

Figure 230–30

Author's Comment: A shunt-trip button does not qualify as a service disconnect because it does not physically interrupt the service conductors.

230.79 Rating of Disconnect. The service disconnecting means for a building or structure must have an ampere rating not less than the calculated load according to Article 220, and in no case less than:

(A) One-Circuit Installation. For installations consisting of a single branch circuit, the disconnecting means must have a rating not less than 15A.

(B) Two-Circuit Installation. For installations consisting of two 2-wire branch circuits, the disconnecting means must have a rating not less than 30A.

(C) One-Family Dwelling. For a one-family dwelling, the disconnecting means must have a rating not less than 100A, 3-wire.

(D) All Others. For all other installations, the disconnecting means must have a rating not less than 60A.

230.82 Equipment Connected to the Supply Side of the Service Disconnect. Electrical equipment must not be connected to the supply side of the service-disconnect enclosure, except for:

(2) Meters and meter sockets are permitted ahead of the service disconnecting means.

(3) Meter disconnect switches that have a short-circuit current rating equal to or greater than the available short-circuit current can be installed ahead of the service disconnecting means. Figure 230–31

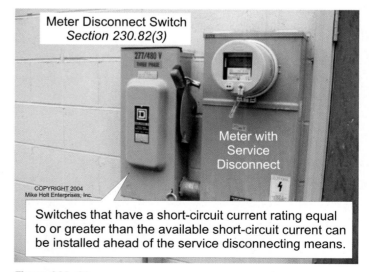

Meter Disconnect Switch
Section 230.82(3)

277/480 V
THREE PHASE

Meter with
Service
Disconnect

COPYRIGHT 2004
Mike Holt Enterprises, Inc.

Switches that have a short-circuit current rating equal to or greater than the available short-circuit current can be installed ahead of the service disconnecting means.

Figure 230–31

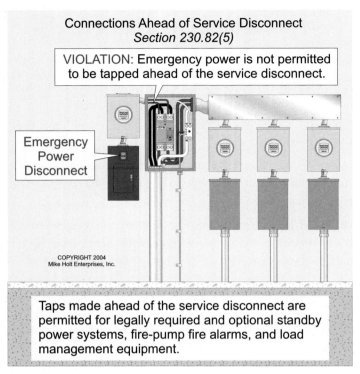

Connections Ahead of Service Disconnect
Section 230.82(5)

VIOLATION: Emergency power is not permitted to be tapped ahead of the service disconnect.

Emergency Power Disconnect

COPYRIGHT 2004 Mike Holt Enterprises, Inc.

Taps made ahead of the service disconnect are permitted for legally required and optional standby power systems, fire-pump fire alarms, and load management equipment.

Figure 230–32

Author's Comment: Electric utilities often require a meter disconnect switch for 277/480V services to enhance safety of utility personnel when they install or remove a meter.

(5) Tap conductors for legally required [701] and optional standby [702] power systems, fire-pump equipment, fire and sprinkler alarms, and load (energy) management devices are permitted ahead of the service disconnecting means.

Author's Comment: Emergency standby power cannot be connected ahead of service equipment. **Figure 230–32**

(6) Solar photovoltaic systems, fuel cell systems, or interconnected electric power production sources are permitted ahead of the service disconnecting means.

(7) Control circuits for power-operable service disconnecting means, if suitable overcurrent protection and disconnecting means are provided, are permitted ahead of the service disconnecting means.

(8) Ground-fault protection systems or transient voltage surge suppressors, where installed as part of listed equipment, are permitted ahead of the service disconnecting means if suitable overcurrent protection and disconnecting means are provided.

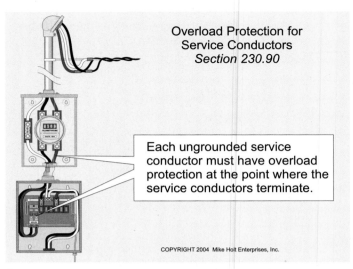

Overload Protection for Service Conductors
Section 230.90

Each ungrounded service conductor must have overload protection at the point where the service conductors terminate.

COPYRIGHT 2004 Mike Holt Enterprises, Inc.

Figure 230–33

PART VII. SERVICE EQUIPMENT OVERCURRENT PROTECTION

Author's Comment: The *NEC* doesn't require service conductors to be provided with short-circuit or ground-fault protection, but the feeder protection device provides overload protection.

230.90 Overload Protection Required. Each ungrounded service conductor must have overload protection at the point where the service conductors terminate [240.21(D)]. **Figure230–33**

(A) Overcurrent Protection Size. The rating of the protection device must not be greater than the ampacity of the conductors.

Exception 2: Where the ampacity of the ungrounded conductors doesn't correspond with the standard rating of overcurrent protection devices as listed in 240.6(A), the next higher protection device can be used, if it doesn't exceed 800A [240.4(B)].

> **Example:** *Two sets of parallel 500 kcmil THHN conductors (each rated 380A at 75°C) can be protected by an 800A overcurrent protection device.* **Figure 230–34**

Exception 3: The combined ratings of two to six service disconnecting means can exceed the ampacity of the service conductors provided the calculated load, in accordance with Article 220, doesn't exceed the ampacity of the service conductors. **Figure 230–35**

Exception 5: Overload protection for 3-wire single-phase 120/240V dwelling unit service conductors can be in accordance with 310.15(B)(6). **Figure 230–36**

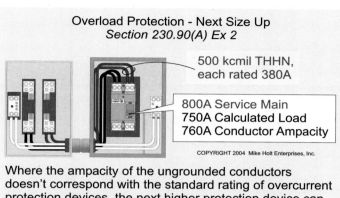

Overload Protection - Next Size Up
Section 230.90(A) Ex 2

500 kcmil THHN,
each rated 380A

800A Service Main
750A Calculated Load
760A Conductor Ampacity

COPYRIGHT 2004 Mike Holt Enterprises, Inc.

Where the ampacity of the ungrounded conductors doesn't correspond with the standard rating of overcurrent protection devices, the next higher protection device can be used, if it doesn't exceed 800A.

Figure 230–34

230.95 Ground-Fault Protection of Equipment.

Ground-fault protection of equipment is required for each service disconnecting means rated 1,000A or more supplied by a 4-wire three-phase 277/480V wye-connected system.

The rating of the service disconnecting means is considered to be the rating of the largest fuse that can be installed or the highest continuous current trip setting for which the actual overcurrent protection device installed in a circuit breaker is rated or can be adjusted.

Exception 2: The ground-fault protection provision of this section does not apply to fire pumps.

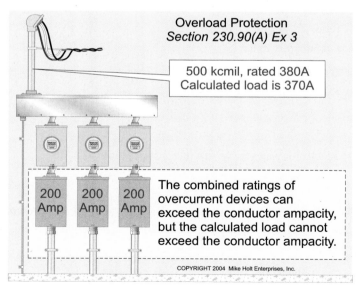

Overload Protection
Section 230.90(A) Ex 3

500 kcmil, rated 380A
Calculated load is 370A

200 Amp | 200 Amp | 200 Amp

The combined ratings of overcurrent devices can exceed the conductor ampacity, but the calculated load cannot exceed the conductor ampacity.

COPYRIGHT 2004 Mike Holt Enterprises, Inc.

Figure 230–35

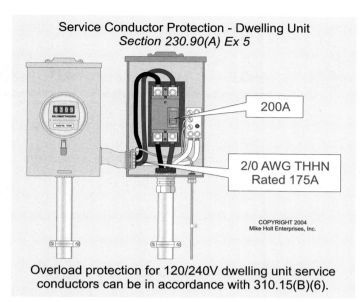

Service Conductor Protection - Dwelling Unit
Section 230.90(A) Ex 5

200A

2/0 AWG THHN
Rated 175A

COPYRIGHT 2004
Mike Holt Enterprises, Inc.

Overload protection for 120/240V dwelling unit service conductors can be in accordance with 310.15(B)(6).

Figure 230–36

Author's Comments:

- Article 100 defines "Ground-Fault Protection of Equipment" as a system intended to provide protection of equipment from ground faults by opening the circuit-protection device at current levels less than those required to protect conductors from damage. This type of protective system isn't intended to protect people, only connected utilization equipment. See 215.10 for similar requirements for feeders.

- Ground-fault protection of equipment isn't required for emergency power systems [700.26] or legally required standby power systems [701.17], because the goal is to have the alternate power supply operating as long as possible.

(• Indicates that 75% or fewer exam takers get the question correct.)

1. A building or structure must be supplied by a maximum of _____ service(s).

   (a) one   (b) two   (c) three   (d) as many as desired

2. •Conductors other than service conductors must not be installed in the same _____.

   (a) service raceway   (b) service cable   (c) enclosure   (d) a or b

3. Service-drop conductors must have _____.

   (a) sufficient ampacity to carry the current for the load   (b) adequate mechanical strength
   (c) a or b   (d) a and b

4. The minimum point of attachment of the service-drop conductors to a building must in no case be less than _____ above finished grade.

   (a) 8 ft   (b) 10 ft   (c) 12 ft   (d) 15 ft

5. Service conductors must be sized at no less than _____ percent of the continuous load, plus 100 percent of the noncontinuous load.

   (a) 100   (b) 115   (c) 125   (d) 150

6. Service cables, where subject to physical damage, must be protected.

   (a) True   (b) False

7. To prevent water from entering service equipment, service-entrance conductors must _____.

   (a) be connected to service-drop conductors below the level of the service head
   (b) have drip loops formed on the service-entrance conductors
   (c) a or b
   (d) a and b

8. Each service disconnecting means must be permanently marked to identify it as a service disconnecting means.

   (a) True   (b) False

9. The additional service disconnecting means permitted by 230.2 for fire pumps or for emergency, legally required standby, or optional standby services, must be installed remote from the one to six service disconnecting means for normal service to minimize the possibility of _____ interruption of supply.

   (a) accidental   (b) intermittent   (c) simultaneous   (d) prolonged

10. Meter disconnect switches that have a short-circuit current rating equal to, or greater than, the available short-circuit current are permitted ahead of the service-disconnecting means.

    (a) True   (b) False

ARTICLE 240 Overcurrent Protection

Introduction

This article provides the requirements for selecting and installing overcurrent protection devices. A review of the basic concept of overcurrent protection will help you avoid confusion as we move forward. Overcurrent exists when current exceeds the rating of conductors or equipment. This can be due to overload, short circuit, or ground fault.

Overload. An overload is a condition where equipment or conductors carry current exceeding their rated ampacity. An example of an overload is plugging two 12.5A (1,500W) hair dryers into a 20A branch circuit.

Ground Fault. A ground fault is an unintentional, electrically conducting connection between an ungrounded conductor of an electrical circuit and the normal noncurrent-carrying conductors, metallic enclosures, metallic raceways, metallic equipment, or earth. During the period of a ground fault, dangerous voltages and larger than normal currents exist.

Short Circuit. A short circuit is the unintentional electrical connection between any two normally current-carrying conductors of an electrical circuit, either line-to-line or line-to-neutral.

Overcurrent protection devices protect conductors and equipment. But it's important to note that they protect conductors differently than equipment.

An overcurrent protection device protects a *circuit* by opening when current reaches a value that would cause an excessive temperature rise in conductors. The overcurrent protection device's interrupting rating must be sufficient for the maximum possible fault current available on the line-side terminals of the equipment [110.9]. You'll find the standard ratings for fuses and *fixed-trip* circuit breakers in 240.6. Using a water analogy, current rises like water in a tank—at a certain level, the overcurrent protection device shuts off the faucet. Think in terms of normal operating conditions that just get too far out of normal range.

An overcurrent protection device protects *equipment* by opening when it detects a short circuit or ground fault. Every piece of electrical equipment must have a short-circuit current rating that permits the overcurrent protection devices (for that equipment) to clear short circuits or ground faults without extensive damage to the electrical components of the circuit [110.10]. Using a water analogy, a pipe bursts—creating a sudden rise in level—and the overcurrent protection device shuts off the supply to the pipe. Short circuits and faults aren't normal operating conditions. Thus, the overcurrent protection devices for equipment will have different characteristics than overcurrent protection devices for conductors.

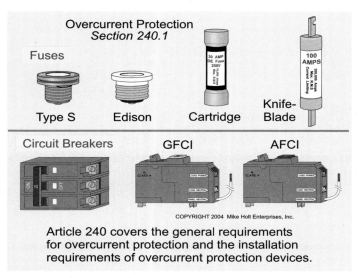

Overcurrent Protection
Section 240.1

Fuses

Type S   Edison   Cartridge   Knife-Blade

Circuit Breakers   GFCI   AFCI

COPYRIGHT 2004 Mike Holt Enterprises, Inc.

Article 240 covers the general requirements for overcurrent protection and the installation requirements of overcurrent protection devices.

Figure 240–1

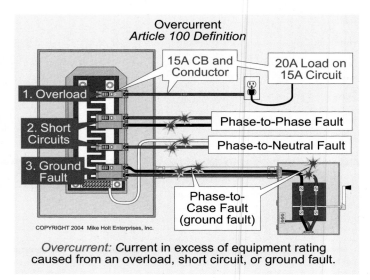

Overcurrent
Article 100 Definition

15A CB and Conductor   20A Load on 15A Circuit

1. Overload
2. Short Circuits   Phase-to-Phase Fault
                     Phase-to-Neutral Fault
3. Ground Fault

Phase-to-Case Fault (ground fault)

COPYRIGHT 2004 Mike Holt Enterprises, Inc.

Overcurrent: Current in excess of equipment rating caused from an overload, short circuit, or ground fault.

Figure 240–2

PART I. GENERAL

240.1 Scope. Article 240 covers the general requirements for overcurrent protection and the installation requirements of overcurrent protection devices. **Figure 240–1** This article is divided into seven parts:

- **PART I. GENERAL**
- **PART II. LOCATION**
- **PART III. ENCLOSURES**
- **PART IV. DISCONNECTING AND GROUNDING**
- **PART V. PLUG FUSES, FUSEHOLDERS, AND ADAPTERS**
- **PART VI. CARTRIDGE FUSES AND FUSEHOLDERS**
- **PART VII. CIRCUIT BREAKERS**

Author's Comment: Overcurrent is a condition where the current exceeds the rating of conductors or equipment due to overload, short circuit, or ground fault [Article 100]. **Figure 240–2**

FPN: An overcurrent protection device protects the circuit by opening the device when the current reaches a value that will cause excessive or dangerous temperature rise (overheating) in conductors. Overcurrent protection devices must have an interrupting rating sufficient for the maximum possible fault current available on the line-side terminals of the equipment [110.9]. Electrical equipment must have a short-circuit current rating that permits the circuit's overcurrent protection device to clear short circuits or ground faults without extensive damage to the circuit's electrical components [110.10].

240.2 Definitions.

Current-Limiting Overcurrent Protective Device. An overcurrent protective device (typically a fast-acting fuse) that reduces the

fault current to a magnitude substantially less than that obtainable in the same circuit if the current-limiting device was not used. See 240.40 and 240.60(B). **Figure 240–3**

Author's Comment: A current-limiting fuse is a type of fuse designed for operations related to short circuits only. When a fuse operates in its current-limiting range, it will clear a bolted short circuit in less than half a cycle. This type of fuse limits the instantaneous peak let-through current to a value substantially less than what would occur in the same circuit if the fuse were replaced with a solid conductor of equal impedance. If the available short-circuit current exceeds the equipment/conductor short-circuit current rating, then the thermal and magnetic forces can cause the equipment circuit conductors, as well as the equipment grounding (bonding) conductors, to vaporize. The only solutions to the problem of excessive available fault current are to:

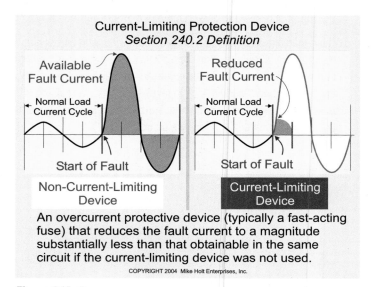

Current-Limiting Protection Device
Section 240.2 Definition

Available Fault Current   Reduced Fault Current

Normal Load Current Cycle   Normal Load Current Cycle

Start of Fault   Start of Fault

Non-Current-Limiting Device   Current-Limiting Device

An overcurrent protective device (typically a fast-acting fuse) that reduces the fault current to a magnitude substantially less than that obtainable in the same circuit if the current-limiting device was not used.

COPYRIGHT 2004 Mike Holt Enterprises, Inc.

Figure 240–3

(1) Install equipment with a higher short-circuit rating, or

(2) Protect the components of the circuit by a current-limiting protection device such as a fast-clearing fuse, which can reduce the let-through energy.

A breaker or a fuse does limit current, but it may not be listed as a current-limiting device. A thermal-magnetic circuit breaker will typically clear fault current in less than three to five cycles when subjected to a short circuit or ground fault of twenty times its rating. A standard fuse would clear the fault in less than one cycle and a current-limiting fuse should clear the same fault in less than one-quarter of one cycle.

Tap Conductors. A conductor, other than a service conductor, that has overcurrent protection rated greater than the ampacity of a conductor. See 240.21 for details. **Figure 240–4**

> **Author's Comment:** This definition is needed so that we can properly apply the tap rules in 240.21.

240.3 Protection of Equipment. The following equipment and their conductors are protected against overcurrent in accordance with the article that covers the type of equipment:

- Air Conditioning and Refrigeration, 440.22
- Appliances, Article 422
- Audio Circuits, 640.9
- Branch Circuits, 210.20
- Class 1, 2, and 3 Circuits, Article 725
- Feeder Conductors, 215.3
- Flexible Cords, 240.5(B)(1)
- Fire Alarms, Article 760
- Fire Pumps, Article 695

- Fixed Electric Heating, 424.3(B)
- Fixture Wire, 240.5(B)(2)
- Panelboards, 408.36(A)
- Service Conductors, 230.90(A)
- Transformers, 450.3

240.4 Protection of Conductors. Except as permitted by (A) through (G), conductors must be protected against overcurrent in accordance with their ampacity after ampacity adjustment, as specified in 310.15.

(A) Power Loss Hazard. Conductor overload protection is not required, but short-circuit protection is required where the interruption of the circuit would create a hazard; such as in a material-handling electromagnet circuit or fire pump circuit.

(B) Overcurrent Protection Not Over 800A. The next higher standard rating overcurrent device (above the ampacity of the ungrounded conductors being protected) is permitted, provided all of the following conditions are met:

(1) The conductors do not supply multioutlet receptacle branch circuits.

(2) The ampacity of a conductor, after ampacity adjustment and/or correction, doesn't correspond with the standard rating of a fuse or circuit breaker in 240.6(A).

(3) The protection device rating doesn't exceed 800A.

> *Example:* A 400A protection device can protect 500 kcmil conductors, where each conductor has an ampacity of 380A at 75°C, in accordance with Table 310.16. **Figure 240–5**

> **Author's Comment:** This rule "next size up" doesn't apply to feeder tap conductors [240.21(B)], or secondary transformer conductors [240.21(C)].

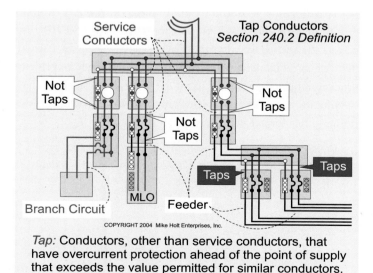

Tap: Conductors, other than service conductors, that have overcurrent protection ahead of the point of supply that exceeds the value permitted for similar conductors.

Figure 240–4

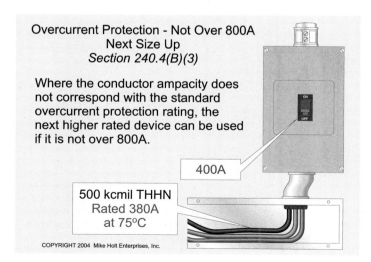

Figure 240–5

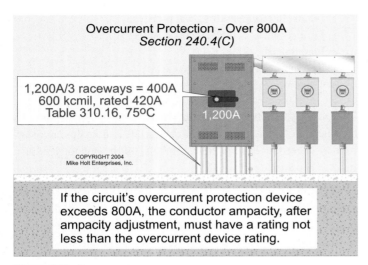

Overcurrent Protection - Over 800A
Section 240.4(C)

1,200A/3 raceways = 400A
600 kcmil, rated 420A
Table 310.16, 75°C

1,200A

COPYRIGHT 2004
Mike Holt Enterprises, Inc.

If the circuit's overcurrent protection device exceeds 800A, the conductor ampacity, after ampacity adjustment, must have a rating not less than the overcurrent device rating.

Figure 240–6

(C) Overcurrent Protection Over 800A. If the circuit's overcurrent protection device exceeds 800A, the conductor ampacity, after ampacity adjustment and/or correction, must have a rating not less than the rating of the overcurrent device.

> *Example: A 1,200A protection device can protect three sets of 600 kcmil conductors per phase, where each conductor has an ampacity of 420A at 75°C, in accordance with Table 310.16.* Figure 240–6

(D) Small Conductors. Unless specifically permitted in 240.4(E) or (G), overcurrent protection must not exceed 15A for 14 AWG, 20A for 12 AWG, and 30A for 10 AWG copper, or 15A for 12 AWG and 25A for 10 AWG aluminum, after ampacity adjustment and/or correction. Figure 240–7

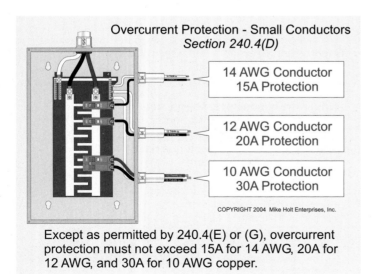

Overcurrent Protection - Small Conductors
Section 240.4(D)

14 AWG Conductor
15A Protection

12 AWG Conductor
20A Protection

10 AWG Conductor
30A Protection

COPYRIGHT 2004 Mike Holt Enterprises, Inc.

Except as permitted by 240.4(E) or (G), overcurrent protection must not exceed 15A for 14 AWG, 20A for 12 AWG, and 30A for 10 AWG copper.

Figure 240–7

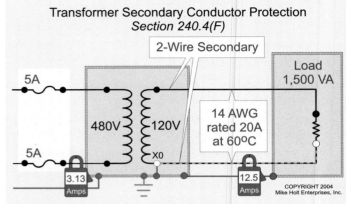

Transformer Secondary Conductor Protection
Section 240.4(F)

2-Wire Secondary

5A

5A

480V 120V

X0

14 AWG
rated 20A
at 60°C

Load
1,500 VA

3.13 Amps

12.5 Amps

COPYRIGHT 2004
Mike Holt Enterprises, Inc.

The primary overcurrent protection device can protect the secondary conductors of a 2-wire system if the primary device does not exceed the value determined by multiplying the secondary conductor ampacity by the secondary-to-primary voltage ratio.

Figure 240–8

(E) Tap Conductors. Tap conductors must be protected against overcurrent as follows:

(1) Household Ranges and Cooking Appliances and Other Loads, 210.19(A)(3) and (4)

(2) Fixture Wire, 240.5(B)(2)

(3) Location in Circuit, 240.21

(4) Reduction in Ampacity Size of Busway, 368.17(B)

(5) Feeder or Branch Circuits (busway taps), 368.17(C)

(6) Single Motor Taps, 430.53(D)

(F) Transformer Secondary Conductors. The primary overcurrent protection device sized in accordance with 450.3(B) can protect the secondary conductors of a 2-wire system or a 3-wire three-phase delta/delta connected system, provided the primary protection device does not exceed the value determined by multiplying the secondary conductor ampacity by the secondary-to-primary transformer voltage ratio.

> *Question: What is the minimum size secondary conductor required for a 2-wire 480V to 120V transformer rated 1.5 kVA?* Figure 240–8
>
> *(a) 16 AWG   (b) 14 AWG   (c) 12 AWG   (d) 10 AWG*
>
> ***Answer:** (b) 14 AWG*
>
> *Primary Current = VA/E*
> *VA = 1,500 VA*
> *E = 480V*

Primary Current = 1,500 VA/480V
Primary Current = 3.13A
Primary Protection [450.3(B)] = 3.13A x 1.67
Primary Protection = 5.22A or 5A Fuse

Secondary Current = 1,500 VA/120V
Secondary Current = 12.5A
Secondary Conductor = 14 AWG, rated 20A at 60C,
                          Table 310.16

The 5A primary protection device can be used to protect 14 AWG secondary conductors because it doesn't exceed the value determined by multiplying the secondary conductor ampacity by the secondary-to-primary transformer voltage ratio (5A = 20A x 120V/480V)

(G) Overcurrent for Specific Applications. Overcurrent protection for specific equipment and conductors must comply with that referenced in Table 240.4(G).

Air-Conditioning or Refrigeration [Article 440]. Air-conditioning and refrigeration equipment and circuit conductors must be protected against overcurrent in accordance with 440.22.

> **Author's Comment:** Typically, the branch-circuit conductor and protection size is marked on the equipment nameplate [440.4(A)].

> **Question:** *What size branch-circuit protection device is required for an air conditioner when the nameplate indicates that the minimum circuit ampacity (MCA) is 23A, and the running load is 18A?* **Figure 240–9**

> *(a) 12 AWG, 40A protection   (b) 12 AWG, 50A protection*
> *(c) 12 AWG, 60A protection   (d) 12 AWG, 70A protection*

> **Answer:** *(a) 12 AWG, 40A fuses*

Step 1. Branch-Circuit Conductor Size [440.32]
        18A x 1.25 = 22.5A, 12 AWG rated 25A at 60ºC

Step 2. Branch-Circuit Protection Size [440.22(A)]
        18A x 2.25 = 40.5A, 40A maximum fuse size (fuses in accordance with the manufacturer's instructions) [110.3(B) and 240.6(A)]

Motors [Article 430]. Motor circuit conductors must be protected against short circuits and ground faults in accordance with 430.52 and 430.62 [430.51].

> **Question:** *What size branch-circuit conductor and protection device (circuit breaker) is required for a 7½ hp, 230V three-phase motor?* **Figure 240–10**

> *(a) 10 AWG, 50A breaker   (b) 10 AWG, 60A breaker*
> *(c) a or b   (d) none of these*

> **Answer:** *(c) 10 AWG, 50A or 60A breaker*

Step 1. Branch-Circuit Conductor Size [Table 310.16, 430.22, and Table 430.250]
        22A x 1.25 = 28A, 10 AWG, rated 30A at 60°C

Step 2. Branch-Circuit Protection Size [240.6(A), 430.52(C)(1) Exception 1, Table 430.250]
        Inverse-Time Breaker: 22A x 2.5 = 55A
        Next size up = 60A

Motor Control [Article 430]. Motor control circuit conductors must be sized and protected in accordance with 430.72.

Remote-Control, Signaling, and Power-Limited Circuits [Article 725]. Remote-control, signaling, and power-limited circuit conductors must be protected against overcurrent according to 725.23 and 725.41.

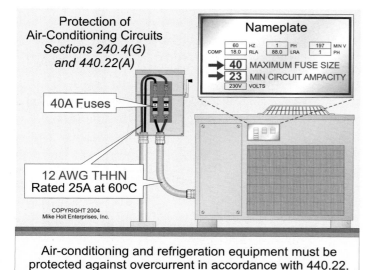

Figure 240–9

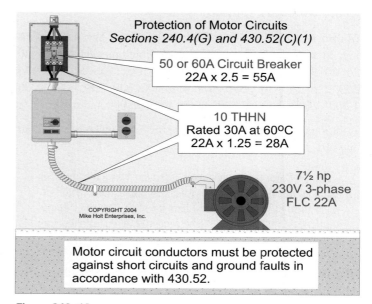

Figure 240–10

240.5 Protection of Flexible Cords and Fixture Wires.

(A) Ampacities. Flexible cord must be protected by an overcurrent device in accordance with its ampacity as specified in Table 400.5(A) and Table 400.5(B). Fixture wire must be protected against overcurrent in accordance with its ampacity as specified in Table 402.5. Supplementary overcurrent protection, as in 240.10, is permitted to provide this protection.

(B) Branch-Circuit Overcurrent Protection.

(1) Cords for Listed Appliance or Portable Lamps. Where flexible cord is used with a specific listed appliance or portable lamp, the conductors are considered protected against overcurrent when used within the appliance or portable lamp listing requirements.

> **Author's Comment:** The *NEC* only applies to premise wiring, not to the supply cords of listed appliances and portable lamps.

(2) Fixture Wire. Fixture wire is considered protected against overcurrent as follows:
- (1) 20A circuit—18 AWG, up to 50 ft of run length
- (2) 20A circuit—16 AWG, up to 100 ft of run length
- (3) 20A circuit—14 AWG and larger
- (4) 30A circuit—14 AWG and larger
- (5) 40A circuit—12 AWG and larger
- (6) 50A circuit—12 AWG and larger

(3) Extension Cord Sets. Where flexible cord is used in listed extension cord sets, the conductors are considered protected against overcurrent when used within the extension cord's listing requirements.

240.6 Standard Ampere Ratings.

(A) Fuses and Fixed-Trip Circuit Breakers. The standard ratings in amperes for fuses and inverse-time breakers are: 15, 20, 25, 30, 35, 40, 45, 50, 60, 70, 80, 90, 100, 110, 125, 150, 175, 200, 225, 250, 300, 350, 400, 450, 500, 600, 700, 800, 1,000, 1,200, 1,600, 2,000, 2,500, 3,000, 4,000, 5,000 and 6,000.

Additional standard ampere ratings for fuses include 1, 3, 6, 10, and 601. Figure 240–11

> **Author's Comment:** Fuses rated less than 15A are sometimes required for the protection of fractional horsepower motor circuits [430.52], motor control circuits [430.72], small transformers [450.3(B)], and remote-control circuit conductors [725.23].

(B) Adjustable Circuit Breakers. The ampere rating of an adjustable circuit breaker is equal to its maximum long-time pickup current setting.

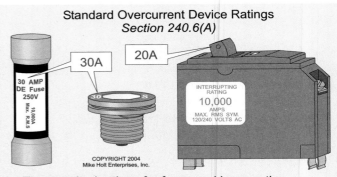

Standard Overcurrent Device Ratings
Section 240.6(A)

The standard ratings for fuses and inverse-time breakers include: 15, 20, 25, 30, 35, 40, 45, 50, 60, 70, 80, 90, 100, 110, 125, 150, 175, 200, 225, 250, 300, 350, 400, 450, 500, 600, 700, 800, 1000, 1200, 1600, 2000, 2500, 3000, 4000, 5000, and 6000A.

Figure 240–11

(C) Restricted-Access, Adjustable-Trip Circuit Breakers. The ampere rating of adjustable-trip circuit breakers that have restricted access to the adjusting means is equal to their adjusted long-time pickup current settings.

240.10 Supplementary Overcurrent Protection.

Supplementary overcurrent protection devices (usually an internal fuse), often used in luminaires, appliances, and equipment for internal circuits and components, cannot be used as the required circuit overcurrent protection device.

> **Author's Comment:** See Article 100 for the definition of "Supplementary Overcurrent Protection Device."

A supplementary overcurrent device isn't required to be readily accessible [240.24(A)(2)]. Figure 240–12

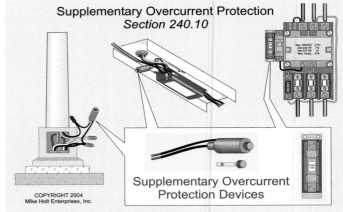

Supplementary Overcurrent Protection
Section 240.10

Supplementary Overcurrent Protection Devices

Supplementary overcurrent protection devices cannot be used as the required circuit protection device.

Figure 240–12

Author's Comment: Article 100 defines a "Supplementary Overcurrent Protection Device" as a device intended to provide limited overcurrent protection for specific applications and utilization equipment. This limited protection is in addition to the protection provided in the required branch circuit by the branch-circuit overcurrent protective device.

240.13 Ground-Fault Protection of Equipment. Each circuit rated 1,000A or more, supplied from a 4-wire three-phase 277/480V wye-connected system must be protected against ground-fault in accordance with 230.95.

The requirement for ground-fault protection of equipment doesn't apply to:

(1) Continuous industrial processes where a nonorderly shutdown would introduce additional or increased hazards.

(2) Installations where ground-fault protection of equipment is already provided.

(3) Fire pumps installed in accordance with Article 695.

Author's Comments:

• Article 100 defines "Ground-Fault Protection of Equipment" as a system intended to provide protection of equipment from ground faults by opening the circuit-protection device at current levels less than those required to protect conductors from damage. This type of protective system isn't intended to protect people, only connected utilization equipment. See 215.10 for similar requirements for feeders.

• Ground-fault protection of equipment isn't permitted for fire pumps [695.6(H)], and ground-fault protection isn't required for emergency power systems [700.26] or legally required standby power systems [701.17].

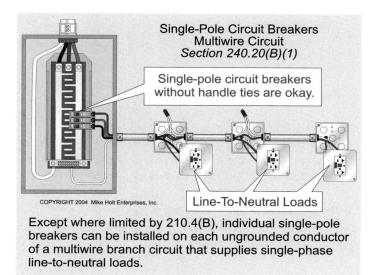

Single-Pole Circuit Breakers
Multiwire Circuit
Section 240.20(B)(1)

Single-pole circuit breakers without handle ties are okay.

Line-To-Neutral Loads

COPYRIGHT 2004 Mike Holt Enterprises, Inc.

Except where limited by 210.4(B), individual single-pole breakers can be installed on each ungrounded conductor of a multiwire branch circuit that supplies single-phase line-to-neutral loads.

Figure 240–13

PART II. LOCATION

240.20 Ungrounded Conductors.

(B) Circuit Breaker as an Overcurrent Device. Circuit breakers must open all ungrounded conductors of the circuit, unless otherwise permitted in 240.20(B)(1), (B)(2), and (B)(3).

(1) Multiwire Branch Circuit. Except where limited by 210.4(B), individual single-pole breakers can be installed on each ungrounded conductor of a multiwire branch circuit that supplies line-to-neutral loads. **Figure 240–13**

Author's Comments:

• Multiwire branch circuits that supply switches, receptacles, or equipment on the same yoke must be provided with a means to disconnect simultaneously all ungrounded conductors that supply those devices or equipment at the point where the branch circuit originates [210.4(B) and 210.7(B)]. This can be accomplished by single-pole circuit breakers with handle ties identified for the purpose or a 2- or 3-pole breaker with common internal trip. **Figure 240–14**

• Single-pole AFCI or GFCI circuit breakers are not suitable for protecting multiwire branch circuits. Multiwire branch circuits protected by AFCI or GFCI circuit breakers must be of the 2-pole type.

(2) Single-Phase, Line-to-Line Loads. Individual single-pole circuit breakers with handle ties identified for the purpose are permitted on each ungrounded conductor of a branch circuit that supplies single-phase, line-to-line loads. **Figure 240–15**

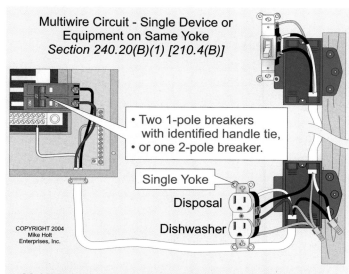

Multiwire Circuit - Single Device or
Equipment on Same Yoke
Section 240.20(B)(1) [210.4(B)]

• Two 1-pole breakers with identified handle tie,
• or one 2-pole breaker.

Single Yoke

Disposal

Dishwasher

COPYRIGHT 2004 Mike Holt Enterprises, Inc.

Multiwire branch circuits that supply devices/equipment on the same yoke must be provided with a means to disconnect simultaneously all ungrounded conductors.

Figure 240–14

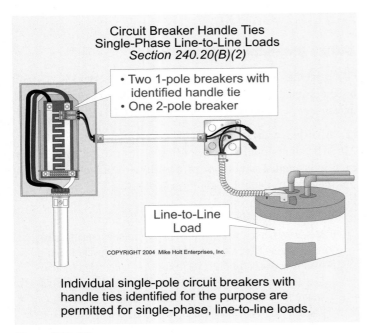

Circuit Breaker Handle Ties
Single-Phase Line-to-Line Loads
Section 240.20(B)(2)

• Two 1-pole breakers with identified handle tie
• One 2-pole breaker

Line-to-Line Load

COPYRIGHT 2004 Mike Holt Enterprises, Inc.

Individual single-pole circuit breakers with handle ties identified for the purpose are permitted for single-phase, line-to-line loads.

Figure 240–15

(3) Three-Phase, Line-to-Line Loads. Individual single-pole breakers with handle ties identified for the purpose are permitted on each ungrounded conductor of a branch circuit that serves three-phase, line-to-line loads. **Figure 240–16**

Author's Comment: Handle ties must be identified for the purpose. This means that handle ties made from nails, screws, wires, or other nonconforming methods are not permitted. **Figure 240–17**

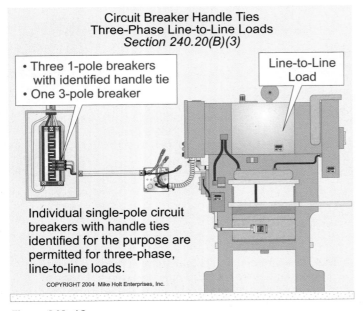

Circuit Breaker Handle Ties
Three-Phase Line-to-Line Loads
Section 240.20(B)(3)

• Three 1-pole breakers with identified handle tie
• One 3-pole breaker

Line-to-Line Load

Individual single-pole circuit breakers with handle ties identified for the purpose are permitted for three-phase, line-to-line loads.

COPYRIGHT 2004 Mike Holt Enterprises, Inc.

Figure 240–16

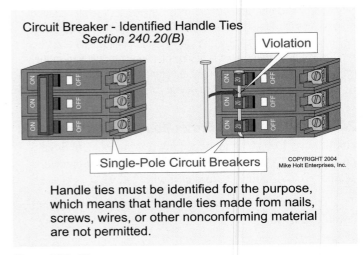

Circuit Breaker - Identified Handle Ties
Section 240.20(B)

Violation

Single-Pole Circuit Breakers

COPYRIGHT 2004 Mike Holt Enterprises, Inc.

Handle ties must be identified for the purpose, which means that handle ties made from nails, screws, wires, or other nonconforming material are not permitted.

Figure 240–17

240.21 Overcurrent Protection Location in Circuit.

Except as permitted by (A) through (G), overcurrent protection devices must be placed at the point where the branch or feeder conductors receive their power.

A tap conductor cannot supply another tap conductor. In other words, you cannot make a tap from a tap.

(A) Branch-Circuit Taps. Branch-circuit taps installed in accordance with 210.19 are permitted.

(B) Feeder Tap Conductors. Conductors can be tapped from a feeder if they are installed in accordance with (1) through (5). The "next size up protection rule" for conductors contained in 240.4(B) is not permitted to be used for feeder tap conductors.

> **Question:** *What size tap conductor would be required for a 150A circuit breaker if the calculated continuous load was 100A?* **Figure 240–18**
>
> *(a) 3 AWG, rated 100A*   *(b) 2 AWG, rated 115A*
> *(c) 1 AWG, rated 130A*   *(d) 1/0 AWG, rated 150A*
>
> **Answer:** *(d) 1/0 AWG tap conductors would be required to supply the circuit breaker.*

Author's Comment: A 150A protection device is permitted to protect a 1 AWG conductor, which is rated 130A [Table 310.16], on the load side of the 150A circuit breaker [240.4(B)].

(1) 10-Foot Feeder Tap. Feeder tap conductors up to 10 ft long are permitted without overcurrent protection if installed as follows: **Figure 240–19**

(1) The ampacity of the tap conductor must not be less than:

  a. The calculated load in accordance with Article 220, and

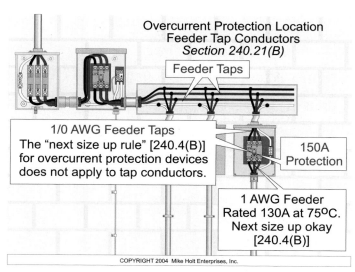

Overcurrent Protection Location
Feeder Tap Conductors
Section 240.21(B)

Feeder Taps

1/0 AWG Feeder Taps
The "next size up rule" [240.4(B)]
for overcurrent protection devices
does not apply to tap conductors.

150A
Protection

1 AWG Feeder
Rated 130A at 75ºC.
Next size up okay
[240.4(B)]

COPYRIGHT 2004 Mike Holt Enterprises, Inc.

Figure 240–18

b. The rating of the device supplied by the tap conductors or the overcurrent protective device at the termination of the tap conductors.

(2) The tap conductors must not extend beyond the equipment they supply.

(3) The tap conductors must be installed in a raceway if they leave the enclosure.

(4) The tap conductors must have an ampacity not less than 10 percent of the ampacity of the overcurrent protection device that protects the feeder.

(2) 25-Foot Feeder Tap. Feeder tap conductors up to 25 ft long are permitted without overcurrent protection if installed as follows: Figure 240–20

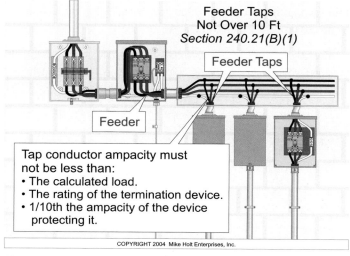

Feeder Taps
Not Over 10 Ft
Section 240.21(B)(1)

Feeder Taps

Feeder

Tap conductor ampacity must
not be less than:
• The calculated load.
• The rating of the termination device.
• 1/10th the ampacity of the device
  protecting it.

COPYRIGHT 2004 Mike Holt Enterprises, Inc.

Figure 240–19

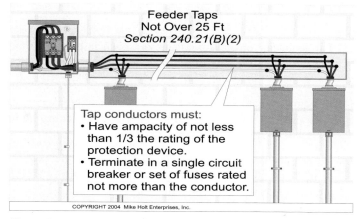

Feeder Taps
Not Over 25 Ft
Section 240.21(B)(2)

Tap conductors must:
• Have ampacity of not less
  than 1/3 the rating of the
  protection device.
• Terminate in a single circuit
  breaker or set of fuses rated
  not more than the conductor.

COPYRIGHT 2004 Mike Holt Enterprises, Inc.

Figure 240–20

(1) The ampacity of the tap conductors must not be less than one-third the ampacity of the overcurrent protection device that protects the feeder.

(2) The tap conductors terminate in a single circuit breaker, or set of fuses rated no greater than the tap conductor ampacity in accordance with 310.15 [Table 310.16].

(3) The tap conductors must be protected from physical damage by being enclosed in a manner approved by the authority having jurisdiction, such as within a raceway.

(3) Taps Supplying a Transformer. Feeder tap conductors that supply a transformer must be installed as follows:

(1) The primary tap conductors must have an ampacity not less than one-third the ampacity of the overcurrent protection device.

(2) The secondary conductors must have an ampacity that, when multiplied by the ratio of the primary-to-secondary voltage, is at least one-third the rating of the overcurrent device that protects the feeder conductors.

(3) The total length of the primary and secondary conductors must not exceed 25 ft.

(4) Primary and secondary conductors must be protected from physical damage by being enclosed in a manner approved by the authority having jurisdiction, such as within a raceway.

(5) Secondary conductors terminate in a single circuit breaker, or set of fuses rated no greater than the tap conductor ampacity in accordance with 310.15 [Table 310.16].

(4) 100 Ft Tap. Feeder tap conductors in a high bay manufacturing building (over 35 ft high at walls) can be run up to 100 ft without overcurrent protection if installed as follows:

(1) Supervision ensures that only qualified persons service the systems.

(2) Tap conductors aren't over 25 ft long horizontally and not over 100 ft in total length.

(3) The ampacity of the tap conductors must not be less than one-third the ampacity of the overcurrent protection device that protects the feeder.

(4) The tap conductors terminate in a single circuit breaker or set of fuses rated no greater than the tap conductor ampacity in accordance with 310.15 [Table 310.16].

(5) Tap conductors must be protected from physical damage by being enclosed in a manner approved by the authority having jurisdiction, such as within a raceway.

(6) Tap conductors contain no splices.

(7) Tap conductors are 6 AWG copper or 4 AWG aluminum or larger.

(8) Tap conductors do not penetrate walls, floors, or ceilings.

(9) The tap is made no less than 30 ft from the floor.

(5) Outside Feeder Tap of Unlimited Length Rule. Outside feeder tap conductors can be of unlimited length without overcurrent protection at the point they receive their supply if installed as follows: Figure 240–21

(1) The tap conductors must be suitably protected from physical damage in a raceway or manner approved by the authority having jurisdiction.

(2) The tap conductors terminate at a single circuit breaker or a single set of fuses that limit the load to the ampacity of the conductors.

(3) The overcurrent device for the tap conductors must be an integral part of the disconnecting means or it must be located immediately adjacent to it.

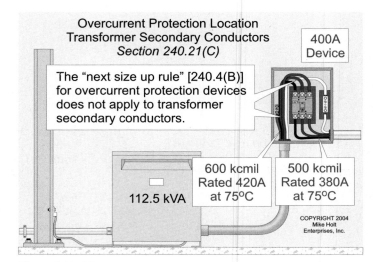

Figure 240–22

(4) The disconnecting means must be located at a readily accessible location either outside the building or structure, or nearest the point of entry of the conductors.

(C) Transformer Secondary Conductors. Each set of conductors feeding separate loads can be connected to a transformer secondary, without overcurrent protection at the secondary, in accordance with (1) through (6).

The "next size up protection rule" for conductors contained in 240.4(B) is not permitted to be used for transformer secondary conductors. Figure 240–22

(1) Protection by Primary Overcurrent Device. The primary overcurrent protection device sized in accordance with 450.3(B) can protect the secondary conductors of a 2-wire system or a 3-wire three-phase, delta/delta connected system, provided the primary protection device does not exceed the value determined by multiplying the secondary conductor ampacity by the secondary-to-primary transformer voltage ratio.

Question: What is the minimum size secondary conductor required for a 2-wire 480V to 120V transformer rated 1.5 kVA? Figure 240–23

(a) 16 AWG   (b) 14 AWG   (c) 12 AWG   (d) 10 AWG

Answer: (b) 14 AWG

Primary Current = VA/E
VA = 1,500 VA
E = 480V

Primary Current = 1,500 VA/480V
Primary Current = 3.13A
Primary Protection [450.3(B)] = 3.13A x 1.67 = 5.22A or 5A Fuse

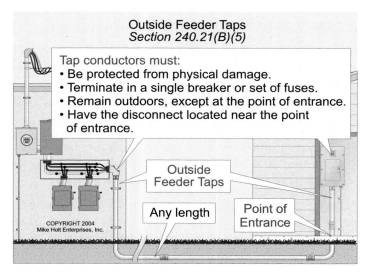

Figure 240–21

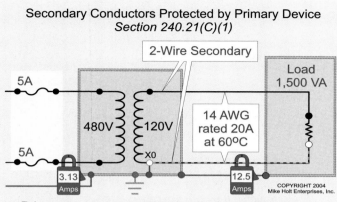

Secondary Conductors Protected by Primary Device
Section 240.21(C)(1)

2-Wire Secondary

5A

480V 120V

14 AWG
rated 20A
at 60°C

Load
1,500 VA

5A

X0

3.13
Amps

12.5
Amps

COPYRIGHT 2004
Mike Holt Enterprises, Inc.

Primary overcurrent protection device can protect
the secondary conductors of a 2-wire system if the
primary device does not exceed the value determined
by multiplying the secondary conductor ampacity by
the secondary-to-primary voltage ratio.

Figure 240–23

Secondary Current = 1,500 VA/120V
Secondary Current = 12.5A
Secondary Conductor = 14 AWG, rated 20A at 60°C, Table 310.16

The 5A primary protection device can be used to protect 14
AWG secondary conductors because it doesn't exceed the value
determined by multiplying the secondary conductor ampacity by
the secondary-to-primary transformer voltage ratio (5A = 20A x
120V/480V).

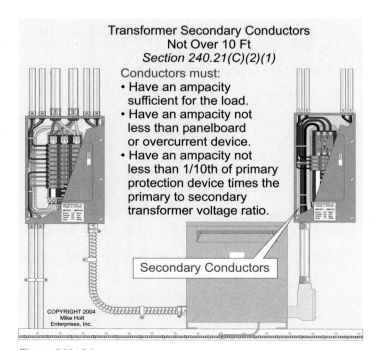

Transformer Secondary Conductors
Not Over 10 Ft
Section 240.21(C)(2)(1)

Conductors must:
• Have an ampacity
sufficient for the load.
• Have an ampacity not
less than panelboard
or overcurrent device.
• Have an ampacity not
less than 1/10th of primary
protection device times the
primary to secondary
transformer voltage ratio.

Secondary Conductors

COPYRIGHT 2004
Mike Holt
Enterprises, Inc.

Figure 240–24

(2) 10 Ft Secondary Conductor. Secondary conductors can be
run up to 10 ft without overcurrent protection if installed as follows: **Figure 240–24**

(1) The ampacity of the secondary conductor must not be less than:

    a. The calculated load in accordance with Article 220,

    b. The rating of the device supplied by the secondary con-
    ductors or the overcurrent protective device at the termina-
    tion of the secondary conductors, and

    c. Not less than one-tenth the rating of the overcurrent device
    protecting the primary of the transformer, multiplied by
    the primary-to-secondary transformer voltage ratio.

(2) The secondary conductors must not extend beyond the
switchboard, panelboard, disconnecting means, or control
devices they supply.

(3) The secondary conductors must be enclosed in a raceway.

> **Author's Comment:** Lighting and appliance branch-circuit pan-
> elboards must have overcurrent protection located on the sec-
> ondary side of the transformer [408.36(D)]. **Figure 240–25**

**(3) Industrial Installation Secondary Conductors not Over 25
Ft.** For industrial installations, secondary conductors can be run
up to 25 ft without overcurrent protection if installed as follows:

(1) The secondary conductor ampacity isn't less than:
    • The secondary current rating of the transformer, and
    • The sum of the ratings of the overcurrent devices.

(2) Secondary overcurrent devices are grouped.

(3) Secondary conductors must be protected from physical
damage by being enclosed in a raceway or manner approved
by the authority having jurisdiction.

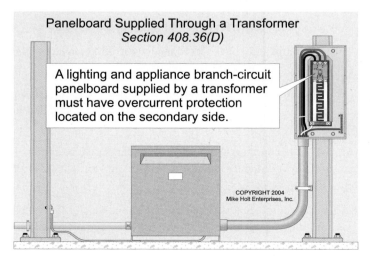

Panelboard Supplied Through a Transformer
Section 408.36(D)

A lighting and appliance branch-circuit
panelboard supplied by a transformer
must have overcurrent protection
located on the secondary side.

COPYRIGHT 2004
Mike Holt Enterprises, Inc.

Figure 240–25

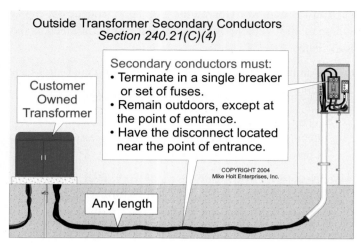

Figure 240–26

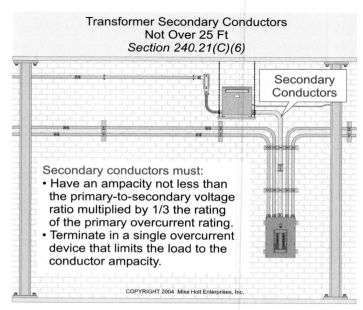

Figure 240–27

(4) Outside Secondary Conductors of Unlimited Length. Outside secondary conductors can be of unlimited length without overcurrent protection at the point they receive their supply if they are installed as follows: Figure 240–26

(1) The conductors must be suitably protected from physical damage in a raceway or manner approved by the authority having jurisdiction.

(2) The conductors terminate at a single circuit breaker or a single set of fuses that limit the load to the ampacity of the conductors.

(3) The overcurrent device for the ungrounded conductors must be an integral part of a disconnecting means or it must be located immediately adjacent thereto.

(4) The disconnecting means must be located at a readily accessible location that complies with one of the following:

   a. Outside of a building or structure.

   b. Inside, nearest the point of entrance of the conductors.

   c. Where installed in accordance with 230.6, nearest the point of entrance of the conductors.

(5) Secondary Conductors from a Feeder Tapped Transformer. Transformer secondary conductors must be installed in accordance with 240.21(B)(3).

(6) 25-Foot Secondary Conductor. Secondary conductors can be run up to 25 ft without overcurrent protection if installed as follows: Figure 240–27

(1) The secondary conductors must have an ampacity that when multiplied by the ratio of the primary-to-secondary voltage isn't less than one-third the rating of the overcurrent device that protects the primary of the transformer.

(2) Secondary conductors terminate in a single circuit breaker or set of fuses rated no greater than the tap conductor ampacity in accordance with 310.15 [Table 310.16].

(3) The secondary conductors must be protected from physical damage by being enclosed in a manner approved by the authority having jurisdiction, such as within a raceway.

Question: True or False. A 112.5 kVA, 120/208V three-phase transformer would be required to terminate in a 400A protection device, with 600 kcmil conductors from the secondary to the line side of the disconnect, but 500 kcmil conductors could be used on the load side!

(a) True   (b) False

Answer: *(a) True*

Secondary Current = VA/(E x 1.732)
Secondary Current = 112,500 VA/208 x 1.732
Secondary Current = 313A

Secondary Overcurrent Protection Device Size = 313 x 1.25
[215.3]
Secondary Overcurrent Protection Device Size = 391
Secondary Overcurrent Protection Device Size = 400A [240.6]

Secondary Conductor Size = 600 kcmil rated 420A, Table 310.16 at 75°C

Conductors leaving the 400A protection device can be 500 kcmil. See 240.4(B).

(D) Service Conductors. Service-entrance conductors must be protected against overload in accordance with 230.91.

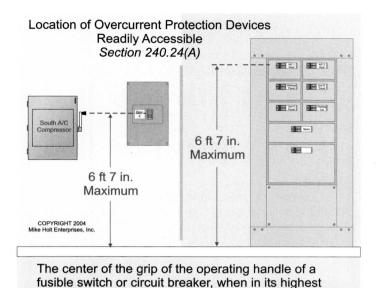

The center of the grip of the operating handle of a fusible switch or circuit breaker, when in its highest position, cannot be more than 6 ft 7 in. above the floor.

Figure 240–28

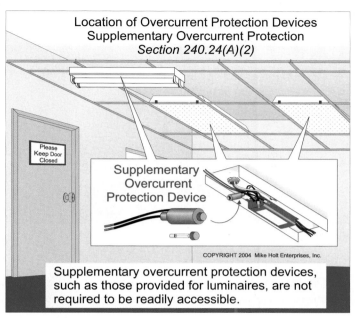

Supplementary overcurrent protection devices, such as those provided for luminaires, are not required to be readily accessible.

Figure 240–30

240.24 Location of Overcurrent Protection Devices.

(A) Readily Accessible. Circuit breakers and fuses must be readily accessible and they must be installed so the center of the grip of the operating handle of the fuse switch or circuit breaker, when in its highest position, isn't more than 6 ft 7 in. above the floor or working platform, unless the installation is for: **Figure 240–28**

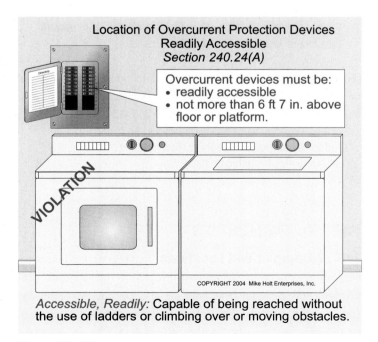

Accessible, Readily: Capable of being reached without the use of ladders or climbing over or moving obstacles.

Figure 240–29

Author's Comment: "Readily accessible" means located so it can be reached quickly without having to climb over or remove obstacles, or use a portable ladder. **Figure 240–29**

(1) Busways, as provided in 368.17(C).

(2) Supplementary overcurrent protection devices [240.10]. **Figure 240–30**

(3) For overcurrent devices, as described in 225.40 and 230.92.

(4) Overcurrent protection devices located next to equipment can be mounted above 6 ft 7 in., if accessible by portable means [404.8(A) Exception 2]. **Figure 240–31**

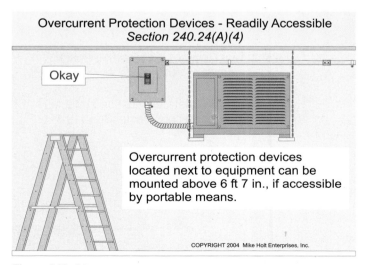

Overcurrent protection devices located next to equipment can be mounted above 6 ft 7 in., if accessible by portable means.

Figure 240–31

Figure 240–32

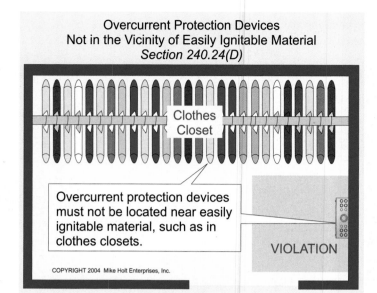

Figure 240–34

(B) Occupancy. Each occupant must have ready access to all overcurrent protection devices that protect the ungrounded conductors that supply that occupancy.

Exception 1: Service and feeder overcurrent protection devices can be accessible only to authorized management personnel, which means they aren't required to be accessible to the occupants in the following buildings:

(1) Multiple-occupancy buildings

(2) Guest rooms or guest suites of hotels and motels that are intended for transient occupancy

Exception 2: Branch-circuit overcurrent protection devices aren't required to be accessible to occupants of guest rooms or guest suites of hotels and motels if electric maintenance is provided in a facility that is under continuous building management.

(C) Not Exposed to Physical Damage. Overcurrent protection devices must not be exposed to physical damage. Figure 240–32

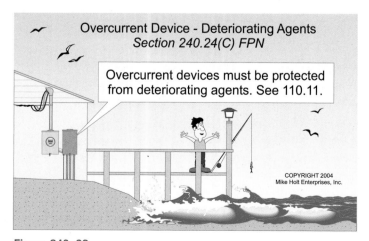

Figure 240–33

FPN: Electrical equipment must be suitable for the environment and consideration must be given to the presence of corrosive gases, fumes, vapors, liquids, or chemicals that have a deteriorating effect on conductors or equipment [110.11]. Figure 240–33

(D) Not in Vicinity of Easily Ignitible Material. Overcurrent protection devices must not be located near easily ignitible material, such as in clothes closets. Figure 240–34

Author's Comment: The purpose of keeping overcurrent protection devices away from easily ignitible material is to prevent fires, not to keep them out of clothes closets.

(E) Not in Bathrooms. Overcurrent protection devices must not be located in the bathrooms of dwelling units, or guest rooms or guest suites of hotels or motels. Figure 240–35

Author's Comment: The service disconnecting means must not be located in any bathroom [230.70(A)(2)].

PART III. ENCLOSURES

240.32 Damp or Wet Locations. In damp or wet locations, enclosures containing overcurrent protection devices must prevent moisture or water from entering or accumulating within the enclosure. When the enclosure is surface mounted in a wet location, it must be mounted with not less than ¼ in. of air space between it and the mounting surface. See 312.2(A).

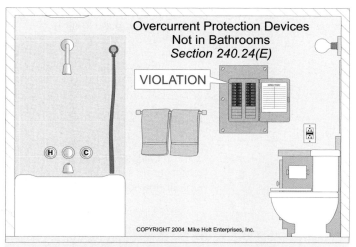

Overcurrent protection devices must not be located in the bathrooms of dwelling units, or in guest rooms or guest suites of hotels or motels.

Figure 240–35

240.33 Vertical Position. Enclosures containing overcurrent protection devices must be mounted in a vertical position unless this isn't practical. Circuit breaker enclosures are permitted horizontally if the circuit breaker is installed in accordance with 240.81. Figure 240–36

> **Author's Comment:** 240.81 specifies that where circuit breaker handles are operated vertically, the "up" position of the handle must be in the "on" position. So, in effect, an enclosure that contains one circuit breaker can be mounted horizontally, but an enclosure that contains a panelboard/loadcenter with multiple circuit breakers on opposite sides of each other would have to be mounted vertically. See Figure 240–36.

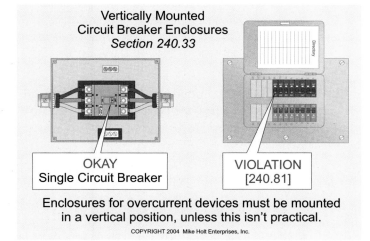

Enclosures for overcurrent devices must be mounted in a vertical position, unless this isn't practical.

Figure 240–36

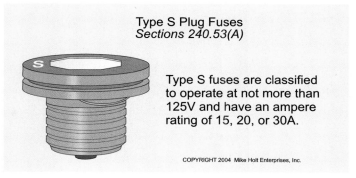

Figure 240–37

PART V. PLUG FUSES, FUSEHOLDERS, AND ADAPTERS

240.51 Edison-Base Fuse.

(A) Classification. Edison-base fuses are classified to operate at not more than 125V and have an ampere rating of not more than 30A.

(B) Replacement Only. Edison-base fuses are permitted only for replacement in an existing installation where there's no evidence of tampering or overfusing.

240.53 Type S Fuses.

(A) Classification. A Type S fuse is classified to operate at not more than 125V and have an ampere rating of 0-15A, 16-20A, and 21-30A. Figure 240–37

(B) Not Interchangeable. Type S fuses are made so that different ampere ratings aren't interchangeable.

240.54 Type S Fuses, Adapters, and Fuseholders.

(A) Type S Adapters. Type S adapters are designed to fit Edison-base fuseholders.

(B) Prevent Edison-Base Fuses. Type S fuseholders and adapters are designed for Type S fuses only.

(C) Nonremovable Adapters. Type S adapters are designed so that they cannot be removed once installed.

PART VI. CARTRIDGE FUSES AND FUSEHOLDERS

> **Author's Comment:** There are two basic designs of cartridge fuses, the ferrule type with a maximum rating of 60A and the knife-blade type with ampere ratings over 60A. The fuse length and diameter varies with the voltage and current rating. Figure 240–38

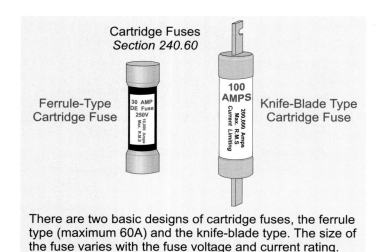

Figure 240-38

240.60 General.

(A) Maximum Voltage—300V Type. Cartridge fuses and fuse-holders of the 300V type can only be used for:

- Circuits not exceeding 300V between conductors.
- Circuits not exceeding 300V from any ungrounded conductor to the grounded neutral conductor.

(B) Noninterchangeable Fuseholders. Fuseholders must be designed to make it difficult to interchange fuses for different voltages and current ratings.

Fuseholders for current-limiting fuses must be designed so that only current-limiting fuses can be inserted. Figure 240-39

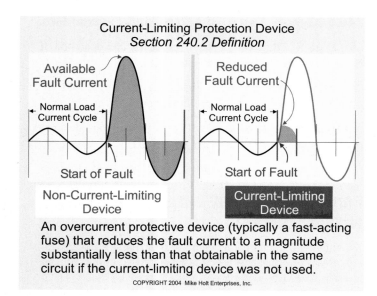

Figure 240-39

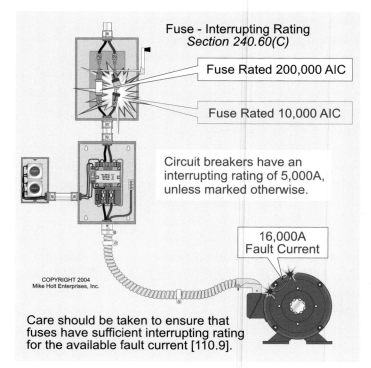

Figure 240-40

Author's Comment: A current-limiting fuse is a fast-clearing overcurrent protective device that reduces the fault current to a magnitude substantially lower than that obtainable in the same circuit if the current-limiting device were not used [240.2].

(C) Marking. Cartridge fuses have an interrupting rating of 10,000A unless marked otherwise.

WARNING: *Take care to ensure that fuses have an interrupting rating sufficient for the short-circuit current that is available at the line terminals of the equipment. Using a fuse with inadequate interrupting current rating could cause equipment to be destroyed from a line-to-line or ground fault, and result in death or serious injury. See 110.9 for more details.* Figure 240-40

240.61 Classification. Cartridge fuses and fuseholders are classified according to voltage and amperage ranges. Fuses rated 600V, nominal, or less are permitted for voltages at or below their ratings.

PART VII. CIRCUIT BREAKERS

240.80 Method of Operation. Circuit breakers must be capable of being opened and closed by hand. Nonmanual means

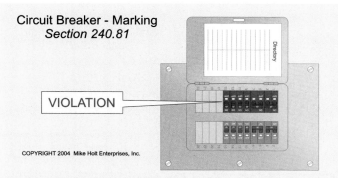

Circuit Breaker - Marking
Section 240.81

VIOLATION

COPYRIGHT 2004 Mike Holt Enterprises, Inc.

Circuit breakers must indicate whether they are in the "off" or "on" position. When the handle is operated vertically, the "up" position of the handle must be the "on" position.

Figure 240–41

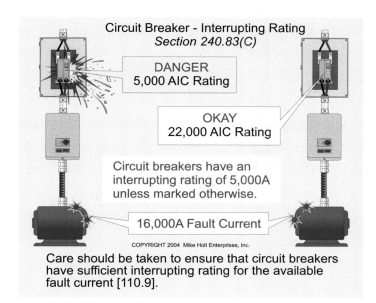

Circuit Breaker - Interrupting Rating
Section 240.83(C)

DANGER
5,000 AIC Rating

OKAY
22,000 AIC Rating

Circuit breakers have an interrupting rating of 5,000A unless marked otherwise.

16,000A Fault Current

COPYRIGHT 2004 Mike Holt Enterprises, Inc.

Care should be taken to ensure that circuit breakers have sufficient interrupting rating for the available fault current [110.9].

Figure 240–42

of operating a circuit breaker, such as electrical shunt trip or pneumatic operation, are permitted as long as the circuit breaker can also be manually operated.

240.81 Indicating. Circuit breakers must clearly indicate whether they are in the open "off" or closed "on" position. When the handle of a circuit breaker is operated vertically, the "up" position of the handle must be the "on" position. See 240.33 and 404.6(C). **Figure 240–41**

240.83 Markings.

(C) Interrupting Rating. Circuit breakers have an interrupting rating of 5,000A unless marked otherwise.

> **WARNING:** *Take care to ensure that the circuit breaker has an interrupting rating sufficient for the short-circuit current that is available at the line terminals of the equipment. Using a circuit breaker with inadequate interrupting current rating could cause equipment to be destroyed from a line-to-line or ground fault, and result in death or serious injury. See 110.9 for more details.* **Figure 240–42**

(D) Used as Switches. Circuit breakers used to switch 120V or 277V fluorescent lighting circuits must be listed and marked SWD or HID. Circuit breakers used to switch high-intensity discharge lighting circuits must be listed and marked HID. **Figure 240–43**

> **Author's Comment:** UL 489, *Standard for Molded Case Circuit Breakers*, permits "HID" breakers to be rated up to 50A, whereas an "SWD" breaker may only be rated to 20A. The tests for "HID" breakers include an endurance test at 75 percent power factor, whereas "SWD" breakers are endurance-tested at 100 percent power factor. The contacts and the spring of an

"HID" breaker are of a heavier duty material to dissipate the increased heat caused by the greater current flow in the circuit, resulting from the "HID" luminaire taking a minute or two to ignite the lamp.

(E) Voltage Markings. Circuit breakers must be marked with a voltage rating that corresponds with their interrupting rating. See 240.85.

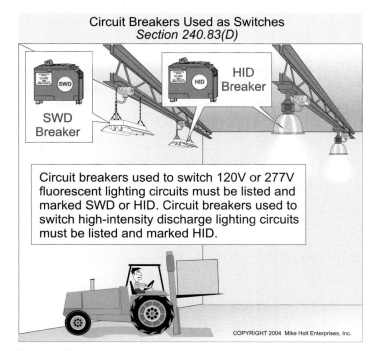

Circuit Breakers Used as Switches
Section 240.83(D)

HID
Breaker

SWD
Breaker

Circuit breakers used to switch 120V or 277V fluorescent lighting circuits must be listed and marked SWD or HID. Circuit breakers used to switch high-intensity discharge lighting circuits must be listed and marked HID.

COPYRIGHT 2004 Mike Holt Enterprises, Inc.

Figure 240–43

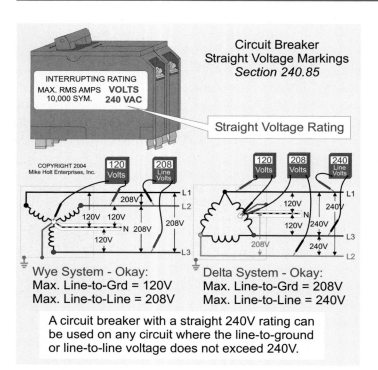

Figure 240–44

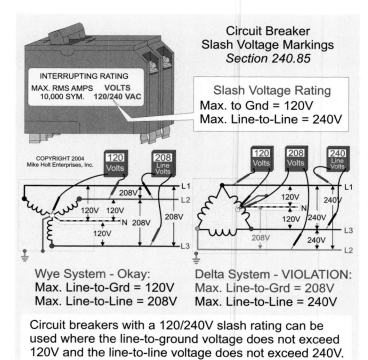

Figure 240–45

240.85 Applications

Straight Voltage Rating. A circuit breaker with a straight voltage rating, such as 240V or 480V, is permitted on a circuit where the nominal voltage between any two conductors (line-to-neutral or line-to-line) doesn't exceed the circuit breaker's voltage rating. Figure 240–44

Slash Voltage Rating. A circuit breaker with a slash rating, such as 120/240V or 277/480V, is permitted on a solidly grounded system where the nominal voltage of any one conductor to metal case doesn't exceed the lower of the two values and the nominal voltage between any two conductors doesn't exceed the higher value.

CAUTION: *A circuit breaker with a slash rating, such as 120/240V, can only be used on a circuit where the nominal voltage of any one ungrounded conductor to metal case doesn't exceed the lower of the two values. A 120/240V slash circuit breaker cannot be used on the high-leg of a solidly grounded 120/240V delta system, because the line-to-ground voltage of the high-leg is 208V, which exceeds the 120V line-to-ground voltage rating of the breaker.* Figure 240–45

FPN: When installing circuit breakers on corner-bonded delta systems, consideration needs to be given to the circuit breakers' individual pole-interrupting capability.

Article 240 Questions

1. Overcurrent protection for conductors and equipment is designed to _____ the circuit if the current reaches a value that will cause an excessive or dangerous temperature in conductors or conductor insulation.

   (a) open   (b) close   (c) monitor   (d) record

2. Flexible cords approved for and used with a specific listed appliance or portable lamp are considered to be protected when _____.

   (a) not more than 6 ft in length   (b) 20 AWG and larger
   (c) applied within the listing requirements   (d) 16 AWG and larger

3. Supplementary overcurrent devices used in luminaires or appliances are not required to be readily accessible.

   (a) True   (b) False

4. Overcurrent protection for tap conductors not over 25 ft is not required at the point where the conductors receive their supply providing the _____.

   (a) ampacity of the tap conductors is not less than one-third the rating of the overcurrent device protecting the feeder conductors being tapped
   (b) tap conductors terminate in a single circuit breaker or set of fuses that limit the load to the ampacity of the tap conductors
   (c) tap conductors are suitably protected from physical damage
   (d) all of these

5. Circuit breakers and fuses must be readily accessible and they must be installed so the center of the grip of the operating handle of the fuse switch or circuit breaker, when in its highest position, isn't more than _____ above the floor or working platform.

   (a) 6 ft 7 in.   (b) 2 ft   (c) 5 ft   (d) 4 ft 6 in.

6. Plug fuses of 15A or less must be identified by a(n) _____ configuration of the window, cap, or other prominent part to distinguish them from fuses of higher ampere ratings.

   (a) octagonal   (b) rectangular   (c) hexagonal   (d) triangular

7. Type _____ fuse adapters must be designed so that once inserted in a fuseholder they cannot be removed.

   (a) A   (b) E   (c) S   (d) P

8. Cartridge fuses and fuseholders must be classified according to _____ ranges.

   (a) voltage   (b) amperage   (c) voltage or amperage   (d) voltage and amperage

9. Circuit breakers rated at _____ amperes or less and _____ volts or less must have the ampere rating molded, stamped, etched, or similarly marked into their handles or escutcheon areas.

   (a) 100, 600   (b) 600, 100   (c) 1,000, 6,000   (d) 6,000, 1,000

10. A circuit breaker with a slash rating (120/240V or 277/480V) can be used on a solidly-grounded circuit where the nominal voltage of any conductor to _____ does not exceed the lower of the two values, and the nominal voltage between any two conductors does not exceed the higher value.

   (a) another conductor   (b) an enclosure   (c) earth   (d) ground

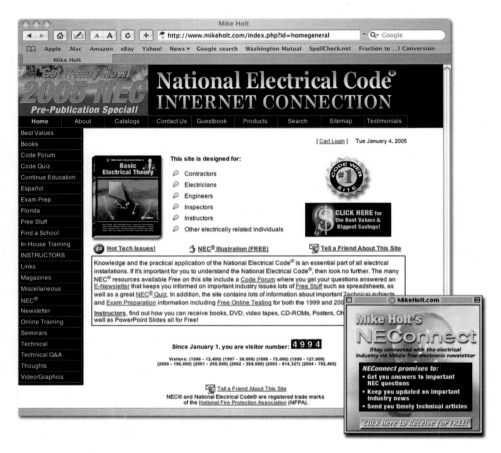

ARTICLE 250

Grounding and Bonding

Introduction

The purpose of the *National Electrical Code* is the practical safeguarding of persons and property from hazards arising from the use of electricity [90.1(A)]. In addition, the *NEC* contains provisions that are considered necessary for safety. Compliance with the *NEC*, combined with proper maintenance, must result in an installation that is essentially free from hazard [90.1(B)].

No other article can match Article 250 for misapplication, violation, and misinterpretation. People often insist on completing installations in a manner that results in violations of this Article. For example, many industrial equipment manuals require violating 250.4(A)(5) as a condition of warranty. The manuals insist on installing an "isolated grounding electrode," which is an electrode without a low-impedance fault-current path back to the electrical supply source, typically the X0 terminal of a transformer, other than through the earth itself. That means the ground-fault current return path to the electrical supply source, typically the X0 terminal (utility transformer), is on the order of several ohms rather than the fraction of an ohm that the typical *NEC*-compliant installation would provide.

If you apply basic physics and basic electrical theory, you can clearly see Article 250 is right and equipment manuals that require isolated grounding are wrong, and other references agree. IEEE-142 and *Soares Book on Grounding* use the same physics and electrical theory as Article 250. This article isn't a "preferred design specification." As with the rest of the *NEC*, it serves the purpose stated in Article 90 to be sure the installation is, and remains, SAFE!

Article 250 covers the requirements for providing paths to divert high voltage to the earth, requirements for the low-impedance fault-current path to facilitate the operation of overcurrent protection devices, and how to remove dangerous voltage potentials between conductive parts of building components and electrical systems.

Over the past two *Code* cycles, this article was extensively revised to make it better organized and easier to implement. It's arranged in a logical manner, so it's a good idea to just read through Article 250 to get a big picture view—after you review the definitions. Then study the article closely so you understand the details. The illustrations will help you understand the key points.

> **Author's Comment:** When the *NEC* uses the word "ground" or "grounding" where the intent is the connection to the earth, this textbook will add "(earth)." If the *NEC* used the word "ground" or "grounding" where the intent is bonding metal parts to the supply source, I will add "(bonding)" to the text. It's unfortunate that the word "grounding" is used interchangeably for grounding and bonding, which are two different things.

PART I. GENERAL

250.1 Scope. Article 250 contains the following grounding and bonding requirements for electrical installations:

(1) Systems and equipment required to be grounded

(2) Circuit conductor to be grounded on grounded systems

(3) Location of grounding connections

(4) Types and sizes of grounding and bonding conductors and electrodes

(5) Methods of grounding and bonding

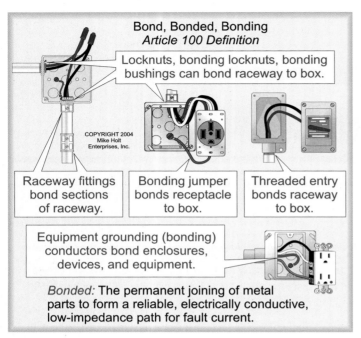

Figure 250–1

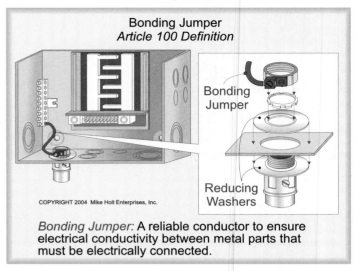

Bonding Jumper: A reliable conductor to ensure electrical conductivity between metal parts that must be electrically connected.

Figure 250–2

250.2 Definitions.

> **Author's Comment:** Why is grounding so difficult to understand? One reason is because many do not understand the definition of many important terms. So before we get too deep into this subject, let's review a few important definitions contained in Articles 100 and 250.

Bonding [100]. The permanent joining of metal parts together to form an electrically conductive path that has the capacity to conduct safely any fault current likely to be imposed on it. Figure 250–1

> **Author's Comment:** Bonding is accomplished by the use of conductors, metallic raceways, connectors, couplings, metallic-sheathed cables with fittings, and other devices recognized for this purpose [250.118].

Bonding Jumper [100]. A conductor properly sized in accordance with Article 250 that ensures electrical conductivity between metal parts of the electrical installation. Figure 250–2

Effective Ground-Fault Current Path [250.2]. An intentionally constructed, permanent, low-impedance conductive path designed to carry fault current from the point of a ground fault on a wiring system to the electrical supply source. Figure 250–3

The effective ground-fault current path is intended to help remove dangerous voltage from a ground fault by opening the circuit overcurrent protective device. Figure 250–4

Equipment Grounding Conductor [100]. The low-impedance fault-current path used to bond metal parts of electrical equipment, raceways, and enclosures to the effective ground-fault-current path at service equipment or the source of a separately derived system.

> **Author's Comments:**
>
> • The purpose of the equipment grounding (bonding) conductor is to provide the low-impedance fault-current path to the electrical supply source to facilitate the operation of circuit

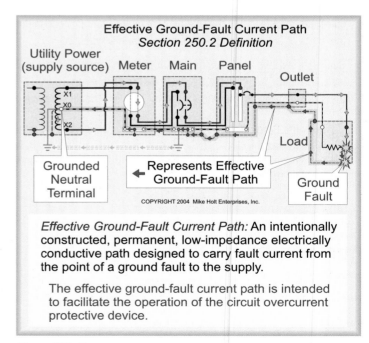

Effective Ground-Fault Current Path: An intentionally constructed, permanent, low-impedance electrically conductive path designed to carry fault current from the point of a ground fault to the supply.

The effective ground-fault current path is intended to facilitate the operation of the circuit overcurrent protective device.

Figure 250–3

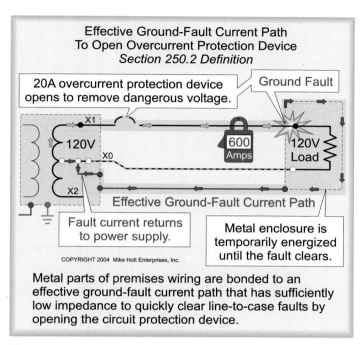

Figure 250–4

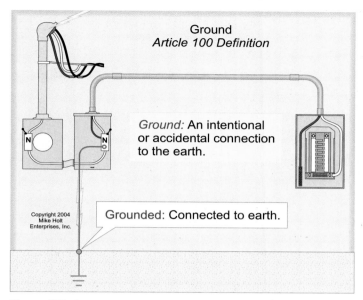

Figure 250–6

overcurrent protection devices in order to remove dangerous ground-fault voltage on conductive parts [250.4(A)(3)]. Fault current returns to the power supply (source), not the earth!

- According to 250.118, the equipment grounding (bonding) conductor must be one or a combination of the following: **Figure 250–5**

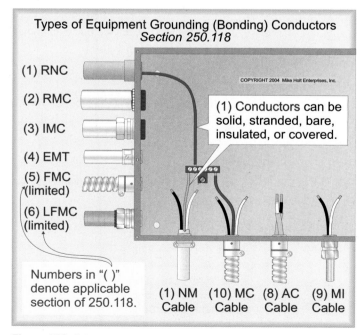

Figure 250–5

- Wire Type. A bare or insulated conductor [250.118(1)]
- Rigid Metal Conduit [250.118(2)]
- Intermediate Metal Conduit [250.118(3)]
- Electrical Metallic Tubing [250.118(4)]
- Listed Flexible Metal Conduit as limited by 250.118(5)
- Listed Liquidtight Flexible Metal Conduit as limited by 250.118(6)
- Armor of Type AC cable [250.118(8)]
- Armor of Type MC cable as limited by 250.118(10)
- Metallic Cable Trays as limited by 250.118(11) and 392.7
- Electrically continuous metal raceways listed for grounding [250.118(13)]
- Surface Metal Raceways listed for grounding [250.118(14)]

Ground (Earth) [100]. Earth or a conductive body that is connected to earth. **Figure 250–6**

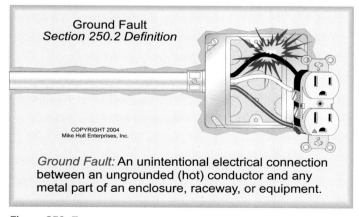

Figure 250–7

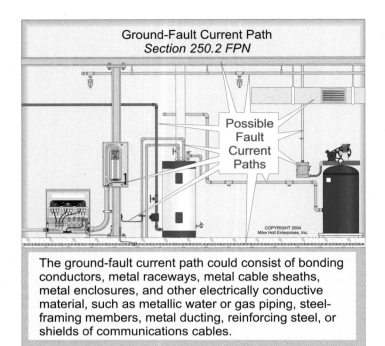

Ground-Fault Current Path
Section 250.2 FPN

Possible Fault Current Paths

The ground-fault current path could consist of bonding conductors, metal raceways, metal cable sheaths, metal enclosures, and other electrically conductive material, such as metallic water or gas piping, steel-framing members, metal ducting, reinforcing steel, or shields of communications cables.

Figure 250–8

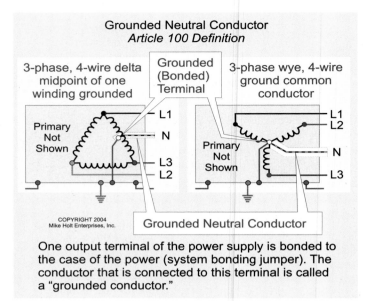

Grounded Neutral Conductor
Article 100 Definition

3-phase, 4-wire delta midpoint of one winding grounded

Grounded (Bonded) Terminal

3-phase wye, 4-wire ground common conductor

Primary Not Shown

L1
N
L3
L2

Primary Not Shown

L1
L2
N
L3

Grounded Neutral Conductor

One output terminal of the power supply is bonded to the case of the power (system bonding jumper). The conductor that is connected to this terminal is called a "grounded conductor."

Figure 250–9

Grounded [100]. Connected to earth.

Ground Fault [100]. An unintentional connection between an ungrounded conductor and metal parts of enclosures, raceways, or equipment. **Figure 250–7**

Ground-Fault Current Path [250.2]. An electrically conductive path from a ground fault to the electrical supply source.

> **Author's Comment:** The fault-current path of a ground fault is not to the earth! It's to the electrical supply source, typically the XO terminal of a transformer.

> **FPN:** The ground-fault current path could be metal raceways, cable sheaths, electrical equipment, or other electrically conductive materials, such as metallic water or gas piping, steel-framing members, metal ducting, reinforcing steel, or the shields of communications cables. **Figure 250–8**

> **Author's Comment:** The difference between an "effective ground-fault current path" and " fault-current path" is that the <u>effective ground-fault current</u> path is "intentionally" constructed to provide the low-impedance fault-current path to the electrical supply source for the purpose of clearing the ground fault. A <u>ground-fault current path</u> is simply all of the available conductive paths over which fault current flows on its return to the electrical supply source during a ground fault.

Grounded (Earthed) [100]. Connected to earth.

Grounded Neutral Conductor [100]. The conductor that terminates to the terminal that is intentionally grounded to the earth. **Figure 250–9**

Grounding (Earthing) Conductor [100]. The conductor that connects equipment to the earth via a grounding electrode.

> **Author's Comment:** An example would be the conductor used to connect equipment to a supplementary grounding electrode [250.56]. **Figure 250–10**

Grounding (Earthing) Electrode [100]. A device that establishes an electrical connection to the earth. **Figure 250–11**

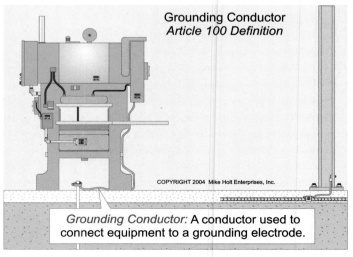

Grounding Conductor
Article 100 Definition

Grounding Conductor: A conductor used to connect equipment to a grounding electrode.

Figure 250–10

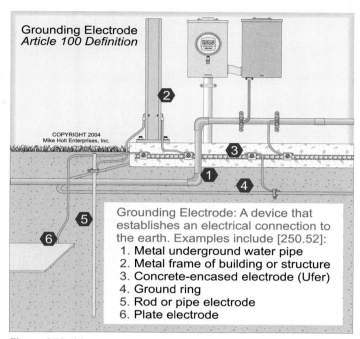

Figure 250–11

Grounding Electrode: A device that establishes an electrical connection to the earth. Examples include [250.52]:
1. Metal underground water pipe
2. Metal frame of building or structure
3. Concrete-encased electrode (Ufer)
4. Ground ring
5. Rod or pipe electrode
6. Plate electrode

Author's Comment: See 250.50 through 250.70

Grounding Electrode (Earth) Conductor [100]. The conductor that connects the grounded neutral conductor at service equipment [250.24(A)], the building or structure disconnecting means enclosure [250.32(A)], or separately derived systems enclosure [250.30(A)] to an electrode (earth). Figure 250–12

Main Bonding Jumper [100]. A conductor, screw, or strap that bonds the equipment grounding (bonding) conductor at service equipment to the grounded neutral service conductor in accordance with 250.24(B). Figure 250–13

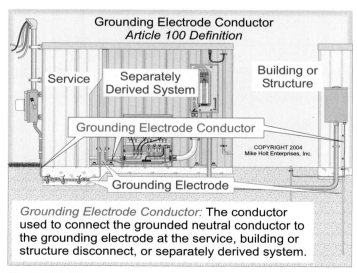

Grounding Electrode Conductor
Article 100 Definition

Service   Separately Derived System   Building or Structure

Grounding Electrode Conductor

Grounding Electrode

Grounding Electrode Conductor: The conductor used to connect the grounded neutral conductor to the grounding electrode at the service, building or structure disconnect, or separately derived system.

Figure 250–12

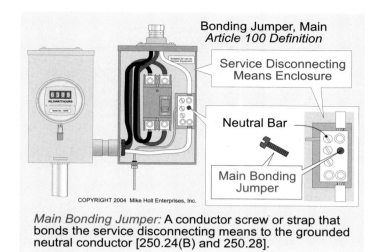

Bonding Jumper, Main
Article 100 Definition

Service Disconnecting Means Enclosure

Neutral Bar

Main Bonding Jumper

Main Bonding Jumper: A conductor screw or strap that bonds the service disconnecting means to the grounded neutral conductor [250.24(B) and 250.28].

Figure 250–13

Author's Comment: For more details, see 250.24(A)(4), 250.28, and 408.3(C).

Solidly Grounded [100]. The intentional electrical connection of one system terminal to the equipment grounding (bonding) conductor in accordance with 250.30(A)(1).

Author's Comment: The industry calls a system that has one terminal bonded to its metal case a solidly grounded system. Figure 250–14

System Bonding Jumper [100]. The conductor, screw, or strap that bonds the metal parts of a separately derived system to a system winding in accordance with 250.30(A)(1). Figure 250–15

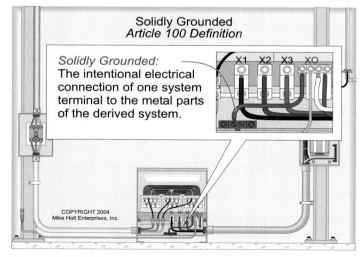

Solidly Grounded
Article 100 Definition

Solidly Grounded: The intentional electrical connection of one system terminal to the metal parts of the derived system.

X1 X2 X3 X0

Figure 250–14

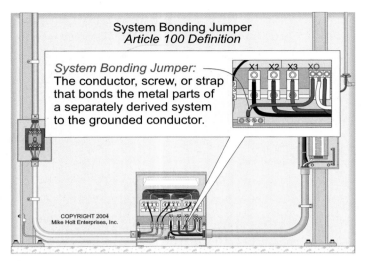

Figure 250–15

Author's Comment: The system bonding jumper provides the low-impedance fault-current path to the electrical supply source for the purpose of clearing the ground fault. For more information see 250.4(A)(5), 250.28, and 250.30(A)(1).

250.4 General Requirements for Grounding and Bonding.

(A) Solidly Grounded Systems.

(1) Grounding Electrical Systems to the Earth. High-voltage system windings are grounded to the earth to help limit high voltage imposed on the system windings from lightning, unintentional contact with higher-voltage lines, or line surges. Figure 250–16

(2) Grounding Electrical Equipment to the Earth. Metal parts of electrical equipment must be grounded to the earth by electrically connecting the building or structure disconnecting means [225.31 or 230.70] with a grounding electrode conductor [250.64(A)] to a grounding electrode [250.52, 250.24(A), and 250.32(A)]. Figure 250–17

Author's Comments:

- Metal parts of the electrical installation are grounded to the earth to reduce voltage on the metal parts from lightning so as to prevent fires from a surface arc within the building or structure. Grounding electrical equipment to earth doesn't serve the purpose of providing a low-impedance fault-current path to clear ground faults. In fact, the *Code* prohibits the use of the earth as the effective ground-fault current path [250.4(A)(5) and 250.4(B)(4)].

- Grounding metal parts to the earth is often necessary in areas where the discharge (arcing) of the voltage buildup (static) could cause dangerous or undesirable conditions.

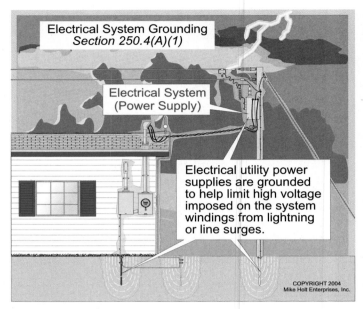

Figure 250–16

Such an occurrence might be the failure of electronic equipment being assembled on a production line, or a fire and explosion in a hazardous (classified) area. See 500.4 FPN 3.

- Grounding metal parts to the earth doesn't protect electrical or electronic equipment from lightning voltage transients (high-frequency voltage impulses) on the circuit conductors. To protect electrical equipment from high-voltage transients, proper transient voltage surge-protection devices must be installed in accordance with Article 280 at service equipment, and Article 285 at panelboards and other locations.

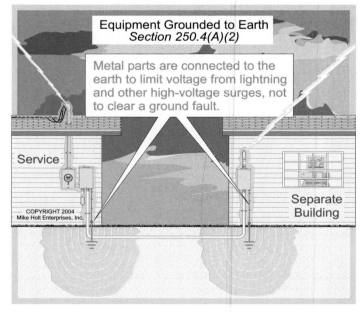

Figure 250–17

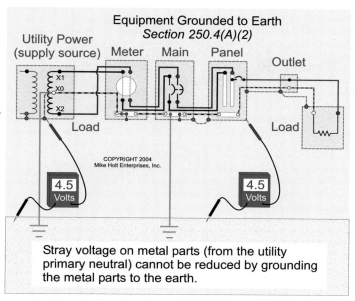

Equipment Grounded to Earth
Section 250.4(A)(2)

Stray voltage on metal parts (from the utility primary neutral) cannot be reduced by grounding the metal parts to the earth.

Figure 250–18

- Grounding metal parts to the earth does not create a zero reference point, nor does it reduce the difference of potential (voltage) between the metal parts and the earth. For example, if the voltage on metal parts from the utility primary neutral is 4.5 (stray voltage), grounding metal parts to the earth will not reduce this value. **Figure 250–18**

(3) Bonding Electrical Equipment to an Effective Ground-Fault Current Path. To remove dangerous voltage from ground faults, metal parts of electrical raceways, cables, enclosures, and equipment must be bonded to an effective ground-fault current path with an equipment grounding (bonding) conductor of a type specified in 250.118. **Figure 250–19**

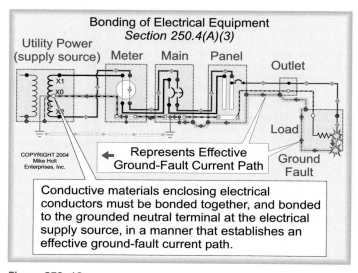

Bonding of Electrical Equipment
Section 250.4(A)(3)

← Represents Effective Ground-Fault Current Path

Conductive materials enclosing electrical conductors must be bonded together, and bonded to the grounded neutral terminal at the electrical supply source, in a manner that establishes an effective ground-fault current path.

Figure 250–19

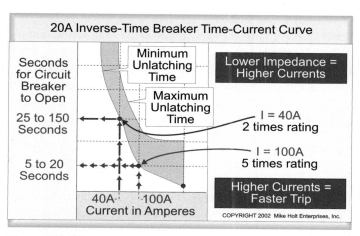

20A Inverse-Time Breaker Time-Current Curve

Figure 250–20

Author's Comment: To protect against electric shock from dangerous voltages on metal parts, a ground fault must quickly be removed by opening the circuit's overcurrent protection device. To quickly remove dangerous touch voltage on metal parts from a ground fault, the fault-current path must have sufficiently low impedance to allow the fault current to quickly rise to a level that will open the branch-circuit overcurrent protection device.

The time it takes for an overcurrent protection device to open is inversely proportional to the magnitude of the fault current. This means that the higher the ground-fault current value, the less time it will take for the protection device to open and clear the fault. For example, a 20A circuit with an overload of 40A (two times the rating) would take 25 to 150 seconds to open the protection device. At 100A (five times the rating) the 20A breaker would trip in 5 to 20 seconds. **Figure 250–20**

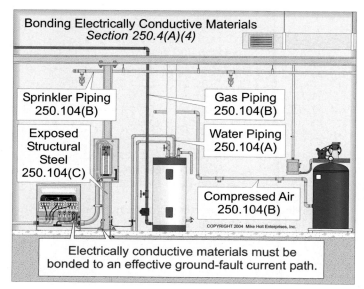

Bonding Electrically Conductive Materials
Section 250.4(A)(4)

Sprinkler Piping 250.104(B)

Gas Piping 250.104(B)

Exposed Structural Steel 250.104(C)

Water Piping 250.104(A)

Compressed Air 250.104(B)

Electrically conductive materials must be bonded to an effective ground-fault current path.

Figure 250–21

(4) Bonding Conductive Materials to an Effective Ground-Fault Current Path. To remove dangerous voltage from ground faults, electrically conductive metal water piping systems, metal sprinkler piping, metal gas piping, and other metal-piping systems, as well as exposed structural steel members that are likely to become energized, must be bonded to an effective ground-fault current path. **Figure 250–21**

> **Author's Comment:** The phrase "likely to become energized" is subject to interpretation by the authority having jurisdiction.

(5) Effective Ground-Fault Current Path. Metal raceways, cables, enclosures, and equipment, as well as other electrically conductive materials that are likely to become energized, must be installed in a manner that creates a permanent, low-impedance fault-current path that facilitates the operation of the circuit overcurrent device. **Figure 250–22**

> **Author's Comment:** To assure a low-impedance ground-fault current path, all circuit conductors must be grouped together in the same raceway, cable, or trench [300.3(B), 300.5(I), and 300.20(A)]. **Figure 250–23**

The earth is not considered an effective ground-fault current path.

> **DANGER:** *Because the resistance of the earth is so high, very little current returns to the electrical supply source via the earth. If a ground rod is used as the ground-fault current path, the circuit overcurrent protection device will not open and metal parts will remain energized.*
>
> *For example, the maximum current flow to the power supply from a 120V ground fault to a 25 ohm ground rod would only be 4.8A.* **Figure 250–24**

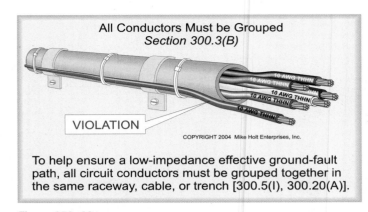

All Conductors Must be Grouped
Section 300.3(B)

VIOLATION

To help ensure a low-impedance effective ground-fault path, all circuit conductors must be grouped together in the same raceway, cable, or trench [300.5(I), 300.20(A)].

Figure 250–23

$$I = E/R$$
$$I = 120V/25\Omega$$
$$I = 4.8A$$

To understand how a ground rod is useless in reducing touch voltage to a safe level, let's answer the following questions:

- What is touch voltage?
- At what level is touch voltage hazardous?
- How do earth surface voltage gradients operate?

Touch/Step Voltage: The IEEE definition of touch/step voltage is "the potential (voltage) difference between a bonded metallic structure and a point on the earth 3 ft from the structure."

Hazardous Level: NFPA 70E, *Standard for Electrical Safety in the Workplace*, cautions that death and/or severe electric shock can occur whenever touch/step voltage exceeds 30V.

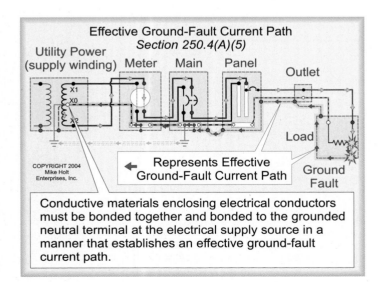

Effective Ground-Fault Current Path
Section 250.4(A)(5)

Utility Power (supply winding)   Meter   Main   Panel   Outlet

Load

← Represents Effective Ground-Fault Current Path

Ground Fault

Conductive materials enclosing electrical conductors must be bonded together and bonded to the grounded neutral terminal at the electrical supply source in a manner that establishes an effective ground-fault current path.

Figure 250–22

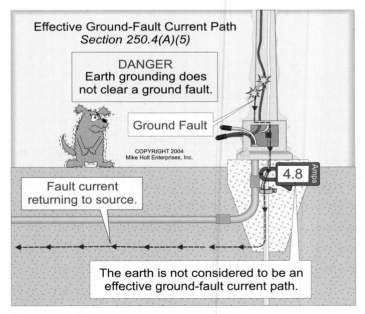

Effective Ground-Fault Current Path
Section 250.4(A)(5)

DANGER
Earth grounding does not clear a ground fault.

Ground Fault

Fault current returning to source.

4.8 Amps

The earth is not considered to be an effective ground-fault current path.

Figure 250–24

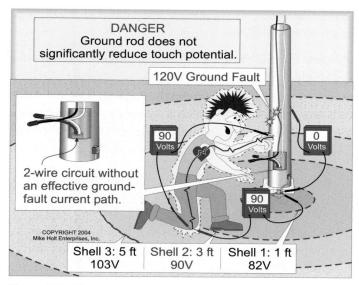

Figure 250–25

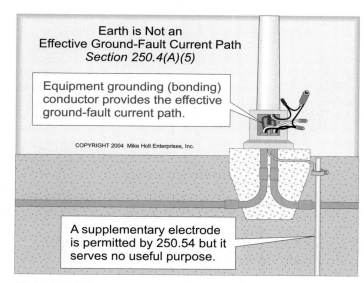

Figure 250–26

Surface Voltage Gradients: According to ANSI/IEEE 142, *Recommended Practice for Grounding of Industrial and Commercial Power Systems* (Green Book) [4.1.1], the resistance of the soil outward from a ground rod is equal to the sum of the series resistances of the earth shells. The shell nearest the rod has the highest resistance and each successive shell has progressively larger areas and progressively lower resistances.

Don't worry if you don't understand the above statement; just review the table below with **Figure 250–25**.

Distance from Rod	Resistance	Touch Voltage
1 Ft (Shell 1)	68%	82V
3 Ft (Shells 1 and 2)	75%	90V
5 Ft (Shells 1, 2, and 3)	86%	103V

Many think a ground rod can reduce touch voltage to a safe value. However, as the above table shows, the voltage gradient of the earth drops off so rapidly that a person in contact with an energized object can receive a lethal electric shock one foot away from an energized object that is grounded to the earth.

The generally accepted grounding practice for street lighting and traffic signaling for many parts of the United States is to ground all metal parts to a ground rod as the only fault-current return path. Studies by some electric utilities indicate that about one-half of one percent of all their metal poles had dangerous touch voltage.

Author's Comment: The common practice of installing a ground rod at a metal pole supporting a luminaire serves no useful purpose. **Figure 250–26**

(B) Ungrounded Systems.

Author's Comment: According to IEEE 242, *Recommended Practice for Protection and Coordination of Industrial and Commercial Power Systems* (Buff Book), if a ground fault is intermittent, or allowed to continue on an ungrounded system, the system wiring could be subjected to severe system overvoltage, which can be as high as six or eight times the phase voltage. This excessive system voltage can puncture conductor insulation and result in additional ground faults. System overvoltage can be caused by repetitive charging of the system capacitance or by resonance between the system capacitance and the inductances of equipment in the system [7.2.5].

In addition, ANSI/IEEE 142, *Recommended Practice for Grounding of Industrial and Commercial Power Systems* (Green Book) states, "One of the dangers of an ungrounded system is that system overvoltages can occur during arcing, resonant or near-resonant ground faults [1.4.2]." And, "Field experience and theoretical studies have shown that arcing, restriking, or vibrating ground faults on ungrounded systems can, under certain conditions, produce surge voltages as high as six times normal.

The conditions necessary for producing overvoltage require that the dielectric strength of the arc path build up at a higher rate after each extinction of the arc than it did after the preceding extinction. This phenomenon is unlikely to take place in open air between stationary contacts because such an arc path is not likely to develop sufficient dielectric recovery strength. It may occur in confined areas where the pressure may increase after each conduction period.

Neutral grounding is effective in reducing transient voltage buildup from such intermittent ground faults by reducing neutral displacement from ground potential and reducing destructive effectiveness of any high-frequency voltage oscillations following each arc initiation or restrike [1.2.14]." **Figure 250–27**

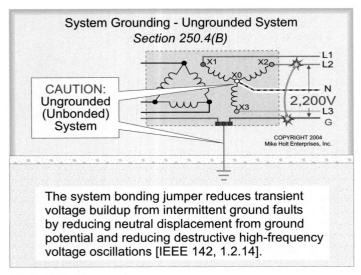

The system bonding jumper reduces transient voltage buildup from intermittent ground faults by reducing neutral displacement from ground potential and reducing destructive high-frequency voltage oscillations [IEEE 142, 1.2.14].

Figure 250–27

(1) Grounding Electrical Equipment to the Earth. Metal parts of electrical equipment must be grounded to the earth by electrically connecting the building or structure disconnecting means [225.31 or 230.70] with a grounding electrode conductor [250.64(A)] to a grounding electrode [250.52, 250.24(D), and 250.32(A)].

Author's Comments:

- Metal parts of the electrical installation are grounded to the earth to reduce voltage on the metal parts from lightning so as to prevent fires from surface arcs within the building or structure. Grounding equipment to the earth doesn't provide a low-impedance fault-current path to the source to clear ground faults. In fact, the *Code* prohibits the use of the earth as the effective ground-fault current path [250.4(A)(5) and 250.4(B)(4)].

- Grounding metal parts to the earth doesn't protect electrical or electronic equipment from lightning voltage transients on the circuit conductors. To protect electrical equipment from high-voltage transients, proper transient voltage surge-protection devices must be installed in accordance with Article 280 at service equipment, and in accordance with Article 285 at panelboards and other locations.

(2) Bonding Wiring Methods to the Metal Enclosure of the System. To remove dangerous voltage from a second ground fault, metal parts of electrical raceways, cables, enclosures, or equipment must be bonded together and to the metal enclosure of the system.

(3) Bonding Conductive Materials to the Metal Enclosure of the System. Electrically conductive materials that are likely to become energized must be bonded together and to the metal enclosure containing the system.

(4) Fault-Current Path. Electrical equipment, wiring, and other electrically conductive material likely to become energized must be installed in a manner that creates a permanent, low-impedance fault-current path from any point on the wiring system to the electrical supply source to facilitate the operation of overcurrent devices should a second ground fault occur on the wiring system.

> **Author's Comment:** A single ground fault cannot be cleared on an ungrounded system because there's no low-impedance fault-current path to the power source. However, in the event of a second ground fault (line-to-line short circuit), the bonding path provides a low-impedance fault-current path so that the circuit-protection device will open to clear the fault.

250.6 Objectionable Current.

(A) Preventing Objectionable Current. To prevent a fire, electric shock, or improper operation of circuit-protection devices or sensitive equipment, electrical systems and equipment must be installed in a manner that prevents objectionable current from flowing on conductive materials, electrical equipment, or grounding and bonding paths.

> **Author's Comment:** Objectionable current occurs because of improper neutral-to-case bonds and wiring errors.

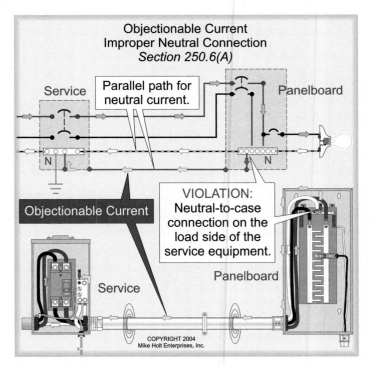

Figure 250–28

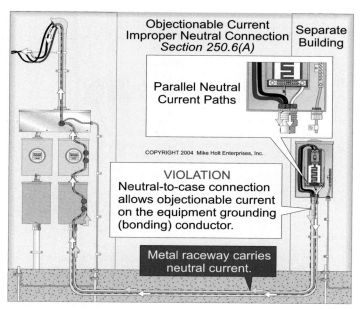

Figure 250–29

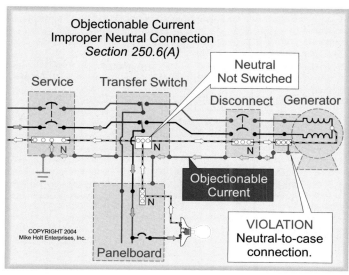

Figure 250–31

Improper Neutral-to-Case Bond [250.142]

Panelboards. Objectionable current will flow on metal parts when the grounded neutral conductor is bonded to the metal case of a panelboard that is not part of service equipment. **Figure 250–28**

Disconnects. Objectionable current will flow on metal parts when the grounded neutral conductor is bonded to the metal case of a disconnecting means that is not part of service equipment. **Figure 250–29**

Separately Derived Systems. Objectionable current will flow on metal parts when the grounded neutral conductor is bonded at the transformer as well as to the metal case on the load side of the transformer. **Figures 250–30 and 250–31**

Wiring Errors

Objectionable current will flow on metal parts when the grounded neutral conductor from one system is connected to a circuit of a different system. **Figure 250–32**

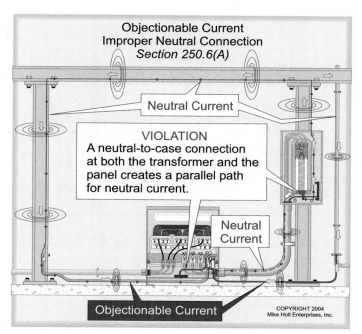

Figure 250–30

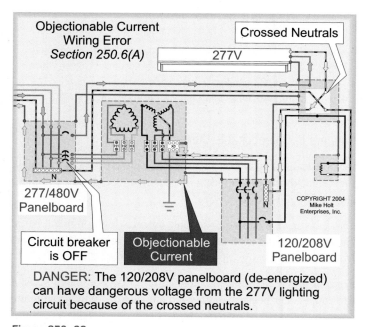

Figure 250–32

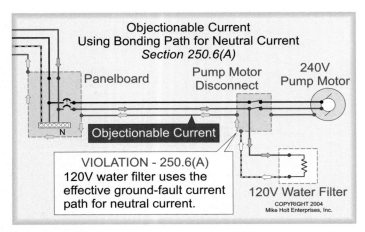

Figure 250–33

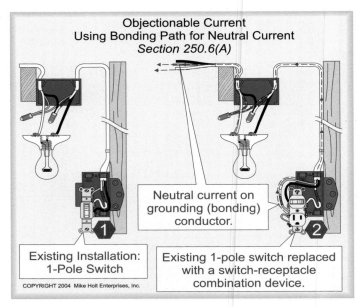

Figure 250–34

Objectionable current will flow on metal parts when the equipment grounding (bonding) conductor is used as a grounded neutral conductor.

Example: A 240V time-clock motor is replaced with a 120V time-clock motor and the equipment grounding (bonding) conductor is used to feed one side of the 120V time clock. Another example is a 120V water filter wired to a 240V well-pump motor circuit, with the equipment grounding (bonding) conductor used for the neutral. Figure 250–33

Using the equipment grounding (bonding) conductor for the neutral is also seen in ceiling fan installations where the bare equipment grounding (bonding) conductor is used as a neutral and the white wire is used as the switch leg for the light, or where a receptacle is added to a switch outlet that doesn't have a neutral conductor. Figure 250–34

> **Author's Comment:** Neutral currents always flow on a community metal underground water piping system because the grounded neutral conductor from each service is grounded to the underground metal water pipe. Figure 250–35

Dangers of Objectionable Current

Objectionable current on metal parts can cause electric shock, fires, and improper operation of sensitive electronic equipment and circuit-protection devices.

Shock Hazard. When objectionable current flows on metal parts, electric shock and even death can occur (ventricular fibrillation) from elevated voltage on the metal parts. Figure 250–36

Fire Hazard. When objectionable current flows on metal parts, a fire could occur because of elevated temperature, which can ignite adjacent combustible material. Heat is generated whenever current flows, particularly over high-resistive parts. In addition, arcing at

loose connections is especially dangerous in areas containing easily ignitible and explosive gases, vapors, or dust. Figure 250–37

Improper Operation of Sensitive Electronic Equipment. Objectionable current flowing on metal parts of electrical equipment and building parts can cause disruptive as well as annoying electromagnetic fields which can negatively affect the performance of sensitive electronic devices, particularly video monitors and medical equipment. For more information, visit www.MikeHolt.com, click on the Technical link, then on Power Quality. Figure 250–38

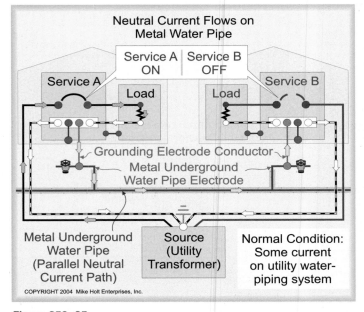

Figure 250–35

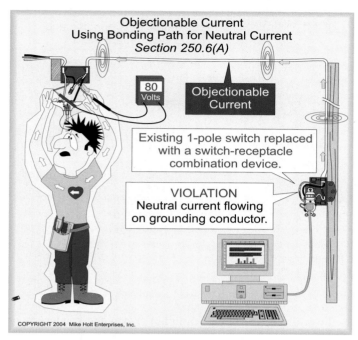

Figure 250-36

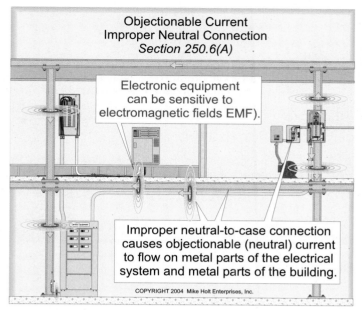

Figure 250-38

In addition, when objectionable current travels on metal parts, a difference of potential will exist between all metal parts, which can cause some sensitive electronic equipment to operate improperly (this is sometimes called a ground loop).

Improper Operation of Circuit-Protection Devices. When objectionable current travels on the metal parts of electrical equipment, nuisance tripping of electronic protection devices equipped with ground-fault protection can occur because some neutral current flows on the equipment grounding (bonding) conductor instead of the grounded neutral conductor.

(C) Temporary Currents Not Classified as Objectionable Currents. Temporary fault current on the effective ground-fault current path isn't classified as objectionable current. Figure 250–39

(D) Electromagnetic Interference (Electrical Noise). Currents that cause noise or data errors in electronic equipment aren't considered objectionable currents. Figure 250–40

> **Author's Comment:** Some sensitive electronic equipment manufacturers require isolation between the metal parts of their equipment and the electrical system, yet they require their equipment to be connected to an independent ground (like a ground

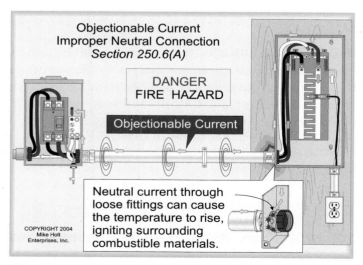

Figure 250-37

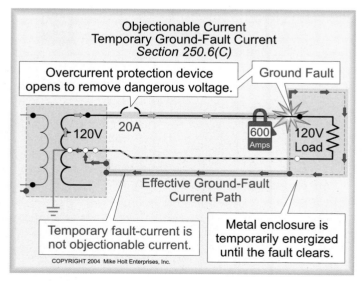

Figure 250-39

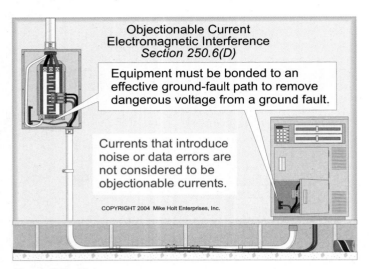

Figure 250–40

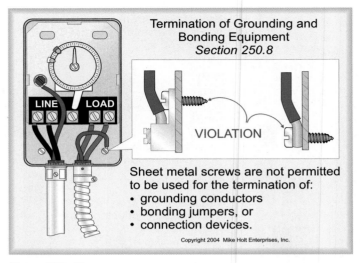

Figure 250–42

rod[s]). This practice violates 250.4(A)(5) and is very dangerous because the earth doesn't provide the low-impedance fault-current path necessary to clear a ground fault. **Figure 250–41**

250.8 Termination of Grounding and Bonding Conductors. The termination of equipment grounding and bonding conductors must be by exothermic welding, listed pressure connectors of the set screw or compression type, listed clamps, or other listed fittings. Sheet-metal screws cannot be

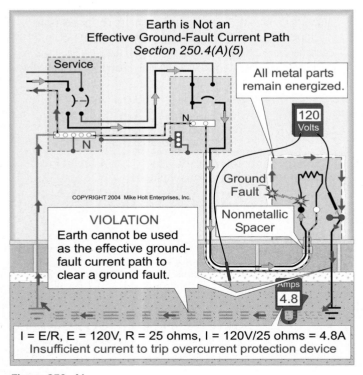

Figure 250–41

used to connect grounding (earthing) or bonding conductors or connection devices to enclosures. **Figure 250–42**

Author's Comment: The rule still doesn't prohibit drywall screws or wood screws from being used for this purpose, just "sheet-metal" screws!

250.10 Protection of Fittings. Ground clamps and other grounding and bonding fittings must be protected from physical damage by:

(1) Locating the fittings so that they aren't likely to be damaged.

(2) Enclosing the fittings in metal, wood, or equivalent protective covering.

Author's Comment: Grounding and bonding fittings can be buried or encased in concrete if installed in accordance with 250.53(G), 250.68(A) Ex. 1, and 250.70.

250.12 Clean Surface. Nonconductive coatings, such as paint, must be removed to ensure good electrical continuity, or the termination fittings must be designed so as to make such removal unnecessary [250.53(A) and 250.96(A)].

Author's Comment: The "tarnish" on copper water pipe need not be removed before making a termination.

PART II. SYSTEM GROUNDING AND BONDING

250.20 Systems Required to be Grounded and Bonded. Alternating-current systems (power supplies) must be grounded and bonded as provided in (A), (B), (C), or (D).

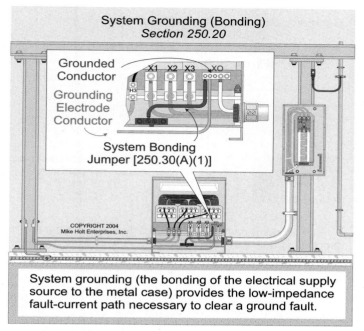

Figure 250–43

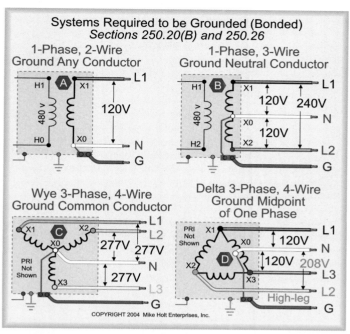

Figure 250–45

Author's Comment: System grounding, the intentional bonding of the electrical supply source to the metal case, provides the low-impedance fault-current path necessary to clear a ground fault. **Figure 250–43**

(A) Alternating-Current Systems Below 50V. Alternating-current systems operating below 50V aren't required to have the system bonded to the metal case unless: **Figure 250–44**

(1) The primary is supplied from a 277V or 480V circuit.

(2) The primary is supplied from an ungrounded power supply.

(3) Where installed as overhead conductors outside of buildings.

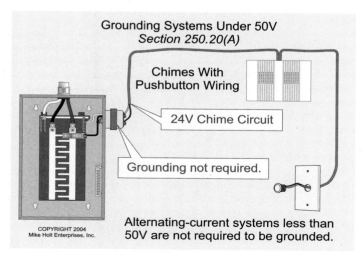

Figure 250–44

(B) Alternating-Current Systems Over 50V. Alternating-current systems over 50V that require a grounded neutral conductor must have the grounded neutral terminal [250.26] of the power supply bonded to the system metal parts in accordance with 250.30(A)(1). Such systems include: **Figure 250–45**

- 2- or 3-wire, single-phase 120V or 120/240V systems
- 4-wire, three-phase wye-connected 120/208V or 277/480V systems
- 4-wire, three-phase delta-connected 120/240V systems

(D) Separately Derived Systems. Separately derived systems that are required to be grounded (bonded) by 250.20(A) or (B), must be grounded and bonded in accordance with 250.30(A).

Author's Comment: A separately derived system is a premises wiring system with no direct electrical connection to conductors originating from another system [Article 100 definition and 250.20(D)]. All transformers except autotransformers are separately derived because the primary circuit conductors do not have any direct electrical connection to the secondary circuit conductors. **Figure 250–46**

Generators that supply a transfer switch that opens the grounded neutral conductor are also considered separately derived. **Figure 250–47**

FPN 1: A generator isn't a separately derived system if the grounded neutral conductor from the generator is solidly connected to the *electrical supply source*.

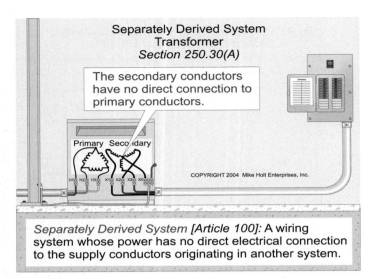

Figure 250–46

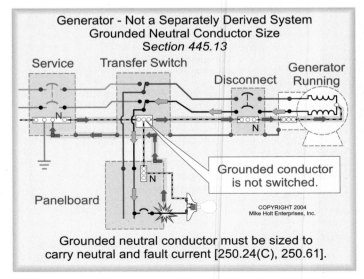

Figure 250–48

Author's Comment: In other words, if the transfer switch doesn't open the grounded neutral conductor, then the generator isn't a separately derived system. Therefore, a *neutral-to-case* bond (system bonding jumper) cannot be made at the generator because it will cause objectionable current to flow on metal parts in violation of 250.6(A). **Figure 250–48**

FPN 2: If a grounded neutral conductor is supplied at a transfer switch, and the transfer switch doesn't open the grounded neutral conductor, then the grounded neutral conductor must be sized:

- To carry fault current back to the generator in accordance with 445.13. **Figure 250–49**
- Not smaller than required to carry the unbalanced load in accordance with 220.61.

(E) Impedance Grounded Neutral Systems. Impedance grounded (bonded) neutral systems must be grounded and bonded in accordance with 250.36.

> **Author's Comment:** The *NEC* refers to the practice of deliberately placing resistance between the system winding and the metal case as an impedance grounded neutral system. **Figure 250–50**

250.24 Grounding and Bonding at Service Equipment.

(A) Grounding. Services supplied from a utility transformer that is grounded to the earth must have the grounded neutral con-

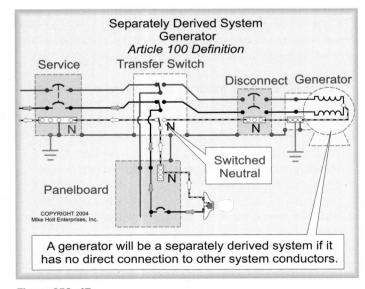

Figure 250–47

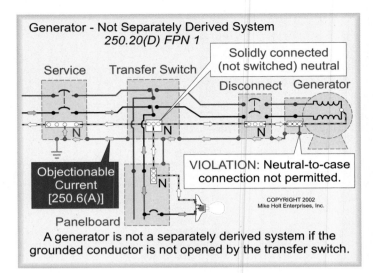

Figure 250–49

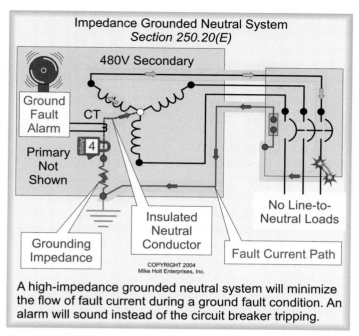

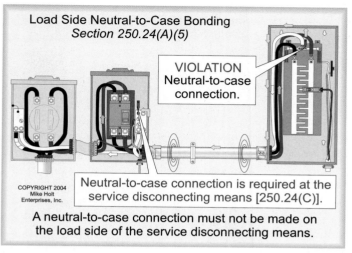

Impedance Grounded Neutral System
Section 250.20(E)

480V Secondary

Ground Fault Alarm

CT

Amps 4

Primary Not Shown

Grounding Impedance

Insulated Neutral Conductor

No Line-to-Neutral Loads

Fault Current Path

COPYRIGHT 2004 Mike Holt Enterprises, Inc.

A high-impedance grounded neutral system will minimize the flow of fault current during a ground fault condition. An alarm will sound instead of the circuit breaker tripping.

Figure 250–50

Load Side Neutral-to-Case Bonding
Section 250.24(A)(5)

VIOLATION Neutral-to-case connection.

COPYRIGHT 2004 Mike Holt Enterprises, Inc.

Neutral-to-case connection is required at the service disconnecting means [250.24(C)].

A neutral-to-case connection must not be made on the load side of the service disconnecting means.

Figure 250–52

ductor grounded to a suitable grounding electrode [250.50] in accordance with the following:

(1) Accessible Location. A grounding electrode conductor must connect the grounded neutral conductor to the grounding electrode and this connection can be made at any accessible location, from the load end of the service drop or service lateral, up to and including the service disconnecting means. **Figure 250–51**

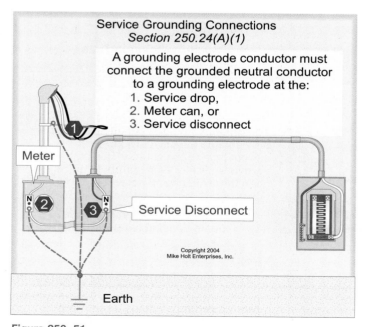

Service Grounding Connections
Section 250.24(A)(1)

A grounding electrode conductor must connect the grounded neutral conductor to a grounding electrode at the:
1. Service drop,
2. Meter can, or
3. Service disconnect

Meter

N 2

N 3

Service Disconnect

Copyright 2004 Mike Holt Enterprises, Inc.

Earth

Figure 250–51

Author's Comment: Some inspectors require the grounding electrode conductor to terminate at the meter enclosure, while other inspectors insist that the grounding electrode conductor terminate at the service disconnect.

The *Code* allows this grounding (earthing) connection to be made at either of these locations.

(4) Main Bonding Jumper. When the grounded neutral conductor is bonded to the service disconnecting means [250.24(B)] by a wire or busbar [250.28], the grounding electrode conductor can terminate to either the grounded neutral terminal or the equipment grounding terminal within the service disconnect.

(5) Load-Side Neutral-to-Case Bonding. A neutral-to-case bond cannot be made on the load side of the service disconnecting means, except as permitted for separately derived systems [250.30(A)(1)] or separate buildings [250.32(B)(2)] in accordance with 250.142. **Figure 250–52**

Author's Comment: If an improper neutral-to-case bond is made on the load side of service equipment, dangerous objectionable current will flow on conductive metal parts of electrical equipment in violation of 250.6(A). Objectionable current on metal parts of electrical equipment can cause electric shock and even death from ventricular fibrillation. **Figure 250–53**

(B) Main Bonding Jumper. An unspliced main bonding jumper complying with 250.28 must be installed between the grounded neutral terminal and the metal parts of the service disconnecting means enclosure in accordance with 250.24(C).

(C) Grounded Neutral Conductor Required. Because electric utilities aren't required to provide an equipment grounding (bonding) conductor to service equipment, a grounded neutral service conductor must be run from the electric utility transformer

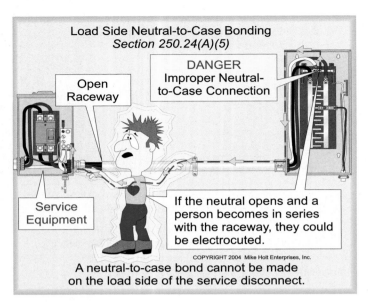

Load Side Neutral-to-Case Bonding
Section 250.24(A)(5)

Open Raceway

DANGER Improper Neutral-to-Case Connection

Service Equipment

If the neutral opens and a person becomes in series with the raceway, they could be electrocuted.

COPYRIGHT 2004 Mike Holt Enterprises, Inc.

A neutral-to-case bond cannot be made on the load side of the service disconnect.

Figure 250–53

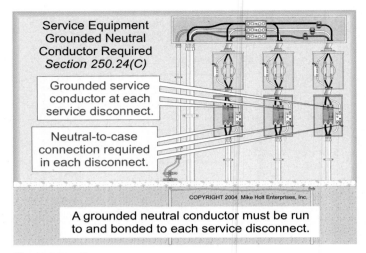

Service Equipment Grounded Neutral Conductor Required
Section 250.24(C)

Grounded service conductor at each service disconnect.

Neutral-to-case connection required in each disconnect.

COPYRIGHT 2004 Mike Holt Enterprises, Inc.

A grounded neutral conductor must be run to and bonded to each service disconnect.

Figure 250–55

to each service disconnecting means and it must be bonded to each service disconnecting means as required by 250.24(B) [250.130(A)]. **Figures 250–54 and 250–55**

Author's Comment: The grounded neutral service conductor provides the effective ground-fault current path to the power source to ensure that dangerous voltage from a ground fault will be quickly removed by opening the circuit-protection device [250.4(A)(3) and 250.4(A)(5)]. **Figure 250–56**

DANGER: *Because the resistance of the earth, when used as a ground-fault current path is so great, often as much as 500 ohms, very little fault current returns to the power source if it is the only fault-current return path. The result—the circuit over-current protection device will not open and clear the ground fault and all metal parts associated with the electrical installation, metal piping, and structural building steel will become and remain energized by circuit voltage.* **Figure 250–57**

Author's Comments:

• A ground-fault cannot be cleared to remove dangerous voltage on the metal parts, metal piping, and structural steel if the service disconnecting means enclosure is not bonded to the grounded neutral service conductor. **Figure 250–58**

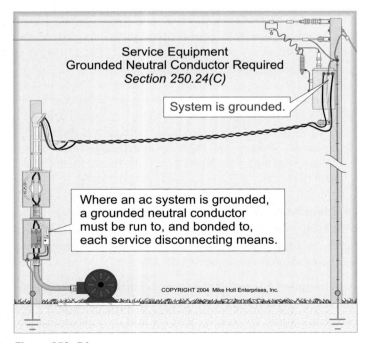

Service Equipment Grounded Neutral Conductor Required
Section 250.24(C)

System is grounded.

Where an ac system is grounded, a grounded neutral conductor must be run to, and bonded to, each service disconnecting means.

COPYRIGHT 2004 Mike Holt Enterprises, Inc.

Figure 250–54

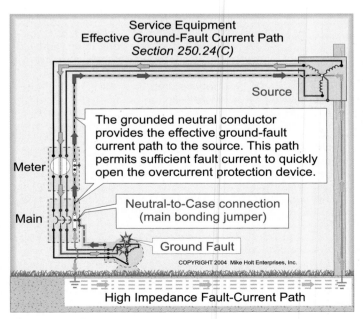

Service Equipment Effective Ground-Fault Current Path
Section 250.24(C)

Source

Meter

The grounded neutral conductor provides the effective ground-fault current path to the source. This path permits sufficient fault current to quickly open the overcurrent protection device.

Neutral-to-Case connection (main bonding jumper)

Main

Ground Fault

COPYRIGHT 2004 Mike Holt Enterprises, Inc.

High Impedance Fault-Current Path

Figure 250–56

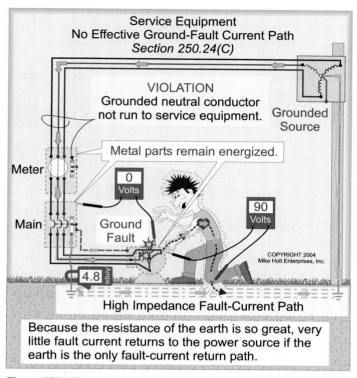

Figure 250–57

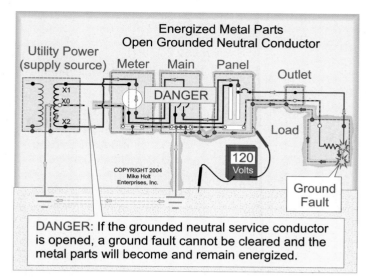

DANGER: If the grounded neutral service conductor is opened, a ground fault cannot be cleared and the metal parts will become and remain energized.

Figure 250–59

• If the grounded neutral service conductor is open, a ground fault cannot be cleared and metal parts will become and remain energized. **Figure 250–59**

In addition, if the grounded neutral conductor is opened, dangerous voltage will be present on metal parts under normal conditions, providing the potential for electric shock. For example: If the earth's ground resistance is 25Ω and the load's resistance is 25Ω, the voltage drop across each of these resistors would be ½ of the voltage source. Since the grounded neutral is bonded to the service disconnect, all metal parts will be elevated to 60V above the earth's potential for a 120/240V system. **Figure 250–60**

This dangerous condition is of particular concern in buildings containing pools, spas, or hot tubs.

To determine the actual voltage on the metal parts from an open grounded neutral service conductor, you need to do some complex math calculations with a spreadsheet to accommodate the variable conditions. Visit www.NECcode.com and go to the Free Stuff link to download a spreadsheet for this purpose.

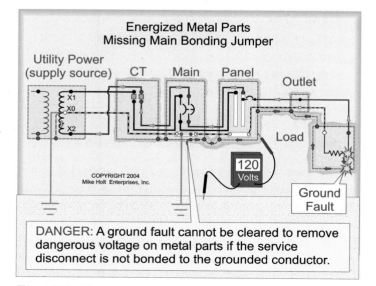

DANGER: A ground fault cannot be cleared to remove dangerous voltage on metal parts if the service disconnect is not bonded to the grounded conductor.

Figure 250–58

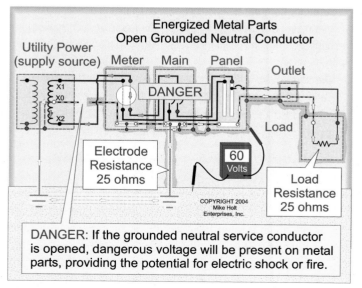

DANGER: If the grounded neutral service conductor is opened, dangerous voltage will be present on metal parts, providing the potential for electric shock or fire.

Figure 250–60

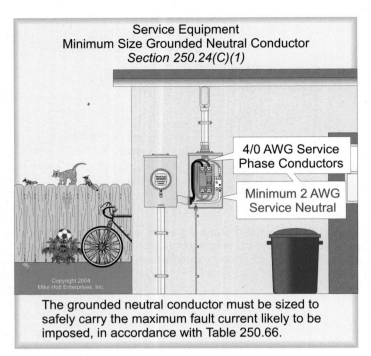

Service Equipment
Minimum Size Grounded Neutral Conductor
Section 250.24(C)(1)

4/0 AWG Service Phase Conductors

Minimum 2 AWG Service Neutral

The grounded neutral conductor must be sized to safely carry the maximum fault current likely to be imposed, in accordance with Table 250.66.

Figure 250–61

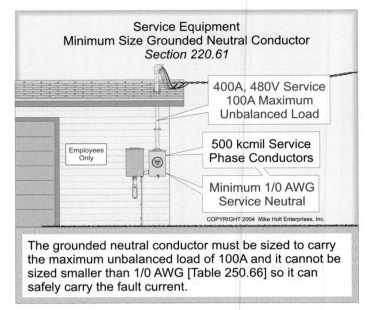

Service Equipment
Minimum Size Grounded Neutral Conductor
Section 220.61

400A, 480V Service
100A Maximum Unbalanced Load

Employees Only

500 kcmil Service Phase Conductors

Minimum 1/0 AWG Service Neutral

COPYRIGHT 2004 Mike Holt Enterprises, Inc.

The grounded neutral conductor must be sized to carry the maximum unbalanced load of 100A and it cannot be sized smaller than 1/0 AWG [Table 250.66] so it can safely carry the fault current.

Figure 250–62

(1) Minimum Size Grounded Neutral Conductor. Because the grounded neutral service conductor is required to serve as the effective ground-fault current path, it must be sized so that it can safely carry the maximum fault current likely to be imposed on it [110.10 and 250.4(A)(5)]. This is accomplished by sizing the grounded neutral conductor in accordance with Table 250.66, based on the total area of the largest ungrounded conductor. Figure 250–61

Author's Comment: In addition, the grounded neutral conductors must have the capacity to carry the maximum unbalanced neutral current in accordance with 220.61.

Question: *What is the minimum size grounded neutral service conductor required for a 480V, three-phase service where the ungrounded service conductors are sized at 500 kcmil and the maximum unbalanced load is 100A?* Figure 250–62

(a) 3 AWG   (b) 2 AWG   (c) 1 AWG   (d) 1/0 AWG

Answer: *(d) 1/0 AWG [Table 250.66]*

The unbalanced load requires a 3 AWG grounded neutral service conductor, which is rated 100A at 75°C in accordance with Table 310.16 [220.61]. However, the grounded neutral service conductor cannot be smaller than 2 AWG in accordance with Table 250.66 to ensure that it will accommodate the maximum fault current likely to be imposed on it.

(2) Parallel Grounded Neutral Conductor. Where service conductors are paralleled, a grounded neutral conductor must be installed in each raceway and it must be sized in accordance with Table 250.66, based on the total area of the largest ungrounded conductor in the raceway. In no case must the grounded neutral conductor in each parallel service raceway be less than 1/0 AWG [310.4].

Author's Comment: In addition, the grounded neutral conductors must have the capacity to carry the maximum unbalanced neutral current in accordance with 220.61.

Question: *What is the minimum size grounded neutral service conductor required for a 480V, three-phase service installed in two raceways where the ungrounded service conductors in each of the raceways are 350 kcmil and the maximum unbalanced load is 100A?* Figure 250–63

(a) 3 AWG   (b) 2 AWG   (c) 1 AWG   (d) 1/0 AWG

Answer: *(d) 1/0 AWG per raceway, Table 250.66 and 310.4*

The unbalanced load only requires a 3 AWG grounded neutral service conductor in accordance with Table 310.16 [220.61]. However, the grounded neutral service conductor in each raceway cannot be smaller than 1 AWG in accordance with Table 250.66 to ensure that it will accommodate the maximum fault current likely to be imposed on it. But ungrounded service conductors run in parallel are not permitted to be sized smaller than 1/0 AWG.

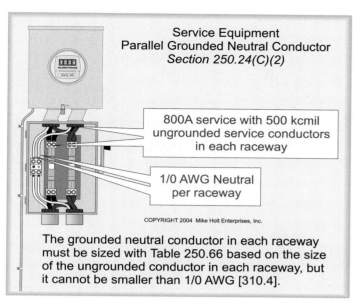

Service Equipment
Parallel Grounded Neutral Conductor
Section 250.24(C)(2)

800A service with 500 kcmil
ungrounded service conductors
in each raceway

1/0 AWG Neutral
per raceway

COPYRIGHT 2004 Mike Holt Enterprises, Inc.

The grounded neutral conductor in each raceway
must be sized with Table 250.66 based on the size
of the ungrounded conductor in each raceway, but
it cannot be smaller than 1/0 AWG [310.4].

Figure 250–63

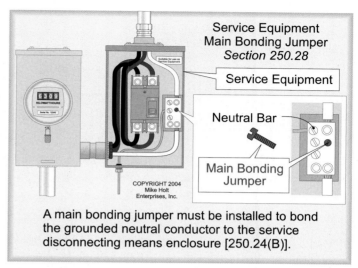

Service Equipment
Main Bonding Jumper
Section 250.28

Service Equipment

Neutral Bar

Main Bonding
Jumper

COPYRIGHT 2004
Mike Holt
Enterprises, Inc.

A main bonding jumper must be installed to bond
the grounded neutral conductor to the service
disconnecting means enclosure [250.24(B)].

Figure 250–64

(D) Grounding Electrode Conductor. A grounding electrode conductor must connect the metal parts of service equipment enclosures to a suitable grounding electrode in accordance with Part III of Article 250.

250.28 Main Bonding Jumper and System Bonding Jumper.

Main Bonding Jumper. At service equipment, a main bonding jumper must be installed to electrically connect the grounded neutral conductor to the metal service disconnecting means enclosure [250.24(B)].

Author's Comment:

- Where equipment is listed for use as service equipment [230.66], the main bonding jumper will be supplied by the equipment manufacturer [408.3(C)]. **Figure 250–64**

- The main bonding jumper provides the low-impedance path for fault current to travel back to the power source to clear a ground fault [250.24(C)]. **Figure 250–65**

DANGER: *Metal parts of the electrical installation, as well as metal piping and structural steel, could become and remain energized with dangerous voltage from a ground fault if a main bonding jumper isn't installed at service equipment.* **Figure 250–66**

System Bonding Jumper. A system bonding jumper is the conductor that is installed between the grounded neutral terminal of a separately derived system and the equipment grounding (bonding) conductor of the secondary system [Article 100 definition and 250.30(A)(1)].

Author's Comment: Some transformers and most generators have the system bonding jumper installed at the factory.

DANGER: *Metal parts of the electrical installation as well as metal piping and structural steel, will remain energized with dangerous voltage from a ground fault if a system bonding jumper isn't installed at a separately derived system.*

(A) Material. The bonding jumper can be a wire, bus, or screw.

(B) Construction. If a bonding jumper is a screw, it must be identified with a green finish that is visible with the screw installed.

(C) Attachment. Main and system bonding jumpers must be attached by exothermic welding, listed pressure connectors of the set screw or compression type, listed clamps, or other listed fittings [250.8].

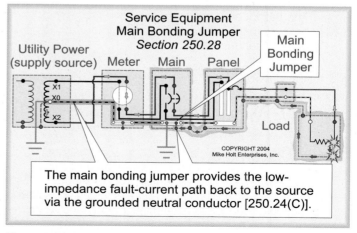

Service Equipment
Main Bonding Jumper
Section 250.28

Utility Power
(supply source)   Meter   Main   Panel

Main
Bonding
Jumper

X1
X0
X2

Load

COPYRIGHT 2004
Mike Holt Enterprises, Inc.

The main bonding jumper provides the low-
impedance fault-current path back to the source
via the grounded neutral conductor [250.24(C)].

Figure 250–65

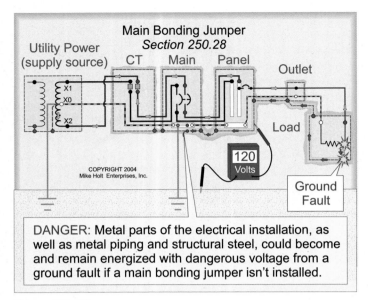

Main Bonding Jumper
Section 250.28

DANGER: Metal parts of the electrical installation, as well as metal piping and structural steel, could become and remain energized with dangerous voltage from a ground fault if a main bonding jumper isn't installed.

Figure 250–66

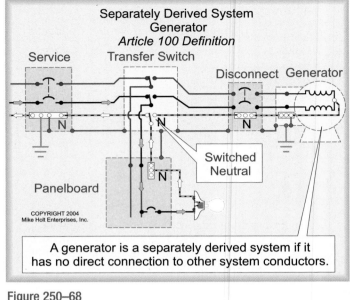

Separately Derived System
Generator
Article 100 Definition

A generator is a separately derived system if it has no direct connection to other system conductors.

Figure 250–68

(D) Size. The main and system bonding jumpers must not be smaller than the sizes shown in Table 250.66, based on the area of the largest ungrounded conductor. Where the service-entrance ungrounded conductors are larger than 1,100 kcmil copper or 1,750 kcmil aluminum, the bonding jumper must have an area that isn't less than 12½ percent of the area of the largest ungrounded conductor, except that, where the ungrounded conductors and the bonding jumper are of different materials (copper or aluminum), the minimum size of the bonding jumper must be based on the assumed use of ungrounded conductors of the same material as the bonding jumper and with an ampacity equivalent to that of the installed ungrounded conductors.

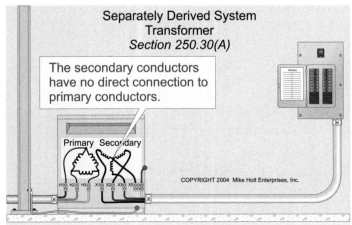

Separately Derived System
Transformer
Section 250.30(A)

The secondary conductors have no direct connection to primary conductors.

Separately Derived System [Article 100]: A wiring system whose power has no direct electrical connection to the supply conductors originating in another system.

Figure 250–67

250.30 Grounding and Bonding of Separately Derived AC Systems.

Author's Comment: A separately derived system is a premises wiring system with no direct electrical connection to conductors originating from another system [Article 100 definition and 250.20(D)].

All transformers, except autotransformers, are separately derived because the primary circuit conductors do not have any direct electrical connection to the secondary circuit conductors. **Figure 250–67**

Generators that supply a transfer switch that opens the grounded neutral conductor would be considered separately derived [250.20(D) FPN 1]. **Figure 250–68**

(A) Grounded Systems. Separately derived systems must be system bonded and grounded in accordance with the following:

A neutral-to-case bond must not be on the load side of the system bonding jumper, except as permitted by 250.142(B).

(1) System Bonding Jumper. Bonding the metal parts of the separately derived system to the secondary grounded neutral terminal by the installation of a system bonding jumper ensures that dangerous voltage from a secondary ground fault can be quickly removed by opening the secondary circuit's overcurrent protection device [250.2(A)(3)]. **Figure 250–69**

DANGER: *During a ground fault, metal parts of electrical equipment, as well as metal piping and structural steel, will become and remain energized providing the potential for electric shock and fire if the system bonding jumper is not installed.* **Figure 250–70**

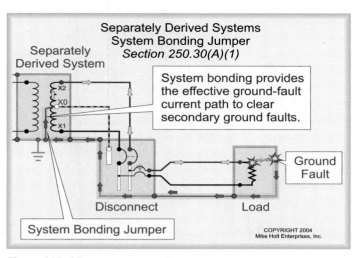

Figure 250–69

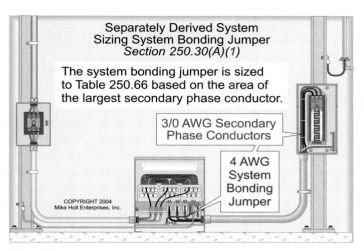

Figure 250–71

The system bonding jumper must be sized in accordance with Table 250.66, based on the area of the largest ungrounded secondary conductor [250.28(D)].

Question: *What size system bonding jumper is required for a 45 kVA transformer, where the secondary conductors are 3/0 AWG?* **Figure 250–71**

(a) 4 AWG   (b) 3 AWG   (c) 2 AWG   (d) 1 AWG

Answer: *(a) 4 AWG, Table 250.66*

The system bonding jumper can be installed at the separately derived system, the first system disconnecting means, or any point in between the separately derived system and the first disconnecting means, but not at both locations. **Figure 250–72**

In addition, the system bonding jumper must be installed at the same location where the grounding electrode conductor terminates to the grounded neutral terminal of the separately derived system, which can be at the separately derived system, the first system disconnecting means, or any point in between, but not at more than one location [250.30(A)(3)]. **Figure 250–73**

CAUTION: *Dangerous objectionable current will flow on conductive metal parts of electrical equipment as well as metal piping and structural steel, in violation of 250.6(A), if the system bonding jumper is not located where the grounding electrode conductor terminates to the grounded neutral conductor.*

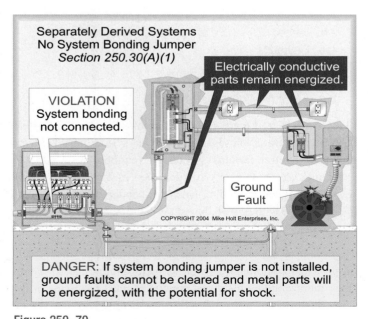

Figure 250–70

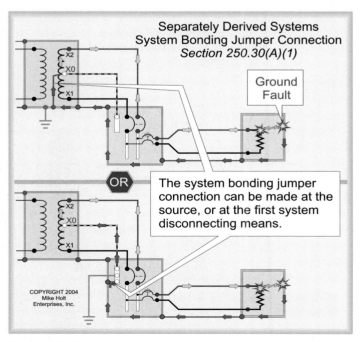

Figure 250–72

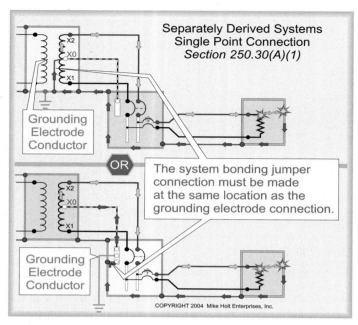

Figure 250–73

Exception 2: *A system bonding jumper can be installed at both the separately derived system and the secondary system disconnecting means where doing so doesn't establish a parallel path for neutral current.*

CAUTION: *Dangerous objectionable current will flow on conductive metal parts of electrical equipment as well as metal piping and structural steel, in violation of 250.6(A), if the*

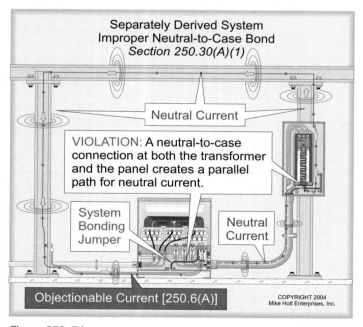

Figure 250–74

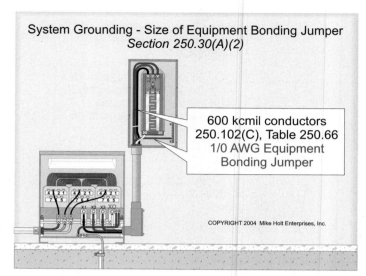

Figure 250–75

system bonding jumper is installed at the separately derived system and the secondary system disconnecting means. Figure 250–74

Author's Comment: For all practical purposes, this isn't possible except in a wood frame building that doesn't have any conductive metal parts.

(2) Equipment Bonding Jumper Size. Where an equipment bonding jumper is run to the secondary system disconnecting means, it must be sized in accordance with Table 250.66, based on the area of the largest ungrounded secondary conductor.

Question: What size equipment bonding jumper is required for a nonmetallic raceway containing 500 kcmil secondary conductors? Figure 250–75

(a) 1 AWG   (b) 1/0 AWG   (c) 2/0 AWG   (d) 3/0 AWG

Answer: (b) 1/0 AWG, Table 250.66

(3) Grounding Electrode Conductor, Single Separately Derived System. Each separately derived system must have the grounded neutral terminal grounded (earthed) to a suitable grounding electrode of a type identified in 250.30(A)(7). The secondary system grounding electrode conductor must be sized in accordance with 250.66, based on the total area of the largest ungrounded secondary conductor. Figure 250–76

To prevent objectionable current from flowing onto metal parts of electrical equipment, as well as metal piping and structural steel, the grounding electrode conductor must terminate at the same point on the separately derived system where the system bonding jumper is installed.

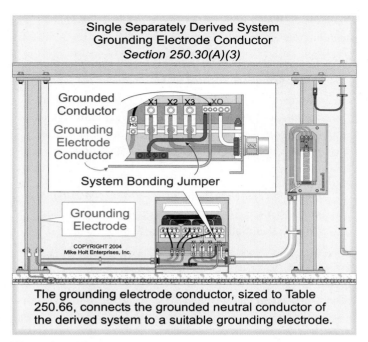

The grounding electrode conductor, sized to Table 250.66, connects the grounded neutral conductor of the derived system to a suitable grounding electrode.

Figure 250–76

Exception 1: Where the system bonding jumper [250.30(A)(1)] is a wire or busbar, the grounding electrode conductor can terminate to the equipment grounding terminal, bar, or bus on the metal enclosure of the separately derived system. Figure 250–77

Exception 3: Separately derived systems rated 1 kVA (1,000 VA) or less are not required to be grounded (earthed); however, to ensure ground faults can be cleared, a system bonding jumper must be installed in accordance with 250.30(A)(1).

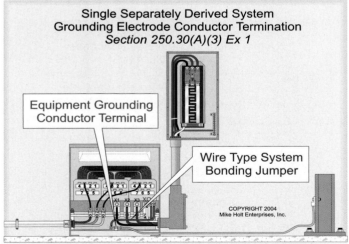

Where the system bonding jumper is a wire or busbar, the grounding electrode conductor can terminate to the equipment grounding terminal of the derived system.

Figure 250–77

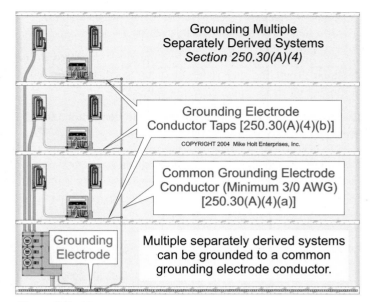

Figure 250–78

(4) Grounding Electrode Conductor, Multiple Separately Derived Systems. Where there are multiple separately derived systems, the grounded neutral terminal of each derived system can be grounded (earthed) to a common grounding electrode conductor. The grounding electrode conductor and grounding electrode tap must comply with (a) through (c). **Figure 250–78**

Exception 1: Where the system bonding jumper [250.30(A)(1)] is a wire or busbar, the grounding electrode tap can terminate to the equipment grounding terminal, bar, or bus on the metal enclosure of the separately derived system.

Exception 2: Separately derived systems rated 1 kVA (1,000 VA) or less are not required to be grounded (earthed); however, to ensure ground faults can be cleared, a system bonding jumper must be installed in accordance with 250.30(A)(1).

(a) Common Grounding Electrode Conductor Size. The common grounding electrode conductor must not be smaller than 3/0 AWG copper or 250 kcmil aluminum.

(b) Tap Conductor Size. Each grounding electrode tap must be sized in accordance with 250.66, based on the largest separately derived ungrounded conductor of the separately derived system.

(c) Connections. All grounding electrode tap connections must be made at an accessible location by:

(1) Listed connector.

(2) Listed connections to aluminum or copper busbars not less than ¼ in. x 2 in. Where aluminum busbars are used, the installation must comply with 250.64(A).

(3) By the exothermic welding process.

Author's Comment: See Article 100 for the definition of "Accessible" as it applies to wiring methods.

Grounding electrode tap conductors must be connected to the common grounding electrode conductor so that the common grounding electrode conductor isn't spliced.

(5) Installation. The grounding electrode conductor must be installed in accordance with 250.64.

Author's Comment: The grounding electrode conductor must comply with the following:

- Be of copper where within 18 in. of earth [250.64(A)].
- Securely fastened to the surface on which it's carried [250.64(B)].
- Adequately protected if exposed to physical damage [250.64(B)].
- Metal enclosures enclosing a grounding electrode conductor must be made electrically continuous from the point of attachment to cabinets or equipment to the grounding electrode [250.64(E)].

(6) Bonding. To ensure that dangerous voltage from a ground fault is removed quickly, structural metal and metal piping in the area served by a separately derived system must be bonded to the grounded neutral conductor at the separately derived system in accordance with 250.104(D).

(7) Grounding (Earthing) Electrode. The grounding electrode conductor must terminate to a grounding electrode that is located as close as possible, and preferably in the same area as, the system bonding jumper. The grounding electrode must be the nearest one of the following: **Figure 250–79**

(1) Metal water pipe electrode as specified in 250.52(A)(1).

(2) Structural metal electrode as specified in 250.52(A)(2).

Exception 1: Where none of the electrodes listed in (1) or (2) is available, one of the following is permitted:

- *Concrete-encased electrode encased by not less than 2 in. of concrete, located within and near the bottom of a concrete foundation or footing that is in direct contact with earth, consisting of not less than 20 ft of electrically conductive steel reinforcing bars or rods not less than ½ in. in diameter [250.52(A)(3)].*

- *A ground ring encircling the building or structure, buried not less than 30 in. below grade, consisting of not less than 20 ft of bare copper conductor not smaller than 2 AWG [250.52(A)(4) and 250.53(F)].*

- *A ground rod having not less than 8 ft of contact with the soil [250.52(A)(5) and 250.53(G)].*

- *Other metal underground systems, piping systems, or underground tanks [250.52(A)(7)].*

FPN: To ensure that dangerous voltage from a ground fault is quickly removed, metal water piping (including structural metal) in the area served by a separately derived system must be bonded to the grounded neutral conductor at the separately derived system in accordance with 250.104(D).

Author's Comment: This FPN makes no sense, since the requirement is contained in 250.30(A)(6).

(8) Grounded Neutral Conductor. Where the system bonding jumper is installed at the secondary system disconnecting means instead of at the source of the separately derived system, the following requirements apply: **Figure 250–80**

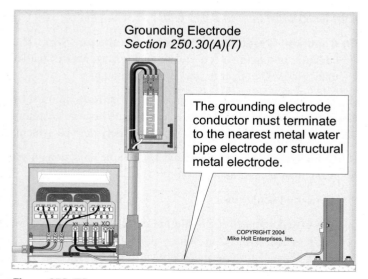

Grounding Electrode
Section 250.30(A)(7)

The grounding electrode conductor must terminate to the nearest metal water pipe electrode or structural metal electrode.

COPYRIGHT 2004
Mike Holt Enterprises, Inc.

Figure 250–79

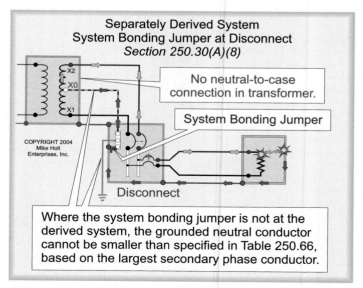

Separately Derived System
System Bonding Jumper at Disconnect
Section 250.30(A)(8)

No neutral-to-case connection in transformer.

System Bonding Jumper

COPYRIGHT 2004
Mike Holt
Enterprises, Inc.

Disconnect

Where the system bonding jumper is not at the derived system, the grounded neutral conductor cannot be smaller than specified in Table 250.66, based on the largest secondary phase conductor.

Figure 250–80

(a) Routing and Sizing. Because the grounded neutral conductor serves as the effective ground-fault current path, the grounded neutral conductor must be routed with the secondary conductors, and it must be sized not smaller than specified in Table 250.66, based on the largest ungrounded conductor for the separately derived system.

(b) Parallel Conductors. If the secondary conductors are installed in parallel, the grounded neutral secondary conductor in each raceway or cable must be sized based on the area of the largest ungrounded secondary conductor in the raceway. But the grounded neutral secondary conductor can not be smaller than 1/0 AWG [310.4].

250.32 Buildings or Structures Supplied by a Feeder or Branch Circuit.

(A) Grounding Electrode. To provide a path to earth for lightning, each building or structure must have its disconnecting means [225.31] grounded (earthed) to one of the following electrodes [250.50 and 250.52(A)]:

- Underground metal water pipe [250.52(A)(1)]
- Metal frame of the building or structure [250.52(A)(2)]
- Concrete-encased steel [250.52(A)(3)]
- Ground ring [250.52(A)(4)]

Author's Comment: See Article 100 for the definitions of "Building" and "Structure."

Where none of the above grounding electrodes are available at a building or structure, then one or more of the following must be used:

- Ground rod [250.52(A)(5)]
- Metal underground systems [250.52(A)(7)]

Author's Comment: Grounding the building or structure disconnecting means to the earth:

- Is intended to limit elevated voltages on the metal parts from lightning [250.4(A)(1)]. **Figure 250–81**
- It doesn't serve as a low-impedance fault-current path to clear ground faults. In fact, the *Code* prohibits the use of the earth as the sole return path since it's such a poor conductor of current [250.4(A)(5) and 250.4(B)(4)].
- It doesn't protect electrical or electronic equipment from lightning voltage transients.

Exception: A grounding electrode isn't required where only one branch circuit serves the building or structure. For the purpose of this section, a multiwire branch circuit is considered to be a single branch circuit. **Figure 250–82**

(B) Bonding Requirements. To quickly clear a ground fault and remove dangerous voltage from metal parts, the building or structure disconnecting means must be grounded (bonded) to an effective ground-fault current path in accordance with (1) or (2) [250.4(A)(3)]. **Figure 250–83**

(1) Equipment Grounding (Bonding) Conductor. The building or structure disconnecting means can be bonded to an equipment grounding (bonding) conductor, as described in 250.118, installed with the feeder conductors. **Figure 250–84**

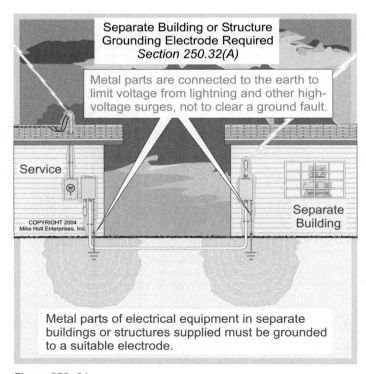

Separate Building or Structure Grounding Electrode Required
Section 250.32(A)

Metal parts are connected to the earth to limit voltage from lightning and other high-voltage surges, not to clear a ground fault.

Service

COPYRIGHT 2004
Mike Holt Enterprises, Inc.

Separate Building

Metal parts of electrical equipment in separate buildings or structures supplied must be grounded to a suitable electrode.

Figure 250–81

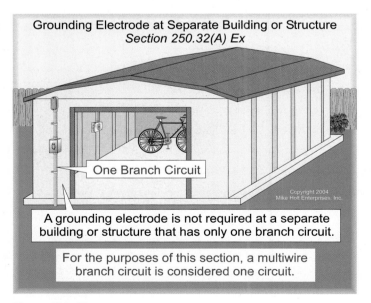

Grounding Electrode at Separate Building or Structure
Section 250.32(A) Ex

One Branch Circuit

Copyright 2004
Mike Holt Enterprises, Inc.

A grounding electrode is not required at a separate building or structure that has only one branch circuit.

For the purposes of this section, a multiwire branch circuit is considered one circuit.

Figure 250–82

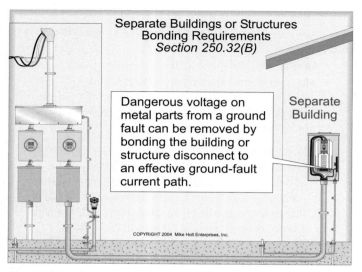

Figure 250-83

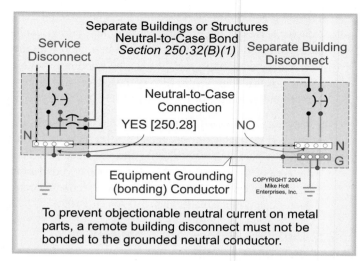

Figure 250-85

The equipment grounding (bonding) conductor, if of the wire type, must be sized in accordance with 250.122, based on the rating of the feeder protection device.

> **CAUTION:** *To prevent dangerous objectionable current from flowing onto metal parts of the electrical installation, as well as metal piping and structural steel [250.6(A)], a building or structure disconnecting means supplied by a feeder must not have the grounded neutral conductor bonded to the building or structure disconnecting means.* **Figures 250-85** *and* **250-86**

(2) Grounded Neutral Conductor. When an equipment grounding (bonding) conductor is not run to the building or structure disconnecting means, the building or structure disconnecting means can be bonded to a grounded neutral conductor installed with the feeder conductors. This is only permitted where there's no continuous metallic path between buildings and structures, and ground-fault protection of equipment isn't installed on the supply side of the feeder.

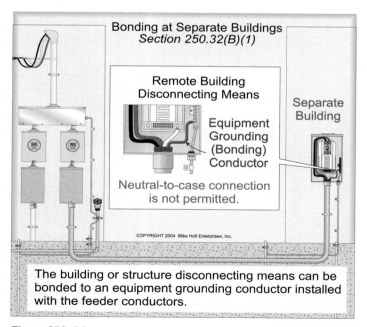

Figure 250-84

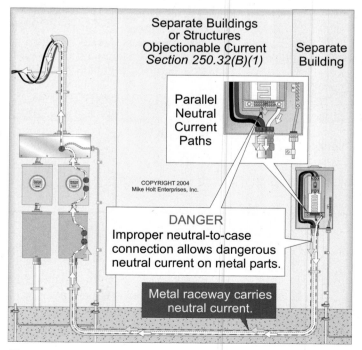

Figure 250-86

Where the grounded neutral feeder conductor serves as the effective ground-fault current path, it must be sized no smaller than the larger of:

(1) The maximum unbalanced neutral load in accordance with 220.61.

(2) The available fault current in accordance with 250.122.

> **CAUTION:** *Using the grounded neutral conductor as the effective ground-fault current path poses potentially dangerous consequences and should only be done after careful consideration. Even if the initial installation doesn't result in dangerous objectionable current on metal parts, there remains the possibility that a future installation of metal piping or cables between the buildings or structures could create unwanted parallel neutral current paths.*

> **Author's Comment:** The preferred practice (or at least my preferred practice) is to not use the grounded neutral conductor as the effective ground-fault current path, but to install an equipment grounding (bonding) conductor with the feeder conductors to the building or structure in accordance with 250.32(B)(1).

(E) Grounding Electrode Conductor. The grounding electrode conductor for a separate building or structure disconnecting means must terminate to the grounding terminal of the disconnecting means and it must be sized in accordance with 250.66, based on the largest ungrounded feeder conductor.

> **Question:** *What size grounding electrode conductor is required for a building disconnect that is supplied with 3/0 AWG?* **Figure 250–87**
>
> *(a) 4 AWG   (b) 3 AWG   (c) 2 AWG   (d) 1 AWG*
>
> **Answer:** *(a) 4 AWG, Table 250.66*

> **Author's Comment:** Where the grounding electrode conductor is connected to a ground rod, that portion of the conductor that is the sole connection to the ground rod isn't required to be larger than 6 AWG copper [250.66(A)]. Where the grounding electrode conductor is connected to a concrete-encased electrode, that portion of the conductor that is the sole connection to the concrete-encased electrode isn't required to be larger than 4 AWG copper [250.66(B)].

250.34 Generators—Portable and Vehicle-Mounted.

(A) Portable Generators. The frame of a portable generator isn't required to be grounded to the earth if: **Figure 250–88**

(1) The generator only supplies equipment or receptacles mounted on the generator, and

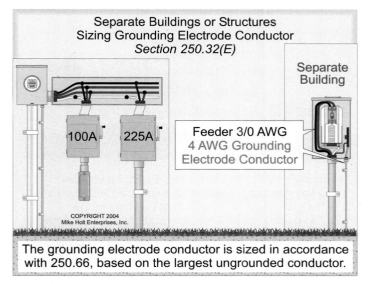

The grounding electrode conductor is sized in accordance with 250.66, based on the largest ungrounded conductor.

Figure 250–87

(2) The metal parts of the generator and the receptacle grounding terminal are bonded to the generator frame.

(B) Vehicle-Mounted Generators. The frame of a vehicle-mounted generator isn't required to be grounded to the earth if: **Figure 250–89**

(1) The generator frame is bonded to the vehicle frame,

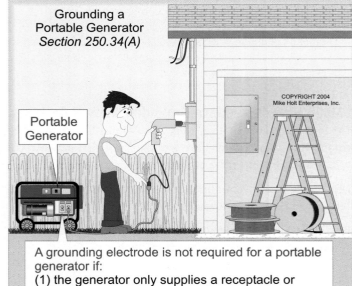

A grounding electrode is not required for a portable generator if:
(1) the generator only supplies a receptacle or equipment mounted on the generator, and
(2) the metal parts of the generator and receptacle terminals for the equipment grounding (bonding) conductor are bonded to the generator frame.

Figure 250–88

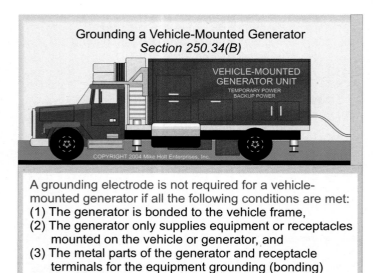

Grounding a Vehicle-Mounted Generator
Section 250.34(B)

A grounding electrode is not required for a vehicle-mounted generator if all the following conditions are met:
(1) The generator is bonded to the vehicle frame,
(2) The generator only supplies equipment or receptacles mounted on the vehicle or generator, and
(3) The metal parts of the generator and receptacle terminals for the equipment grounding (bonding) conductor are bonded to the generator frame.

Figure 250–89

(2) The generator only supplies equipment or receptacles mounted on the vehicle or generator, and

(3) The metal parts of the generator and the receptacle grounding terminal are bonded to the generator frame.

(C) Grounded Neutral Conductor Bonding. If the portable generator is a separately derived system (transfer switch opens the grounded neutral conductor), then the portable generator must be grounded and bonded in accordance with 250.30.

> **Author's Comment:** When a generator provides the sole power for a building or structure, it's a separately derived system even though no transfer switch is present.

250.36 High-Impedance Grounded Neutral Systems.

To limit fault current to a very low value, high-impedance grounded neutral systems have a resistor installed between the system and the metal parts of the system. **Figure 250–90**

This is only permitted for three-phase ac systems (typically rated 480V) where all the following conditions are met:

(1) Conditions of maintenance and supervision ensure that only qualified persons service the installation.

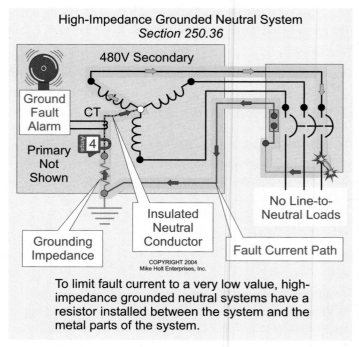

High-Impedance Grounded Neutral System
Section 250.36

To limit fault current to a very low value, high-impedance grounded neutral systems have a resistor installed between the system and the metal parts of the system.

Figure 250–90

(2) Continuity of power is required.

(3) Ground detectors are installed on the system.

(4) Line-to-neutral loads aren't served.

> **FPN:** The system bonding resistor is selected to limit fault current to a value slightly greater than or equal to the capacitive charging current of the system. This value of impedance will also limit transient overvoltages to safe values. For guidance, refer to criteria for limiting transient overvoltages in ANSI/IEEE 142, *Recommended Practice for Grounding of Industrial and Commercial Power Systems* (Green Book).

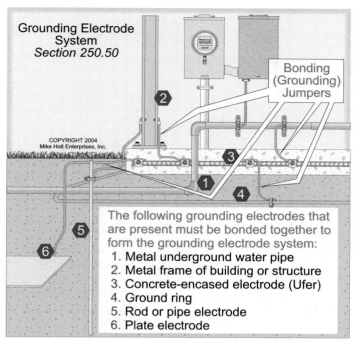

Figure 250–91

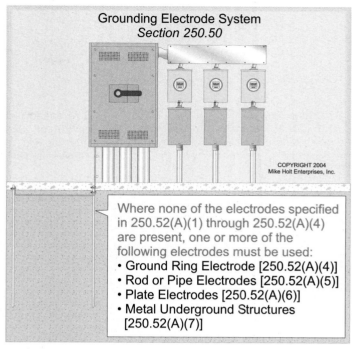

Figure 250–92

PART III. GROUNDING ELECTRODE SYSTEM AND GROUNDING ELECTRODE CONDUCTOR

Grounding metal parts of the electrical installation to earth is intended to reduce voltage on the metal parts from lightning so as to prevent fires from surface arcing (side-flashes) within the building or structure [250.4(A)(2)].

250.50 Grounding Electrode System. All grounding electrodes as described in 250.52(A)(1) through (A)(6) that are present at each building or structure must be bonded together to form the grounding electrode (earthing) system. **Figure 250–91**

- Underground metal water pipe [250.52(A)(1)]
- Metal frame of the building or structure [250.52(A)(2)]
- Concrete-encased steel [250.52(A)(3)]
- Ground ring [250.52(A)(4)]
- Ground rod [250.52(A)(5)]
- Grounding plate [250.52(A)(6)]

Exception: Concrete-encased electrodes are not required for existing buildings or structures where the conductive steel reinforcing bars aren't accessible without disturbing the concrete.

Where an underground metal water pipe electrode, metal building or structure frame electrode, or concrete-encased electrode is not present, one or more of the following electrodes specified in 250.52(A)(4) through (A)(7) must be installed to create the grounding electrode (earthing) system. **Figure 250–92**

- Ground rod [250.52(A)(5)]
- Grounding plate [250.52(A)(6)]
- Metal underground systems [250.52(A)(7)]

250.52 Grounding (Earthing) Electrodes.

(A) Electrodes Permitted for Grounding.

(1) Underground Metal Water Pipe Electrode. Underground metal water pipe in direct contact with earth for 10 ft or more can serve as a grounding electrode. **Figure 250–93**

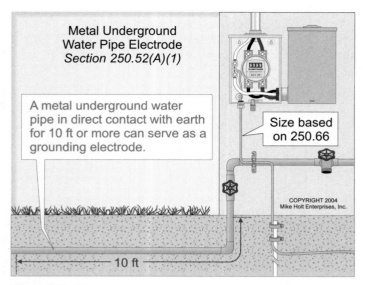

Metal Underground
Water Pipe Electrode
Section 250.52(A)(1)

A metal underground water pipe in direct contact with earth for 10 ft or more can serve as a grounding electrode.

Size based on 250.66

COPYRIGHT 2004
Mike Holt Enterprises, Inc.

|← 10 ft →|

Figure 250–93

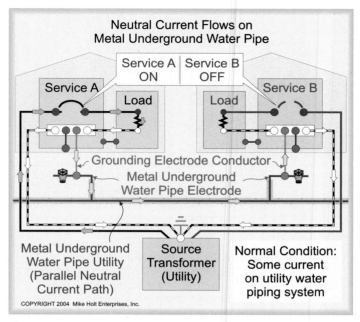

Neutral Current Flows on
Metal Underground Water Pipe

Service A ON | Service B OFF

Service A | Load | Load | Service B

Grounding Electrode Conductor
Metal Underground Water Pipe Electrode

Metal Underground
Water Pipe Utility
(Parallel Neutral
Current Path)

Source
Transformer
(Utility)

Normal Condition:
Some current
on utility water
piping system

COPYRIGHT 2004 Mike Holt Enterprises, Inc.

Figure 250–94

Author's Comment: The grounding electrode conductor to the water pipe electrode must be sized in accordance with Table 250.66.

If the underground metal water pipe electrode is interrupted, such as with a water meter, it must be made electrically continuous with a bonding jumper sized according to 250.66 before it can serve as a grounding (earthing) electrode [250.68(B)].

Interior metal water piping located more than 5 ft from the point of entrance to the building or structure cannot be used to interconnect electrodes that are part of the grounding electrode (earthing) system.

Exception: In industrial and commercial buildings where conditions of maintenance and supervision ensure that only qualified persons service the installation, the entire length of the metal water pipe can be used for the grounding if it is exposed, provided the entire length, other than short sections passing through walls, floors, or ceilings, is exposed.

Author's Comment: Controversy about using the metal underground water supply piping as a grounding electrode has existed since the early 1900s. The water supply industry feels that neutral current flowing on the metal water pipe system corrodes the metal. For more information, contact the American Water Works Association about their report "Effects of Electrical Grounding on Pipe Integrity and Shock Hazard," Catalog No. 90702 1-800-926-7337. **Figure 250–94**

(2) Metal Frame of the Building or Structure Electrode. The metal frame of the building or structure can serve as a grounding electrode, where any of the following methods exist:

(a) 10 ft or more of a single structural metal member is in direct contact with the earth or encased in concrete that is in direct contact with the earth.

(b) The structural metal is bonded to an electrode as defined in 250.52(A)(1), (3), or (4). **Figure 250–95**

(c) The structural metal is bonded to two ground rods if the ground resistance of a single ground rod exceeds 25 ohms [250.52(A)(5) and 250.56].

(d) Other means approved by the authority having jurisdiction.

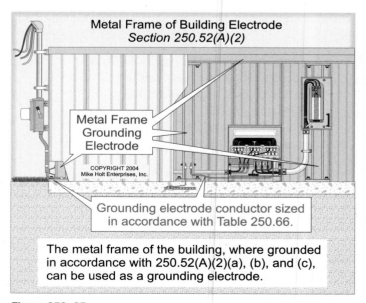

Metal Frame of Building Electrode
Section 250.52(A)(2)

Metal Frame
Grounding
Electrode

COPYRIGHT 2004
Mike Holt Enterprises, Inc.

Grounding electrode conductor sized in accordance with Table 250.66.

The metal frame of the building, where grounded in accordance with 250.52(A)(2)(a), (b), and (c), can be used as a grounding electrode.

Figure 250–95

Author's Comments:

- The intent is that where structural metal is to be used as an electrode, it must be of substantial cross-sectional area.

- The grounding electrode conductor to the metal frame of a building or structure must be sized in accordance with Table 250.66.

(3) Concrete-Encased Grounding Electrode (Ufer). Electrically conductive steel reinforcing bars not smaller than ½ in. in diameter or 4 AWG copper conductor can serve as a grounding electrode if the steel or copper conductor:

- Has a total conductive length of 20 ft,
- Is encased in not less than 2 in. of concrete, and
- Is located near the bottom of a foundation or footer that is in direct contact with earth.

The steel rebar isn't required to be one continuous length and the usual steel tie wires can be used to conductively tie multiple sections together to create a 20 ft concrete-encased grounding electrode. **Figure 250–96**

Author's Comments:

- The grounding electrode conductor required for a concrete-encased grounding electrode isn't required to be larger than 4 AWG copper [250.66(B)].

- The concrete-encased grounding electrode is also called a "Ufer Ground," named after Herb Ufer, the person who determined its usefulness as a grounding electrode in the 1960s. This type of grounding electrode generally offers the lowest ground resistance for the cost and it's the grounding electrode of choice for many where new concrete foundations are available.

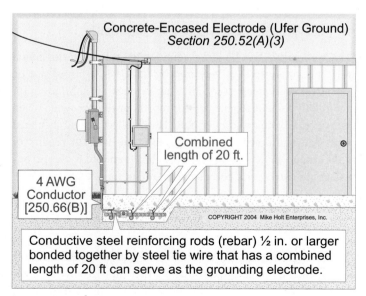

Concrete-Encased Electrode (Ufer Ground)
Section 250.52(A)(3)

Combined length of 20 ft.

4 AWG Conductor [250.66(B)]

COPYRIGHT 2004 Mike Holt Enterprises, Inc.

Conductive steel reinforcing rods (rebar) ½ in. or larger bonded together by steel tie wire that has a combined length of 20 ft can serve as the grounding electrode.

Figure 250–96

(4) Ground Ring Electrode. A ground ring encircling a building or structure, in direct contact with earth consisting of not less than 20 ft of bare copper conductor not smaller than 2 AWG copper, can serve as a grounding (earthing) electrode.

Author's Comment: The ground ring must be buried at a depth below the earth's surface of not less than 30 in. [250.53(F)]. The grounding electrode conductor for the ground ring isn't required to be larger than the conductor used for the ground ring [250.66(C)].

(5) Ground Rod Electrodes. Ground rod electrodes must not be less than 8 ft long and must have not less than 8 ft of length in contact with the soil [250.53(G)].

(b) Rod. Unlisted ground rods must have a diameter of at least ⅝ in. Nonferrous ground rods smaller than ⅝ in. must be <u>listed</u> and must not be less than ½ in. in diameter. **Figure 250–97**

Author's Comment:

- The grounding electrode conductor that is the sole connection to a ground rod isn't required to be larger than 6 AWG copper [250.66(A)].

- The diameter of a ground rod has an insignificant effect on the ground resistance of the ground rod. However, larger diameter ground rods (¾ in. and 1 in.) are sometimes installed where mechanical strength is required or where necessary to compensate for the loss of the electrode's metal due to corrosion.

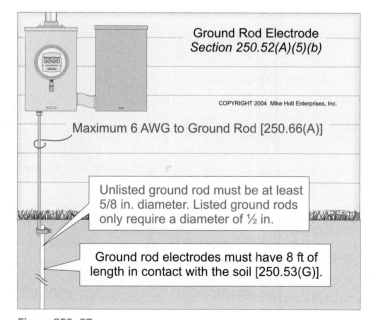

Ground Rod Electrode
Section 250.52(A)(5)(b)

COPYRIGHT 2004 Mike Holt Enterprises, Inc.

Maximum 6 AWG to Ground Rod [250.66(A)]

Unlisted ground rod must be at least 5/8 in. diameter. Listed ground rods only require a diameter of ½ in.

Ground rod electrodes must have 8 ft of length in contact with the soil [250.53(G)].

Figure 250–97

(6) Ground Plate Electrode. A buried iron or steel plate with not less than ¼ in. of thickness, or a nonferrous (copper) metal plate not less than 0.06 in. of thickness, with an exposed surface area not less than 2 sq ft can be used as a grounding electrode.

> **Author's Comment:** The grounding electrode conductor that is the sole connection to a ground plate electrode isn't required to be larger than 6 AWG copper [250.66(A)].

(7) Metal Underground Systems Electrode. Metal underground systems such as piping systems, underground tanks, or an underground metal well casing that isn't effectively bonded to a metal water pipe system, can be used as a grounding electrode.

> **Author's Comment:** The grounding electrode conductor to the metal underground systems must be sized in accordance with Table 250.66.

(B) Electrodes Not Permitted.

(1) Underground Metal Gas Piping System. Underground metal gas piping systems and structures cannot be used as a grounding electrode. **Figure 250–98**

> **FPN:** See 250.104(B) for the bonding requirements for gas piping.

> **Author's Comment:** According to 250.104(B), metal gas piping that is likely to become energized must be bonded to the service equipment enclosure, the grounded neutral service conductor, or the grounding electrode or grounding electrode conductor where the grounding electrode conductor is of sufficient size [250.104(B)]. The equipment grounding (bonding) conductor for

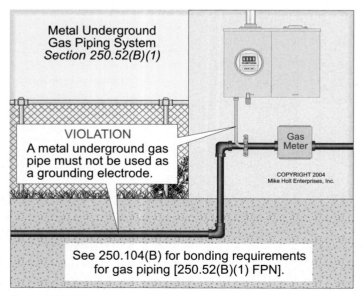

Metal Underground
Gas Piping System
Section 250.52(B)(1)

VIOLATION
A metal underground gas pipe must not be used as a grounding electrode.

Gas
Meter

COPYRIGHT 2004
Mike Holt Enterprises, Inc.

See 250.104(B) for bonding requirements
for gas piping [250.52(B)(1) FPN].

Figure 250–98

the circuit that may energize the piping can serve as the bonding means. So effectively, this means that no action is actually required by the electrical installer!

(2) Aluminum Electrodes. Aluminum cannot be used as a grounding electrode because it corrodes more quickly than copper.

250.53 Installation of Grounding Electrode System.

(A) Ground Rod Electrodes. Where practicable, ground rods must be embedded below permanent moisture level and must be free from nonconductive coatings such as paint or enamel [250.12].

> **Author's Comment:** See 250.53(G) for additional details.

(B) Electrode Spacing. Where more than one grounding electrode system exists at a building or structure, they must be separated by at least 6 ft.

(C) Grounding Electrode Bonding Jumper. Where within 18 in. of earth, the conductor used to bond grounding electrodes together to form the grounding electrode system must be copper [250.64(A)], securely fastened to the surface on which it's carried, and be protected if exposed to physical damage [250.64(B)]. The bonding jumper to each electrode must be sized in accordance with 250.66.

In addition, the grounding electrode bonding jumpers must terminate to the grounding electrode by exothermic welding, listed lugs, listed pressure connectors, listed clamps, or other listed means [250.8]. When the termination is encased in concrete or buried, the termination fittings must be listed and identified for this purpose [250.70].

(D) Underground Metal Water Pipe Electrode.

(1) Continuity. The bonding connection to the interior metal water piping system, as required by 250.104(A), must not be dependent on water meters, filtering devices, or similar equipment likely to be disconnected for repairs or replacement. When necessary, a bonding jumper must be installed around insulated joints and equipment likely to be disconnected for repairs or replacement to assist in clearing and removing dangerous voltage on metal parts because of a ground fault. **Figure 250–99**

> **Author's Comment:** See 250.68(B) and 250.104 for additional details.

(2) Underground Metal Water Pipe Supplemental Electrode Required. The underground metal water pipe grounding electrode, if present [250.52(A)(1)], must be supplemented by one of the following electrodes:

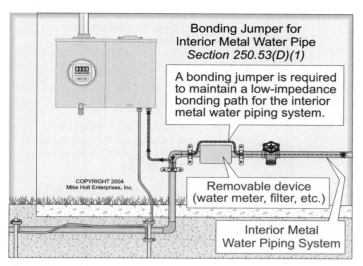

Bonding Jumper for
Interior Metal Water Pipe
Section 250.53(D)(1)

A bonding jumper is required
to maintain a low-impedance
bonding path for the interior
metal water piping system.

Removable device
(water meter, filter, etc.)

Interior Metal
Water Piping System

COPYRIGHT 2004
Mike Holt Enterprises, Inc.

Figure 250–99

- Metal frame of the building or structure [250.52(A)(2)]
- Concrete-encased steel [250.52(A)(3)] **Figure 250–100**
- Ground ring [250.52(A)(4)]

Where none of the above electrodes are available, one of the following electrodes must be used:

- Ground rod in accordance with 250.56 [250.52(A)(5)]
- Grounding plate [250.52(A)(6)]
- Metal underground systems [250.52(A)(7)]

The underground water pipe supplemental electrode must terminate to one of the following:

- Grounding electrode conductor
- Grounded neutral service conductor
- Metal service raceway
- Service equipment enclosure **Figure 250–101**

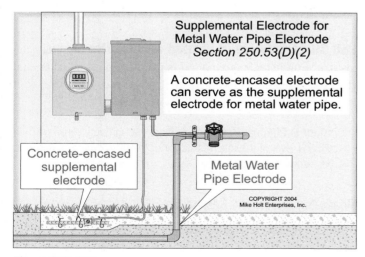

Supplemental Electrode for
Metal Water Pipe Electrode
Section 250.53(D)(2)

A concrete-encased electrode
can serve as the supplemental
electrode for metal water pipe.

Concrete-encased
supplemental
electrode

Metal Water
Pipe Electrode

COPYRIGHT 2004
Mike Holt Enterprises, Inc.

Figure 250–100

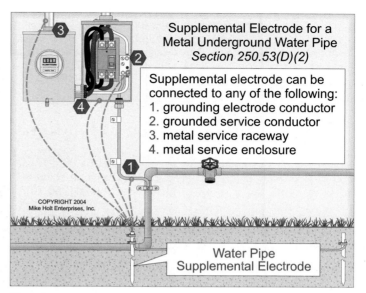

Supplemental Electrode for a
Metal Underground Water Pipe
Section 250.53(D)(2)

Supplemental electrode can be
connected to any of the following:
1. grounding electrode conductor
2. grounded service conductor
3. metal service raceway
4. metal service enclosure

Water Pipe
Supplemental Electrode

COPYRIGHT 2004
Mike Holt Enterprises, Inc.

Figure 250–101

(E) Underground Metal Water Pipe Supplemental Electrode Bonding Jumper. Where the supplemental electrode is a ground rod, that portion that is the sole connection to a ground rod isn't required to be larger than 6 AWG copper.

> **Author's Comment:** The bonding jumper for the underground metal water pipe supplemental electrode is sized in accordance with 250.66, including Table 250.66, where applicable.

(F) Ground Ring. A ground ring encircling the building or structure, consisting of at least 20 ft of bare copper conductor not smaller than 2 AWG, must be buried at a depth of not less than 30 in. See 250.52(A)(4) for additional details.

(G) Ground Rod Electrodes. Ground rod electrodes must be installed so that not less than 8 ft of length is in contact with the soil. Where rock bottom is encountered, the ground rod must be driven at an angle not to exceed 45 degrees from vertical. If rock bottom is encountered at an angle up to 45 degrees from vertical, the ground rod can be buried in a minimum 30 in. deep trench. **Figure 250–102**

The upper end of the ground rod must be flush with or underground unless the grounding electrode conductor attachment is protected against physical damage as specified in 250.10.

> **Author's Comments:**
> - See 250.52(A)(5) and 250.53(A) for additional details.
> - When the grounding electrode attachment fitting is located underground, it must be listed for direct soil burial [250.68(A) Ex. 1, and 250.70].

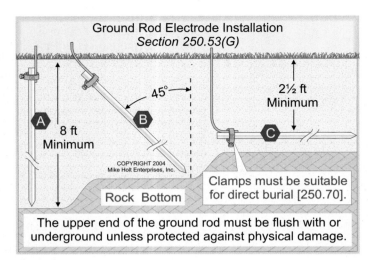

Ground Rod Electrode Installation
Section 250.53(G)

45°

8 ft Minimum

2½ ft Minimum

Clamps must be suitable for direct burial [250.70].

Rock Bottom

COPYRIGHT 2004 Mike Holt Enterprises, Inc.

The upper end of the ground rod must be flush with or underground unless protected against physical damage.

Figure 250–102

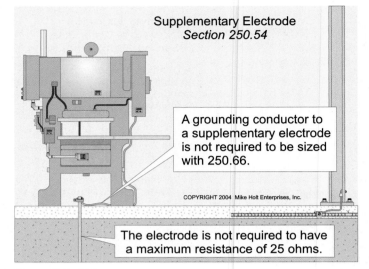

Supplementary Electrode
Section 250.54

A grounding conductor to a supplementary electrode is not required to be sized with 250.66.

COPYRIGHT 2004 Mike Holt Enterprises, Inc.

The electrode is not required to have a maximum resistance of 25 ohms.

Figure 250–104

(H) Ground Plate Electrode. A plate electrode with not less than 2 sq ft of surface exposed to exterior soils must be installed so that it's at least 30 in. below the surface of the earth [250.52(A)(6)].

250.54 Supplementary Electrodes. A supplementary electrode is an electrode that is not required by the *NEC*. This electrode is not required to be bonded to the building or structure grounding electrode (earthing) system. **Figure 250–103**

The supplementary electrode is not required to be sized to 250.66, and it is not required to comply with the 25 ohm resistance requirement of 250.56. **Figure 250–104**

The earth cannot be used as an effective ground-fault current path as required by 250.4(A)(4).

Author's Comment: Because the resistance of the earth is so high, very little current will return to the electrical supply source via the earth. If a ground rod is used as the ground-fault current path, the circuit overcurrent protection device will not open and metal parts will remain energized.

CAUTION: *The requirements contained in 250.54 for a "supplementary" electrode should not be confused with the requirements contained in 250.53(D)(2) for the underground metal water pipe "supplemental" electrode.*

Author's Comment: Typically, a supplementary electrode serves no useful purpose, and in some cases it may actually create equipment or performance failure. However, in a few cases, the supplementary electrode is used to help reduce static charges on metal parts. For information on protection against static electricity in hazardous (classified) locations, see NFPA 77, *Recommended Practice on Static Electricity.*

250.56 Resistance of Ground Rod Electrode. When the resistance of a single ground rod is over 25 ohms, an additional electrode is required to augment the ground rod electrode, and it must be installed not less than 6 ft away. **Figure 250–105**

Author's Comment: No more than two ground rods are required, even if the total resistance of the two parallel ground rods exceeds 25 ohms.

Measuring the Ground Resistance

A ground resistance clamp meter, or a three-point fall of potential ground resistance meter, can measure the resistance of a grounding electrode.

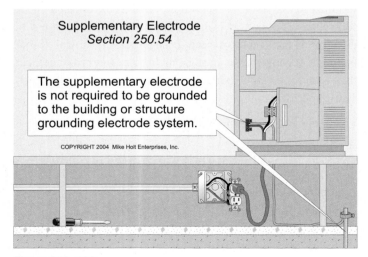

Supplementary Electrode
Section 250.54

The supplementary electrode is not required to be grounded to the building or structure grounding electrode system.

COPYRIGHT 2004 Mike Holt Enterprises, Inc.

Figure 250–103

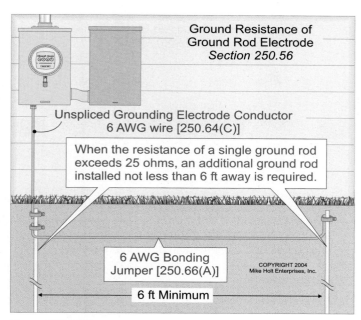

Figure 250–105

Ground Clamp Meter. The ground resistance clamp meter measures the resistance of the grounding (earthing) system by injecting a high-frequency signal via the grounded neutral conductor to the utility ground, and then measuring the strength of the return signal through the earth to the grounding electrode being measured. **Figure 250–106**

Fall of Potential Ground Resistance Meter. The three-point fall of potential ground resistance meter determines the ground resistance by using Ohm's Law: R=E/I. This meter divides the voltage difference between the electrode to be measured and a driven potential test stake (P) by the current flowing between the electrode to be measured and a driven current test stake (C). The test stakes are typically made of ¼ in. diameter steel rods, 24 in. long, driven two-thirds of their length into earth.

The distance and alignment between the potential and current test stakes, and the electrode, is extremely important to the validity of the ground resistance measurements. For an 8 ft ground rod, the accepted practice is to space the current test stake (C) 80 ft from the electrode to be measured.

The potential test stake (P) is positioned in a straight line between the electrode to be measured and the current test stake (C). The potential test stake should be located at approximately 62 percent of the distance that the current test stake is located from the electrode. Since the current test stake (C) is located 80 ft from the grounding (earthing) electrode, the potential test stake (P) will be about 50 ft from the electrode to be measured.

> **Question:** *If the voltage between the ground rod and the potential test stake (P) is 3V and the current between the ground rod and the current test stake (C) is 0.2A, then the ground resistance is _____.* **Figure 250–107**
>
> *(a) 5 ohms   (b) 10 ohms   (c) 15 ohms   (d) 25 ohms*
>
> **Answer:** *(c) 15 ohms*
>
> *Resistance = Voltage/Current*
> *E (Voltage) = 3V*
> *I (Current) = 0.2A*
> *R = E/I*
> *Resistance = 3V/0.2A*
> *Resistance = 15 ohms*

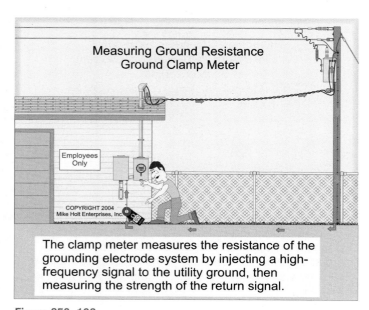

Figure 250–106

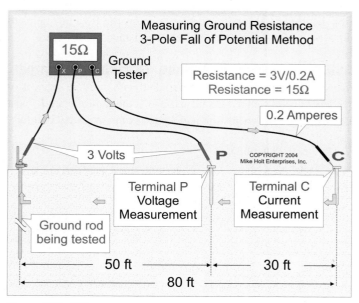

Figure 250–107

Author's Comment: The three-point fall of potential meter can only be used to measure one electrode at a time. Two electrodes bonded together cannot be measured until they have been separated. The total resistance for two separate electrodes is calculated as if they were two resistors in parallel. For example, if the ground resistance of each electrode were 50 ohms, the total resistance of two electrodes bonded together is about 25 ohms.

CAUTION: *If the electrode to be measured is connected to the electrical utility ground via the grounded neutral service conductor, the ohmmeter will give an erroneous reading. To measure the ground resistance of electrodes that aren't isolated from the electric utility (such as at industrial facilities, commercial buildings, cell phone sites, broadcast antennas, data centers, and telephone central offices), a clamp-on ground resistance tester would better serve the purpose.*

Author's Comment: The resistance of the grounding electrode can be lowered by bonding multiple grounding (earthing) electrodes that are properly spaced apart or by chemically treating the earth around the grounding (earthing) electrode. There are many readily available commercial products for this purpose.

Soil Resistivity

The earth's ground resistance is directly impacted by the soil's resistivity, which varies throughout the world. Soil resistivity is influenced by the soil's electrolytes, which consist of moisture, minerals, and dissolved salts. Because soil resistivity changes with moisture content, the resistance of any grounding (earthing) system will vary with the seasons of the year. Since moisture becomes more stable at greater distances below the surface of the earth, grounding (earthing) systems appear to be more effective if the grounding electrode can reach the water table. In addition, having the grounding electrode below the frost line helps to ensure less deviation in the system's resistance year round.

250.58 Common Grounding (Earthing) Electrode.
Where a building or structure is supplied with multiple services or feeders as permitted by 225.30 and 230.2, the same electrode must be used to ground enclosures and equipment in or on that building.

Author's Comments:

* Metal parts of the electrical installation are grounded to the earth to reduce voltage on the metal parts from lightning so as to prevent fires from a surface arc within the building or structure. Grounding electrical equipment to earth doesn't serve the purpose of providing a low-impedance fault-current path to clear ground faults.

* The most practical method of meeting this requirement is to ground each of the disconnecting means to a common concrete-encased grounding electrode [250.52(A)(3)]. Figure 250–108

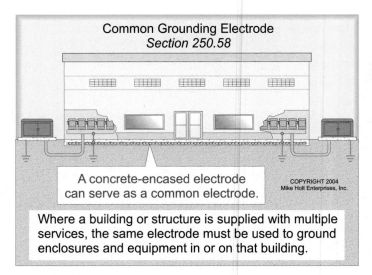

Figure 250–108

CAUTION: *Potentially dangerous objectionable current flows on the grounding electrode conductor when multiple service disconnecting means are grounded to the same electrode. This is because neutral current from each service can return to the utility via the common grounding electrode and its conductors. This is especially a problem if one of the grounded neutral service conductors is opened.* Figure 250–109

250.60 Lightning Protection System Grounding (Earthing) Electrode.
The grounding electrode for a lightning protection system cannot be used for the building or structure grounding electrode system as required by 250.50. Figure 250–110

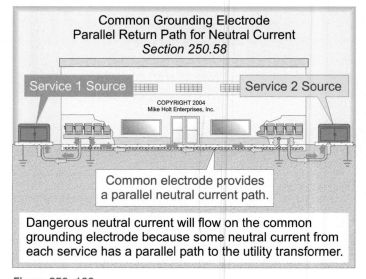

Figure 250–109

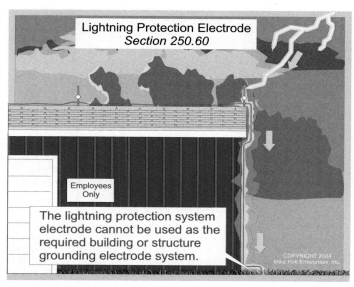

Figure 250–110

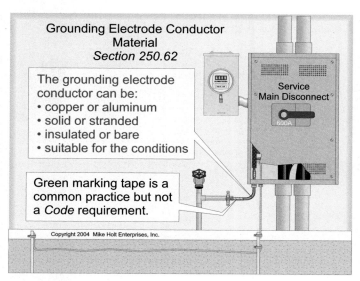

Figure 250–111

Author's Comments:

• Where a lightning protection system is installed, it must be bonded to the building or structure grounding electrode (earthing) system [250.106].

• A lightning protection system installed in accordance with NFPA 780, *Standard for the Installation of Lightning Protection Systems,* is intended to protect the building structure from lightning damage. It isn't intended to protect the electrical wiring or equipment within the structure from lightning damage. To protect from high-voltage transients, proper transient voltage surge-protection devices must be installed in accordance with Articles 280 and 285.

• GROUNDING ELECTRODE CONDUCTOR. Grounding metal parts of the electrical installation to earth is intended to reduce voltage on the metal parts from lightning so as to prevent fires from surface arcing (side-flashes) within the building or structure [250.4(A)(2)]. In addition, high-voltage distribution systems (typically electric utility wiring) are grounded to the earth to help limit high voltage on system windings from lightning, unintentional contact with higher-voltage lines, or line surges [250.4(A)(1)].

250.62 Grounding Electrode Conductor—Material.

The grounding electrode conductor, or its jumpers, can be solid or stranded, insulated or bare, and it must be copper, except aluminum is permitted if it is not subjected to corrosive conditions and not within 18 in. of the earth [250.64(A)]. **Figure 250–111**

Author's Comment: The *NEC* doesn't require the identification of the grounding electrode conductor or grounding electrode bonding jumpers. Common practice is to either not identify the conductor at all, or to apply green marking tape.

250.64 Grounding Electrode Conductor Installation.

(A) Aluminum Grounding Electrode Conductor. Aluminum grounding electrode conductors cannot be in contact with earth, masonry, or subjected to corrosive conditions. When used outdoors, the termination to the electrode must not be within 18 in. of earth.

(B) Grounding Electrode Conductor Protection. Where exposed, grounding electrode conductors sized 8 AWG and smaller must be installed in rigid metal conduit, intermediate metal conduit, rigid nonmetallic conduit, or electrical metallic tubing.

Author's Comment: Ferrous metal raceways containing the grounding electrode conductors must be made electrically continuous by bonding each end of the ferrous metal raceway to the grounding electrode conductor [250.64(E)].

Grounding electrode conductors 6 AWG copper and larger can be run exposed along the surface if securely fastened to the construction and not subject to physical damage.

(C) Continuous Run. The grounding electrode conductor, which runs to any convenient grounding electrode [250.64(F)], must not be spliced, except as permitted in (1) through (3): **Figure 250–112**

(1) Splicing is permitted by irreversible compression-type connectors listed for grounding or by exothermic welding.

(2) Sections of busbars can be connected together to form a grounding electrode conductor.

(3) Bonding and grounding electrode conductors are permitted to terminate to a busbar that is sized not smaller than ¼ x 2 in.,

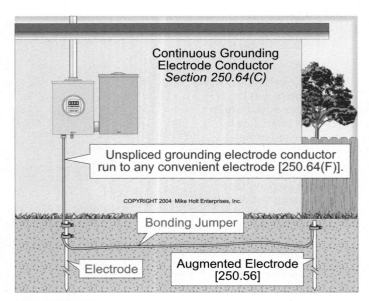

Figure 250–112

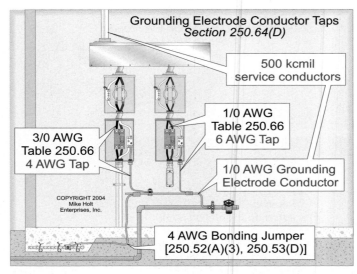

Figure 250–114

and the busbar must be securely fastened in place at an accessible location. Connections must be made by a listed connector or by the exothermic welding process. **Figure 250–113**

(D) Grounding Electrode Tap Conductors. When a service consists of multiple disconnecting means as permitted in 230.71(A), a grounding electrode tap from each disconnect to a common grounding electrode conductor is permitted.

The grounding electrode tap must be sized in accordance with 250.66, based on the largest ungrounded conductor serving that disconnect.

The common grounding electrode conductor for the grounding electrode taps is also sized in accordance with 250.66, based on the service conductors feeding all the service disconnects.

Each grounding electrode tap must terminate to the common grounding electrode conductor in such a manner that there will be no splices or joints in the common grounding electrode conductor. **Figure 250–114**

(E) Enclosures for Grounding Electrode Conductor. Ferrous (iron/steel) raceways, boxes, and enclosures containing the grounding electrode conductors must have each end of the ferrous metal raceway, box, and enclosure bonded to the grounding electrode conductor [250.92(A)(3)]. **Figure 250–115**

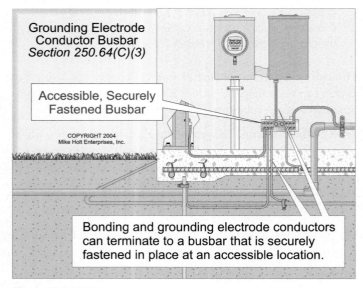

Figure 250–113

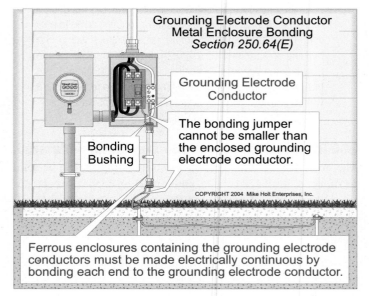

Figure 250–115

Author's Comment: "Nonferrous" metal raceways, such as aluminum rigid metal conduit, enclosing the grounding electrode conductor aren't required to meet the "bonding each end of the raceway to the grounding electrode conductor" provisions of this section.

The bonding jumper must be sized no smaller than the enclosed grounding electrode conductor.

CAUTION: *The effectiveness of the grounding electrode can be significantly reduced if a ferromagnetic raceway containing a grounding electrode conductor isn't bonded to the grounding electrode conductor at both ends. This is because a single conductor carrying high-frequency lightning current in a ferrous raceway causes the raceway to act as an inductor, which severely limits (chokes) the current flow through the grounding electrode conductor. ANSI/IEEE 142, Recommended Practice for Grounding of Industrial and Commercial Power Systems (Green Book) states, "An inductive choke can reduce the current flow by 97 percent."*

Author's Comment: To save a lot of time and effort, simply run the grounding electrode conductor exposed if not subject to physical damage [250.64(B)], or enclose it in a nonmetallic conduit that is suitable for the application.

(F) To Electrode(s). The grounding electrode conductor can be run to any convenient grounding electrode available in the grounding electrode (earthing) system. The grounding electrode conductor must be sized for the largest grounding electrode conductor required among all the electrodes connected to it.

Author's Comment: It is not necessary to run the grounding electrode conductor to all of the electrodes unbroken, just to the first electrode.

Table 250.66—Grounding Electrode Conductor

Ungrounded Conductor or Area of Parallel Conductors	Copper Grounding Electrode Conductor
12 through 2 AWG	8 AWG
1 or 1/0 AWG	6 AWG
2/0 or 3/0 AWG	4 AWG
Over 3/0 through 350 kcmil	2 AWG
Over 350 through 600 kcmil	1/0 AWG
Over 600 through 1,100 kcmil	2/0 AWG
Over 1,100 kcmil	3/0 AWG

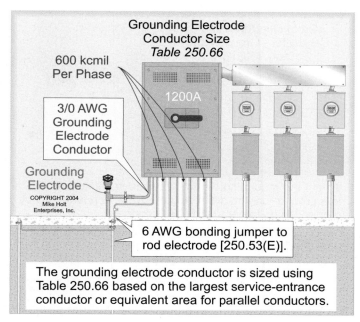

Figure 250–116

250.66 Grounding Electrode Conductor—Size.

Except for a ground rod electrode [250.66(A)], a concrete-encased electrode [250.66(B)], or a ground ring electrode [250.66(C)], the grounding electrode conductor must be sized based on the largest service-entrance conductor or equivalent area for parallel conductors in accordance with Table 250.66

Question: *What size grounding electrode conductor is required for a 1,200A service that is supplied with three parallel sets of 600 kcmil conductors per phase?* **Figure 250–116**

(a) 1 AWG   (b) 1/0 AWG   (c) 2/0 AWG   (d) 3/0 AWG

Answer: *(d) 3/0 AWG*

The equivalent area of three parallel 600 kcmil conductors is 1,800 kcmil per phase [Table 250.66].

FPN: Because the grounded neutral service conductor is required to serve as the low-impedance ground-fault current path back to the source, it must be sized no smaller than that shown in Table 250.66 [250.24(C)(1)]. Of course, it must be sized to carry the maximum unbalanced load as calculated by 220.61.

(A) Ground Rod. Where the grounding electrode conductor is connected to a ground rod, that portion of the grounding electrode conductor that is the sole connection to the ground rod isn't required to be larger than 6 AWG copper. **Figure 250–117**

Author's Comment: See 250.52(A)(5) for the installation requirements of a ground rod electrode.

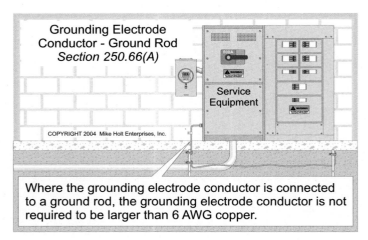

Figure 250–117

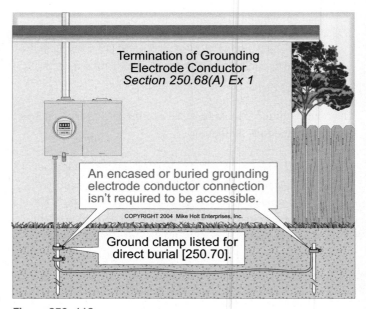

Figure 250–119

(B) Concrete-Encased Grounding Electrode (Ufer Ground). Where the grounding electrode conductor is connected to a concrete-encased electrode, that portion of the grounding electrode conductor that is the sole connection to the concrete-encased electrode isn't required to be larger than 4 AWG copper. Figure 250–118

> **Author's Comment:** See 250.52(A)(3) for the installation requirements of a concrete-encased electrode.

(C) Ground Ring. Where the grounding electrode conductor is connected to a ground ring, that portion of the conductor that is the sole connection to the ground ring isn't required to be larger than the conductor used for the ground ring.

> **Author's Comment:** A ground ring encircling the building or structure in direct contact with earth must consist of not less than 20 ft of bare copper conductor not smaller than 2 AWG [250.52(A)(4)].

250.68 Grounding Electrode Conductor Termination.

(A) Attachment Fitting. The grounding electrode attachment fitting must be accessible.

Exception 1: The grounding electrode attachment fitting to an encased or buried grounding electrode isn't required to be accessible. Figure 250–119

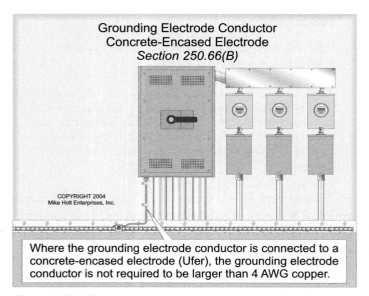

Where the grounding electrode conductor is connected to a concrete-encased electrode (Ufer), the grounding electrode conductor is not required to be larger than 4 AWG copper.

Figure 250–118

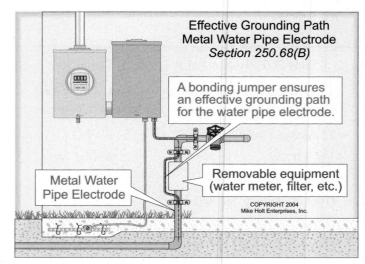

Figure 250–120

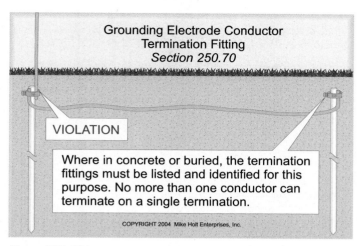

Figure 250–121

Author's Comment: When the grounding electrode attachment fitting is encased in concrete or buried, it must be listed and identified for this purpose [250.70].

Exception 2: An exothermic or irreversible compression connection to fireproofed structural metal isn't required to be accessible.

(B) Effective Ground-Fault Current Path. To ensure a permanent and effective grounding path for an underground metal water pipe electrode, a bonding jumper must be installed around insulated joints and equipment likely to be disconnected for repairs or replacement. **Figure 250–120**

Author's Comment: Continuity of the conductive bonding path for metal water piping as required by 250.104(A) cannot rely on water meters, filtering devices, or similar equipment [250.53(D)(1)].

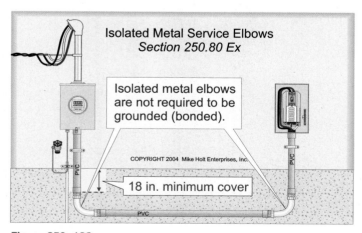

Figure 250–122

250.70 Grounding Electrode Conductor Termination Fitting.
The grounding electrode conductor must terminate to the grounding electrode by exothermic welding, listed lugs, listed pressure connectors, listed clamps, or other listed means. In addition, termination fittings must be listed for the materials of the grounding electrode.

When the termination to a grounding electrode is encased in concrete or buried, the termination fitting must be listed and identified for this purpose. No more than one conductor can terminate on a single clamp or fitting unless the clamp or fitting is listed for multiple connections. **Figure 250–121**

PART IV. GROUNDING (BONDING) OF ENCLOSURE, RACEWAY, AND SERVICE CABLE

250.80 Service Enclosures. Metal raceways and enclosures containing service conductors must be grounded (bonded) to an effective ground-fault current path [250.4(A)(5)].

Exception: Metal elbows having 18 in. or more of earth cover aren't required to be grounded (bonded) to an effective ground-fault current path. **Figure 250–122**

250.86 Other Enclosures. Metal raceways and enclosures containing electrical conductors that operate at over 50V [250.112(I)] must be grounded (bonded) to an effective ground-fault current path [250.4(A)(5)].

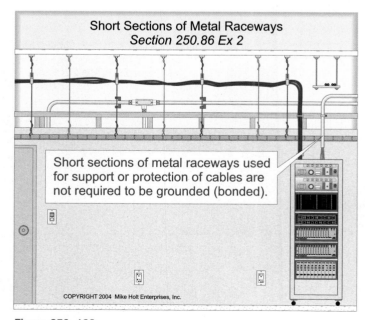

Figure 250–123

Author's Comment: Metal raceways and metal enclosures containing circuit conductors from electrical supply sources that operate at 50V or less aren't required to be grounded [250.86]. For example, metal boxes used with power-limited fire alarm circuits operating at 50V or less aren't required to be grounded.

Exception 2: Short sections of metal raceways used for the support or physical protection of cables aren't required to be grounded (bonded) to an effective ground-fault current path. **Figure 250–123**

Author's Comment: NM cable on a wall of an unfinished basement installed, in a listed metal raceway, must be grounded to an effective ground-fault current path [334.15(C)].

Exception 3: Metal elbows having 18 in. or more of earth cover aren't required to be grounded (bonded) to an effective ground-fault current path. **Figure 250–124**

PART V. BONDING

250.92 Service Bonding.

(A) Equipment and Raceways. The following metal parts must be service bonded to an effective ground-fault current path in accordance with 250.92(B): **Figure 250–125**

(1) Metal raceways containing service conductors.

(2) Enclosures containing service conductors.

Author's Comment: Metal raceways and enclosures containing service conductors must be effectively bonded in accordance with 250.92(B).

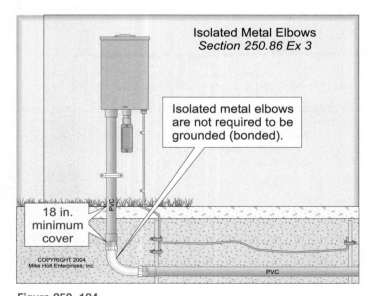

Figure 250–124

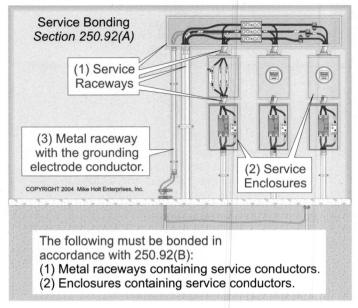

Figure 250–125

(3) A metal raceway containing the grounding electrode conductor.

Author's Comments:

- The metal raceway containing the grounding electrode conductor must be effectively bonded in accordance with 250.64(E).

- Raceways or enclosures containing feeder and branch-circuit conductors are not required to be service bonded in accordance with 250.92(B). **Figure 250–126**

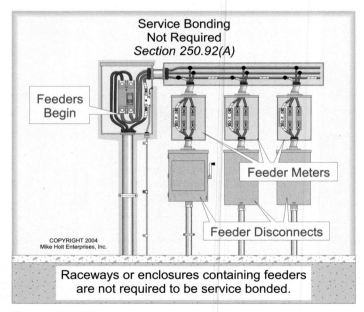

Figure 250–126

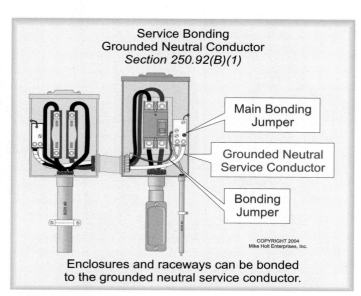

Service Bonding
Grounded Neutral Conductor
Section 250.92(B)(1)

Main Bonding Jumper

Grounded Neutral Service Conductor

Bonding Jumper

COPYRIGHT 2004
Mike Holt Enterprises, Inc.

Enclosures and raceways can be bonded
to the grounded neutral service conductor.

Figure 250–127

(B) Methods of Bonding. Enclosures and raceways containing service conductors must be bonded to an effective ground-fault current path by one of the following methods:

(1) Grounded Neutral Conductor. Enclosures and raceways containing service conductors are considered bonded to an effective ground-fault current path by bonding to the grounded neutral service conductor via the main bonding jumper. Figure 250–127

The bonding must be by exothermic welding, listed pressure connectors, listed clamps, or other listed fittings [250.8].

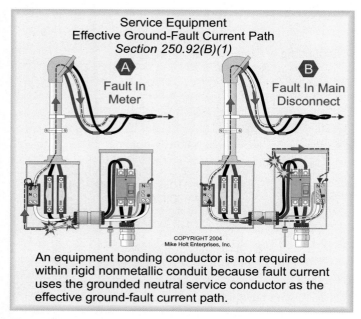

Service Equipment
Effective Ground-Fault Current Path
Section 250.92(B)(1)

A Fault In Meter

B Fault In Main Disconnect

COPYRIGHT 2004
Mike Holt Enterprises, Inc.

An equipment bonding conductor is not required within rigid nonmetallic conduit because fault current uses the grounded neutral service conductor as the effective ground-fault current path.

Figure 250–128

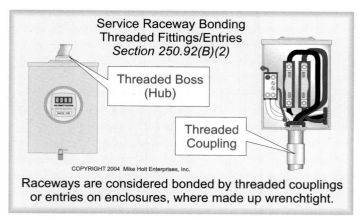

Service Raceway Bonding
Threaded Fittings/Entries
Section 250.92(B)(2)

Threaded Boss (Hub)

Threaded Coupling

COPYRIGHT 2004 Mike Holt Enterprises, Inc.

Raceways are considered bonded by threaded couplings
or entries on enclosures, where made up wrenchtight.

Figure 250–129

Author's Comments:

- A main bonding jumper is required to bond the service disconnect to the grounded neutral service conductor [250.24(B) and 250.28].

- At service equipment, the grounded neutral service conductor is used to provide the effective ground-fault current path to the power source [250.24(C)]. Therefore, an equipment grounding (bonding) conductor isn't required to be installed within a nonmetallic raceway containing service-entrance conductors [250.142(A)(1) and 352.60 Ex. 2]. Figure 250–128

(2) Threaded Fittings or Entries. Raceways containing service conductors are considered bonded to an effective ground-fault current path by threaded couplings or threaded entries on enclosures where made up wrenchtight. Figure 250–129

(3) Threadless Fitting. Raceways containing service conductors are considered bonded to an effective ground-fault current path by threadless raceway couplings and connectors where made up tight. Figure 250–130

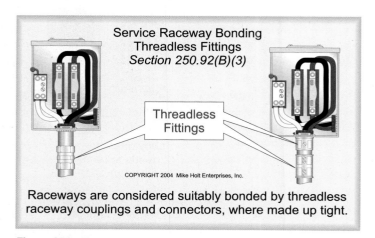

Service Raceway Bonding
Threadless Fittings
Section 250.92(B)(3)

Threadless Fittings

COPYRIGHT 2004 Mike Holt Enterprises, Inc.

Raceways are considered suitably bonded by threadless
raceway couplings and connectors, where made up tight.

Figure 250–130

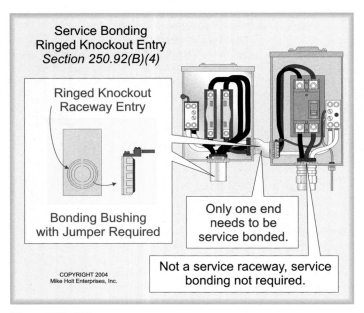

Service Bonding
Ringed Knockout Entry
Section 250.92(B)(4)

Ringed Knockout
Raceway Entry

Bonding Bushing
with Jumper Required

Only one end
needs to be
service bonded.

Not a service raceway, service
bonding not required.

COPYRIGHT 2004
Mike Holt Enterprises, Inc.

Figure 250–131

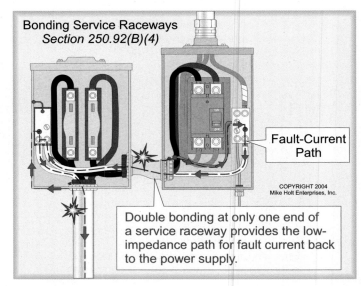

Bonding Service Raceways
Section 250.92(B)(4)

Fault-Current
Path

Double bonding at only one end of
a service raceway provides the low-
impedance path for fault current back
to the power supply.

COPYRIGHT 2004
Mike Holt Enterprises, Inc.

Figure 250–133

(4) Bonding Fitting. When a metal service raceway terminates to an enclosure with a ringed knockout, a listed bonding device, such as a bonding wedge or bushing, must bond one end of the service raceway with a bonding jumper sized in accordance with Table 250.66 [250.92(B)(4) and 250.102(C)]. **Figure 250–131**

Author's Comments:

- When a metal raceway containing service conductors terminates to an enclosure without a ringed knockout, a bonding-type locknut can be used instead of a bonding wedge or bushing. **Figure 250–132**

- A bonding-type locknut differs from a standard-type locknut in that it has a bonding screw with a sharp point that drives into the metal enclosure to ensure a solid termination.

- Bonding one end of a service raceway in accordance with 250.92(B) provides the low-impedance fault-current path to the utility electrical supply source. **Figure 250–133**

250.94 Grounding (Bonding) of Communications Systems. An accessible bonding point must be provided at service equipment or the disconnecting means of separate buildings or structures for communications systems. The point can be any one of the following:

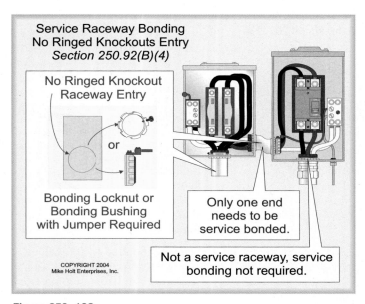

Service Raceway Bonding
No Ringed Knockouts Entry
Section 250.92(B)(4)

No Ringed Knockout
Raceway Entry

or

Bonding Locknut or
Bonding Bushing
with Jumper Required

Only one end
needs to be
service bonded.

Not a service raceway, service
bonding not required.

COPYRIGHT 2004
Mike Holt Enterprises, Inc.

Figure 250–132

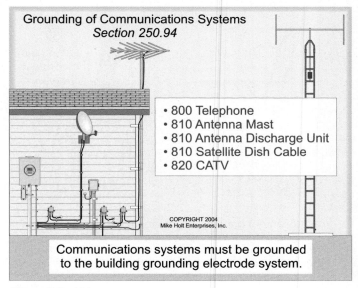

Grounding of Communications Systems
Section 250.94

• 800 Telephone
• 810 Antenna Mast
• 810 Antenna Discharge Unit
• 810 Satellite Dish Cable
• 820 CATV

COPYRIGHT 2004
Mike Holt Enterprises, Inc.

Communications systems must be grounded
to the building grounding electrode system.

Figure 250–134

(1) An exposed, nonflexible metallic raceway.

(2) An exposed grounding electrode conductor.

(3) An external connection approved by the authority having jurisdiction.

> **FPN No. 2:** Communications systems must be bonded together. **Figure 250–134**
>
> • Antennas/Satellite Dishes, 810.21
> • CATV, 820.100
> • Telephone Circuits, 800.100

Author's Comment: The bonding of all external communications systems to a single point minimizes the possibility of damage to the systems from potential (voltage) differences between the systems. **Figure 250–135**

250.96 Bonding Other Enclosures.

(A) Maintaining Effective Ground-Fault Current Path. All metal parts intended to serve as the effective ground-fault current path, such as raceways, cables, equipment, and enclosures must be bonded together to ensure they have the capacity to conduct safely any fault current likely to be imposed on them [110.10, 250.4(A)(5), and Note to Table 250.122].

Author's Comment: Bonding jumpers are required around reducing washers (donuts) because they do not provide an effective ground-fault current path [250.4(A)(5)]. **Figure 250–136**

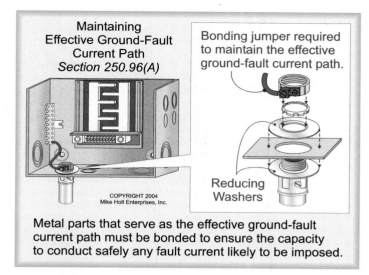

Maintaining
Effective Ground-Fault
Current Path
Section 250.96(A)

Bonding jumper required to maintain the effective ground-fault current path.

Reducing Washers

COPYRIGHT 2004
Mike Holt Enterprises, Inc.

Metal parts that serve as the effective ground-fault current path must be bonded to ensure the capacity to conduct safely any fault current likely to be imposed.

Figure 250–136

Nonconductive coatings such as paint, lacquer, and enamel on equipment must be removed to ensure an effective ground-fault current path, or the termination fittings must be designed so as to make such removal unnecessary [250.12].

Author's Comment: The practice of driving a locknut tight with a screwdriver and pliers is considered sufficient in removing paint and other nonconductive finishes to ensure an effective ground-fault current path.

(B) Isolated Ground Circuit. A metal raceway containing circuit conductors for sensitive electronic equipment can be electrically isolated from the sensitive equipment it supplies by a nonmetallic raceway fitting located at the equipment. However, the metal raceway must contain an insulated equipment grounding (bonding) conductor to provide the effective ground-fault current path to the power source in accordance with 250.145(D). **Figure 250–137**

> **DANGER:** *Some digital equipment manufacturers insist that their equipment be electrically isolated from the building or structure's grounding (earthing) system. This is a dangerous practice, and it violates 250.4(A)(5), which prohibits the earth from being used as an effective ground-fault current path. If the metal enclosures of sensitive electronic equipment are isolated or floated as required by some sensitive equipment manufacturers, dangerous voltage on metal parts can remain from a ground fault.* **Figure 250–138**

Author's Comment: For more information on how to properly ground sensitive electronic equipment, visit www.MikeHolt.com, go to "Technical," then to the "Grounding" link.

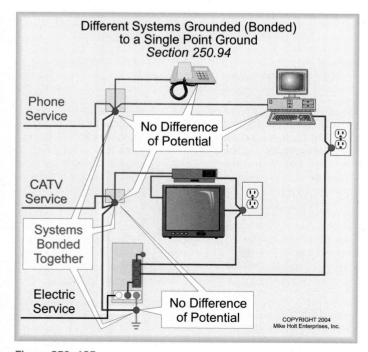

Different Systems Grounded (Bonded)
to a Single Point Ground
Section 250.94

Phone Service

No Difference of Potential

CATV Service

Systems Bonded Together

Electric Service

No Difference of Potential

COPYRIGHT 2004
Mike Holt Enterprises, Inc.

Figure 250–135

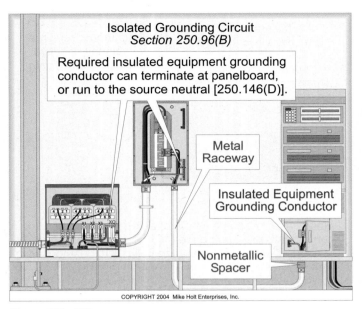

Figure 250–137

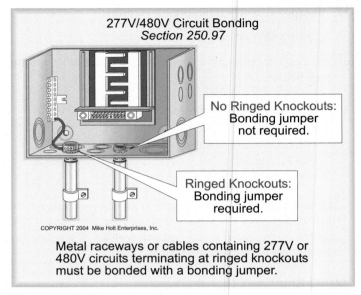

Figure 250–139

250.97 Bonding 277V/480V. Metal raceways or cables, containing 277V or 480V circuits, terminating at ringed knockouts must be bonded to the metal enclosure with a bonding jumper sized in accordance with Table 250.122, based on the rating of the circuit overcurrent protection device [250.102(D)].
Figure 250–139

Author's Comments:

- Bonding jumpers for raceways and cables containing 277V or 480V circuits are required at ringed knockout terminations to ensure that the effective ground-fault current path has the

capacity to safely conduct the maximum ground-fault current likely to be imposed on it back to the electrical supply source, in accordance with 110.10 and 250.4(A)(5).

- Ringed knockouts aren't listed to withstand the heat generated by a 277V ground fault because a 277V ground fault generates five times as much heat as a 120V ground fault.
Figure 250–140

Exception: A bonding jumper isn't required where ringed knockouts aren't encountered, or where the box is listed to provide a permanent and reliable electrical bond. Figure 250–141

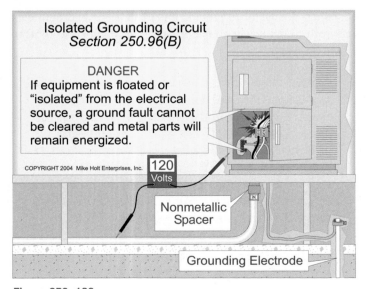

Figure 250–138

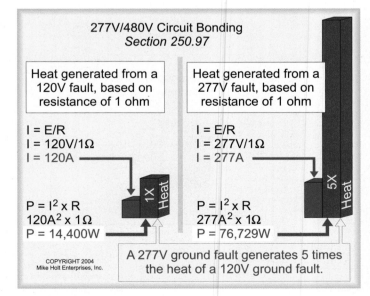

Figure 250–140

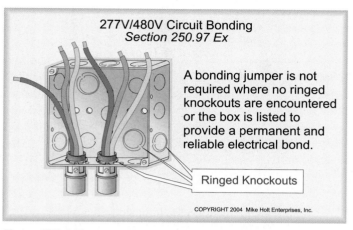

277V/480V Circuit Bonding
Section 250.97 Ex

A bonding jumper is not required where no ringed knockouts are encountered or the box is listed to provide a permanent and reliable electrical bond.

Ringed Knockouts

COPYRIGHT 2004 Mike Holt Enterprises, Inc.

Figure 250–141

250.100 Bonding—Hazardous (Classified) Locations.

Because of the explosive conditions associated with electrical installations in hazardous (classified) locations, electrical continuity of the effective ground-fault current path (metal parts of equipment and raceways) must be ensured by one of the methods specified in 250.92(B)(2) through (4): **Figure 250–142**

Author's Comments:

- The methods of bonding metal raceways that extend into a hazardous (classified) location include:
  - Threaded couplings or threaded entries on enclosures [250.92(B)(2)].
  - Threadless raceway couplings and connectors where permitted in Articles 501, 502, and 503 [250.92(B)(3)].

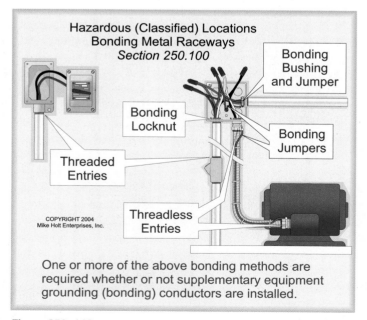

Hazardous (Classified) Locations
Bonding Metal Raceways
Section 250.100

Bonding Bushing and Jumper

Bonding Locknut

Bonding Jumpers

Threaded Entries

Threadless Entries

COPYRIGHT 2004 Mike Holt Enterprises, Inc.

One or more of the above bonding methods are required whether or not supplementary equipment grounding (bonding) conductors are installed.

Figure 250–142

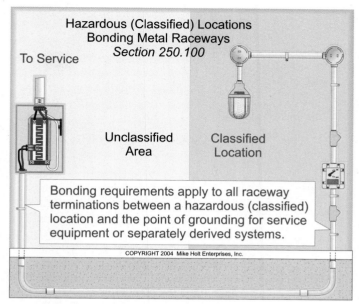

Hazardous (Classified) Locations
Bonding Metal Raceways
Section 250.100

To Service

Unclassified Area

Classified Location

Bonding requirements apply to all raceway terminations between a hazardous (classified) location and the point of grounding for service equipment or separately derived systems.

COPYRIGHT 2004 Mike Holt Enterprises, Inc.

Figure 250–143

- Bonding the raceway with a listed bonding wedge or bushing [250.92(B)(4)].
- Bonding the raceway with a bonding-type locknut if the metal raceway terminates to an enclosure without a ringed knockout [250.92(B)(4)].

- Hazardous (classified) location bonding requirements apply to all intervening raceways, fittings, boxes, and enclosures between the hazardous (classified) location and the point of grounding and bonding at service equipment and separately derived systems. See 501.100 for Class I locations, 502.100 for Class II locations, and 503.100 for Class III locations for specific requirements. **Figure 250–143**

250.102 Bonding Jumper.

(A) Bonding Material. Bonding jumpers must be of copper.

(B) Bonding Jumper Attachment. Bonding jumpers must terminate by exothermic welding, listed pressure connectors, listed clamps, or other listed means. Sheet-metal screws cannot be used for termination of bonding conductors or connection devices [250.8].

(C) Supply Side of Service—Bonding Jumper. Bonding jumpers for service raceways must be sized in accordance with Table 250.66, based on the ungrounded service conductors within the service raceway. Where service conductors are paralleled in two or more raceways or cables, the bonding jumper for each raceway or cable must be sized on the ungrounded service conductors in each raceway or cable.

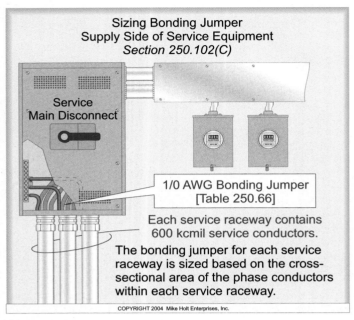

Figure 250–144

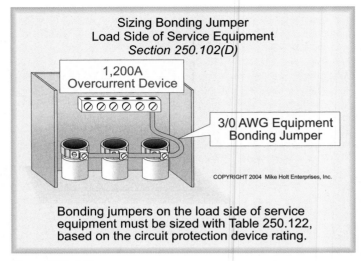

Figure 250–146

Question: *What size bonding jumper is required for a metal raceway containing 600 kcmil service conductors?* **Figure 250–144**

(a) 1 AWG   (b) 1/0 AWG   (c) 2/0 AWG   (d) 3/0 AWG

Answer: *(b) 1/0 AWG, Table 250.66*

(D) Load Side of Service—Bonding Jumper. Bonding jumpers on the load side of service equipment must be sized in accordance with Table 250.122, based on the rating of the circuit-protection device.

Question: *What size bonding jumper is required for a metal raceway where the circuit conductors are protected by a 1,200A protection device?* **Figure 250–145**

(a) 1 AWG   (b) 1/0 AWG   (c) 2/0 AWG   (d) 3/0 AWG

Answer: *(d) 3/0 AWG, Table 250.122*

A single bonding jumper sized in accordance with 250.122 can be used to bond multiple raceways or cables. **Figure 250–146**

(E) Installation. Where the equipment bonding jumper is installed outside a raceway, its length must not exceed 6 ft and it must be routed with the raceway. **Figure 250–147**

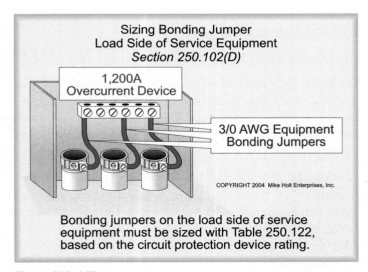

Figure 250–145

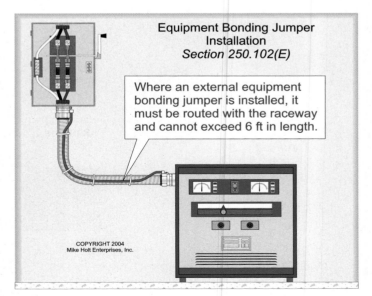

Figure 250–147

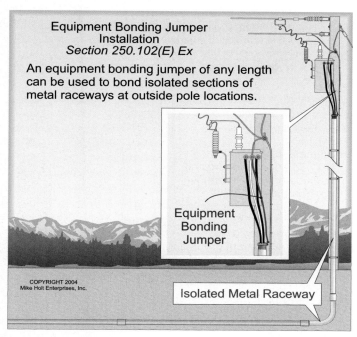

Figure 250–148

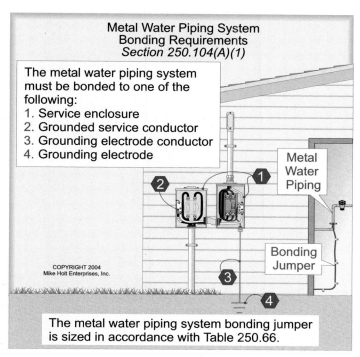

The metal water piping system bonding jumper is sized in accordance with Table 250.66.

Figure 250–149

Exception: An equipment bonding jumper of any length can be used to bond isolated sections of metal raceways at outside pole locations. **Figure 250–148**

250.104 Bonding of Piping Systems and Exposed Structural Metal.

Author's Comment: To remove dangerous voltage on metal parts from a ground fault, electrically conductive metal water piping systems, metal sprinkler piping, metal gas piping, and other metal piping systems, as well as exposed structural steel members that are likely to become energized, must be bonded to an effective ground-fault current path [250.4(A)(4)].

(A) Metal Water Piping System. Metal water piping systems must be bonded in accordance with (1), (2), or (3).

Author's Comment: Bonding isn't required for isolated sections of metal water piping connected to a nonmetallic water piping system.

(1) Building Supplied by a Service. The metal water piping system of a building or structure must be bonded to one of the following: **Figure 250–149**

- Service equipment enclosure
- Grounded neutral service conductor
- Grounding electrode conductor where the grounding electrode conductor is sized in accordance with Table 250.66

- One of the electrodes of the grounding electrode system

The metal water pipe bonding jumper must be sized in accordance with Table 250.66, based on the largest ungrounded service conductor.

> *Question: What size bonding jumper is required for the metal water piping system if the service conductors are 4/0 AWG?* **Figure 250–150**
>
> *(a) 6 AWG   (b) 4 AWG   (c) 2 AWG   (d) 1/0 AWG*
>
> *Answer: (c) 2 AWG, Table 250.66*

Author's Comment: Where hot and cold water pipes are electrically connected, only one bonding jumper is required, either to the cold or hot water pipe. Otherwise, a single bonding jumper, sized in accordance with 250.104(A)(1), must be used to bond the hot and cold water piping together.

(2) Multiple Occupancy Building. When the metal water piping system in an individual occupancy is metallically isolated from other occupancies, the metal water piping system for that occupancy can be bonded to the equipment grounding terminal of the occupancy's panelboard. **Figure 250–151**

The bonding jumper used for this purpose must be sized to the ampere rating of the occupancy's feeder overcurrent protection device in accordance with Table 250.122.

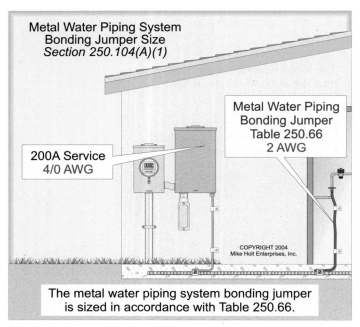

The metal water piping system bonding jumper
is sized in accordance with Table 250.66.

Figure 250–150

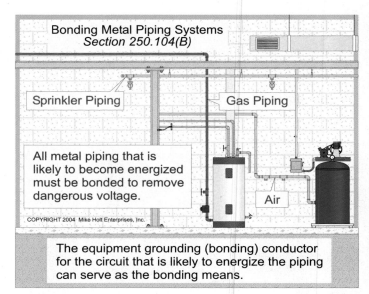

The equipment grounding (bonding) conductor
for the circuit that is likely to energize the piping
can serve as the bonding means.

Figure 250–152

(3) Building or Structure Supplied by a Feeder. The metal
water piping system of a building or structure that is supplied by
a feeder must be bonded to:

- The equipment grounding terminal of the building
  disconnect enclosure,
- The feeder equipment grounding (bonding) conductor, or
- One of the electrodes of the grounding electrode system.

The bonding jumper for the metal water piping system must be
sized to the feeder circuit conductors that supply the building or
structure in accordance with Table 250.66. The bonding jumper is
not required to be larger than the ungrounded feeder conductors.

> **Author's Comment:** It makes no sense to size the metal water
> piping bonding jumper in accordance with Table 250.66, based
> on the feeder circuit conductor size. The bonding jumper should
> be sized in accordance with Table 250.122, based on the
> building or structure feeder overcurrent protection device.

(B) Other Metal Piping Systems. Metal piping systems such as
gas or air that are likely to become energized must be bonded to
an effective ground-fault current path. The equipment grounding
(bonding) conductor for the circuit that may energize the piping
can serve as the bonding means.

> **Author's Comments:**
> - This exact text is contained in NFPA 54, *National Fuel Gas
>   Code*.
> - Because the equipment grounding (bonding) conductor for
>   the circuit that may energize the piping can serve as the
>   bonding means, no action is required by the electrical
>   installer. **Figure 250–152**

> **FPN:** Bonding of all metal piping and metal ducts within the
> building provides an additional degree of safety, but this isn't
> an *NEC* requirement.

(C) Structural Metal. Exposed structural metal that forms a
metal building frame that is likely to become energized must be
bonded to: **Figure 250–153**

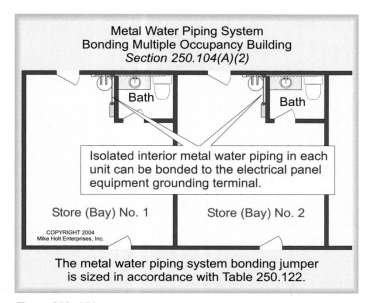

The metal water piping system bonding jumper
is sized in accordance with Table 250.122.

Figure 250–151

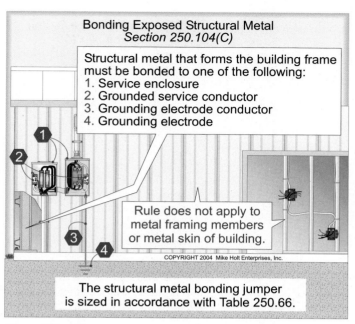

Bonding Exposed Structural Metal
Section 250.104(C)

Structural metal that forms the building frame must be bonded to one of the following:
1. Service enclosure
2. Grounded service conductor
3. Grounding electrode conductor
4. Grounding electrode

Rule does not apply to metal framing members or metal skin of building.

COPYRIGHT 2004 Mike Holt Enterprises, Inc.

The structural metal bonding jumper is sized in accordance with Table 250.66.

Figure 250–153

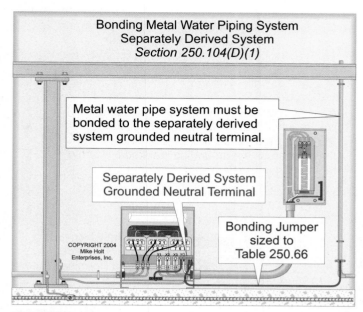

Bonding Metal Water Piping System
Separately Derived System
Section 250.104(D)(1)

Metal water pipe system must be bonded to the separately derived system grounded neutral terminal.

Separately Derived System Grounded Neutral Terminal

COPYRIGHT 2004 Mike Holt Enterprises, Inc.

Bonding Jumper sized to Table 250.66

Figure 250–154

- Service equipment enclosure
- Grounded neutral service conductor
- Grounding electrode conductor where the grounding electrode conductor is sized in accordance with Table 250.66
- One of the electrodes of the grounding electrode system

Author's Comment: This rule doesn't require the bonding of sheet metal framing members (studs) or the metal skin of a wood frame building, but it would be a good practice.

The bonding jumper for the structural metal must be sized to the feeder or service conductors that supply the building or structure in accordance with Table 250.66, and the bonding jumper must be:

- Copper where within 18 in. of earth [250.64(A)].
- Securely fastened and adequately protected if exposed to physical damage [250.64(B)].
- Installed without a splice or joint, unless spliced by irreversible compression connectors listed for the purpose or by the exothermic welding process [250.64(C)].

(D) Separately Derived Systems. Metal water pipe systems and structural metal that forms a building frame must be bonded in accordance with (1), (2), and (3).

(1) Metal Water Pipe. In the area served by a separately derived system, the nearest available point of the metal water piping system must be bonded to the grounded neutral terminal of the separately derived system. The bonding at the separately derived system must be at the same location where the grounding electrode conductor and system bonding jumper terminate [250.32(A)]. **Figure 250–154**

The metal water piping bonding jumper must be sized in accordance with Table 250.66, based on the largest ungrounded conductor of the separately derived system.

Exception 1: A water pipe bonding jumper isn't required if the water pipe is used as the grounding electrode for the separately derived system.

Exception 2: A water pipe bonding jumper isn't required if the metal water pipe is bonded to the structural metal building frame that is used as the grounding electrode for the separately derived system. **Figure 250–155**

(2) Structural Metal. Where exposed structural metal is interconnected to form the building frame, it must be bonded to the grounded neutral conductor of each separately derived system. This connection must be made at the same point on the separately derived system where the grounding electrode conductor is connected.

Each bonding jumper must be sized in accordance with 250.66, based on the largest ungrounded conductor of the separately derived system.

Exception 1: A structural metal bonding jumper isn't required if the metal structural frame is used as the grounding electrode for the separately derived system.

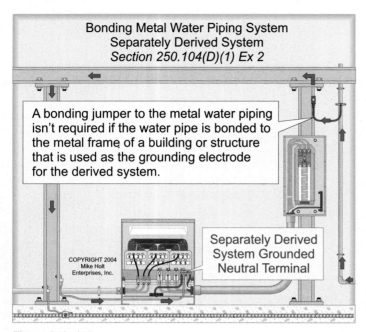

Bonding Metal Water Piping System
Separately Derived System
Section 250.104(D)(1) Ex 2

A bonding jumper to the metal water piping isn't required if the water pipe is bonded to the metal frame of a building or structure that is used as the grounding electrode for the derived system.

COPYRIGHT 2004
Mike Holt
Enterprises, Inc.

Separately Derived System Grounded Neutral Terminal

Figure 250–155

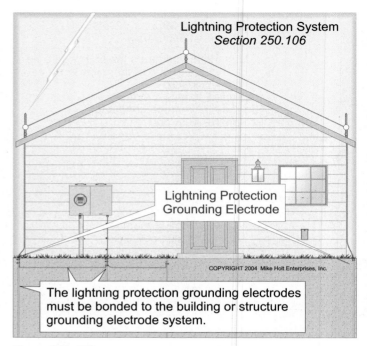

Lightning Protection System
Section 250.106

Lightning Protection Grounding Electrode

COPYRIGHT 2004 Mike Holt Enterprises, Inc.

The lightning protection grounding electrodes must be bonded to the building or structure grounding electrode system.

Figure 250–156

Exception 2: A structural metal bonding jumper isn't required if the structural metal frame is bonded to metal water piping that is used as the grounding electrode for the separately derived system.

(3) Common Grounding Electrode Conductor. Where a common grounding electrode conductor is installed for multiple separately derived systems as permitted by 250.30(A)(4), exposed structural metal and interior metal piping in the area served by the separately derived system must be bonded to the common grounding electrode conductor.

Exception: A separate bonding jumper from each derived system to metal water piping and to structural metal members isn't required where the metal water piping and the structural metal members in the area served by the separately derived system are bonded to the common grounding electrode conductor.

250.106 Lightning Protection System. Where a lightning protection system is installed, the lightning protection system must be bonded to the building or structure grounding electrode system. Figure 250–156

Author's Comment: The grounding electrode for a lightning protection system cannot be used for the building or structure grounding electrode [250.60]. Figure 250–157

FPN No. 1: See NFPA 780, *Standard for the Installation of Lightning Protection Systems* for additional details on grounding and bonding requirements for lightning protection.

FPN No. 2: Metal raceways, enclosures, frames, and other metal parts of electrical equipment may require bonding or spacing from the lightning protection conductors in accordance with NFPA 780, *Standard for the Installation of Lightning Protection Systems*. Separation from lightning protection conductors is typically 6 ft through air, or 3 ft through dense materials, such as concrete, brick, or wood. Figure 250–158

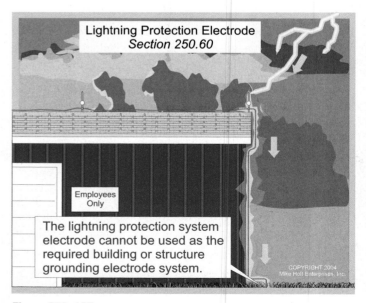

Lightning Protection Electrode
Section 250.60

Employees Only

The lightning protection system electrode cannot be used as the required building or structure grounding electrode system.

COPYRIGHT 2004
Mike Holt Enterprises, Inc.

Figure 250–157

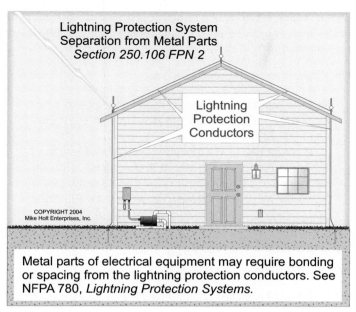

Lightning Protection System
Separation from Metal Parts
Section 250.106 FPN 2

Lightning Protection Conductors

COPYRIGHT 2004
Mike Holt Enterprises, Inc.

Metal parts of electrical equipment may require bonding or spacing from the lightning protection conductors. See NFPA 780, *Lightning Protection Systems*.

Figure 250–158

PART VI. EQUIPMENT GROUNDING (BONDING) AND EQUIPMENT GROUNDING (BONDING) CONDUCTORS

250.110 Fixed Equipment Connected by Permanent Wiring Methods—General.
Exposed metal parts of fixed equipment likely to become energized must be grounded (bonded) to an effective ground-fault current path where the fixed equipment is:

(1) Within 8 ft vertically or 5 ft horizontally of earth or grounded metal objects

(2) Located in a wet or damp location

(3) In electrical contact with metal

(4) In a hazardous (classified) location [Articles 500 through 517]

(5) Supplied by a wiring method that provides an equipment grounding (bonding) conductor of a type specified in 250.118

(6) Supplied by a 277V or 480V circuit

Exception 3: Double-insulated equipment isn't required to be grounded (bonded) to an effective ground-fault current path.

250.112 Fixed Equipment Connected by Permanent Wiring Methods—Specific.
Exposed metal parts of equipment and enclosures described in (A) through (M) must be grounded (bonded) to an effective ground-fault current path.

(A) Switchboard Frames and Structures. Switchboard frames and structures supporting switching equipment must be grounded (bonded) to an effective ground-fault current path.

(C) Motor Frames. Motor frames in accordance with 430.242.

(D) Enclosures for Motor Controllers.

(I) Power-Limited Remote-Control, Signaling, and Fire Alarm Circuits Rated Not Over 50V. Metal raceways and metal enclosures containing circuit conductors from an *electrical supply source* that operate at 50V or less aren't required to be grounded to an effective ground-fault current path [250.86].

> **Author's Comment:** For example, metal boxes used with power-limited fire alarm circuits operating at 50V or less aren't required to be bonded to an effective ground-fault path.

(J) Luminaires (Lighting Fixtures). Luminaires in accordance with 410.17 through 410.21.

(K) Skid Mounted Equipment. Electrical equipment permanently on skids (such as generators).

(M) Metal Well Casings. Where a submersible pump is used in a metal well casing, the well casing must be bonded to the pump circuit equipment grounding (bonding) conductor.

250.114 Equipment Connected by Cord and Plug.
Under any of the conditions described in (1) through (4), exposed, metal parts must be grounded (bonded) to an effective ground-fault current path [250.4(A)(3)].

(1) Equipment in a hazardous (classified) location [Articles 500 through 517]

(2) Equipment supplied by a 277V or 480V circuit

(3) The following equipment in residential occupancies must be grounded (bonded) to an effective ground-fault current path:

 a. Refrigerators, freezers, and air conditioners

 b. Clothes-washing, clothes-drying, dish-washing machines, kitchen waste disposers, information technology equipment, sump pumps, and electric aquarium equipment

 c. Hand-held motor-operated tools, stationary and fixed motor-operated tools, light industrial motor-operated tools

 d. Motor-operated appliances of the following types: hedge clippers, lawn mowers, snow blowers, and wet scrubbers

 e. Portable handlamps

(4) In other than residential occupancies, the following must be grounded (bonded) to an effective ground-fault current path:

a. Refrigerators, freezers, and air conditioners

b. Clothes-washing, clothes-drying, dish-washing machines, information technology equipment, sump pumps, and electric aquarium equipment

c. Hand-held motor-operated tools, stationary and fixed motor-operated tools, light industrial motor operated tools

d. Motor-operated appliances of the following types: hedge clippers, lawn mowers, snow blowers, and wet scrubbers

e. Portable handlamps

f. Cord-and-plug connected appliances used in damp or wet locations or by persons standing on the ground or on metal floors or working inside of metal tanks or boilers

g. Tools likely to be used in wet or conductive locations

Exception: Double-insulated tools, appliances, and equipment covered in (2) through (4) aren't required to be grounded (bonded) to an effective ground-fault current path.

Author's Comment: The equipment grounding (bonding) conductor (actually it's an equipment bonding conductor) provides the low-impedance fault-current path necessary to facilitate the operation of overcurrent protection devices in order to remove dangerous voltage potentials between conductive parts of building components and electrical systems [250.4(A)(3)].

250.118 Types of Equipment Grounding (Bonding) Conductors. The equipment grounding (bonding) con-

ductor, which serves as the effective ground-fault current path to the source, must be one or a combination of the following: **Figure 250–159**

(1) A bare or insulated conductor.

> **Author's Comment:** The equipment grounding (bonding) conductor can be copper or aluminum and must be sized in accordance with 250.122.
> To ensure that it has a low-impedance path, the equipment grounding (bonding) conductors must be installed within the same raceway, cable, or trench with the circuit conductors in accordance with 250.134(B) [300.3(b), 300.5(l), and 300.20(A)].

(2) Rigid metal conduit.

(3) Intermediate metal conduit.

(4) Electrical metallic tubing.

(5) Listed flexible metal conduit meeting the following [348.60]: **Figure 250–160**

a. The conduit terminates in fittings listed for grounding.

b. The circuit conductors are protected by overcurrent devices rated 20A or less.

c. The combined length of the conduit in the same fault return path doesn't exceed 6 ft. **Figure 250–161**

d. Where flexibility is necessary after installation, an equipment grounding (bonding) conductor must be installed in accordance with 250.102(E).

Types of Equipment Grounding (Bonding) Conductors
Section 250.118

(1) RNC
(2) RMC
(3) IMC
(4) EMT
(5) FMC (limited)
(6) LFMC (limited)

COPYRIGHT 2004 Mike Holt Enterprises, Inc.

(1) Conductors can be solid, stranded, bare, insulated, or covered.

Numbers in "( )" denote applicable section of 250.118.

(1) NM Cable   (10) MC Cable   (8) AC Cable   (9) MI Cable

Figure 250–159

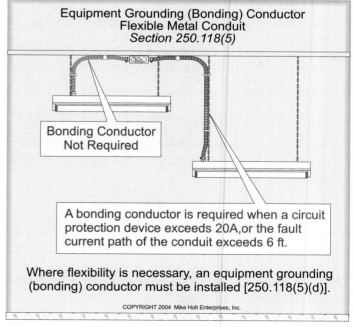

Equipment Grounding (Bonding) Conductor
Flexible Metal Conduit
Section 250.118(5)

Bonding Conductor
Not Required

A bonding conductor is required when a circuit protection device exceeds 20A, or the fault current path of the conduit exceeds 6 ft.

Where flexibility is necessary, an equipment grounding (bonding) conductor must be installed [250.118(5)(d)].

COPYRIGHT 2004 Mike Holt Enterprises, Inc.

Figure 250–160

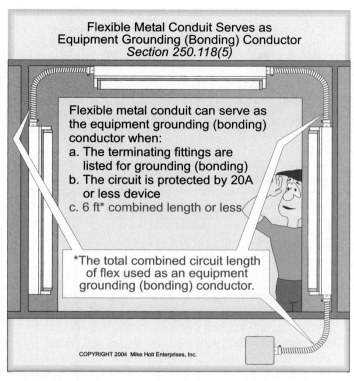

Flexible Metal Conduit Serves as
Equipment Grounding (Bonding) Conductor
Section 250.118(5)

Flexible metal conduit can serve as the equipment grounding (bonding) conductor when:
a. The terminating fittings are listed for grounding (bonding)
b. The circuit is protected by 20A or less device
c. 6 ft* combined length or less

*The total combined circuit length of flex used as an equipment grounding (bonding) conductor.

COPYRIGHT 2004 Mike Holt Enterprises, Inc.

Figure 250–161

(6) Listed liquidtight flexible metal conduit meeting the following [350.60]: **Figure 250–162**

   a. The conduit terminates in fittings listed for grounding.

   b. For ⅜ in. through ½ in., the circuit conductors are protected by overcurrent devices rated 20A or less.

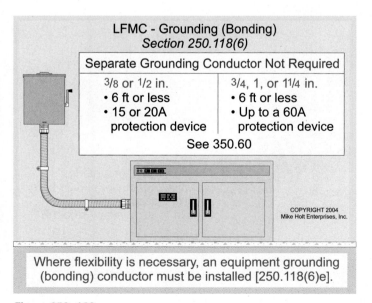

LFMC - Grounding (Bonding)
Section 250.118(6)

Separate Grounding Conductor Not Required	
3/8 or 1/2 in. • 6 ft or less • 15 or 20A protection device	3/4, 1, or 11/4 in. • 6 ft or less • Up to a 60A protection device
See 350.60	

COPYRIGHT 2004
Mike Holt Enterprises, Inc.

Where flexibility is necessary, an equipment grounding (bonding) conductor must be installed [250.118(6)e].

Figure 250–162

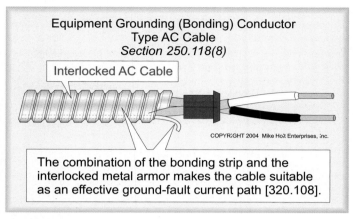

Equipment Grounding (Bonding) Conductor
Type AC Cable
Section 250.118(8)

Interlocked AC Cable

COPYRIGHT 2004 Mike Holt Enterprises, Inc.

The combination of the bonding strip and the interlocked metal armor makes the cable suitable as an effective ground-fault current path [320.108].

Figure 250–163

   c. For ¾ through 1 ¼ in., the circuit conductors are protected by overcurrent devices rated 60A or less.

   d. The combined length of the conduit in the same ground return path doesn't exceed 6 ft.

   e. Where flexibility is necessary after installation, an equipment grounding (bonding) conductor must be installed in accordance with 250.102(E) regardless of the circuit rating or the length of the flexible metal conduit.

(8) Type AC cable as provided in 320.108.

Author's Comment: Interlocked Type AC cable is manufactured with an internal bonding strip that is in direct contact with the interlocked metal armor. The combination of the bonding strip and the interlocked metal armor makes the cable suitable as an effective ground-fault current path [320.108]. **Figure 250–163**

(9) The copper metal sheath of Type MI cable.

(10) Type MC cable where listed and identified for grounding as follows:

   a. Interlocked Type MC cable containing an equipment grounding (bonding) conductor within the cable.

Author's Comment: The metal armor of interlocked Type MC cable isn't suitable as an effective ground-fault current path because it doesn't have an internal bonding strip like Type AC cable. **Figure 250–164**

   b. Smooth or corrugated-tube Type MC cable.

Author's Comment: The sheath of smooth or corrugated-tube Type MC cable is suitable as the effective ground-fault current path, therefore an internal equipment grounding (bonding) conductor isn't required within the cable.

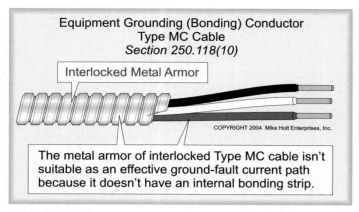

Equipment Grounding (Bonding) Conductor
Type MC Cable
Section 250.118(10)

Interlocked Metal Armor

COPYRIGHT 2004 Mike Holt Enterprises, Inc.

The metal armor of interlocked Type MC cable isn't suitable as an effective ground-fault current path because it doesn't have an internal bonding strip.

Figure 250–164

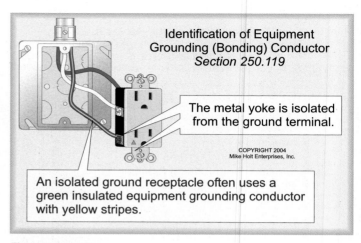

Identification of Equipment
Grounding (Bonding) Conductor
Section 250.119

The metal yoke is isolated from the ground terminal.

COPYRIGHT 2004
Mike Holt Enterprises, Inc.

An isolated ground receptacle often uses a green insulated equipment grounding conductor with yellow stripes.

Figure 250–166

(11) Metallic cable trays where continuous maintenance and supervision ensure that qualified persons service the cable tray [392.3(C)] if all the following are met [392.7]:

• Cable tray and fittings are identified for grounding.

• Cable tray, fittings, and raceways are bonded in accordance with 250.96 using bolted mechanical connectors or bonding jumpers sized in accordance with 250.102.

(13) Other electrically continuous metal raceways listed for bonding, such as metal wireways.

(14) Surface metal raceways listed for grounding.

250.119 Identification of Equipment Grounding (Bonding) Conductor. Unless required to be insulated, equipment grounding (bonding) conductors can be bare.

Author's Comment: Equipment grounding (bonding) conductors must be insulated for patient care equipment [517.13(B)] and permanently installed pool, outdoor spa, and outdoor hot tub equipment [680.21(A)(1), 680.23(F)(2), and 680.25(B)].

Individually covered or insulated equipment grounding (bonding) conductors must have a continuous outer finish that is either green or green with one or more yellow stripes.

Author's Comment: Isolated ground circuits [250.96(B)] and isolated ground receptacles [250.146(D)] frequently use a green insulated conductor for equipment and a green insulated conductor with yellow stripes for the sensitive electronic equipment. Figures 250–165 and 250–166

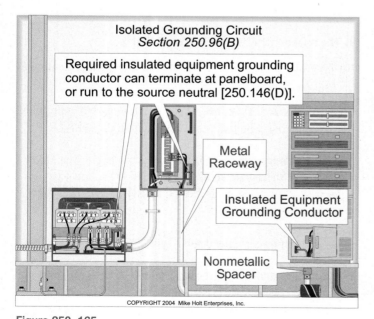

Isolated Grounding Circuit
Section 250.96(B)

Required insulated equipment grounding conductor can terminate at panelboard, or run to the source neutral [250.146(D)].

Metal
Raceway

Insulated Equipment
Grounding Conductor

Nonmetallic
Spacer

COPYRIGHT 2004 Mike Holt Enterprises, Inc.

Figure 250–165

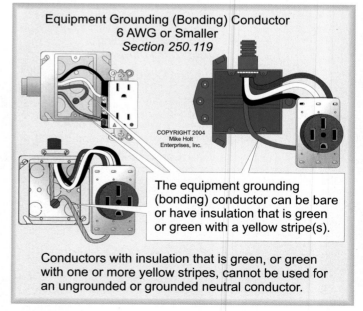

Equipment Grounding (Bonding) Conductor
6 AWG or Smaller
Section 250.119

COPYRIGHT 2004
Mike Holt
Enterprises, Inc.

The equipment grounding (bonding) conductor can be bare or have insulation that is green or green with a yellow stripe(s).

Conductors with insulation that is green, or green with one or more yellow stripes, cannot be used for an ungrounded or grounded neutral conductor.

Figure 250–167

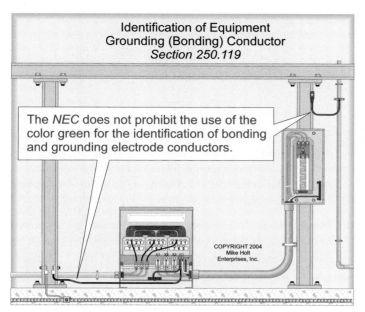

Figure 250–168

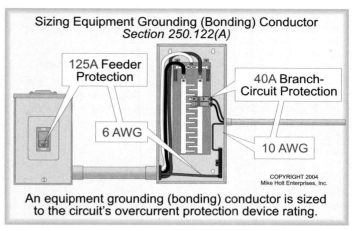

Figure 250–169

ampere rating of the circuit-protection device, but in no case is it required to be larger than the circuit conductors. **Figure 250–169**

Author's Comment: Equipment grounding (bonding) conductors must be capable of safely conducting any ground-fault current likely to be imposed on them [110.10]. If the equipment grounding (bonding) conductor isn't sized to withstand the ground-fault current, the conductor may burn clear before the protective device responds.

Conductors with insulation that is green, or green with one or more yellow stripes cannot be used for an ungrounded or grounded neutral conductor. **Figure 250–167**

Author's Comment: The *NEC* neither requires nor prohibits the use of the color green for the identification of bonding and grounding electrode conductors. **Figure 250–168**

(A) Conductors Larger Than 6 AWG

(1) Identified Where Accessible. Equipment grounding (bonding) conductors larger than 6 AWG that are insulated can be permanently reidentified at the time of installation at every point where the conductor is accessible.

Exception: Identification of equipment grounding (bonding) conductors larger than 6 AWG in conduit bodies is not required.

(2) Identification Method. Equipment grounding (bonding) conductor identification must encircle the conductor and be accomplished by one of the following:

(a) Stripping exposed insulation

(b) Coloring exposed insulation green

(c) Marking exposed insulation with green tape or green adhesive labels

250.122 Sizing Equipment Grounding (Bonding) Conductor.

(A) General. The equipment grounding (bonding) conductor must be sized in accordance with Table 250.122, based on the

Table 250.122—Minimum Size Equipment Grounding (Bonding) Conductor

Protection Rating	Copper Conductor
15A	14 AWG
20A	12 AWG
30—60A	10 AWG
70—100A	8 AWG
110—200A	6 AWG
225—300A	4 AWG
350—400A	3 AWG
450—500A	2 AWG
600A	1 AWG
700—800A	1/0 AWG
1,000A	2/0 AWG
1,200A	3/0 AWG

(B) Increased in Size. When ungrounded circuit conductors are increased in size for any reason, the equipment grounding (bonding) conductor must be proportionately increased in size.

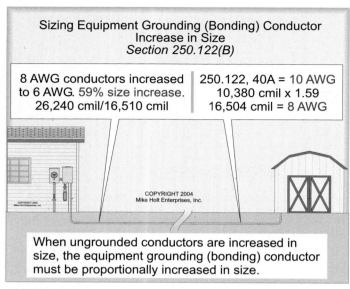

Figure 250–170

The equipment grounding (bonding) conductor for a 40A protection device can be 10 AWG (10,380 cmil) [Table 250.122], but it must be increased in size by a multiplier of 1.59.

Conductor Size = 10,380 cmil x 1.59
Conductor Size = 16,504 cmil
Conductor Size = 8 AWG, Chapter 9, Table 8

(C) Multiple Circuits. When multiple circuits are installed in the same raceway or cable, only one equipment grounding (bonding) conductor is required. This conductor must be sized in accordance with Table 250.122, based on the largest overcurrent device protecting the circuit conductors. **Figure 250–171**

(F) Parallel Runs. When circuit conductors are run in parallel [310.4], an equipment grounding (bonding) conductor must be installed with each parallel conductor set and it must be sized in accordance with (1) or (2).

(1) Based on the ampere rating of the circuit-protection device in accordance with Table 250.122. **Figure 250–172**

(2) Based on the ampere rating of the ground-fault protection in accordance with Table 250.122 where ground-fault protection of equipment is installed if: **Figure 250–173**

(1) Maintenance and supervision ensure that only qualified persons will service the installation.

(2) Ground-fault protection is set to trip at not more than the ampacity of a single ungrounded conductor.

(G) Feeder Tap Conductors. Equipment grounding (bonding) conductors for feeder taps must be sized in accordance with Table 250.122, based on the ampere rating of the circuit-protection device ahead of the feeder, but in no case is it required to be larger than the circuit conductors. **Figure 250–174**

Author's Comment: Ungrounded conductors could be increased in size to accommodate voltage drop, because of excessive heating from harmonic currents, fault-current studies, or future capacity.

Question: If the ungrounded conductors for a 40A circuit are increased in size from 8 AWG to 6 AWG, the equipment grounding (bonding) conductor must be increased in size from 10 AWG to _____. Figure 250–170

(a) 10 AWG   (b) 8 AWG   (c) 6 AWG   (d) 4 AWG

Answer: (b) 8 AWG

The circular mil area of 6 AWG is 59 percent greater than 8 AWG (26,240 cmil/16,510 cmil) [Chapter 9, Table 8].

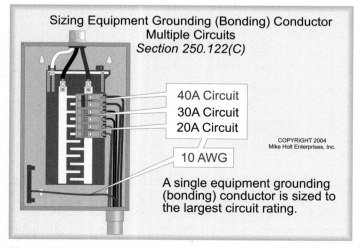

Figure 250–171

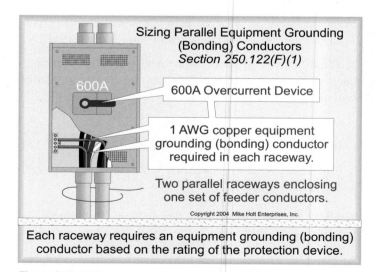

Figure 250–172

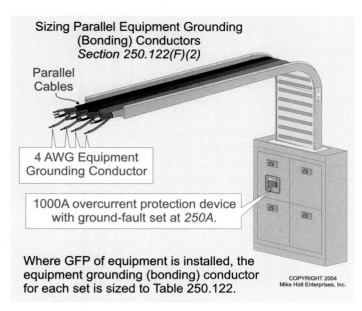

Figure 250-173

250.126 Identification of Wiring Device Terminals.

The terminal for the equipment grounding (bonding) conductor must be a:

(1) Green screw, hexagonal head, and not readily removable. Figure 250-175

(2) Green nut, hexagonal head, and not readily removable.

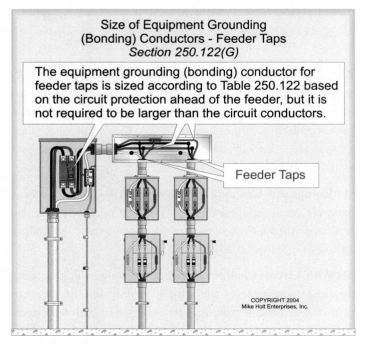

Figure 250-174

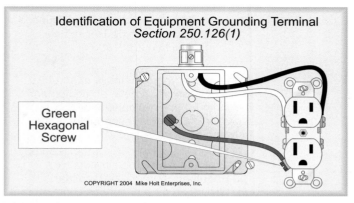

Figure 250-175

(3) Green pressure wire connector. If the terminal for the equipment grounding (bonding) conductor isn't visible, the conductor entrance hole must be marked with the word green or ground, the letters G or GR, a grounding (earthing) symbol, or otherwise identified by a distinctive green color.

PART VII. METHODS OF EQUIPMENT GROUNDING (BONDING)

250.130 Equipment Grounding (Bonding) Conductor Connections.

(A) For Grounded Systems. The service disconnecting means supplied by a grounded utility system must have the grounded neutral service conductor bonded to the service-disconnect enclosure in accordance with 250.24(C). The grounded neutral service conductor must also be bonded to the grounding (earthing) electrodes [250.24(A)] and the equipment grounding (bonding) conductors.

(B) For Ungrounded Systems. The service disconnecting means supplied by an ungrounded utility system must have the service-disconnect enclosure grounded to a grounding electrode in accordance with 250.24(E). The equipment grounding (bonding) conductors must also be bonded to the grounding (earthing) electrodes [250.130(B)].

(C) Nongrounding Receptacle Replacement or Branch-Circuit Extensions. Where a nongrounding receptacle is replaced with a grounding-type receptacle, or when a branch-circuit extension is made from an outlet box that doesn't contain an equipment grounding (bonding) conductor, the grounding contacts of the receptacle must be grounded (bonded) to one of the following (effective ground-fault current path): Figure 250-176

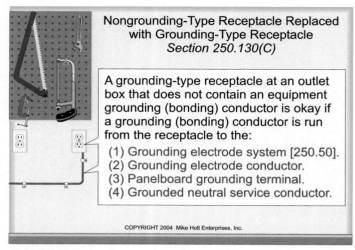

Figure 250–176

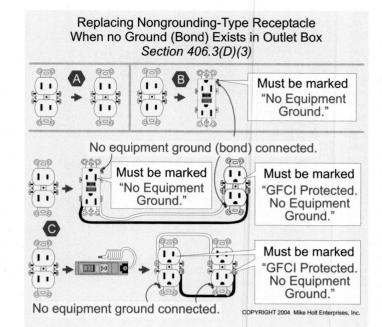

Figure 250–178

(1) Grounding electrode system [250.50] (which is bonded to the grounded neutral service conductor).

(2) Grounding electrode conductor (which is bonded to the grounded neutral service conductor).

(3) Panelboard equipment grounding terminal.

(4) Grounded neutral service conductor.

Author's Comment: A branch-circuit extension is not permitted to be made from an outlet box that doesn't have an equipment grounding (bonding) conductor, unless an equipment grounding (bonding) conductor for the circuit extension is run as specified in 250.130(C)(1) through (4). **Figure 250–177**

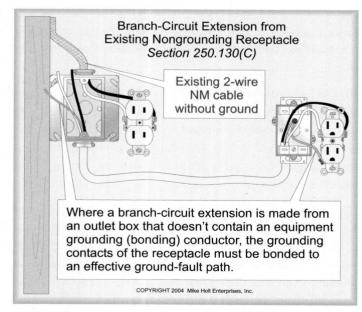

Figure 250–177

FPN: A grounding-type receptacle can replace a nongrounding type receptacle, without having the grounding terminal grounded (bonded) to an effective ground-fault current path, if the receptacle is GFCI protected and marked in accordance with 406.3(D)(3). **Figure 250–178**

Author's Comment: But this rule does not apply to new grounding-type receptacles placed on a branch-circuit extension from an existing outlet box that does not have an equipment grounding (bonding) conductor.

250.134 Grounding (Bonding)—Fixed Equipment.

Metal parts of fixed electrical equipment, raceways, and enclosures must be grounded (bonded) to an effective ground-fault current path in accordance with (A) or (B), except where they are grounded (bonded) to the grounded neutral conductor as permitted by 250.142.

(A) Equipment Grounding (Bonding) Types. Metal parts of fixed electrical equipment, raceways, and enclosures can be grounded (bonded) by any of the equipment grounding (bonding) types specified in 250.118.

(B) With Circuit Conductors. Where an equipment grounding (bonding) conductor of the wire type provides the effective ground-fault current path, it must be installed with the circuit conductors in the same raceway, cable tray, trench, cable, or cord. **Figures 250–179 and 250–180**

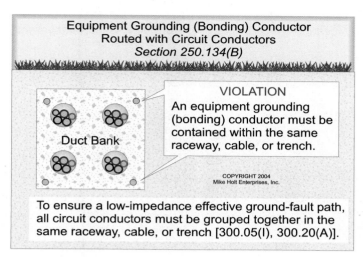

Figure 250–179

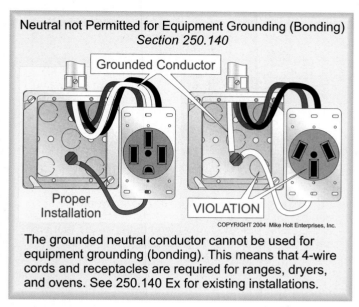

The grounded neutral conductor cannot be used for equipment grounding (bonding). This means that 4-wire cords and receptacles are required for ranges, dryers, and ovens. See 250.140 Ex for existing installations.

Figure 250–181

Exception 1: The equipment grounding (bonding) conductor can be run separately from the circuit conductors when necessary to comply with 250.130(C).

FPN No. 1: Where the equipment bonding jumper is installed outside a raceway, its length must not exceed 6 ft and it must be routed with the raceway [250.102(E)].

250.138 Grounding (Bonding)—Cord-and-Plug Connected Equipment.

(A) Equipment Grounding (Bonding) Conductor. Metal parts of cord-and-plug connected equipment must be grounded (bonded) to an effective ground-fault current path by an equipment grounding (bonding) conductor within the flexible cord that terminates in a grounding-type attachment plug.

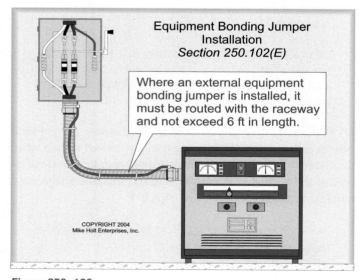

Figure 250–180

250.140 Grounding (Bonding)—Ranges, Ovens, and Clothes Dryers.

The frames of electric ranges, wall-mounted ovens, counter-mounted cooking units, clothes dryers, and outlet boxes that are part of the circuit for these appliances must be grounded (bonded) to an effective ground-fault current path by an equipment grounding (bonding) conductor of a type specified in 250.118 [250.134(A)]. Figure 250–181

> **CAUTION:** *Ranges, dryers, and ovens have their metal cases bonded to the grounded neutral conductor at the factory. This neutral-to-case bond must be removed when these appliances are installed in new construction, and a 4-wire cord and receptacle must be used.*

Exception: For existing branch-circuit installations, where an equipment grounding (bonding) conductor isn't present in the outlet box, the frames of electric ranges, wall-mounted ovens, counter-mounted cooking units, clothes dryers, and outlet boxes that are part of the circuit for these appliances can be grounded (bonded) to the grounded neutral conductor if all the following conditions are met: Figure 250–182

(1) The circuit is 120/240V or 120/208V.

(2) The grounded neutral conductor is not smaller than 10 AWG copper or 8 AWG aluminum.

(3) The grounded neutral conductor is insulated, or it's uninsulated and part of a Type SE cable and the branch circuit originates at service equipment.

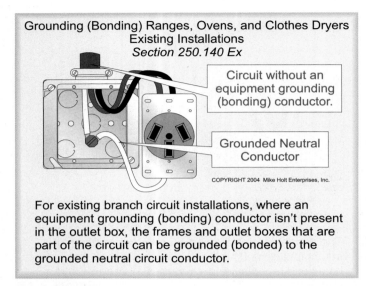

Figure 250–182

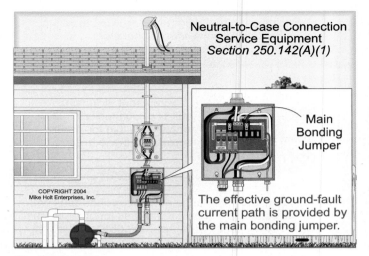

Figure 250–184

250.142 Use of Grounded Neutral Conductor for Equipment Grounding (Bonding).

Author's Comment: To remove dangerous voltage on metal parts from a ground fault, the metal parts of electrical raceways, cables, enclosures, and equipment must be bonded to an effective ground-fault current path in accordance with 250.4(A)(3).

(A) Supply Side of Service Equipment. A grounded neutral conductor can be used as the effective ground-fault current path for metal parts of equipment, raceways, and other enclosures.

(1) Service Equipment. Because an equipment grounding (bonding) conductor isn't run from the utility to electrical services, the grounded neutral service conductor can serve as the effective ground-fault current path to the utility power source. Figure 250–183

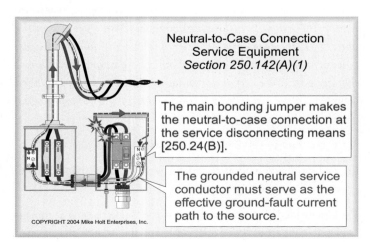

Figure 250–183

Author's Comment: The effective ground-fault current path for service equipment is provided by the installation of the main bonding jumper at service equipment in accordance with 250.24(B) [250.28]. **Figure 250–184**

(2) Separate Buildings and Structures. Where no equipment grounding (bonding) conductor is run to a building or structure disconnect, the grounded neutral conductor can serve as the effective ground-fault current path to the power source.

Author's Comment: This is accomplished by bonding the grounded neutral conductor to the equipment grounding (bonding) conductor at the separate building or structure building disconnecting means in accordance with 250.32(B)(2).

CAUTION: *Using the grounded neutral conductor as the effective ground-fault current path poses potentially dangerous consequences and should only be done after careful consideration.*

Author's Comment: The safest practice is to install an equipment grounding (bonding) conductor with the feeder conductors to the building or structure to serve as the effective ground-fault current path, as provided by 250.32(B)(1).

(3) Separately Derived Systems. The effective ground-fault current path is established for a separately derived system when the system bonding jumper is installed between the metal enclosure of the separately derived system and the grounded neutral terminal in accordance with 250.30(A)(1).

DANGER: *Failure to install the system bonding jumper as required by 250.30(A)(1) will create a condition where dangerous touch voltage from a ground fault will remain on the metal parts of electrical equipment.*

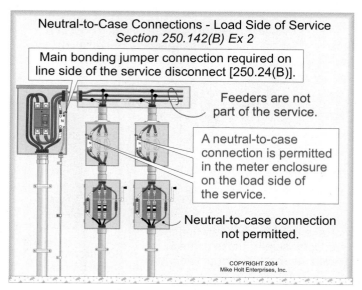

Figure 250–185

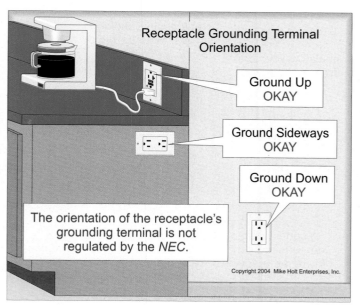

Figure 250–187

(B) Load-Side Equipment. To prevent dangerous voltage on metal parts, the grounded neutral conductor must not be bonded to the equipment grounding (bonding) conductor on the load side of service equipment, except as permitted by 250.142(A).

Exception 1: The grounded neutral conductor can serve as the effective ground-fault current path for existing ranges, dryers, and ovens [250.140 Ex].

Exception 2: The grounded neutral conductor can be bonded to the meter enclosure on the load side of the service disconnecting means if: Figure 250–185

(a) No service ground-fault protection is installed,

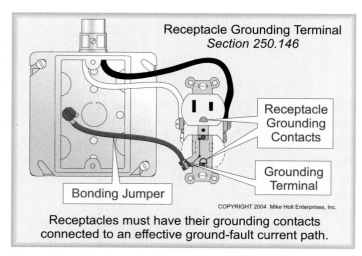

Receptacles must have their grounding contacts connected to an effective ground-fault current path.

Figure 250–186

(b) Meter enclosures are located immediately adjacent to the service disconnecting means, and

(c) The grounded neutral conductor is sized not smaller than specified in Table 250.122.

250.146 Connecting Receptacle Grounding Terminal to Box.
Receptacles must have their grounding contacts connected to an effective ground-fault current path by bonding the receptacle's grounding terminal to a metal box, unless the receptacle's grounding terminal is grounded (bonded) to an effective ground-fault current path by one of the methods provided in (A) through (D). See 406.3 for additional details. Figure 250–186

> **Author's Comment:** The *NEC* does not restrict the position of the receptacle grounding terminal; it can be up, down, or sideways. All *Code* proposals to specify the mounting position of receptacles have been rejected. Figure 250–187

(A) Surface-Mounted Box. Where the box is mounted on the surface, direct metal-to-metal contact between the device yoke and the box can serve as the effective ground-fault current path. To ensure an effective ground-fault current path between the receptacle and metal box, at least one of the insulating retaining washers on the yoke screw must be removed. Figure 250–188

Receptacles secured to a metal cover [406.4(C)] must have the receptacle's grounding terminal bonded to the box, unless the box and cover are listed as providing continuity between the box and the receptacle. Figure 250–189

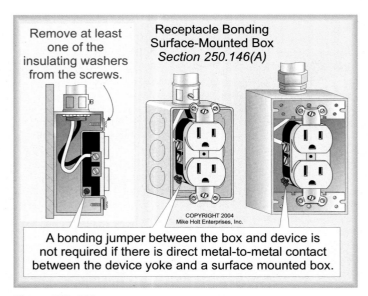

Remove at least one of the insulating washers from the screws.

Receptacle Bonding
Surface-Mounted Box
Section 250.146(A)

COPYRIGHT 2004
Mike Holt Enterprises, Inc.

A bonding jumper between the box and device is not required if there is direct metal-to-metal contact between the device yoke and a surface mounted box.

Figure 250–188

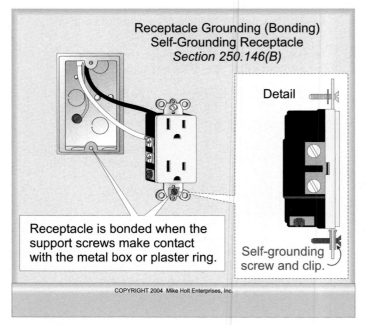

Receptacle Grounding (Bonding)
Self-Grounding Receptacle
Section 250.146(B)

Detail

Receptacle is bonded when the support screws make contact with the metal box or plaster ring.

Self-grounding screw and clip.

COPYRIGHT 2004 Mike Holt Enterprises, Inc.

Figure 250–190

(B) Self-Grounding Receptacles. Receptacle yokes designed and listed as self-grounding can be used to establish the effective ground-fault current path between the device yoke and a metal outlet box. Figure 250–190

> **Author's Comment:** Outlet boxes cannot be set back more than ¼ in. from the finished mounting surface [314.20].

(C) Floor Boxes. Listed floor boxes are permitted to establish the bonding path between the device yoke and a grounded (bonded) outlet box.

(D) Isolated Ground Receptacles. Isolated ground receptacles have the grounding terminal insulated from its metal mounting yoke. Therefore, the grounding terminal of an isolated ground receptacle must be connected to an equipment grounding (bonding) conductor that provides the effective ground-fault current path to the power source. Figure 250–191

> **Author's Comment:** Isolated ground receptacles must be identified by an orange triangle located on the face of the receptacle [406.2(D)]. Sometimes the entire receptacle is orange, with the triangle molded into the plastic face in a color other than orange.

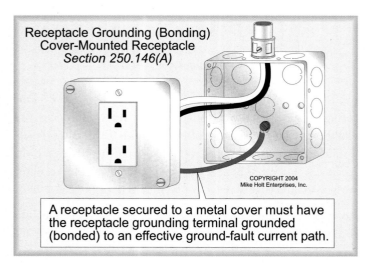

Receptacle Grounding (Bonding)
Cover-Mounted Receptacle
Section 250.146(A)

COPYRIGHT 2004
Mike Holt Enterprises, Inc.

A receptacle secured to a metal cover must have the receptacle grounding terminal grounded (bonded) to an effective ground-fault current path.

Figure 250–189

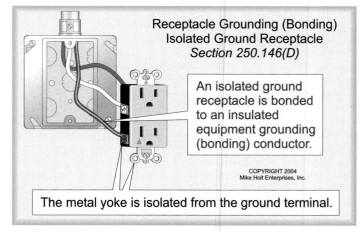

Receptacle Grounding (Bonding)
Isolated Ground Receptacle
Section 250.146(D)

An isolated ground receptacle is bonded to an insulated equipment grounding (bonding) conductor.

COPYRIGHT 2004
Mike Holt Enterprises, Inc.

The metal yoke is isolated from the ground terminal.

Figure 250–191

DANGER: *Some digital equipment manufacturers insist that their equipment be electrically isolated from the building or structure's grounding (earthing) system. This is a dangerous practice, and it violates 250.4(A)(5), which prohibits the earth to be used as an effective ground-fault current path. If the metal enclosures of sensitive electronic equipment were isolated or floated as required by some sensitive equipment manufacturers, dangerous voltage from a ground fault would remain an metal parts.*

Author's Comment: For more information on how to properly ground sensitive electronic equipment, visit www.MikeHolt.com, click on "Technical," then on "Grounding."

WARNING: *The outer metal sheath of interlocked Type MC cable isn't listed as an equipment grounding (bonding) conductor [250.118(10)]; therefore, this wiring method can't be used to supply an isolated ground receptacle, unless the cable contains two equipment grounding (bonding) conductors of the wire type.*

However, interlocked Type AC cable containing an insulated equipment grounding (bonding) conductor of the wire type can be used to supply isolated ground receptacles, because the metal armor of the cable is listed as an equipment grounding (bonding) conductor [250.118(8)]. Figure 250–192

Author's Comment: When should an isolated ground receptacle be installed and how should the isolated ground system be designed? These questions are design issues and cannot be answered based on the *NEC* alone [90.1(C)]. In most cases the use of an isolated ground receptacle is a waste of money. For example, IEEE 1100, *Powering and Grounding Sensitive Electronic Equipment* (Emerald Book) states, "The results from the use of the isolated ground method range from no observable effects, the desired effects, or worse noise conditions than when standard equipment bonding configurations are used to serve electronic load equipment [8.5.3.2]."

In reality, few electrical installations truly require an isolated ground system. For those systems that could benefit from an isolated ground system, engineering opinions differ as to what is a proper design. Making matters worse—of those that are properly designed, few are installed correctly and even fewer are properly maintained. For more information on how to properly ground sensitive electronic equipment, go to: http://mikeholt.com, click on the Technical link, and then visit the Power Quality page.

FPN: Metal raceways and metal enclosures containing an insulated equipment grounding (bonding) conductor must be grounded to an effective ground-fault current path [250.86].

250.148 Continuity and Attachment of Equipment Grounding (Bonding) Conductors to Boxes.

Equipment grounding (bonding) conductors associated with circuit conductors that are spliced or terminated on equipment within a metal outlet box, must be spliced together or joined to the box with devices suitable for the purpose. Figure 250–193

Isolated Ground Receptacle
Wiring Methods
Section 250.146(D)

OKAY - AC Cable with an insulated equipment grounding (bonding) conductor.

OKAY - Spiral interlock MC Cable with two equipment grounding (bonding) conductors.

VIOLATION - Spiral interlock MC Cable with one equipment grounding (bonding) conductor.

COPYRIGHT 2004 Mike Holt Enterprises, Inc.

Figure 250–192

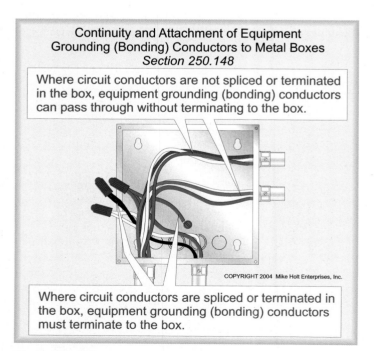

Continuity and Attachment of Equipment Grounding (Bonding) Conductors to Metal Boxes
Section 250.148

Where circuit conductors are not spliced or terminated in the box, equipment grounding (bonding) conductors can pass through without terminating to the box.

Where circuit conductors are spliced or terminated in the box, equipment grounding (bonding) conductors must terminate to the box.

COPYRIGHT 2004 Mike Holt Enterprises, Inc.

Figure 250–193

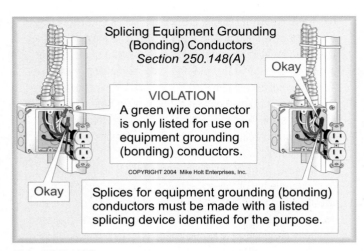

Figure 250–194

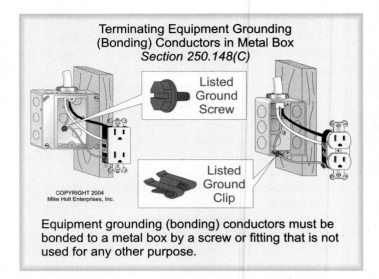

Figure 250–196

Exception: The equipment grounding (bonding) conductor for isolated ground receptacles [250.146(D)] isn't required to terminate to the metal outlet box.

(A) Splicing. Equipment grounding (bonding) conductors must be spliced with a listed splicing device that is identified for the purpose [110.14(B)]. Figure 250–194

> **Author's Comment:** Wire connectors of any color can be used with equipment grounding (bonding) conductor splices, but green wire connectors can only be used with equipment grounding (bonding) conductors.

(B) Grounding (Bonding) Continuity. Equipment grounding (bonding) connections must be made so that the disconnection or

the removal of a receptacle, luminaire, or other device will not interrupt the effective ground-fault current path. Figure 250–195

(C) Metal Boxes. Where equipment grounding (bonding) conductors enter a metal box, they must be bonded to the box by a screw or device (i.e. ground clip) that is not used for any other purpose. Figure 250–196

> **Author's Comment:** Equipment grounding (bonding) conductors aren't permitted to terminate to a cable clamp or screw that secures the plaster (mud) ring to the box. Figure 250–197

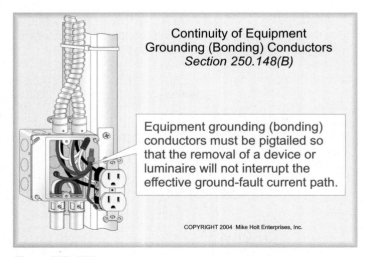

Figure 250–195

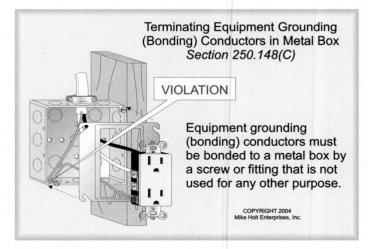

Figure 250–197

1. A ground-fault current path is an electrically conductive path from the point of a line-to-case fault extending to the _____.

   (a) ground   (b) earth   (c) electrical supply source   (d) none of these

2. For grounded systems, non current-carrying conductive materials enclosing electrical conductors or equipment, or forming part of such equipment, must be connected to earth so as to limit the voltage-to-ground on these materials.

   (a) True   (b) False

3. For ungrounded systems, noncurrent-carrying conductive materials enclosing electrical conductors or equipment, or forming part of such equipment, must be connected to earth in a manner that will limit the voltage imposed by lightning or unintentional contact with higher-voltage lines.

   (a) True   (b) False

4. Grounding electrode conductor fittings must be protected from physical damage by being enclosed in _____ where there may be a possibility of physical damage.

   (a) metal   (b) wood   (c) the equivalent of a or b   (d) none of these

5. An alternate ac power source such as an onsite generator is not a separately derived system if the _____ is solidly interconnected to a service-supplied system neutral.

   (a) ignition system   (b) fuel cell   (c) neutral   (d) line conductor

6. When service-entrance conductors exceed 1,100 kcmil for copper, the required grounded neutral conductor for the service must be sized not less than _____ percent of the area of the largest ungrounded service-entrance (phase) conductor.

   (a) 15   (b) 19   (c) 12½   (d) 25

7. For a single separately derived system, the grounding electrode conductor connects the grounding electrode to the grounded neutral conductor of the derived system at the same point on the separately derived system where the _____ is installed.

   (a) metering equipment   (b) transfer switch   (c) bonding jumper   (d) largest circuit breaker

8. When supplying a grounded system at a separate building or structure, if the equipment grounding (bonding) conductor is run with the supply conductors and connected to the building disconnecting means, there must be no connection made between the grounded neutral conductor and the equipment grounding (bonding) conductor at the separate building.

   (a) True   (b) False

9. Where none of the items in 250.52(A)(1) through (A)(6) are present for use as a grounding electrode, one or more of the following must be installed and used as the grounding electrode: _____.

   (a) a ground ring   (b) rod and pipe electrodes or plate electrodes
   (c) local metal underground systems or structures   (d) any of these

10. Electrodes of pipe or conduit must not be smaller than _____ and, where of iron or steel, must have the outer surface galvanized or otherwise metal-coated for corrosion protection.

(a) ½ in.   (b) ¾ in.   (c) 1 in.   (d) none of these

11. When driving a ground rod electrode, if rock bottom is encountered, the rod is allowed to be bent over in a trench and buried or shortened with a hack saw.

(a) True   (b) False

12. Where separate services supply a building and are required to be connected to a grounding electrode, the same grounding electrode must be used. Two or more grounding electrodes that are _____ are considered as a single grounding electrode system in this sense.

(a) effectively bonded together   (b) spaced no more than 6 ft apart
(c) a and b   (d) none of these

13. The grounding electrode conductor must be installed in one continuous length without a splice or joint, unless spliced _____.

(a) by connecting to a busbar
(b) by irreversible compression-type connectors listed as grounding and bonding
(c) by the exothermic welding process
(d) any of these

14. In an ac system, the size of the grounding electrode conductor to a concrete-encased electrode is not required to be larger than _____ copper wire.

(a) 4 AWG   (b) 6 AWG   (c) 8 AWG   (d) 10 AWG

15. The grounding conductor connection to the grounding electrode must be made by _____.

(a) listed lugs   (b) exothermic welding   (c) listed pressure connectors   (d) any of these

16. Bonding must be provided where necessary to ensure _____ and the capacity to conduct safely any fault current likely to be imposed.

(a) electrical continuity   (b) fiduciary responsibility   (c) listing requirements   (d) electrical demand

17. An accessible means external to enclosures for connecting intersystem _____ conductors must be provided at the service equipment and at the disconnecting means.

(a) bonding   (b) grounding   (c) secondary   (d) a and b

18. Regardless of the voltage of the electrical system, the electrical continuity of non-current carrying metal parts of equipment, raceways, and other enclosures in any hazardous (classified) location as defined in Article 500 must be ensured by any of the methods specified in 250.92(B)(2) through (B)(4). One or more of these _____ methods must be used whether or not supplementary equipment grounding (bonding) conductors are installed.

(a) grounded   (b) securing   (c) sealing   (d) bonding

19. What is the minimum size copper bonding jumper for a service raceway containing 4/0 THHN aluminum conductors?

(a) 6 AWG aluminum   (b) 3 AWG copper   (c) 4 AWG aluminum   (d) 4 AWG copper

20. Metal gas piping is considered bonded by the circuit's equipment grounding (bonding) conductor of the circuit that is likely to energize the piping.

    (a) True   (b) False

21. Which of the following appliances installed in residential occupancies need not be grounded?

    (a) Toaster   (b) Aquarium   (c) Dishwasher   (d) Refrigerator

22. An equipment grounding (bonding) conductor must be identified by _____.

    (a) a continuous outer finish that is green
    (b) being bare
    (c) a continuous outer finish that is green with one or more yellow stripes
    (d) any of these

23. When ungrounded conductors are increased in size, the equipment grounding (bonding) conductor is not required to be increased because it is not a current-carrying conductor.

    (a) True   (b) False

24. Cord-and-Plug connected equipment must be grounded by means of _____.

    (a) an equipment grounding (bonding) conductor in the cable assembly   (b) a separate flexible wire or strap   (c) either a or b   (d) none of these

25. Where a metal box is surface-mounted, direct metal-to-metal contact between the device yoke and the box is permitted to ground the receptacle to the box. Unless the receptacle is listed as _____, at least one of the insulating retaining washers must be removed from the receptacle to ensure direct metal-to-metal contact between the device yoke and metal outlet box.

    (a) self-grounding   (b) weatherproof   (c) metal contact sufficient   (d) isolated grounding

Notes

280 Surge Arresters

Introduction

This article covers general requirements, installation requirements, and connection requirements for surge arresters installed on premises wiring systems. Surge arresters are generally installed on the supply side (line side) of the service disconnecting means [280.22]. Some surge arresting devices are dual-listed for use on either the line side or load side of the service disconnecting means.

Transient voltage surges are short-term deviations from a desired voltage level, of high enough magnitude to cause equipment malfunction or damage. Surge arresters are designed to reduce transient voltages present on utility power lines and other line-side equipment. A surge arrester limits surge voltages by discharging or bypassing surge current. It prevents continued flow of follow current without sacrificing itself in the process so it can repeatedly provide the intended protection.

PART I. GENERAL

280.1 Scope. Article 280 covers the installation and connection requirements for surge arresters that are permanently installed on premises wiring systems.

280.2 Definition. A surge arrester is a protective device intended to limit surge voltages by discharging or bypassing surge current. It also prevents continued flow of follow current while remaining capable of repeating these functions.

> **Author's Comment:** The components in the surge protection device are not conductive for normal circuit voltage, but become temporarily conductive for the higher surge voltage. Figure 280–1

280.3 Number Required. Where used, a surge arrester must be connected to each ungrounded conductor of the system.

280.4 Surge Arrester Selection.

(A) Circuits of Less Than 1,000V.

(1) The rating of the surge arrester must be equal to or greater than the maximum phase-to-metal case voltage at the point of connection.

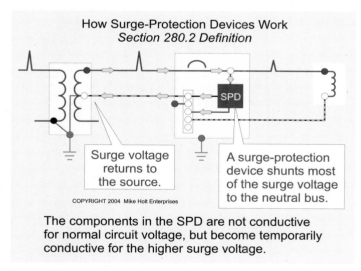

How Surge-Protection Devices Work
Section 280.2 Definition

Surge voltage returns to the source.

A surge-protection device shunts most of the surge voltage to the neutral bus.

COPYRIGHT 2004 Mike Holt Enterprises

The components in the SPD are not conductive for normal circuit voltage, but become temporarily conductive for the higher surge voltage.

Figure 280–1

(2) Surge arresters installed on circuits of less than 1,000V must be listed.

(3) Surge arresters must be marked with a short-circuit current rating and must not be installed at a point on the system where the available fault current exceeds that rating.

(4) Surge arrestors must not be installed on an ungrounded system, impedance grounded system, or corner-grounded delta system, unless listed specifically for use on these systems.

> **FPN 2:** See the manufacturer's application requirements for the selection of an arrester to be used at a particular location.

PART II. INSTALLATION

280.11 Location. Surge arresters can be located indoors or outdoors.

280.12 Routing of Conductors. Surge arrester conductors must not be longer than necessary, and unnecessary bends should be avoided.

PART III. CONNECTING SURGE ARRESTERS

280.21 Installed at Services of Less Than 1,000V.
Surge arrester conductors on the line side of service equipment must not be smaller than 14 AWG copper, and the grounding (earthing) conductor must be connected to one of the following:

(1) Grounded neutral service conductor

(2) Grounding electrode conductor

(3) Grounding electrode for the service

(4) Equipment grounding terminal in the service equipment

280.22 Installed on the Load Side of Services of Less Than 1,000V.
Surge-arrester conductors installed on the load side of service equipment must not be smaller than 14 AWG copper.

Article 280 Questions

1. A surge arrester is a protective device for limiting surge voltages by _____ or bypassing surge current.

    (a) decreasing   (b) discharging   (c) limiting   (d) derating

2. Line and ground-connecting conductors for a surge arrester must not be smaller than _____ AWG copper.

    (a) 14   (b) 12   (c) 10   (d) 8

3. The conductor between a surge arrester and the line and the grounding connection must not be smaller than _____ AWG copper for installations operating at 1 kV or more.

    (a) 4   (b) 6   (c) 8   (d) 2

Notes

285 Transient Voltage Surge Suppressors (TVSSs)

Introduction

This article covers general requirements, installation requirements, and connection requirements for transient voltage surge suppressors (TVSSs) permanently installed on premises wiring systems. It doesn't apply to cord-and-plug connected units such as "computer power strips."

TVSS devices generally apply to the load side of the service disconnecting means [285.21]. Transient voltage surges are short-term deviations from a desired voltage level, of high enough magnitude to cause equipment malfunction or damage. TVSS devices are designed to reduce transient voltages present on premises power distribution wiring and load-side equipment, particularly sensitive electronic equipment such as computers, telecommunications equipment, security systems, and electronic appliances.

A TVSS is designed to *shunt* the current and *clamp* the voltage. This is a key difference between a TVSS and a surge arrester. A TVSS clamps at a given voltage, depending on its rating, regardless of the actual surge current at its input terminals. This allows a TVSS to provide a more precise level of protection than a surge arrester can.

PART I. GENERAL

285.1 Scope. Article 285 covers the installation and connection requirements for transient voltage surge suppressors (TVSSs) that are permanently installed on premises wiring systems. Figure 285–1

285.2 Definition.

Transient voltage surge suppressor (TVSS). A protective device for limiting transient voltages that diverts or limits surge current; it also prevents continued flow of follow current while remaining capable of repeating these functions.

285.3 Uses Not Permitted. A TVSS cannot be used in:

(1) Circuits that exceed 600V.

(2) Ungrounded systems, impedance grounded systems, or corner-grounded delta systems, unless listed specifically for use on these systems.

(3) Where the voltage rating of the TVSS is less than the maximum continuous phase-to-metal case voltage available at the point of connection.

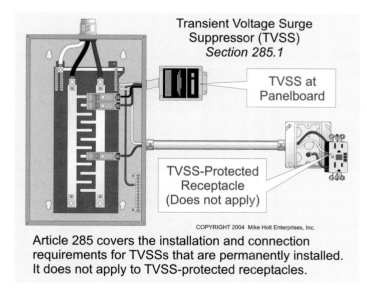

Transient Voltage Surge Suppressor (TVSS)
Section 285.1

TVSS at Panelboard

TVSS-Protected Receptacle (Does not apply)

COPYRIGHT 2004 Mike Holt Enterprises, Inc.

Article 285 covers the installation and connection requirements for TVSSs that are permanently installed. It does not apply to TVSS-protected receptacles.

Figure 285–1

285.4 Number Required. Where used, the TVSS must be connected to each ungrounded conductor of the circuit.

285.5 Listing. A TVSS must be listed.

Author's Comment: According to UL 1449, *Transient Voltage Surge Suppressors,* these units are intended to limit the maximum amplitude of transient voltage surges on power lines to specified values. They aren't intended to function as lightning arresters. The adequacy of the voltage suppression level to protect connected equipment from voltage surges has not been evaluated.

285.6 Short-Circuit Current Rating. TVSS devices must be marked with their short-circuit current rating, and they must not be installed where the available fault current exceeds that rating. This short-circuit current marking requirement does not apply to receptacles containing TVSS protection.

WARNING: *TVSS devices of the series type are susceptible to failure at high fault currents. A hazardous condition is present if the TVSS short-circuit current rating is less than the available fault current.*

PART II. INSTALLATION

285.11 Location. TVSS devices can be located indoors or outdoors.

285.12 Routing of Conductors. TVSS conductors must not be any longer than necessary, and unnecessary bends should be avoided.

PART III. CONNECTING TRANSIENT VOLTAGE SURGE SUPPRESSORS

285.21 Connection. Where a TVSS is installed, it must be connected as follows:

(A) Location.

(1) Service Supplied Building or Structure. A TVSS must be located on the load side of service equipment, unless installed in accordance with 230.82(8). Figure 285–2

Exception: A TVSS that is also listed as a surge arrester [Article 280] can be connected to the line side of the service disconnecting means device.

Author's Comments:
- Only one conductor can be connected to a terminal, unless the terminal is identified for multiple conductors [110.14(A)]. Figure 285–3
- TVSS devices are listed for installation only on the load side of service equipment. They cannot be installed on the line

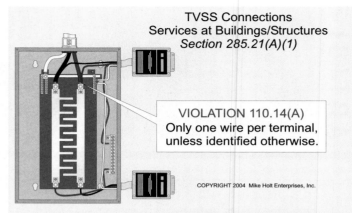

TVSS Connections
Services at Buildings/Structures
Section 285.21(A)(1)

VIOLATION 110.14(A)
Only one wire per terminal, unless identified otherwise.

COPYRIGHT 2004 Mike Holt Enterprises, Inc.

TVSSs must be connected on the load side of the service equipment, unless installed in accordance with 230.82(8).

Figure 285–2

side of the building or structure overcurrent device because of the concern that they might be exposed to lightning-induced surges.

(2) Feeder Supplied Building or Structure. A TVSS can be connected anywhere on the premises wiring system, but not on the line side of the building or structure disconnect overcurrent device.

Author's Comment: TVSS devices cannot be installed on the line side of the building or structure overcurrent device because there's concern that they might be exposed to lightning-induced surges.

(3) Separately Derived System. A TVSS can be connected anywhere on the premises wiring of the separately derived system, but not on the line side of the separately derived system overcurrent device.

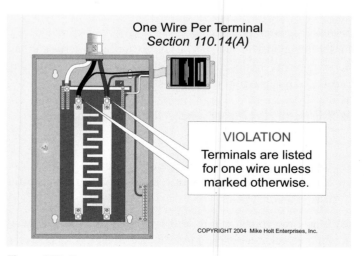

One Wire Per Terminal
Section 110.14(A)

VIOLATION
Terminals are listed for one wire unless marked otherwise.

COPYRIGHT 2004 Mike Holt Enterprises, Inc.

Figure 285–3

1. Article 285 covers surge arresters.

    (a) True   (b) False

2. The scope of Article 285 applies to devices such as cord-and-plug connected TVSS units or receptacles, or appliances that have integral TVSS protection.

    (a) True   (b) False

3. A TVSS is listed to limit transient voltages by diverting or limiting surge current.

    (a) True   (b) False

4. TVSSs must be marked with their short-circuit current rating, and they must not be installed where the available fault current is in excess of that rating.

    (a) True   (b) False

5. The conductors for the TVSS cannot be any longer than _____, and unnecessary bends should be avoided.

    (a) 6 in.   (b) 12 in.   (c) 18 in.   (d) none of these

In-House Electrical Training

Electrical Theory Library

Understanding the NEC Library

Grounding vs. Bonding Library

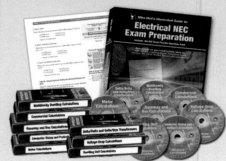

Exam Prep Library

PowerPoint Presentations

Code Change Library

Comprehensive Materials • Proven Training Methods • Learn At Your Own Pace

Our In-House Electrical Training Programs allow your staff to learn at their own pace and in familiar surroundings. We supply the material—you set the schedule. Our in-house training program has everything you'll need at drastically reduced pricing. Programs include 2002 or 2005 NEC Code® cycle textbooks, videos, DVDs, MP3s, and PowerPoint presentations. Well trained employees are your most valuable asset. Call Freddy to find out how he can help you develop an in-house electrical training program to suit your company's needs. Be sure to ask how you can qualify to receive our free PowerPoint presentations.

Contact Our In-House Training Specialist, Freddy Snow 1.888.NEC®CODE

CHAPTER 3
Wiring Methods and Materials

Introduction

Chapter 2 provided the general rules for wiring and protection of conductors and is primarily concerned with correct installation of circuits and the means of protecting them. This differs from the purpose of Chapter 3, which is to correctly size and install the conductors that make up those circuits.

Chapter 2 was a bit of an uphill climb, because many rules had a kind of abstract quality to them. Chapter 3, on the other hand, gets very specific about conductors, cables, boxes, raceways, and fittings. It's also highly detailed about the installation and restrictions involved with wiring methods.

It is because of that detail that many people incorrectly apply Chapter 3 wiring methods. You need to pay careful attention to the details, rather than making the mistake of glossing over something. This is especially true when it comes to applying the Tables.

The type of wiring method you will use depends on several factors: *Code* requirements, environment, need, and cost are among them.

Power quality is a major concern today. The cost of poor power quality runs into the millions of dollars each month in the United States alone. Grounding, actually bonding deficiencies (refer back to Article 250), constitute the number one cause of power quality problems. Violations of Chapter 3 wiring methods constitute the number two cause of power quality problems. *Code* violations in general lead to fire and other hazards. This is particularly true of Chapter 3 violations.

Chapter 3 is really a modular assembly of articles, each detailing a specific area of electrical installation. It starts with *wiring methods* [Article 300], covers *conductors* [Article 310], then *enclosures* [Articles 312 and 314]. The next string [Articles 320–340] addresses specific *types of cables*, with Articles 342 through 390 covering specific *types of raceways*. We close with Article 392, a support system, and the last string [Articles 394–398] for *open wiring*.

Wiring Method Articles

Article 300. Wiring Methods. Article 300 contains the general requirements for all wiring methods included in the *NEC*, except for signaling and communications systems, which are covered in Chapters 7 and 8.

Article 310. Conductors for General Wiring. This article contains the general requirements for conductors, such as insulation markings, ampacity ratings, and conductor use. Article 310 doesn't apply to conductors that are part of cable assemblies, flexible cords, fixture wires, or conductors that are an integral part of equipment [90.6, 300.1(B)].

Article 312. Cabinets, Cutout Boxes, and Meter Socket Enclosures. Article 312 covers the installation and construction specifications for cabinets, cutout boxes, and meter socket enclosures.

Article 314. Outlet, Device, Pull and Junction Boxes, Conduit Bodies, Fittings, and Handhole Enclosures. Article 314 contains installation requirements for outlet boxes, pull and junction boxes, as well as conduit bodies, and handhole enclosures.

Cable Articles

Articles 320 through 340 address specific types of cable. If you take the time to become familiar with the various types of cable, you will:

- Understand what's available for doing the work.
- Recognize cable types that have special *Code* requirements.
- Avoid buying cable you cannot install due to *Code* requirements you cannot meet with that particular installation.

Here's a brief overview of each one:

Article 320. Armored Cable (Type AC). Armored cable is an assembly of insulated conductors, 14 AWG through 1 AWG, that are individually wrapped with waxed paper. The conductors are contained within a flexible spiral metal (steel or aluminum) sheath that interlocks at the edges. Armored cable looks like flexible metal conduit. Many electricians call this metal cable BX®.

Article 330. Metal-Clad Cable (Type MC). Metal-clad cable encloses one or more insulated conductors in a metal sheath of either corrugated or smooth copper or aluminum tubing, or spiral interlocked steel or aluminum. The physical characteristics of type MC cable make it a versatile wiring method that is permitted in almost any location and for almost any application. The most common type of MC cable is the interlocking type, which looks similar to armored cable or flexible metal conduit.

Article 334. Nonmetallic-Sheathed Cable (Type NM). Nonmetallic-sheathed cable encloses two, three, or four insulated conductors, 14 AWG through 2 AWG, within a nonmetallic outer jacket. Because this cable is nonmetallic, it contains a separate equipment grounding (bonding) conductor. Nonmetallic-sheathed cable is a common wiring method used for residential and commercial branch circuits. Most electricians call this plastic cable Romex®.

Article 336. Power and Control Tray Cable (Type TC). Power and control cable tray is a factory assembly of two or more insulated conductors under a nonmetallic sheath for installation in cable trays, in raceways, or where supported by a messenger wire.

Article 338. Service-Entrance Cables (Type SE). Service-entrance cable can be a single-conductor or multiconductor assembly within an overall nonmetallic covering. This cable is used primarily for services not over 600V, but is also permitted for feeders and branch circuits.

Article 340. Underground Feeder and Branch-Circuit Cable (Type UF). Underground feeder cable is a moisture-, fungus-, and corrosion-resistant cable suitable for direct burial in the earth, and it comes in sizes 14 AWG through 4/0 AWG [340.104]. Multiconductor UF cable is covered in molded plastic that encapsulates the insulated conductors.

Raceway Articles

Articles 342 through 392 address specific types of raceways. Refer to Article 100 for the definition of raceway. If you take the time to become familiar with the various types of raceways, you will:

- Understand what's available for doing the work.
- Recognize raceway types that have special *Code* requirements.
- Avoid buying a raceway you cannot install due to *NEC* requirements you cannot meet with that particular installation.

Here's a brief overview of each one:

Article 342. Intermediate Metal Conduit (Type IMC). Intermediate metal conduit is a circular metal raceway with the same outside diameter as rigid metal conduit. The wall thickness of intermediate metal conduit is less than that of rigid metal conduit, so it has a greater interior cross-sectional area. Intermediate metal conduit is lighter and less expensive than rigid metal conduit, but it's permitted in all the same locations as rigid metal conduit. Intermediate metal conduit also uses a different steel alloy, which makes it stronger than rigid metal conduit, even though the walls are thinner.

Article 344. Rigid Metal Conduit (Type RMC). Rigid metal conduit is similar to intermediate metal conduit, except the wall thickness is greater, so it has a smaller interior cross-sectional area. Rigid metal conduit is heavier than intermediate metal conduit and it's permitted to be installed in any location, just like intermediate metal conduit.

Article 348. Flexible Metal Conduit (Type FMC). Flexible metal conduit is a raceway of circular cross section made of a helically wound, interlocked metal strip of either steel or aluminum. It's commonly called "Greenfield" or "Flex."

Article 350. Liquidtight Flexible Metal Conduit (Type LFMC). Liquidtight flexible metal conduit is a listed raceway of circular cross section with an outer liquidtight, nonmetallic, sunlight-resistant jacket over an inner flexible metal core, with associated couplings, connectors, and fittings. It's listed for the installation of electric conductors. Liquidtight flexible metal conduit is commonly called Sealtite® or simply "liquidtight." Liquidtight flexible metal conduit is of similar construction to flexible metal conduit but has an outer thermoplastic covering.

Article 352. Rigid Nonmetallic Conduit (Type RNC). Rigid nonmetallic conduit is a listed nonmetallic raceway of circular cross section with integral or associated couplings, connectors, and fittings. It's listed for the installation of electrical conductors. Typically, it's constructed of polyvinyl chloride (PVC).

Article 354. Nonmetallic Underground Conduit with Conductors (Type NUCC). Nonmetallic underground conduit with conductors is a factory assembly of conductors or cables inside a nonmetallic, smooth wall conduit with a circular cross section. It can also be supplied on reels without damage or distortion and is of sufficient strength to withstand abuse, such as impact or crushing when handled and installed, without damage to conduit or conductors.

Article 356. Liquidtight Flexible Nonmetallic Conduit (Type LFNC). Liquidtight flexible nonmetallic conduit is a listed raceway of circular cross section with an outer liquidtight, nonmetallic, sunlight-resistant jacket over an inner flexible core, with associated couplings, connectors, and fittings. It's listed for the installation of electric conductors. LFNC is available in three types:

- Type LFNC-A (orange). A smooth seamless inner core and cover bonded together. One or more reinforcement layers are inserted between the core and covers.
- Type LFNC-B (gray). A smooth inner surface with integral reinforcement within the conduit wall.
- Type LFNC-C (black). A corrugated internal and external surface without integral reinforcement within the conduit wall.

Article 358. Electrical Metallic Tubing (EMT). Electrical metallic tubing is a listed metallic tubing of circular cross section raceway listed for the installation of electrical conductors. Compared to rigid metal conduit and intermediate metal conduit, electrical metallic tubing is relatively easy to bend, cut, and ream. Because it isn't threaded, all connectors and couplings are of the threadless type.

Article 362. Electrical Nonmetallic Tubing (ENT). Electrical nonmetallic tubing is a pliable, corrugated, circular raceway made of PVC. It's often called "Smurf Pipe" or "Smurf Tube," because it originally came out at the height of popularity of the children's cartoon characters "the Smurfs," and was available only in blue.

Article 376. Metal Wireways. A metal wireway is a sheet metal trough with hinged or removable covers for housing and protecting electric wires and cable, in which conductors are placed after the wireway has been installed as a complete system.

Article 378. Nonmetallic Wireways. A nonmetallic wireway is a flame-retardant trough with hinged or removable covers for housing and protecting electric wires and cable, in which conductors are placed after the wireway has been installed as a complete system.

Article 380. Multioutlet Assembly. A multioutlet assembly is a surface, flush, or freestanding raceway designed to hold conductors and receptacles. It's assembled in the field or at the factory.

Article 384. Strut-Type Channel Raceway. Strut-type channel raceway is a metallic raceway intended to be mounted to the surface or suspended with associated accessories, in which conductors are placed after the raceway has been installed as a complete system.

Article 386. Surface Metal Raceways. A surface metal raceway is a metallic raceway intended to be mounted to the surface with associated accessories, in which conductors are placed after the raceway has been installed as a complete system.

Article 388. Surface Nonmetallic Raceways. A surface nonmetallic raceway is intended to be surface mounted with associated accessories. Conductors are placed after the raceway has been installed as a complete system.

Article 392. Cable Trays. A cable tray system is a unit or assembly of units or sections with associated fittings that form a structural system used to securely fasten or support cables and raceways. A cable tray isn't a raceway, but a support system for raceways, cables, and enclosures.

Notes

ARTICLE 300 Wiring Methods

Introduction

Article 300 contains the general requirements for all wiring methods included in the *NEC*. However, this article doesn't apply to signaling and communications systems, which are covered in Chapters 7 and 8, except as specifically referenced in those chapters.

This article is primarily concerned with how you install, route, splice, protect, and secure conductors and raceways. How well you conform to the requirements of Article 300 will generally be evident in the finished work, because many of the requirements tend to determine the appearance of the installation.

Because of this, it's often easy to spot Article 300 problems if you're looking for *Code* violations. For example, you can easily spot when someone runs an equipment grounding (bonding) conductor outside a raceway instead of grouping all conductors of a circuit together, as required by 300.3(B). From your study of Article 300 you'll learn there's a difference between running an equipment grounding (bonding) conductor outside a raceway, which is illegal, and running an external bonding jumper outside the same raceway.

This is just one of the common points of confusion your studies here will clear up for you. To help achieve that end, be sure to carefully consider the accompanying illustrations and also refer to Article 100 as needed.

PART I. GENERAL REQUIREMENTS

300.1 Scope.

(A) Wiring Installations. Article 300 contains the general requirements for power and lighting wiring methods.

> **Author's Comment:** The requirements contained in Article 300 do not apply to the wiring methods for signaling and communications systems. However, the following sections in Chapter 7 and 8 refer back to Article 300 requirements.
>
> - CATV, 820.3
> - Class 2 and 3 Circuits, 725.3
> - Communications (Telephone), 800.133(A)(2)
> - Fiber Optics, 770.3
> - Fire Alarm Circuits, 760.3

(B) Integral Parts of Equipment. The requirements contained in Article 300 do not apply to the internal parts of electrical equipment.

> **Author's Comment:** Product evaluation by a testing laboratory approved by the authority having jurisdiction avoids the necessity for repetitive examinations by different persons, and the confusion that could result from conflicting test reports [90.7].
> **Figure 300–1**

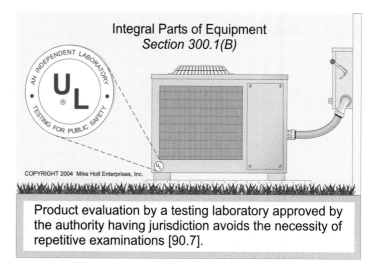

Integral Parts of Equipment
Section 300.1(B)

COPYRIGHT 2004 Mike Holt Enterprises, Inc.

Product evaluation by a testing laboratory approved by the authority having jurisdiction avoids the necessity of repetitive examinations [90.7].

Figure 300–1

(C) Metric Designators and Trade Sizes. Metric designators for raceway trade sizes are given in Table 300.1(C).

Author's Comments:

- Since compliance with either the metric or the trade size measurement system constitutes compliance with the *NEC*, this textbook only uses trade sizes [90.9(D)].

- The industry practice is to refer to raceways in relationship to inches, such as ½ in., 2 in., etc., however the proper reference (2005 *NEC* change) is to use "Trade Size ½," or "Trade Size 2." In this textbook we will generally use the term "Trade Size," except where I think the term "inches" makes more sense. Please forgive our inconsistency, but we only do this to make the reading easier.

Table 300.1(C) Metric Designator and Trade Sizes

Trade Size	Metric
⅜	12
½	16
¾	21
1	27
1 ¼	35
1 ½	41
2	53
2 ½	63
3	78
3 ½	91
4	103
5	129
6	155

300.3 Conductors.

(A) Conductors. Conductors must be installed within a raceway, cable, or enclosure.

Exception: Overhead conductors can be installed in accordance with 225.6.

(B) Circuit Conductors Grouped Together. All conductors of a circuit must be installed in the same raceway, cable, trench, cord, or cable tray, except as permitted by (1) through (4). Figure 300–2

> **Author's Comment:** All conductors of a circuit must be installed in the same raceway, cable, trench, cord, or cable tray to minimize induction heating of metallic raceways and enclosures, and to maintain a low-impedance ground-fault current path.

(1) Paralleled Installations. Conductors can be run in parallel, in accordance with 310.4, and must have all circuit conductors within the same raceway, auxiliary gutter, cable tray, trench, or cable. Figure 300–3

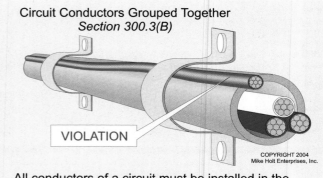

Circuit Conductors Grouped Together
Section 300.3(B)

VIOLATION

All conductors of a circuit must be installed in the same raceway, cable, trench, cord, or cable tray to minimize induction heating of metallic raceways and enclosures, and to maintain a low-impedance ground-fault current path.

Figure 300–2

Exception: Parallel conductors can be run in different raceways (Phase A in raceway 1, Phase B in raceway 2, etc.) if, in order to reduce or eliminate inductive heating, the raceway is nonmetallic or nonmagnetic and the installation complies with 300.20(B). See 300.3(B)(3).

(2) Grounding and Bonding Conductors. Equipment grounding (bonding) conductors can be installed outside a raceway or cable assembly for certain existing installations. See 250.130(C). Equipment grounding (bonding) jumpers can be located outside a flexible raceway if the bonding jumper is installed in accordance with 250.102(E). Figure 300–4

(3) Nonferrous Wiring Methods. Circuit conductors can be run in different raceways (Phase A in raceway 1, Phase B in raceway 2, etc.) if, in order to reduce or eliminate inductive heating, the raceway is nonmetallic or nonmagnetic and the installation complies with 300.20(B). See 300.5(I) Ex. 2.

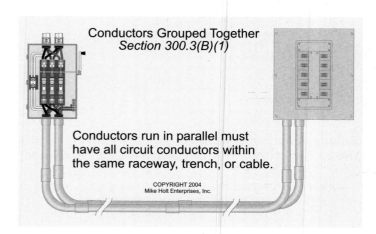

Conductors Grouped Together
Section 300.3(B)(1)

Conductors run in parallel must have all circuit conductors within the same raceway, trench, or cable.

Figure 300–3

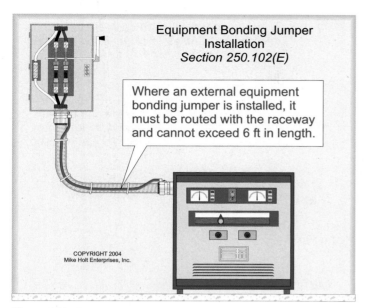

Equipment Bonding Jumper Installation
Section 250.102(E)

Where an external equipment bonding jumper is installed, it must be routed with the raceway and cannot exceed 6 ft in length.

COPYRIGHT 2004
Mike Holt Enterprises, Inc.

Figure 300–4

(C) Conductors of Different Systems.

(1) Mixing. Power conductors of different systems can occupy the same raceway, cable, or enclosure if all conductors have an insulation voltage rating not less than the maximum circuit voltage. **Figure 300–5**

> **Author's Comment:** Control, signal, and communications wiring must be separated from power and lighting circuits so the higher-voltage conductors do not accidentally energize them. The following *Code* sections prohibit the mixing of signaling and communications conductors with power conductors:
>
> • CATV Coaxial Cable, 820.133(A)(1)(2)
> • Class 1, Class 2, and Class 3 Control Circuits, 725.26 and 725.55(A). **Figure 300–6**

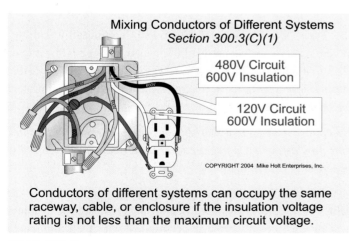

Mixing Conductors of Different Systems
Section 300.3(C)(1)

480V Circuit
600V Insulation

120V Circuit
600V Insulation

COPYRIGHT 2004 Mike Holt Enterprises, Inc.

Conductors of different systems can occupy the same raceway, cable, or enclosure if the insulation voltage rating is not less than the maximum circuit voltage.

Figure 300–5

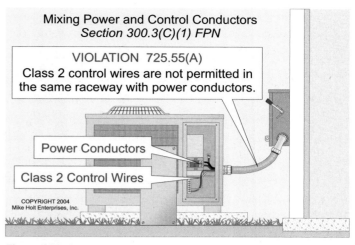

Mixing Power and Control Conductors
Section 300.3(C)(1) FPN

VIOLATION 725.55(A)
Class 2 control wires are not permitted in the same raceway with power conductors.

Power Conductors

Class 2 Control Wires

COPYRIGHT 2004
Mike Holt Enterprises, Inc.

Figure 300–6

• Communications Circuits, 800.133(A)(1)(c)
• Fire Alarm Circuits, 760.55(A)
• Instrument Tray Cable, 727.5
• Sound Circuits, 640.9(C)

Class 2 or Class 3 circuits reclassified as a Class 1 circuit, can be run with associated power conductors [725.26(B)(1)]. Exceptions to the above allow Class 1 or Class 2 reclassified [725.52 Ex. 2] as Class 1 control circuits to be run with associated power conductors. **Figure 300–7**

300.4 Protection Against Physical Damage. Conductors must be protected against physical damage [110.27(B)].

(A) Cables and Raceways Through Wood Members. When the following wiring methods are installed through wood members, they must comply with (1) and (2). **Figure 300–8**

• Armored Cable, Article 320

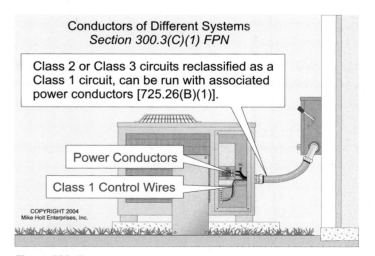

Conductors of Different Systems
Section 300.3(C)(1) FPN

Class 2 or Class 3 circuits reclassified as a Class 1 circuit, can be run with associated power conductors [725.26(B)(1)].

Power Conductors

Class 1 Control Wires

COPYRIGHT 2004
Mike Holt Enterprises, Inc.

Figure 300–7

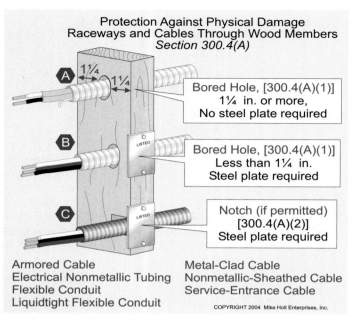

**Protection Against Physical Damage
Raceways and Cables Through Wood Members**
Section 300.4(A)

Ⓐ 1¼ ← 1¼ →

Bored Hole, [300.4(A)(1)]
1¼ in. or more,
No steel plate required

Ⓑ

Bored Hole, [300.4(A)(1)]
Less than 1¼ in.
Steel plate required

Ⓒ

Notch (if permitted)
[300.4(A)(2)]
Steel plate required

Armored Cable
Electrical Nonmetallic Tubing
Flexible Conduit
Liquidtight Flexible Conduit

Metal-Clad Cable
Nonmetallic-Sheathed Cable
Service-Entrance Cable

COPYRIGHT 2004 Mike Holt Enterprises, Inc.

Figure 300–8

- Electrical Nonmetallic Tubing, Article 362
- Flexible Metal Conduit, Article 348
- Liquidtight Flexible Metal Conduit, Article 350
- Metal-Clad Cable, Article 330
- Nonmetallic-Sheathed Cable, Article 334
- Service-Entrance Cable, Article 338

(1) Holes in Wood Members. Holes through wood framing members for the above cables or raceways must be not less than 1¼ in. from the edge of the wood member. If the edge of the hole is less than 1¼ in. from the edge, a ¹⁄₁₆ in. thick steel plate of sufficient length and width must be installed to protect the wiring method from screws and nails.

Exception 1: A steel plate isn't required to protect rigid metal conduit, intermediate metal conduit, rigid nonmetallic conduit, or electrical metallic tubing.

Exception 2: A listed and marked steel plate less than ¹⁄₁₆ in. thick that provides equal or better protection against nail or screw penetration is permitted.

Author's Comment: Hardened steel plates thinner than ¹⁄₁₆ in. have been tested and found to provide better protection from screw and nail penetration than the thicker plates.

(2) Notches in Wood Members. Where notching of wood framing members for cables and raceways are permitted by the building code, a ¹⁄₁₆ in. thick steel plate of sufficient length and width must be installed to protect the wiring method laid in these wood notches from screws and nails.

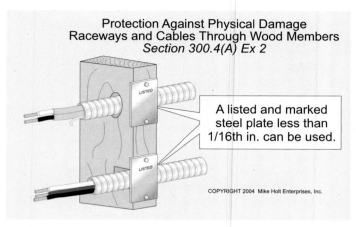

**Protection Against Physical Damage
Raceways and Cables Through Wood Members**
Section 300.4(A) Ex 2

A listed and marked
steel plate less than
1/16th in. can be used.

COPYRIGHT 2004 Mike Holt Enterprises, Inc.

Figure 300–9

CAUTION: *When drilling or notching wood members, be sure to check with the building code inspector to ensure that you do not damage or weaken the structure and violate the building code.*

Exception 1: A steel plate isn't required to protect rigid metal conduit, intermediate metal conduit, rigid nonmetallic conduit, or electrical metallic tubing.

Exception 2: A listed and marked steel plate less than ¹⁄₁₆ in. thick that provides equal or better protection against nail or screw penetration is permitted. Figure 300–9

(B) Nonmetallic-Sheathed Cable and Electrical Nonmetallic Tubing Through Metal Framing Members.

(1) Nonmetallic-Sheathed Cable (NM). Where NM cables pass through factory or field openings in metal framing members, the cable must be protected by listed bushings or listed grommets that cover all metal edges. The protection fitting must be securely fastened in the opening before the installation of the cable. Figure 300–10

(2) Type NM Cable and Electrical Nonmetallic Tubing. Where nails or screws are likely to penetrate Type NM cable or electrical nonmetallic tubing, a steel sleeve, steel plate, or steel clip not less than ¹⁄₁₆ in. in thickness must be installed to protect the cable or tubing.

Exception: A listed and marked steel plate less than ¹⁄₁₆ in. thick that provides equal or better protection against nail or screw penetration is permitted.

(C) Behind Suspended Ceilings. Wiring methods, such as cables or raceways installed behind panels designed to allow access, must be supported in accordance with their applicable article. Figure 300–11

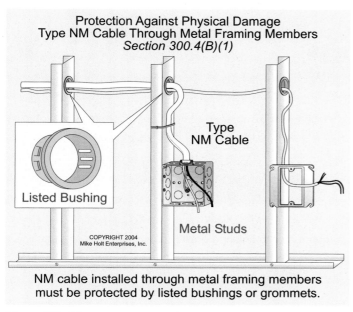

Figure 300–10

NM cable installed through metal framing members must be protected by listed bushings or grommets.

Author's Comment: This rule doesn't apply to control, signal, and communications cables, but similar requirements are contained as follows:

- CATV Coaxial Cable, 820.21 and 820.24
- Communications Cable, 800.3(B) and 800.21
- Control and Signaling Cable, 725.7 and 725.8
- Fire Alarm Cable, 760.7 and 760.8
- Optical Fiber Cable, 770.21 and 770.24
- Sound Cable, 640.5 and 640.6

(D) Cables and Raceways Parallel to Framing Members and Furring Strips. Cables or raceways run parallel to framing members or furring strips must be protected where they are

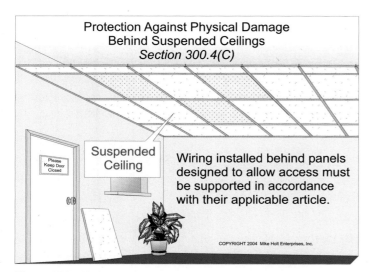

Figure 300–11

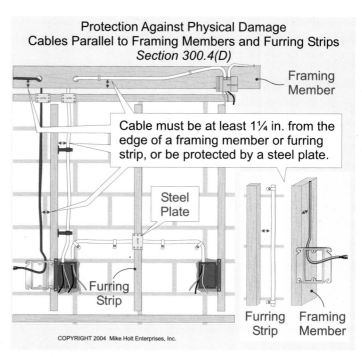

Figure 300–12

likely to be penetrated by nails or screws, by installing the wiring method so it isn't less than 1¼ in. from the nearest edge of the framing member or furring strips. If the edge of the framing member or furring strip is less than 1¼ in. away, a ¹⁄₁₆ in. thick steel plate of sufficient length and width must be installed to protect the wiring method from screws and nails. **Figure 300–12**

Author's Comment: This rule doesn't apply to control, signaling, and communications cables, but similar requirements are contained as follows:

- CATV Cable, 820.24
- Communications Cable, 800.24
- Control and Signaling Cable, 725.8
- Optical Fiber Cable, 770.24
- Fire Alarm Cable, 760.8
- Sound Cable, 640.6

Exception 1: Protection isn't required for rigid metal conduit, intermediate metal conduit, rigid nonmetallic conduit, or electrical metallic tubing.

Exception 2: For concealed work in finished buildings, or finished panels for prefabricated buildings where such supporting is impracticable, the cables can be fished between access points.

Exception 3: A listed and marked steel plate less than ¹⁄₁₆ in. thick that provides equal or better protection against nail or screw penetration is permitted.

(E) Cables and Raceways Installed in Grooves. Cables and raceways installed in a groove must be protected by a ¹⁄₁₆ in. thick steel plate or sleeve, or by 1¼ in. of free space.

Author's Comment: An example is Type NM cable installed in a groove cut into the Styrofoam-type insulation building block structure and then covered with wallboard.

Exception 1: Protection isn't required if the cable is installed in rigid metal conduit, intermediate metal conduit, rigid nonmetallic conduit, or electrical nonmetallic tubing.

Exception 2: A listed and marked steel plate less than ¹⁄₁₆ in. thick that provides equal or better protection against nail or screw penetration is permitted.

(F) Insulating Bushings. Where raceways contain conductors 4 AWG and larger that enter an enclosure, the conductors must be protected from abrasion during and after installation by a fitting that provides a smooth, rounded insulating surface, such as an insulating bushing. Figure 300–13

Author's Comment: Rigid nonmetallic conduit male-adapter termination fittings are sometimes, but not always, considered to provide the required smooth rounded insulating surface. Check with the authority having jurisdiction.

Exception: Insulating bushings aren't required where a raceway terminates in a threaded raceway entry that provides a smooth, rounded, or flared surface for the conductors. An example would be a meter hub fitting or a Meyer's hub-type fitting.

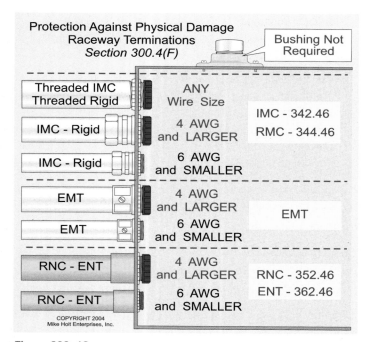

Figure 300–13

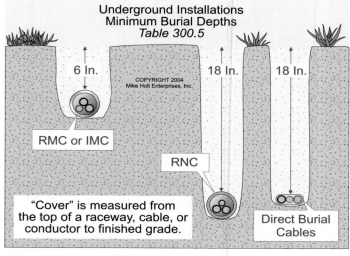

Figure 300–14

300.5 Underground Installations.

(A) Minimum Burial Depths. When cables and raceways are run underground, they must have a minimum "cover" in accordance with Table 300.5.

Author's Comment: Note 1 to Table 300.5 defines "Cover" as the distance from the top of the underground cable or raceway to the surface of finish grade. Figure 300–14

Table 300.5 Minimum Cover Requirements in Inches
Figure 300–15

Location	Buried Cables	Metal Raceway	Nonmetallic Raceway	120V 20A GFCI Circuit
Under Building	0	0	0	0
Dwelling Unit	18	18	18	12
Under Roadway	24	24	24	24
Other Locations	24	6	18	12

Author's Comment: The cover requirements contained in 300.5 do not apply to signaling and communications wiring, see: Figure 300–16

- CATV, 90.3
- Class 2 and 3 Circuits, 725.3
- Communications, 90.3
- Optical Fiber Cables, 770.3
- Fire Alarm Circuits, 760.3

(B) Listing. Cables and insulated conductors installed in enclosures or raceways underground must be listed for use in wet locations.

Underground Installations
Minimum Burial Depths
Table 300.5

	UF or USE Cables or Conductors	RMC or IMC	RNC (PVC) not encased in concrete	Residential 15 & 20A GFCI Branch Circuits
Applications NOT listed below	24	6	18	12
STREET Driveway Parking Lot	24	24	24	24
DRIVEWAYS One - Two Family	18	18	18	12
SOLID ROCK With not less than 2 in. of concrete	Raceway Only			Raceway Only

COPYRIGHT 2004 Mike Holt Enterprises, Inc.

Figure 300–15

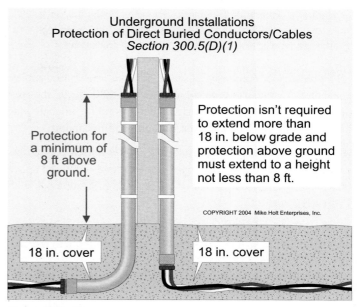

Underground Installations
Protection of Direct Buried Conductors/Cables
Section 300.5(D)(1)

Protection for a minimum of 8 ft above ground.

Protection isn't required to extend more than 18 in. below grade and protection above ground must extend to a height not less than 8 ft.

COPYRIGHT 2004 Mike Holt Enterprises, Inc.

18 in. cover   18 in. cover

Figure 300–17

(C) Cables Under Buildings. Cables run under a building must be installed in a raceway that extends past the outside walls of the building.

(D) Protecting Underground Cables and Conductors from Damage. Direct-buried conductors and cables must be protected from damage in accordance with (1) through (4).

(1) Emerging from Grade. Direct-buried cables or conductors that emerge from the ground must be installed in an enclosure or raceway to protect against physical damage. Protection isn't required to extend more than 18 in. below grade and protection above ground must extend to a height not less than 8 ft. Figure 300–17

(2) Conductors Entering Buildings. Conductors that enter a building must be protected to the point of entrance.

(3) Service Conductors. Service conductors that aren't under the exclusive control of the electric utility, and are buried 18 in. or more below grade, must have their location identified by a warning ribbon placed in the trench not less than 1 ft above the underground installation. Figure 300–18

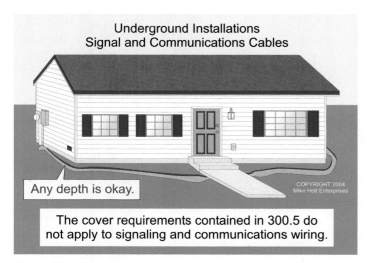

Underground Installations
Signal and Communications Cables

Any depth is okay.

COPYRIGHT 2004 Mike Holt Enterprises

The cover requirements contained in 300.5 do not apply to signaling and communications wiring.

Figure 300–16

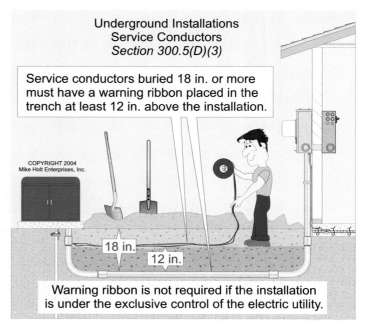

Underground Installations
Service Conductors
Section 300.5(D)(3)

Service conductors buried 18 in. or more must have a warning ribbon placed in the trench at least 12 in. above the installation.

COPYRIGHT 2004 Mike Holt Enterprises, Inc.

18 in.

12 in.

Warning ribbon is not required if the installation is under the exclusive control of the electric utility.

Figure 300–18

Author's Comment: It's impossible to comply with the service conductor identification location requirement when service conductors are installed using directional boring equipment.

(4) Enclosure or Raceway Damage. Where direct-buried cables, enclosures, or raceways are subject to physical damage, the conductors must be installed in rigid metal conduit, intermediate metal conduit, or Schedule 80 rigid nonmetallic conduit.

(E) Splices and Taps Underground. Direct-buried conductors or cables can be spliced or tapped underground without a splice box [300.15(G)], if the splice or tap is made in accordance with 110.14(B). Figure 300–19

(F) Backfill. Backfill material for underground wiring must not damage the underground cable raceway or contribute to the corrosion of the metal raceway.

Author's Comment: Large rocks, chunks of concrete, steel rods, mesh, and other sharp-edged objects cannot be used for backfill material because they can damage the underground conductors, cables, or raceways.

(G) Raceway Seals. Where moisture could enter a raceway and contact energized live parts, seals must be installed at one or both ends of the raceway.

Author's Comment: This is a common problem for equipment located physically downhill from the supply or in underground equipment rooms. See 230.8 for service raceway seals and 300.7(A) for different temperature area seals.

FPN: Hazardous explosive gases or vapors make it necessary to seal underground conduits or raceways that enter the building in accordance with 501.15.

Author's Comment: It isn't the intent of this FPN to imply that seal-offs of the types required in hazardous (classified) locations must be installed in unclassified locations, except as required in Chapter 5. This also doesn't imply that the sealing material provides a watertight seal, but only that it prevents moisture from entering.

(H) Bushing. Raceways that terminate underground must have a bushing or fitting at the end of the raceway to protect emerging cables or conductors.

(I) Conductors Grouped Together. All conductors of the same circuit, including the equipment grounding (bonding) conductor, must be inside the same raceway or in close proximity to each other. See 300.3(B). Figure 300–20

Exception 1: Conductors can be installed in parallel in accordance with 310.4.

Exception 2: Individual sets of parallel circuit conductors can be installed in underground nonmetallic raceways, if inductive heating at raceway terminations is reduced by complying with 300.20(B) [300.3(B)(1) Ex. 1]. Figure 300–21

Author's Comment: Installing phase and neutral wires in different nonmetallic raceways makes it easier to terminate larger parallel sets of conductors, but it will result in higher levels of electromagnetic fields (EMF), which can cause computer monitors to flicker in a distracting manner.

(J) Ground Movement. Direct-buried conductors, cables, or raceways subject to movement by settlement or frost, must be arranged to prevent damage to conductors or equipment connected to the wiring.

Underground Splices - Single Conductors
Section 300.5(E)

Type UF or USE single conductor

COPYRIGHT 2004
Mike Holt Enterprises, Inc.

Single Type UF or USE conductors can be spliced underground with a device that is listed for direct burial.

Figure 300–19

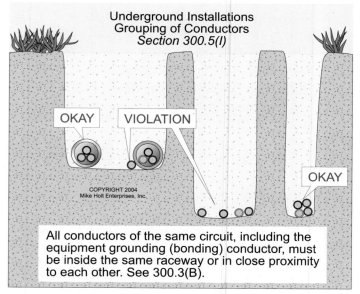

Underground Installations Grouping of Conductors
Section 300.5(I)

OKAY   VIOLATION   OKAY

COPYRIGHT 2004
Mike Holt Enterprises, Inc.

All conductors of the same circuit, including the equipment grounding (bonding) conductor, must be inside the same raceway or in close proximity to each other. See 300.3(B).

Figure 300–20

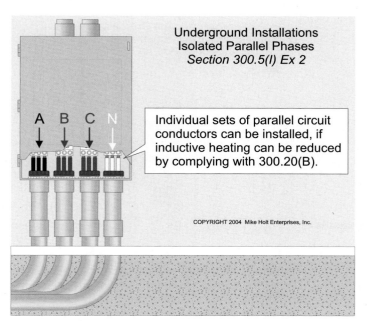

Underground Installations
Isolated Parallel Phases
Section 300.5(I) Ex 2

A B C N

Individual sets of parallel circuit conductors can be installed, if inductive heating can be reduced by complying with 300.20(B).

COPYRIGHT 2004 Mike Holt Enterprises, Inc.

Figure 300–21

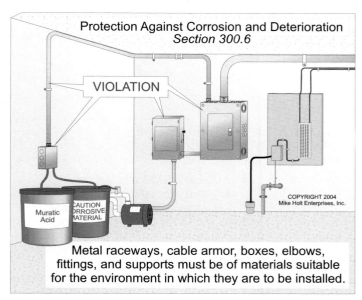

Protection Against Corrosion and Deterioration
Section 300.6

VIOLATION

Muriatic Acid

CAUTION CORROSIVE MATERIAL

COPYRIGHT 2004 Mike Holt Enterprises, Inc.

Metal raceways, cable armor, boxes, elbows, fittings, and supports must be of materials suitable for the environment in which they are to be installed.

Figure 300–22

(K) Directional Boring. Cables or raceways installed using directional boring equipment must be approved by the authority having jurisdiction for this purpose.

Author's Comment: Directional boring technology uses a directional drill, which is steered continuously from point "A" to point "B." When the drill head comes out of the earth at point "B," it's replaced with a back-reamer and the duct or conduit being installed is attached to it. The size of the boring rig (hp, torque, and pull-back power) comes into play along with the type of soil in determining the type of raceways required. For telecom work, multiple poly innerducts are pulled in at one time. If a major crossing is encountered, such as an expressway, railroad, or river, outerduct may be installed to create a permanent sleeve for the innerducts.

"Innerduct" and "outerduct" are terms usually associated with optical fiber cable installations, while unitduct comes with current-carrying conductors factory installed. All of these come in various sizes. Galvanized rigid steel conduit and Schedule 40 and 80 RNC are installed extensively with directional boring installations.

300.6 Protection Against Corrosion and Deterioration.

Raceways, cable trays, cablebus, auxiliary gutters, cable armor, boxes, cable sheathing, cabinets, elbows, couplings, fittings, supports, and support hardware must be suitable for the environment. Figure 300–22

(A) Ferrous Metal Equipment. Ferrous metal raceways, enclosures, cables, cable trays, fittings, and support hardware must be protected against corrosion inside and outside by a coating of listed corrosion-resistant material. Where corrosion protection is necessary, such as underground and in wet locations, and the conduit is threaded in the field, the threads must be coated with an electrically conductive, corrosion-resistant compound approved by the authority having jurisdiction, such as cold zinc.

(1) Protected from Corrosion Solely by Enamel. Where ferrous metal parts are protected from corrosion solely by enamel, they cannot be used outdoors or in wet locations as described in 300.6(D).

(2) Organic Coatings on Boxes or Cabinets. Boxes or cabinets having a system of organic coatings marked "Raintight," "Rainproof," or "Outdoor Type," can be installed outdoors.

(3) In Concrete or in Direct Contact with the Earth. Ferrous metal raceways, cable armor, boxes, cable sheathing, cabinets, elbows, couplings, nipples, fittings, supports, and support hardware can be installed in concrete or in direct contact with the earth, or in areas subject to severe corrosive influences where made of material approved by the authority having jurisdiction for the condition, or where provided with corrosion protection approved by the authority having jurisdiction for the condition.

Author's Comment: Galvanized rigid steel or galvanized steel intermediate metal conduits are generally okay for most locations without requiring additional corrosion protection. Galvanized steel electrical metallic tubing can be installed in concrete at grade level and in direct contact with earth but supplementary corrosion protection is usually required (UL *White Book*). Electrical metallic tubing can be installed in concrete above the ground floor slab generally without supplementary corrosion protection. Figure 300–23

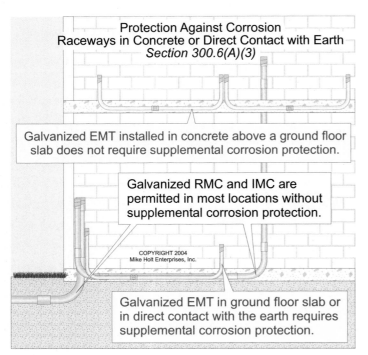

Figure 300–23

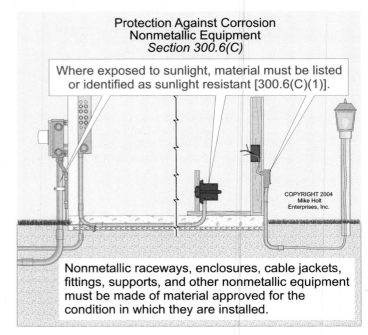

Figure 300–24

(B) Nonferrous Metal Equipment. Nonferrous raceways, cable trays, cablebus, auxiliary gutters, cable armor, boxes, cable sheathing, cabinets, elbows, couplings, nipples, fittings, supports, and support hardware embedded or encased in concrete or in direct contact with the earth must be provided with supplementary corrosion protection.

(C) Nonmetallic Equipment. Nonmetallic raceways, cable trays, cablebus, auxiliary gutters, boxes, cables with a non-metallic outer jacket and internal metal armor or jacket, cable sheathing, cabinets, elbows, couplings, nipples, fittings, supports, and support hardware must be made of material approved by the authority having jurisdiction for the condition and must comply with (1) and (2). Figure 300–24

(1) Exposed to Sunlight. Where exposed to sunlight, the materials must be listed or identified as sunlight resistant.

(2) Chemical Exposure. Where subject to exposure to chemical solvents, vapors, splashing, or immersion, materials or coatings must either be inherently resistant to chemicals based upon their listing or be identified for the specific chemical reagent.

(D) Indoor Wet Locations. In portions of dairy processing facilities, laundries, canneries, and other indoor wet locations, and in locations where walls are frequently washed or where there are surfaces of absorbent materials, such as damp paper or wood, the entire wiring system, where installed exposed, including all boxes, fittings, raceways, and cable used therewith, must be mounted so that there's at least ¼ in. of airspace between it and the wall or supporting surface.

> **Author's Comment:** See Article 100 for the definitions of "Exposed" and "Location, Wet."

Exception: Nonmetallic raceways, boxes, and fittings are permitted without the airspace on a concrete, masonry, tile, or similar surface.

> **FPN:** Areas where acids and alkali chemicals are handled and stored may present such corrosive conditions, particularly when wet or damp. Severe corrosive conditions may also be present in portions of meatpacking plants, tanneries, glue houses, and some stables; in installations immediately adjacent to a seashore or swimming pool, spa, hot tub, and fountain areas; in areas where chemical deicers are used; and in storage cellars or rooms for hides, casings, fertilizer, salt, and bulk chemicals.

300.7 Raceways Exposed to Different Temperatures.

(A) Sealing. Where a raceway is subjected to different temperatures and where condensation is known to be a problem, the raceway must be filled with a material approved by the authority having jurisdiction that will prevent the circulation of warm air to a colder section of the raceway. An explosionproof seal isn't required for this purpose. Figure 300–25

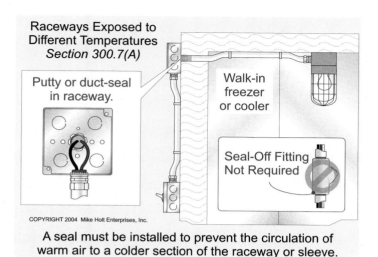

Raceways Exposed to Different Temperatures
Section 300.7(A)

Putty or duct-seal in raceway.

Walk-in freezer or cooler

Seal-Off Fitting Not Required

COPYRIGHT 2004 Mike Holt Enterprises, Inc.

A seal must be installed to prevent the circulation of warm air to a colder section of the raceway or sleeve.

Figure 300–25

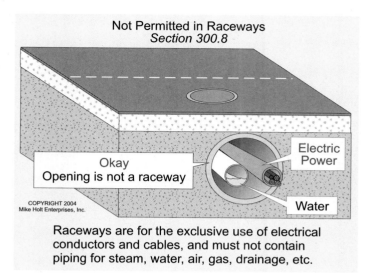

Not Permitted in Raceways
Section 300.8

Okay
Opening is not a raceway

Electric Power

Water

COPYRIGHT 2004 Mike Holt Enterprises, Inc.

Raceways are for the exclusive use of electrical conductors and cables, and must not contain piping for steam, water, air, gas, drainage, etc.

Figure 300–27

(B) Expansion Fittings. Raceways must be provided with expansion fittings where necessary to compensate for thermal expansion and contraction. Figure 300–26

> **FPN:** Table 352.44(A) provides the expansion characteristics for rigid nonmetallic conduit. The expansion characteristics for metal raceways are determined by multiplying the values from Table 352.44(A) by 0.20.

> **Author's Comment:** Where an expansion fitting is used with a metal raceway, a bonding jumper is required to maintain the low-impedance fault-current path [250.98].

300.8 Not Permitted in Raceways. Raceways are designed for the exclusive use of electrical conductors and cables, and cannot contain nonelectrical components, such as pipes or tubes for steam, water, air, gas, drainage, etc. Figure 300–27

300.10 Electrical Continuity. All metal raceways, cable, boxes, fittings, cabinets, and enclosures for conductors must be metallically joined together (bonded) to form a continuous low-impedance fault-current path that is capable of carrying any fault current likely to be imposed on it [110.10, 250.4(A)(3), and 250.122]. Figure 300–28

Metal raceways and cable assemblies must be mechanically secured to boxes, fittings, cabinets, and other enclosures.

Exception 1: Short lengths of metal raceways used for the support or protection of cables aren't required to be electrically continuous, nor are they required to be bonded to an effective ground-fault current path [250.86 Ex. 2 and 300.12 Ex.]. Figure 300–29

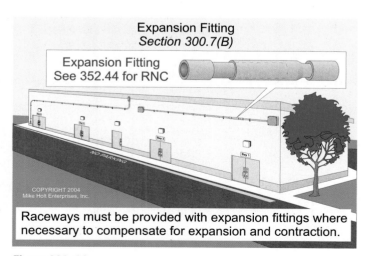

Expansion Fitting
Section 300.7(B)

Expansion Fitting
See 352.44 for RNC

NO PARKING

COPYRIGHT 2004 Mike Holt Enterprises, Inc.

Raceways must be provided with expansion fittings where necessary to compensate for expansion and contraction.

Figure 300–26

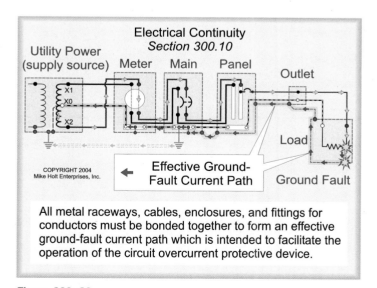

Electrical Continuity
Section 300.10

Utility Power (supply source)   Meter   Main   Panel

Outlet

X1
X0
X2

Load

Effective Ground-Fault Current Path

Ground Fault

COPYRIGHT 2004 Mike Holt Enterprises, Inc.

All metal raceways, cables, enclosures, and fittings for conductors must be bonded together to form an effective ground-fault current path which is intended to facilitate the operation of the circuit overcurrent protective device.

Figure 300–28

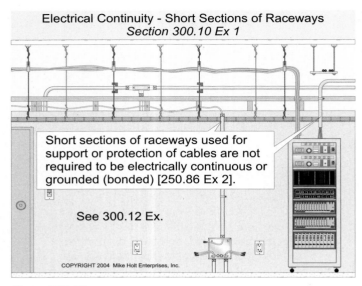

Figure 300-29

300.11 Securing and Supporting.

(A) Secured in Place. Raceways, cable assemblies, boxes, cabinets, and fittings must be securely fastened in place. The ceiling-support wires or ceiling grid cannot be used to support raceways and cables (power, signaling, or communications). However, independent support wires, secured at both ends, that provide secure support are permitted. Figure 300–30

> **Author's Comment:** Outlet boxes [314.23(D)] and luminaires can be secured to the suspended-ceiling grid if securely fastened to the ceiling-framing member by mechanical means such as bolts, screws, or rivets, or by the use of clips or other securing means identified for use with the type of ceiling framing member(s) [410.16(C)]. Figure 300–31

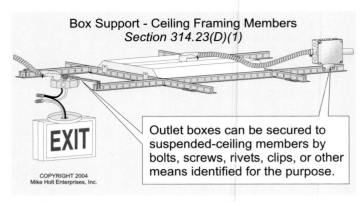

Figure 300–31

(1) Fire-Rated Assembly. Electrical wiring within the cavity of a fire-rated floor-ceiling or roof-ceiling assembly can be supported by independent support wires that are attached to the ceiling assembly. The independent support wires must be distinguishable from the suspended ceiling-support wires by color, tagging, or other effective means.

(2) Nonfire-Rated Assembly. Electrical wiring located within the cavity of a nonfire-rated floor-ceiling or roof-ceiling assembly, can be supported by independent support wires that are secured at both ends. Support wires within nonfire-rated assemblies aren't required to be distinguishable from the suspended-ceiling framing support wires.

(B) Raceways Used for Support. Raceways can only be used as a means of support for other raceways, cables, or nonelectrical equipment under the following conditions: Figure 300–32

(2) Class 2 and 3 Circuits. Class 2 and 3 cables can be supported by the raceway that supplies power to the equipment controlled by the Class 2 or 3 circuit. Figure 300–33

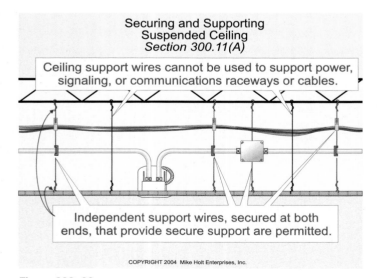

Figure 300–30

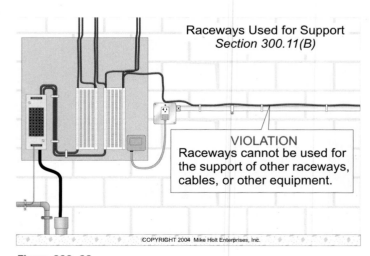

Figure 300–32

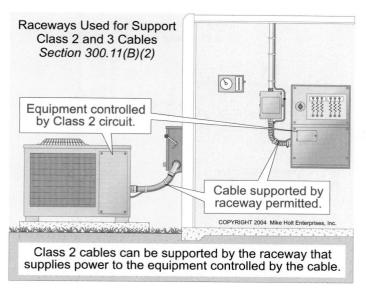

Raceways Used for Support
Class 2 and 3 Cables
Section 300.11(B)(2)

Equipment controlled
by Class 2 circuit.

Cable supported by
raceway permitted.

COPYRIGHT 2004 Mike Holt Enterprises, Inc.

Class 2 cables can be supported by the raceway that
supplies power to the equipment controlled by the cable.

Figure 300–33

Author's Comment: Where a Class 2 or Class 3 circuit is
reclassified as a Class 1 circuit [725.52(A) Ex. 2], it can be run
with the associated power conductors in accordance with
725.55(D)(2)(b).

(3) Boxes Supported by Conduits. Raceways are permitted as
a means of support for threaded boxes and conduit bodies in
accordance with 314.23(E) and (F).

(C) Cables Not Used as Means of Support. Cables cannot be
used to support other cables, raceways, or nonelectrical equipment. Figure 300–34

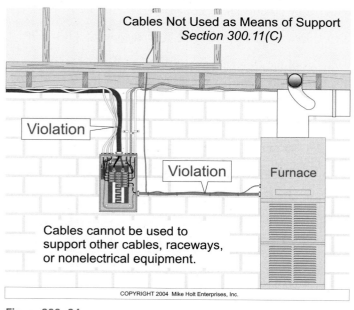

Cables Not Used as Means of Support
Section 300.11(C)

Violation

Violation

Furnace

Cables cannot be used to
support other cables, raceways,
or nonelectrical equipment.

COPYRIGHT 2004 Mike Holt Enterprises, Inc.

Figure 300–34

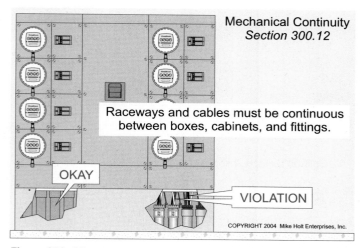

Mechanical Continuity
Section 300.12

Raceways and cables must be continuous
between boxes, cabinets, and fittings.

OKAY

VIOLATION

COPYRIGHT 2004 Mike Holt Enterprises, Inc.

Figure 300–35

300.12 Mechanical Continuity. Raceways and cable sheaths
must be mechanically continuous between boxes, cabinets, and
fittings. Figure 300–35

*Exception: Short sections of raceways used to provide support
or protection of cable from physical damage aren't required to
be mechanically continuous [250.86 Ex. 2 and 300.10 Ex. 1].*
Figure 300–36

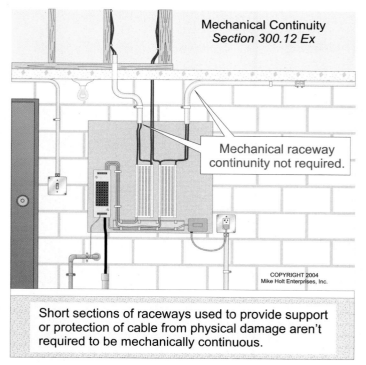

Mechanical Continuity
Section 300.12 Ex

Mechanical raceway
continuity not required.

COPYRIGHT 2004
Mike Holt Enterprises, Inc.

Short sections of raceways used to provide support
or protection of cable from physical damage aren't
required to be mechanically continuous.

Figure 300–36

Figure 300–37

300.13 Splices and Pigtails.

(A) Conductor Splices. Conductors in raceways must be continuous between all points of the system, which means that splices cannot be made in raceways, except as permitted by 376.56, 378.56, 384.56, 386.56, or 388.56. See 300.15. Figure 300–37

(B) Conductor Continuity (Pigtail). Continuity of the grounded neutral conductor of a multiwire branch circuit must not be interrupted by the removal of a wiring device. Therefore, the grounded neutral conductors must be spliced together, and a pigtail must be provided for the wiring device. Figure 300–38

> **Author's Comment:** The opening of the ungrounded conductors, or the grounded neutral conductor of a 2-wire circuit during the replacement of a device doesn't cause a safety hazard, so pigtailing of these conductors isn't required [110.14(B)].

> **CAUTION:** *If the continuity of the grounded neutral conductor of a multiwire circuit is interrupted (open), the resultant over- or undervoltage could cause a fire and/or destruction of electrical equipment.*

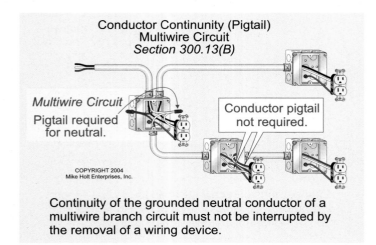

Figure 300–38

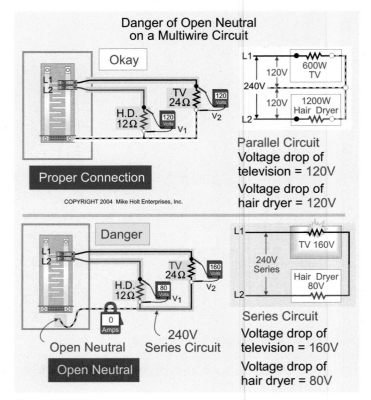

Figure 300–39

Example: A 3-wire single-phase 120/240V multiwire circuit supplies a 1,200W, 120V hair dryer and a 600W, 120V television. If the grounded neutral conductor of the multiwire circuit is interrupted, it will cause the 120V television to operate at 160V and consume 1,067W of power (instead of 600W) for only a few seconds before it burns up. Figure 300–39

Step 1. Determine the resistance of each appliance, $R = E^2/P$.

R of the Hair Dryer $= 120V^2/1,200W$
R of the Hair Dryer $= 12\Omega$
R of the Television $= 120V^2/600W$
R of the Television $= 24\Omega$

Step 2. Determine the current of the circuit, $I = E/R$.

$I = 240V/36\Omega\ (12\Omega + 24\Omega)$
$I = 6.7A$

Step 3. Determine the operating voltage for each appliance, $E = I \times R$.

Voltage of Hair Dryer $= 6.7A \times 12\Omega$
Voltage of Hair Dryer $= 80V$
Voltage of Television $= 6.7A \times 24\Omega$
Voltage of Television $= 160V$

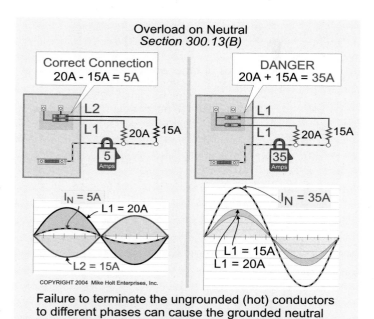

Overload on Neutral
Section 300.13(B)

Failure to terminate the ungrounded (hot) conductors to different phases can cause the grounded neutral conductor to be overloaded, which can cause a fire.

Figure 300–40

WARNING: *Failure to terminate the ungrounded conductors to separate phases could cause the grounded neutral conductor to become overloaded, and the insulation could be damaged or destroyed by excessive heat. Conductor overheating is known to decrease insulating material service life, which creates the potential for arcing faults in hidden locations and could ultimately lead to fires. It isn't known just how long conductor insulation will last, but heat does decrease its life span.* Figure 300–40

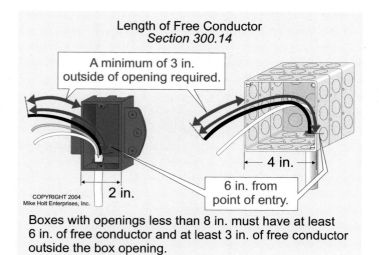

Length of Free Conductor
Section 300.14

A minimum of 3 in. outside of opening required.

4 in.

6 in. from point of entry.

2 in.

Boxes with openings less than 8 in. must have at least 6 in. of free conductor and at least 3 in. of free conductor outside the box opening.

Figure 300–41

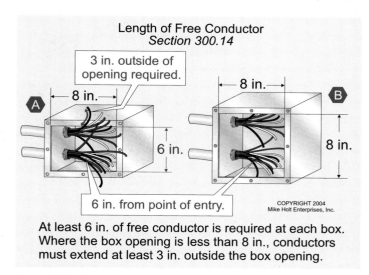

Length of Free Conductor
Section 300.14

3 in. outside of opening required.

8 in.

8 in.

6 in.

8 in.

6 in. from point of entry.

At least 6 in. of free conductor is required at each box. Where the box opening is less than 8 in., conductors must extend at least 3 in. outside the box opening.

Figure 300–42

300.14 Length of Free Conductors. At least 6 in. of free conductor, measured from the point in the box where the conductors enter the enclosure, must be left at each outlet, junction, and switch point for splices or terminations of luminaires or devices. Figure 300–41

Boxes that have openings less than 8 in. in any dimension, must have at least 6 in. of free conductor, measured from the point where the conductors enter the box, and at least 3 in. of free conductor outside the box opening. Figure 300–42

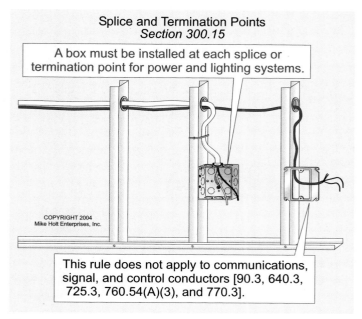

Splice and Termination Points
Section 300.15

A box must be installed at each splice or termination point for power and lighting systems.

This rule does not apply to communications, signal, and control conductors [90.3, 640.3, 725.3, 760.54(A)(3), and 770.3].

Figure 300–43

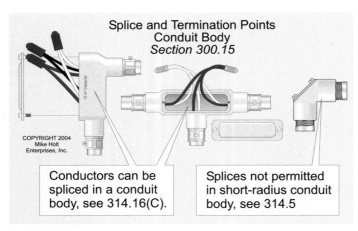

Splice and Termination Points
Conduit Body
Section 300.15

Conductors can be spliced in a conduit body, see 314.16(C).

Splices not permitted in short-radius conduit body, see 314.5

Figure 300–44

Author's Comment: The *NEC* doesn't limit the number of extension rings permitted on an outlet box. However it does stipulate that there must be not less than 3 in. of free conductors outside the opening of the final extension ring. To be sure, check with the authority having jurisdiction.

Exception: This rule doesn't apply to conductors that pass through a box without splice or termination.

300.15 Boxes or Conduit Bodies. A box must be installed at each splice or termination point, except as permitted for: **Figure 300–43**

- Cabinet or Cutout Boxes, 312.8
- Conduit Bodies, 314.16(C) **Figure 300–44**
- Luminaires, 410.31
- Surface Raceways, 386.56 and 388.56
- Wireways, 376.56

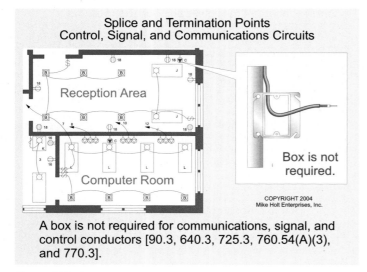

Splice and Termination Points
Control, Signal, and Communications Circuits

Reception Area

Computer Room

Box is not required.

A box is not required for communications, signal, and control conductors [90.3, 640.3, 725.3, 760.54(A)(3), and 770.3].

Figure 300–45

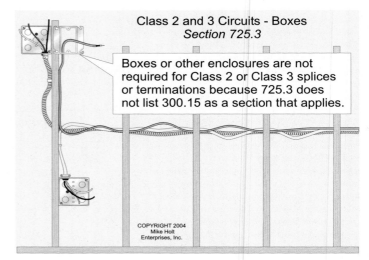

Class 2 and 3 Circuits - Boxes
Section 725.3

Boxes or other enclosures are not required for Class 2 or Class 3 splices or terminations because 725.3 does not list 300.15 as a section that applies.

Figure 300–46

Author's Comment: Boxes aren't required for the following signaling and communications cables: **Figure 300–45**

- CATV, 90.3
- Class 2 and 3 Control and Signaling, 725.3 **Figure 300–46**
- Communications, 90.3
- Optical Fiber Cable, 770.3
- Sound Systems, 640.3(A)

Fittings and Connectors. Fittings can only be used with the specific wiring methods for which they are listed and designed. For example, Type NM cable connectors cannot be used with Type AC cable, and electrical metallic tubing fittings cannot be used with rigid metal conduit or intermediate metal conduit, unless listed for the purpose. **Figure 300–47**

Author's Comment: Rigid nonmetallic conduit couplings and connectors are permitted with electrical nonmetallic tubing if the proper glue is used in accordance with manufacturer's instructions [110.3(B)]. See 362.48.

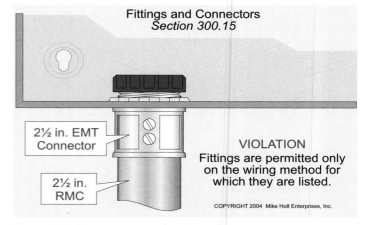

Fittings and Connectors
Section 300.15

2½ in. EMT Connector

2½ in. RMC

VIOLATION
Fittings are permitted only on the wiring method for which they are listed.

Figure 300–47

(C) Raceways for Support or Protection. When a raceway is used for the support or protection of cables, a fitting to reduce the potential for abrasion must be placed at the location the cables enter the raceway. See 250.86 Ex. 2, 300.5(D), 300.10 Ex. 1, and 300.12 Ex. for more details.

(G) Underground Splices. A box or conduit body isn't required where a splice is made underground if the conductors are spliced with a splicing device listed for direct burial. See 110.14(B) and 300.5(E).

> **Author's Comment:** See Article 100 for the definition of "Conduit Body."

(I) Enclosures. A box or conduit body isn't required where a splice is made in a cabinet or in cutout boxes containing switches or overcurrent protection devices if the splices or taps do not fill the wiring space at any cross section to more than 75 percent, and the wiring at any cross section doesn't exceed 40 percent. See 312.8 and 404.3(B). **Figure 300–48**

> **Author's Comment:** See Article 100 for the definition of "Cutout Box."

(J) Luminaires. A box or conduit body isn't required where a luminaire is used as a raceway as permitted in 410.31 and 410.32.

(L) Handhole Enclosures. Where accessible only to qualified persons, a box or conduit body isn't required for conductors in handhole enclosures installed in accordance with 314.30. **Figure 300–49**

Author's Comments:

- See Article 100 for the definition of "Handhole Enclosure."

- Handhole enclosures are often used in conjunction with underground RNC conduit for the installation of landscape lighting, light poles, and other applications. However, the

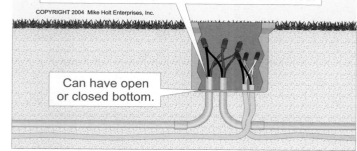

**Splice and Termination Points
Handhole Enclosure
Section 300.15(L)**

Where accessible only to qualified persons, a box isn't required for conductors in handhole enclosures installed in accordance with 314.30.

COPYRIGHT 2004 Mike Holt Enterprises, Inc.

Can have open or closed bottom.

Figure 300–49

Code requirements specify that the conductors can only be accessible to qualified persons!

300.17 Raceway Sizing. Raceways must be large enough to permit the installation and removal of conductors without damaging the conductor's insulation.

> **Author's Comment:** When all conductors in a raceway are the same size and of the same insulation type, the number of conductors permitted can be determined by Annex C.
>
> *Question: How many 12 THHN conductors can be installed in trade size ¾ electrical metallic tubing?* **Figure 300–50**
>
> *(a) 12   (b) 13   (c) 14   (d) 16*
>
> *Answer: (d) 16 conductors, Annex C, Table C1*

> **Author's Comment:** When different size conductors are installed in a raceway, conductor fill is limited to the percentages of Table 1 of Chapter 9. **Figure 300–51**

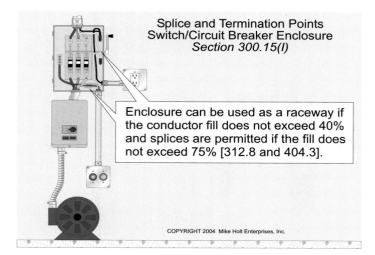

**Splice and Termination Points
Switch/Circuit Breaker Enclosure
Section 300.15(I)**

Enclosure can be used as a raceway if the conductor fill does not exceed 40% and splices are permitted if the fill does not exceed 75% [312.8 and 404.3].

COPYRIGHT 2004 Mike Holt Enterprises, Inc.

Figure 300–48

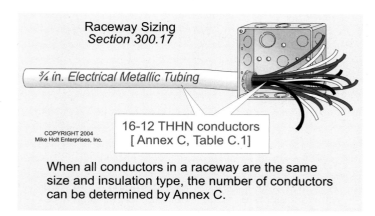

**Raceway Sizing
Section 300.17**

¾ in. Electrical Metallic Tubing

COPYRIGHT 2004
Mike Holt Enterprises, Inc.

16-12 THHN conductors
[Annex C, Table C.1]

When all conductors in a raceway are the same size and insulation type, the number of conductors can be determined by Annex C.

Figure 300–50

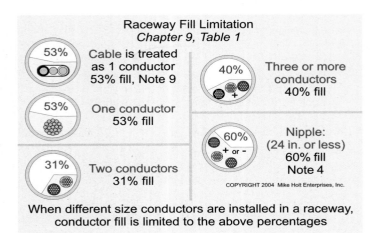

Figure 300–51

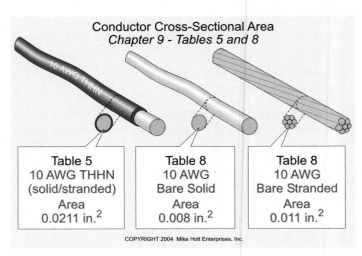

Figure 300–53

Table 1, Chapter 9

Number	Percent Fill
1 Conductor	53%
2 Conductors	31%
3 or more	40%

The above percentages are based on conditions where the length of the conductor and number of raceway bends are within reasonable limits.

Author's Comment: Where a raceway has a maximum length of 24 in., it can be filled to 60 percent of its total cross-sectional area [Chapter 9, Table 1, Note]. **Figure 300–52**

Step 1. When sizing a raceway, first determine the total area of conductors (Chapter 9, Table 5 for insulated conductors and Chapter 9, Table 8 for bare conductors). **Figure 300–53**

Step 2. Select the raceway from Chapter 9, Table 4, in accordance with the percent fill listed in Chapter 9, Table 1. **Figure 300–54**

Question: *What trade size Schedule 40 rigid nonmetallic conduit is required for the following THHN conductors?* **Figure 300–55**

3—500 THHN
1—250 THHN
1—3 THHN

(a) 2   (b) 3   (c) 4   (d) none of these

Answer: *(b) 3*

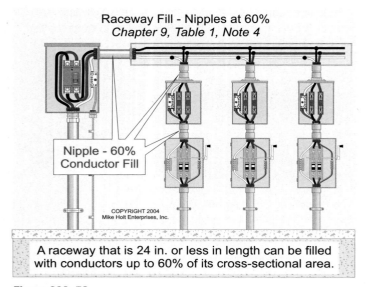

Figure 300–52

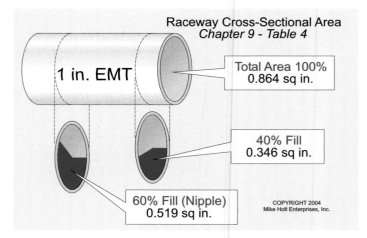

Figure 300–54

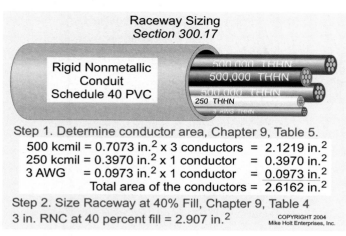

Raceway Sizing
Section 300.17

Rigid Nonmetallic Conduit Schedule 40 PVC

500,000 THHN
500,000 THHN
500,000 THHN
250 THHN
3 AWG THHN

Step 1. Determine conductor area, Chapter 9, Table 5.
500 kcmil = 0.7073 in.² x 3 conductors = 2.1219 in.²
250 kcmil = 0.3970 in.² x 1 conductor = 0.3970 in.²
3 AWG = 0.0973 in.² x 1 conductor = 0.0973 in.²
Total area of the conductors = 2.6162 in.²

Step 2. Size Raceway at 40% Fill, Chapter 9, Table 4
3 in. RNC at 40 percent fill = 2.907 in.²

COPYRIGHT 2004 Mike Holt Enterprises, Inc.

Figure 300–55

Step 1. Determine the total area of conductors [Chapter 9, Table 5]:

500 THHN	*0.7073 x 3*	*=*	*2.1219 in.²*
250 THHN	*0.3970 x 1*	*=*	*0.3970 in.²*
3 THHN	*0.0973 x 1*	*=*	*0.0973 in.²*
Total Area			*2.6162 in.²*

Step 2. Select the raceway at 40 percent fill [Chapter 9, Table 4]: Trade size 3 rigid nonmetallic conduit Schedule 40 = 2.907 sq in. of conductor fill at 40%.

300.18 Inserting Conductors in Raceways.

(A) Complete Runs. To protect conductor insulation from abrasion during installation, raceways must be mechanically completed between the pulling points before conductors are installed. See 300.10 and 300.12. **Figure 300–56**

Exception: Short sections of raceways used for protection of cables from physical damage aren't required to be installed complete between outlet, junction, or splicing points.

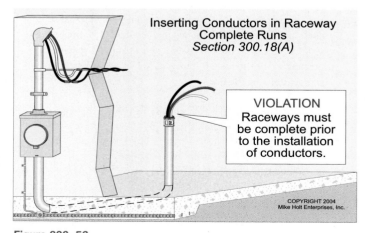

Inserting Conductors in Raceway
Complete Runs
Section 300.18(A)

VIOLATION
Raceways must be complete prior to the installation of conductors.

COPYRIGHT 2004 Mike Holt Enterprises, Inc.

Figure 300–56

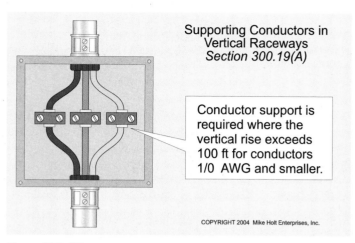

Supporting Conductors in Vertical Raceways
Section 300.19(A)

Conductor support is required where the vertical rise exceeds 100 ft for conductors 1/0 AWG and smaller.

COPYRIGHT 2004 Mike Holt Enterprises, Inc.

Figure 300–57

300.19 Supporting Conductors in Vertical Raceways.

(A) Spacing Intervals. If the vertical rise of a raceway exceeds the values of Table 300.19(A), the conductors must be supported at the top, or as close to the top as practical. **Figure 300–57**

> **Author's Comment:** The weight of long vertical runs of conductors can cause the conductors to actually drop out of the raceway if they aren't properly secured. There have been many cases where conductors in a vertical raceway were released from the pulling basket (at the top) without being secured, and the conductors fell down and out of the raceway, injuring those at the base of the installation.

300.20 Induced Currents in Metal Parts.

(A) Conductors Grouped Together. To minimize induction heating of ferrous metallic raceways and enclosures, and to maintain an effective ground-fault current path, all conductors of a circuit must be installed in the same raceway, cable, trench, cord, or cable tray. See 250.102(E), 300.3(B), 300.5(I), and 392.8(D).

> **Author's Comment:** When alternating current (ac) flows through a conductor, a pulsating or varying magnetic field is created around the conductor. This magnetic field is constantly expanding and contracting with the amplitude of the ac current. In the United States, the frequency is 60 cycles per second. Since ac reverses polarity 120 times per second, the magnetic field that surrounds the conductor also reverses its direction 120 times per second. This expanding and collapsing magnetic field induces eddy currents in the ferrous metal parts that surround the conductors, causing the metal parts to heat up from hysteresis.

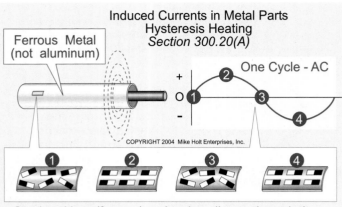

Induced Currents in Metal Parts
Hysteresis Heating
Section 300.20(A)

One Cycle - AC

Steel and iron (ferrous) molecules align to the polarity of the magnetic field and when the field reverses, the molecules reverse their polarity. This back-and-forth alignment of the molecules heats up ferrous metal parts.

Figure 300–58

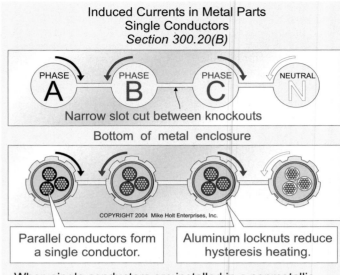

Induced Currents in Metal Parts
Single Conductors
Section 300.20(B)

PHASE **A**   PHASE **B**   PHASE **C**   NEUTRAL **N**

Narrow slot cut between knockouts

Bottom of metal enclosure

Parallel conductors form a single conductor.

Aluminum locknuts reduce hysteresis heating.

When single conductors are installed in a nonmetallic raceway, inductive heating of the metal enclosure can be minimized by cutting a slot between the individual holes through which the conductors pass.

Figure 300–59

Hysteresis heating affects only ferrous metals with magnetic properties, such as steel and iron, but not aluminum. Simply put, the molecules of steel and iron align to the polarity of the magnetic field and when the magnetic field reverses, the molecules reverse their polarity as well. This back-and-forth alignment of the molecules heats up the metal, and the greater the current flow, the greater the heat rises in the ferrous metal parts. **Figure 300–58**

When conductors of the same circuit are grouped together, the magnetic fields of the different conductors tend to cancel each other out, resulting in a reduced magnetic field around the conductors. The lower magnetic field reduces induced currents in the ferrous metal raceways or enclosures, which reduces hysteresis heating of the surrounding metal enclosure.

WARNING: *There has been much discussion in the press on the effects of electromagnetic fields on humans. According to the Institute of Electrical and Electronics Engineers (IEEE), there's insufficient information at this time to define an unsafe electromagnetic field level.*

(B) Single Conductors. When single conductors are installed in nonmetallic raceways as permitted in 300.5(I) Ex. 2, the inductive heating of the metal enclosure can be minimized by the use of aluminum locknuts and by cutting a slot between the individual holes through which the conductors pass. **Figure 300–59**

FPN: Because aluminum is a nonmagnetic metal, aluminum parts do not heat up due to hysteresis.

Author's Comment: Aluminum conduit, locknuts, and enclosures carry eddy currents, but because aluminum is nonferrous, it doesn't heat up [300.20(B) FPN].

300.21 Spread of Fire or Products of Combustion.

Electrical circuits and equipment must be installed in such a way that the spread of fire or products of combustion will not be substantially increased. Openings in fire-rated walls, floors, and ceilings for electrical equipment must be firestopped using methods approved by the authority having jurisdiction to maintain the fire-resistance rating of the fire-rated assembly.

Author's Comment: Firestop material is listed for the specific types of wiring methods and construction structures.

FPN: Directories of electrical construction materials published by qualified testing laboratories contain listing and installation restrictions necessary to maintain the fire-resistive rating of assemblies. Outlet boxes must have a horizontal separation not less than 24 in. when installed in a fire-rated assembly, unless an outlet box is listed for closer spacing or protected by fire-resistant "putty pads" in accordance with manufacturer's instructions.

Author's Comment: This rule applies to control, signal, and communications cables. **Figure 300–60**

- CATV, 820.3(A)
- Communications, 800.3(C)
- Control and Signaling, 725.3(B)
- Fire Alarm, 760.3(A)
- Optical Fiber Cables, 770.3(A)
- Sound Systems, 640.3(A)

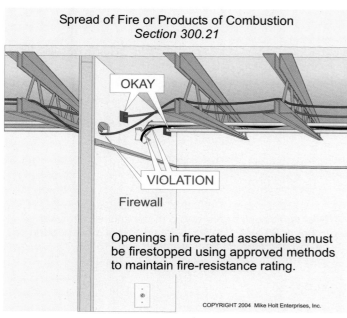

Figure 300–60

300.22 Ducts, Plenums, and Air-Handling Spaces.

(A) Ducts Used for Dust, Loose Stock, or Vapor. Ducts that transport dust, loose stock, or vapors must not have any wiring method installed within them. Figure 300–61

(B) Ducts or Plenums Used for Environmental Air. Where necessary for the direct action upon, or sensing of, the contained air, Type MI cable, Type MC cable that has a smooth or corrugated impervious metal sheath without an overall nonmetallic covering, electrical metallic tubing, flexible metallic tubing, intermediate metal conduit, or rigid metal conduit without an overall nonmetallic covering can be installed in ducts or plenums specifically fabricated to transport environmental air.

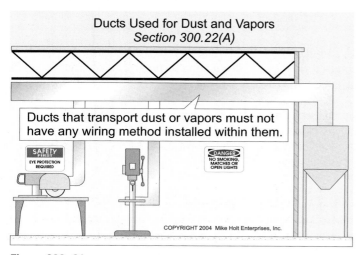

Figure 300–61

Author's Comment: See Article 100 for the definition of "Plenum."

Flexible metal conduit in lengths not exceeding 4 ft can be used to connect physically adjustable equipment and devices, provided any openings are effectively closed.

Where equipment or devices are installed and illumination is necessary to facilitate maintenance and repair, enclosed gasketed-type luminaires are permitted.

(C) Space Used for Environmental Air. Wiring and equipment in spaces used for environmental air-handling purposes must comply with (1) and (2). This requirement doesn't apply to habitable rooms or areas of buildings, the prime purpose of which isn't air handling.

> FPN: The spaces above a suspended ceiling or below a raised floor that are used for environmental air are examples of the type of space to which this section applies. Figure 300–62

(1) Wiring Methods Permitted. Electrical metallic tubing, rigid metal conduit, intermediate metal conduit, armored cable, metal clad cable without a nonmetallic cover, and flexible metal conduit can be installed in other environmental air spaces. Figure 300–63

Where accessible, surface metal raceways, metal wireways with metal covers, or solid bottom metal cable tray with solid metal covers can be installed in other environmental air spaces.

Author's Comments:

- Control, signaling, and communications cables installed in surface metal raceways, metal wireways with solid metal covers, or solid bottom metal cables with solid metal covers are not required to be plenum-rated.

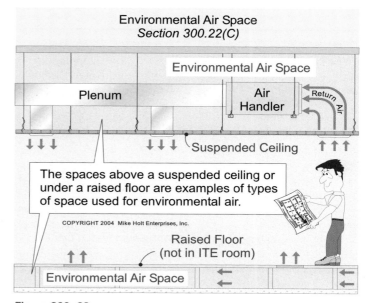

Figure 300–62

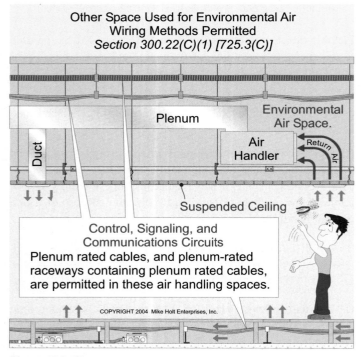

**Other Space Used for Environmental Air
Wiring Methods Permitted
Section 300.22(C)(1) [725.3(C)]**

Plenum

Duct

Environmental
Air Space.

Air
Handler

Return Air

Suspended Ceiling

**Control, Signaling, and
Communications Circuits**
Plenum rated cables, and plenum-rated
raceways containing plenum rated cables,
are permitted in these air handling spaces.

COPYRIGHT 2004 Mike Holt Enterprises, Inc.

Figure 300–63

- Rigid nonmetallic conduit [Article 352], electrical nonmetallic tubing [Article 362], and nonmetallic cables are not permitted to be installed in spaces used for environmental air because they give off deadly toxic fumes when burned or super heated.

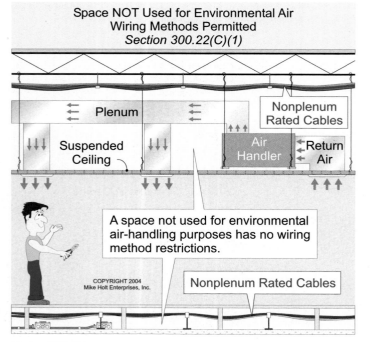

**Space NOT Used for Environmental Air
Wiring Methods Permitted
Section 300.22(C)(1)**

Plenum

Nonplenum
Rated Cables

Suspended
Ceiling

Air
Handler

Return
Air

A space not used for environmental
air-handling purposes has no wiring
method restrictions.

COPYRIGHT 2004
Mike Holt Enterprises, Inc.

Nonplenum Rated Cables

Figure 300–64

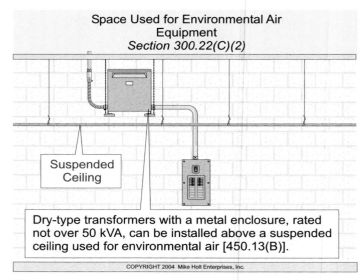

**Space Used for Environmental Air
Equipment
Section 300.22(C)(2)**

Suspended
Ceiling

Dry-type transformers with a metal enclosure, rated
not over 50 kVA, can be installed above a suspended
ceiling used for environmental air [450.13(B)].

COPYRIGHT 2004 Mike Holt Enterprises, Inc.

Figure 300–65

However, control, signaling and communications cables, and nonmetallic raceways installed in spaces used for environmental air must be plenum rated. **See Figure 300–63.**

- CATV, 820.179(A)
- Communications, 800.154(A)
- Control and Signaling, 725.61(A)
- Fire Alarm, 760.61(A)
- Optical Fiber Cables, 770.154(A)
- Sound Systems, 640.9(C) and 725.61(A)

- A space not used for environmental air-handling purposes has no wiring method restrictions. **Figure 300–64**

(2) Equipment. Electrical equipment with a metal enclosure is permitted in other environmental air spaces, unless prohibited elsewhere in this *Code*.

Author's Comment: Dry-type transformers with a metal enclosure, rated not over 50 kVA, can be installed above suspended ceilings used for environmental air [450.13(B)]. **Figure 300–65**

(D) Information Technology Equipment Room. Wiring under a raised floor in an information technology room must comply with 645.5(D). **Figure 300–66**

Author's Comment: Signal and communications cables under a raised floor are not required to be plenum rated [645.5(D)(5)(c)], because ventilation is restricted to that room/space [645.5(D)(3)].

300.23 Panels Designed to Allow Access. Wiring, cables, and equipment installed behind panels must be located so the panels can be removed to give access to electrical equipment. **Figure 300–67**

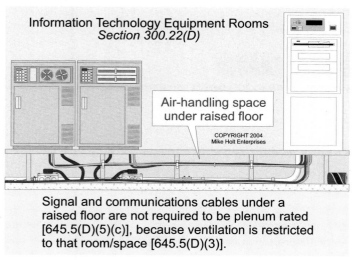

Figure 300–66

Author's Comment: Access to equipment must not be hindered by an accumulation of cables that prevent the removal of suspended-ceiling panels. Control, signaling, and communications cables must be located and supported so that the suspended-ceiling panels can be moved to provide access to electrical equipment.

- CATV Cable, 820.21
- Communications Cable, 800.7
- Control & Signaling Cable, 725.7
- Fire Alarm Cable, 760.5
- Optical Fiber Cable, 770.21
- Sound Cable, 640.5

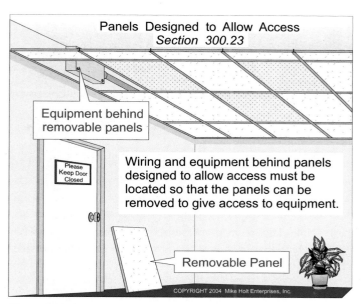

Figure 300–67

Article 300 Questions

1. Single conductors as specified in Table 310.13 are only permitted when installed as part of a recognized wiring method from Chapter _____ of the *NEC*.

    (a) 4   (b) 3   (c) 2   (d) 9

2. Cables laid in wood notches require protection against nails or screws by using a steel plate at least _____ thick, installed before the building finish is applied. A thinner plate that provides equal or better protection may be used if listed and marked.

    (a) ⅟₁₆ in.   (b) ⅛ in.   (c) ½ in.   (d) none of these

3. In both exposed and concealed locations, where a cable or nonmetallic raceway-type wiring method is installed parallel to framing members such as joists, rafters, or studs or furring strips, the nearest outside surface of the cable or raceway must be _____ the nearest edge of the framing member where nails or screws are likely to penetrate.

    (a) not less than 1¼ inch from   (b) immediately adjacent to
    (c) not less than ⅟₁₆ inch from   (d) 90° away from

4. What is the minimum cover requirement in inches for direct burial UF cable installed outdoors that supplies power to a 120V, 30A circuit?

    (a) 6 in.   (b) 12 in.   (c) 18 in.   (d) 24 in.

5. Service conductors that are not encased in concrete and that are buried 18 in. or more below grade must have their location identified by a warning ribbon placed in the trench at least _____ above the underground installation.

    (a) 6 in.   (b) 12 in.   (c) 18 in.   (d) none of these

6. All conductors of the same circuit are required to be _____.

    (a) in the same raceway or cable   (b) in close proximity in the same trench
    (c) the same size   (d) a or b

7. Where corrosion protection is necessary and the conduit is threaded in the field, the threads must be coated with a(n) _____, electrically conductive, corrosion-resistance compound.

    (a) marked   (b) listed   (c) labeled   (d) approved

8. Nonmetallic raceways, cable trays, cablebus, auxiliary gutters, boxes, and cables with nonmetallic outer jackets must be made of material approved for the conditions where they will be installed, and where exposed to chemicals, the materials or coatings must be _____.

    (a) listed as inherently resistant to chemicals   (b) identified for the specific chemical reagent
    (c) both a and b   (d) either a or b

9. •Metal raceways, cable armor, and other metal enclosures for conductors must be _____ joined together to form a continuous electrical conductor.

    (a) electrically   (b) permanently   (c) metallically   (d) none of these

10. Ceiling-support wires used for the support of electrical raceways and cables within nonfire rated assemblies are required to be distinguishable from the suspended-ceiling framing support wires.

    (a) True   (b) False

11. In multiwire circuits, the continuity of the _____ conductor must not be dependent upon the device connections.

    (a) ungrounded   (b) grounded   (c) grounding   (d) a and b

12. A box or conduit body is not required for splices and taps in direct-buried conductors and cables as long as the splice is made with a splicing device that is identified for the purpose.

    (a) True   (b) False

13. The number of conductors permitted in a raceway must be limited to _____.

    (a) permit heat to dissipate   (b) prevent damage to insulation during installation
    (c) prevent damage to insulation during removal of conductors (d) all of these

14. A 100 ft vertical run of 4/0 AWG copper requires the conductors to be supported at _____ locations.

    (a) 4   (b) 5   (c) 2   (d) none of these

15. Openings around electrical penetrations through fire-resistant-rated walls, partitions, floors, or ceilings must _____ to maintain the fire resistance rating.

    (a) be documented   (b) not be allowed
    (c) be firestopped using approved methods (d) be enlarged

16. When equipment or devices are installed in ducts or plenum chambers used to transport environmental air, and illumination is necessary to facilitate maintenance and repair, enclosed _____-type luminaires are permitted.

    (a) screw   (b) plug   (c) gasketed   (d) neon

Notes

ARTICLE 310

Conductors for General Wiring

Introduction

This article contains the general requirements for conductors, such as insulation markings, ampacity ratings, and conditions of use. Article 310 doesn't apply to conductors that are part of cable assemblies, flexible cords, fixture wires, or to conductors that are an integral part of equipment [90.6, 300.1(B)].

People most often make errors in applying the ampacity tables contained in Article 310. If you study the explanations carefully, you'll avoid common errors such as applying Table 310.17 when you should be applying Table 310.16.

But why so many tables? Why does Table 310.17 list the ampacity of 6 THHN as 105A, yet Table 310.16 lists the same conductor as having an ampacity of only 75A? To answer that, go back to Article 100 and review the definition of ampacity. Notice the phrase "conditions of use." What these tables do is set a maximum current value at which you can ensure the installation won't undergo premature failure of the conductor insulation in normal use.

The designations THHN, THHW, RHH, and so on, describe insulation. Every type of insulation has a heat withstand limit. When current flows through a conductor it creates heat. How well the insulation around a conductor can dissipate that heat depends on factors such as whether that conductor is in free air or not. Think what happens to you if you put on a sweater, a jacket, and then a coat—all at the same time. You heat up. Your skin cannot dissipate heat with all this clothing on nearly as well as it dissipates heat in free air.

Conductors fail with age. That's why we conduct cable testing and take other measures to predict failure and replace certain conductors (for example, feeders or critical equipment conductors) while they are still within design specifications. But conductor failure takes decades under normal use and occurs slowly—and it's a maintenance issue. However, if a conductor is forced to exceed the ampacity listed in the appropriate table, failure happens much more rapidly—often catastrophically. Exceeding the allowable ampacity is a safety issue.

310.1 Scope. Article 310 contains the general requirements for conductors, such as insulation markings, ampacity ratings, and their use. This article doesn't apply to conductors that are an integral part of cable assemblies, cords, or equipment. See 90.6 and 300.1(B).

310.2 Conductors

(A) Insulation. All conductors must be insulated, and they must be installed as part of a recognized wiring method listed in Chapter 3. See 110.8 and 300.3(A). Figure 310–1

Exception: Where permitted, bare equipment grounding or bonding conductors are not required to be insulated. See 250.64 and 250.119. Figure 310–2

Author's Comment: Equipment grounding (bonding) conductors must be insulated for patient-care receptacles, switches and equipment [517.13(B)], and for wet-niche pool lights and permanently installed pool, outdoor spa, and outdoor hot tub equipment [680.21(A)(1), 680.23(F)(2), and 680.25(B)].

310.3 Stranded Conductors. Conductors 8 AWG and larger must be stranded when installed in a raceway. Figure 310–3

Author's Comment: Bare solid conductors are often used for the grounding electrode conductor [250.62] and for permanently installed pool, outdoor spa, and outdoor hot tub bonding [680.26(C)].

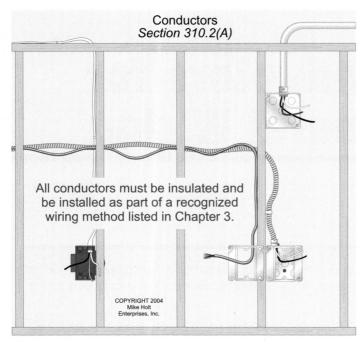

Figure 310–1

Figure 310–3

310.4 Conductors in Parallel.
Ungrounded and grounded neutral conductors sized 1/0 AWG and larger can be connected in parallel *(electrically joined at both ends)*.

When conductors are run in parallel, the current must be evenly distributed between the individual parallel conductors. This is accomplished by ensuring that all ungrounded and grounded neutral conductors within a parallel set are identical. Each conductor of a parallel set must:

(1) Be the same length.
(2) Be made of the same conductor material (copper/aluminum).
(3) Be the same size in circular mil area (minimum 1/0 AWG).
(4) Use the same insulation material (like THHN).
(5) Terminate in the same method (set screw versus compression).

Author's Comment: Each current-carrying conductor of a paralleled set of conductors must be counted as a current-carrying conductor for the purpose of conductor ampacity adjustment, in accordance with Table 310.15(B)(2)(a). **Figure 310–4**

In addition, raceways or cables containing parallel conductors must have the same physical characteristics and the same number of conductors in each raceway or cable, **Figure 310–5**. Conductors for one phase (ungrounded conductor) or the grounded neutral conductor, aren't required to have the same physical characteristics as those of another phase or grounded neutral conductor to achieve balance.

Author's Comment: If one set of parallel conductors is run in a metallic raceway and the other conductors are run in a nonmetallic raceway, the conductors in the metallic raceway will have an increased opposition to current flow (impedance) as compared to the conductors in the nonmetallic raceway. This results in an unbalanced distribution of current between the parallel conductors. Without getting into the details, this isn't good.

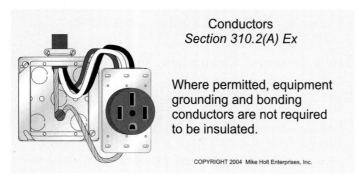

Figure 310–2

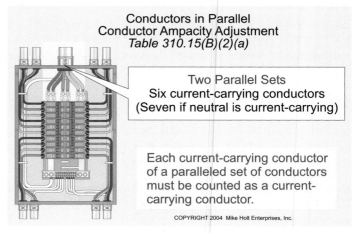

Figure 310–4

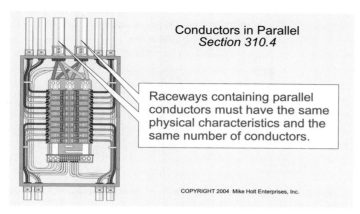

Figure 310–5

Paralleling is done in sets. Parallel sets of conductors aren't required to have the same physical characteristics as those of another set to achieve balance.

For example, a 400A feeder with a neutral load of 240A can be in parallel as follows: **Figure 310–6**

• Phase A, Two—250 kcmil THHN aluminum, 100 ft
• Phase B, Two—3/0 THHN copper, 104 ft
• Phase C, Two—3/0 THHN copper, 102 ft
• Neutral, Two—1/0 THHN aluminum, 103 ft
• Equipment Ground, Two—3 AWG copper, 101 ft

Equipment Grounding (Bonding) Conductors. The equipment grounding (bonding) conductors for circuits in parallel must be identical to each other in length, material, size, insulation, and termination. In addition, each raceway, where required, must have an equipment grounding (bonding) conductor sized in accordance with 250.122. The minimum 1/0 AWG rule of 310.4 doesn't apply to equipment grounding (bonding) conductors [250.122(F)(1)]. **Figure 310–7**

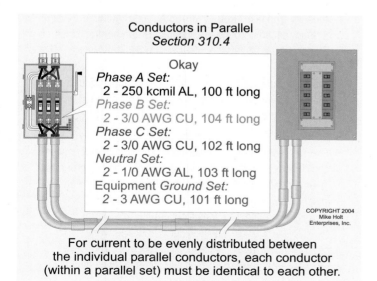

Figure 310–6

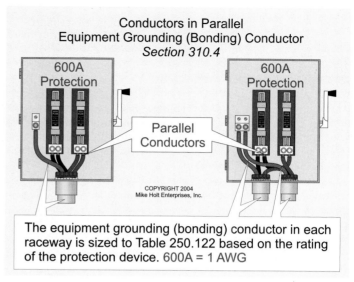

Figure 310–7

Ampacity Adjustment. When more than three current-carrying conductors are run together in a raceway longer than 24 in., the ampacity adjustment factors of Table 310.15(B)(2)(a) must be applied. See 310.10 and 310.15 for details and examples.

310.5 Minimum Size Conductors. The smallest conductor permitted for branch circuits for residential, commercial, and industrial locations is 14 AWG copper, except as permitted elsewhere in this *Code*.

> **Author's Comment:** There's a misconception that 12 AWG copper is the smallest conductor permitted for commercial or industrial facilities. Although this isn't true based on *NEC* rules, it may be a local code requirement.

310.8 Location.

(D) Locations Exposed to Direct Sunlight. Insulated conductors and cables exposed to the direct rays of the sun must comply with one of the following:

(1) Cables must be listed as being sunlight resistant or marked as being sunlight resistant.

> **Author's Comment:** Type SE cable and the conductors contained in the cable are listed as sunlight resistant. However, according to the UL listing standard, the conductors contained in Type SE cable aren't required to be identified or marked as sunlight resistant.

(2) Conductors must be listed as being sunlight resistant or marked as being sunlight resistant. **Figure 310–8**

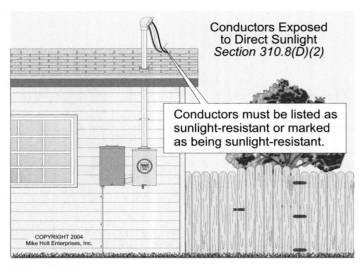

Figure 310–8

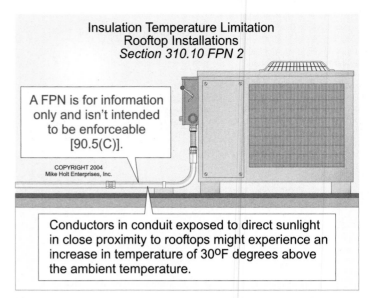

Figure 310–9

Author's Comment: All insulated conductors exposed to the direct rays of the sun must have a "Sunlight Resistant" marking on the conductor insulation. However, this is not a requirement for conductors within a cable that is listed as sunlight resistant [310.8(D)(1)].

(3) Conductors can be covered with insulating material, such as tape or sleeving that is listed as being sunlight resistant or marked as being sunlight resistant.

310.9 Corrosive Conditions. Conductor insulation must be suitable for any substance to which it may be exposed that may have a detrimental effect on the conductor's insulation, such as oil, grease, vapor, gases, fumes, liquids, or other substances. See 110.11.

310.10 Insulation Temperature Limitation. Conductors cannot be used where the operating temperature exceeds that designated for the type of insulated conductor involved.

> **FPN No. 1:** The insulation temperature rating of a conductor (see Table 310.13) is the maximum temperature a conductor can withstand over a prolonged time without serious degradation. The main factors to consider for conductor operating temperature include ambient temperature, heat generated internally from current flow through the conductor, the rate at which heat can dissipate, and adjacent load carrying conductors.

> **FPN No. 2:** Conductors installed in conduit exposed to direct sunlight in close proximity to rooftops have been shown, under certain conditions, to experience an increase in temperature of 30°F above ambient temperature. **Figure 310–9**

Author's Comment: A Fine Print Note (FPN) is for information only and isn't intended to be enforceable [90.5(C)].

310.12 Conductor Identification.

(A) Grounded Neutral Conductor. The grounded neutral conductor must be identified in accordance with 200.6.

(B) Equipment Grounding (Bonding) Conductor. The equipment grounding (bonding) conductor must be identified in accordance with 250.119.

(C) Ungrounded Conductors. Ungrounded conductors must be clearly distinguishable from grounded neutral and equipment grounding (bonding) conductors.

Author's Comments:

• Where the premises wiring system has branch circuits or feeders supplied from more than one nominal voltage system, each ungrounded conductor of the branch circuit or feeder, where accessible, must be identified by system. The means of identification can be by separate color-coding, marking tape, tagging, or other means approved by the authority having jurisdiction. Such identification must be permanently posted at each panelboard. See 210.5(C) and 215.12 for specific requirements. **Figure 310–10**

• The *NEC* doesn't require color-coding of ungrounded conductors, except for the high-leg conductor when a grounded neutral conductor is present [110.15 and 230.56]. However, electricians often use the following color system for power and lighting conductor identification:
  • 120/240V single-phase—black, red, and white
  • 120/208V three-phase—black, red, blue, and white
  • 120/240V three-phase—black, orange, blue, and white
  • 277/480V three-phase—brown, orange, yellow, and gray; or, brown, purple, yellow, and gray

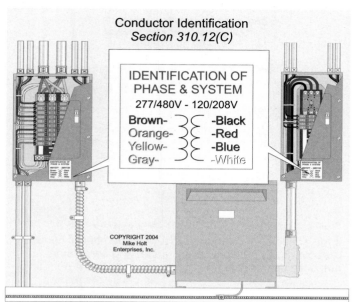

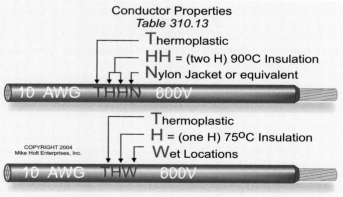

Conductor Identification
Section 310.12(C)

IDENTIFICATION OF
PHASE & SYSTEM
277/480V - 120/208V

Brown-   -Black
Orange-   -Red
Yellow-   -Blue
Gray-   -White

COPYRIGHT 2004
Mike Holt
Enterprises, Inc.

Where the premises wiring system has more than one nominal voltage system, each ungrounded conductor must be identified by system and posted at each panelboard.

Figure 310–10

310.13 Conductor Construction. Table 310.13 contains conductor insulation information, such as operating temperature, applications, sizes, and outer cover. These conductors can be used in any Chapter 3 wiring method.

Author's Comment: The following explains the lettering on conductor insulation: **Figure 310–11**

- F   Fixture wires (solid or 7 strands) [Table 402.3]
- FF   Flexible fixture wire (19 strands) [Table 402.3]
- No H   60°C insulation rating

Conductor Properties
Table 310.13

Thermoplastic
HH = (two H) 90°C Insulation
Nylon Jacket or equivalent

10 AWG THHN 600V

Thermoplastic
H = (one H) 75°C Insulation
Wet Locations

10 AWG THW 600V

COPYRIGHT 2004
Mike Holt Enterprises, Inc.

Table 310.13 contains conductor insulation information, such as operating temperature and applications. These conductors can be used in any Chapter 3 wiring method.

Figure 310–11

- H   75°C insulation rating
- HH   90°C insulation rating
- -2   Suitable for 90°C in wet locations
- N   Nylon outer cover
- R   Thermoset insulation
- T   Thermoplastic insulation
- U   Underground
- W   Wet or damp locations
- X   Cross-linked polyethylene insulation

310.15 Conductor Ampacity.

(A) General Requirements.

(1) Tables for Engineering Supervision. The ampacity of a conductor can be determined either by using the tables in accordance with 310.15(B), or under engineering supervision as provided in 310.15(C).

> **FPN No. 1:** Ampacities provided by this section do not take voltage drop into consideration. See 210.19(A) FPN No. 4, for branch circuits and 215.2(D) FPN No. 2, for feeders.

(2) Conductor Ampacity—Lower Rating. If a single length of conductor is routed in a manner that two or more ampacity ratings apply to a single conductor length, the lower ampacity must be used for the entire circuit. See 310.15(B). **Figure 310–12**

Exception: When different ampacities apply to a length of conductor, the higher ampacity is permitted for the entire circuit if the reduced ampacity length does not exceed 10 ft and its length doesn't exceed 10 percent of the length of the higher ampacity. **Figure 310–13**

(B) Ampacity Table. The allowable conductor ampacities listed in Table 310.16 are based on conditions where the ambient temperature isn't over 86°F and no more than three current-carrying conductors are bundled together. **Figure 310–14**

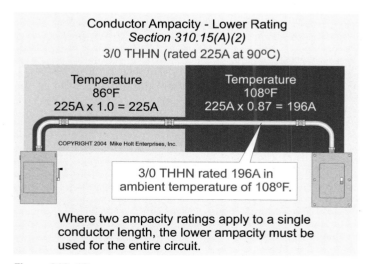

Conductor Ampacity - Lower Rating
Section 310.15(A)(2)
3/0 THHN (rated 225A at 90°C)

Temperature 86°F	Temperature 108°F
225A x 1.0 = 225A	225A x 0.87 = 196A

COPYRIGHT 2004 Mike Holt Enterprises, Inc.

3/0 THHN rated 196A in ambient temperature of 108°F.

Where two ampacity ratings apply to a single conductor length, the lower ampacity must be used for the entire circuit.

Figure 310–12

Conductor Ampacity - Higher Rating
Section 310.15(A)(2) Ex

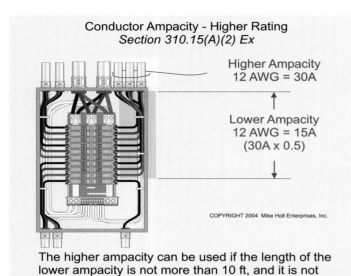

Higher Ampacity
12 AWG = 30A

Lower Ampacity
12 AWG = 15A
(30A x 0.5)

COPYRIGHT 2004 Mike Holt Enterprises, Inc.

The higher ampacity can be used if the length of the lower ampacity is not more than 10 ft, and it is not longer than 10 percent of the higher ampacity length.

Figure 310–13

Author's Comment: When conductors are installed in an ambient temperature other than 78 to 86°F, ampacities listed in Table 310.16 must be corrected in accordance with the multipliers listed in Table 310.16.

Ambient Temperature °F	Ambient Temperature °C	Correction Factor
70–77°F	21-25°C	1.04
78–86°F	26-30°C	1.00
87–95°F	31-35°C	0.96
96–104°F	36-40°C	0.91
105–113°F	41-45°C	0.87
114–122°F	46-50°C	0.82
123–131°F	51-55°C	0.76
132–140°F	56-60°C	0.71
141–158°F	61-70°C	0.58
159–176°F	71-80°C	0.41

Author's Comment: When correcting conductor ampacity for elevated ambient temperature, the correction factor used for THHN conductors is based on the 90°C rating of the conductor, based on the conductor ampacity listed in the 90°C column of Table 310.16 [110.14(C)].

Question: What is the corrected ampacity of 3/0 THHN conductors if the ambient temperature is 108°F?

(a) 173A   (b) 196A   (c) 213A   (d) 241A

Answer: *(b) 196A*

Conductor Ampacity [90°C] = 225A
Correction Factor [Table 310.16] = 0.87

Conductor Ampacity - Table 310.16
Correction and Adjustment
Section 310.15(B)

This raceway contains only 3 current-carrying conductors

Table 310.16 ampacity is based on an ambient temperature of not over 86°F and no more than 3 current-carrying conductors bundled together.

Conductor Ampacity Adjustment

Ambient Temperature

Conductor Bundling

If the ambient temperature is above 86°F, the conductor ampacity decreases.

If the number of current-carrying conductors exceeds 3, the conductor ampacity decreases.

COPYRIGHT 2004 Mike Holt Enterprises, Inc.

Figure 310–14

Corrected Ampacity = 225A x 0.87
Corrected Ampacity = 196A

Author's Comment: When adjusting conductor ampacity, the ampacity is based on the temperature insulation rating of the conductor as listed in Table 310.16, not the temperature rating of the terminal [110.14(C)].

(2) Ampacity Adjustment.

(a) Conductor Bundle. Where the number of current-carrying conductors in a raceway or cable exceeds three, or where single conductors or multiconductor cables are stacked or bundled in lengths exceeding 24 in., the allowable ampacity of each conductor, as listed in Table 310.16, must be adjusted in accordance with the adjustment factors contained in Table 310.15(B)(2)(a).

Each current-carrying conductor of a paralleled set of conductors must be counted as a current-carrying conductor. **Figure 310–15**

Table 310.15(B)(2)(a)

Number of Current–Carrying	Adjustment Factor
1–3 Conductors	1.00
4–6 Conductors	0.80
7–9 Conductors	0.70*
10–20 Conductors	0.50

*Figure 310–16

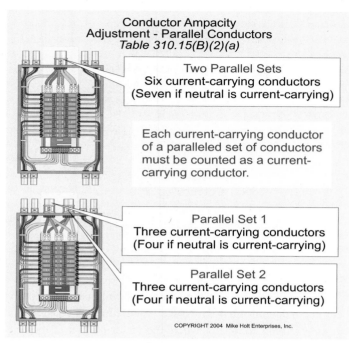

Conductor Ampacity
Adjustment - Parallel Conductors
Table 310.15(B)(2)(a)

Two Parallel Sets
Six current-carrying conductors
(Seven if neutral is current-carrying)

Each current-carrying conductor
of a paralleled set of conductors
must be counted as a current-
carrying conductor.

Parallel Set 1
Three current-carrying conductors
(Four if neutral is current-carrying)

Parallel Set 2
Three current-carrying conductors
(Four if neutral is current-carrying)

COPYRIGHT 2004 Mike Holt Enterprises, Inc.

Figure 310–15

Author's Comment: The grounded neutral conductor is considered a current-carrying conductor, but only under the conditions specified in 310.15(B)(4). Equipment grounding (bonding) conductors are never considered current carrying, but they are not designed to be used for this purpose [310.15(B)(5)].

Question: What is the adjusted ampacity of 3/0 THHN conductors if the raceway contains a total of four current-carrying conductors?

(a) 180A   (b) 196A   (c) 213A   (d) 241A

Answer: (a) 180A

Conductor Ampacity [90°C] = 225A
Adjustment Factor [Table 310.15(B)(2)(a)] = 0.80
Adjusted Ampacity = 225A x 0.80
Adjusted Ampacity = 180A

Author's Comments:

• When adjusting conductor ampacity, the ampacity is based on the temperature insulation rating of the conductor as listed in Table 310.16, not the temperature rating of the terminal [110.14(C)]. See a modified version of Table 310.16 on page 274 of this textbook.

• Where more than three current-carrying conductors are present and the ambient temperature isn't between 78 and 86°F, the ampacity listed in Table 310.16 must be adjusted for both conditions.

Question: What is the adjusted ampacity of 3/0 THHN conductors at an ambient temperature of 108°F if the raceway contains four current-carrying conductors?

(a) 157A   (b) 176A   (c) 199A   (d) 214A

Answer: (a) 157A

Table 310.16 Ampacity 3/0 THHN = 225A
Ambient Temperature Correction [Table 310.16] = 0.87
Conductor Bundle Adjustment [310.15(B)(2)(a)] = 0.80
Adjusted Ampacity = 225A x 0.87 x 0.80 = 157A

Author's Comment: When adjusting conductor ampacity, the ampacity of THHN conductors is based on the 90°C rating of the conductor [110.14(C)].

FPN 2: See 376.22 for conductor ampacity adjustment factors for conductors in metal wireways.

Author's Comment: Conductor ampacity adjustment only applies when more than 30 current-carrying conductors are installed in any cross-sectional area of a metal wireway.

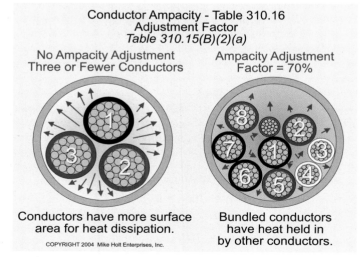

Conductor Ampacity - Table 310.16
Adjustment Factor
Table 310.15(B)(2)(a)

No Ampacity Adjustment
Three or Fewer Conductors

Ampacity Adjustment
Factor = 70%

Conductors have more surface area for heat dissipation.

Bundled conductors have heat held in by other conductors.

COPYRIGHT 2004 Mike Holt Enterprises, Inc.

Figure 310–16

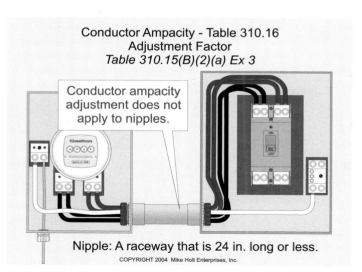

Conductor Ampacity - Table 310.16
Adjustment Factor
Table 310.15(B)(2)(a) Ex 3

Conductor ampacity adjustment does not apply to nipples.

Nipple: A raceway that is 24 in. long or less.

COPYRIGHT 2004 Mike Holt Enterprises, Inc.

Figure 310–17

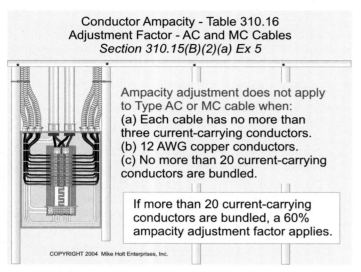

Figure 310–18

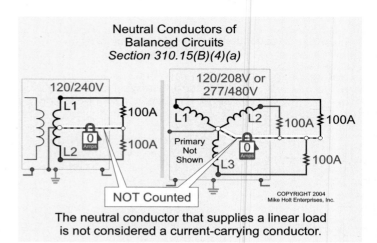

Figure 310–19

Exception 3: The conductor ampacity adjustment factors of Table 310.15(B)(2)(a) do not apply to conductors installed in raceways not exceeding 24 in. in length. **Figure 310–17**

Exception 5: The conductor ampacity adjustment factors of Table 310.15(B)(2)(a) do not apply to Type AC or MC cable when: **Figure 310–18**

    (1) Each cable has not more than three current-carrying conductors,

    (2) The conductors are 12 AWG copper, and

    (3) No more than 20 current-carrying conductors (ten 2-wire cables or six 3-wire cables) are bundled.

Author's Comment (to Exception 5): When eleven or more 2-wire cables or seven or more 3-wire cables (more than 20 current-carrying conductors) are bundled or stacked for more than 24 in., an ampacity adjustment factor of 60 percent must be applied.

(4) Neutral Conductor.

(a) Balanced Circuits. The neutral conductor of a 3-wire single-phase 120/240V system, or 4-wire three-phase 120/208V or 277/480V wye-connected system, isn't considered a current-carrying conductor. **Figure 310–19**

(b) 3-Wire Circuits. The neutral conductor of a 3-wire circuit from a 4-wire three-phase 120/208V or 277/480V wye-connected system is considered a current-carrying conductor.

Author's Comment: When a 3-wire circuit is supplied from a 4-wire three-phase wye-connected system, the neutral conductor carries approximately the same current as the ungrounded conductors. **Figure 310–20**

(c) Wye 4-Wire Circuits That Supply Nonlinear Loads. The neutral conductor of a 4-wire three-phase circuit is considered a current-carrying conductor where the major portion of the neutral load consists of nonlinear loads. This is because harmonic currents will be present in the neutral conductor, even if the loads on each of the three phases are balanced.

Author's Comment: Nonlinear loads supplied by 4-wire three-phase 120/208V or 277/480V wye-connected systems can produce unwanted and potentially hazardous triplen harmonic currents (3rd, 9th, 15th, etc.) that can add on the neutral conductor. To prevent fire or equipment damage from excessive harmonic neutral current, the designer should consider increasing the size of the neutral conductor or installing a separate neutral for each phase. For more information, visit www.MikeHolt.com and see 210.4(A) FPN, 220.61 FPN 2, and 450.3 FPN 2. **Figure 310–21**

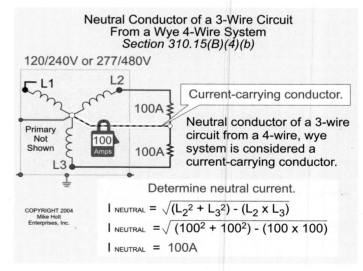

Figure 310–20

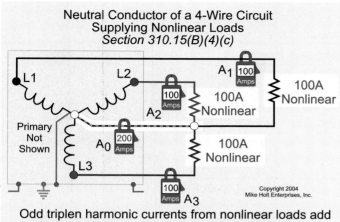

Figure 310–21

(5) Grounding (Earthing) Conductors. Grounding (earthing) and bonding conductors aren't considered current carrying.

(6) Dwelling Unit Feeder/Service Conductors. For individual dwelling units of one-family, two-family, and multifamily dwellings, Table 310.15(B)(6) can be used to size 3-wire single-phase 120/240V service or feeder conductors (including neutral conductors) that serve as the main power feeder. Feeder conductors are not required to have an ampacity rating greater than the service conductors [215.2(A)(3)].

Table 310.15(B)(6)

Amperes	Copper	Aluminum
100	4 AWG	2 AWG
110	3 AWG	1 AWG
125	2 AWG	1/0 AWG
150	1 AWG	2/0 AWG
175	1/0 AWG	3/0 AWG
200	2/0 AWG	4/0 AWG
225	3/0 AWG	250 kcmil
250	4/0 AWG	300 kcmil
300	250 kcmil	350 kcmil
350	350 kcmil	500 kcmil
400	400 kcmil	600 kcmil

WARNING: *Table 310.15(B)(6) doesn't apply to 3-wire single-phase 120/208V systems, because the grounded neutral conductor in these systems carries neutral current even when the load on the phases is balanced [310.15(B)(4)(6)]. For more information on this topic, see 220.61(C)(1).*

Grounded Neutral Conductor Sizing. Table 310.15(B)(6) can be used to size the grounded neutral conductor of a 3-wire single-phase 120/240V service or feeder that serves as the main power feeder, based on the feeder calculated load in accordance with 220.61.

Author's Comment: Because the grounded neutral service conductor is required to serve as the effective ground-fault current path, it must be sized so that it can safely carry the maximum fault current likely to be imposed on it [110.10 and 250.4(A)(5)]. This is accomplished by sizing the grounded neutral conductor in accordance with Table 250.66, based on the total area of the largest ungrounded conductor [250.24(C)(1)].

Question: What size service conductors would be required if the calculated load for a dwelling unit equals 195A and the maximum unbalanced neutral load is 100A? **Figure 310–22**

(a) 1/0 AWG and 6 AWG   (b) 2/0 AWG and 4 AWG
(c) 3/0 AWG and 2 AWG   (d) 4/0 AWG and 1 AWG

Answer: (b) 2/0 AWG and 4 AWG

Service Conductor: 2/0 AWG rated 200A [Table 310.15(B)(6)]

Grounded Neutral Conductor: 4 THHN AWG is rated 100A in accordance with Table 310.15(B)(6). In addition, 250.24(C) requires the grounded neutral conductor to be sized no smaller than 4 AWG based on 2/0 AWG service conductors in accordance with Table 250.66.

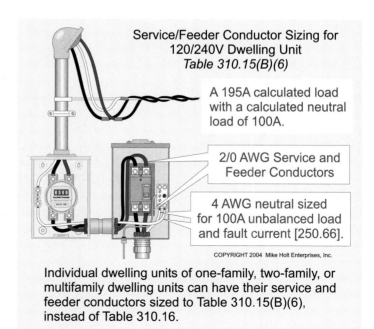

Individual dwelling units of one-family, two-family, or multifamily dwelling units can have their service and feeder conductors sized to Table 310.15(B)(6), instead of Table 310.16.

Figure 310–22

Table 310.16   *Understanding the National Electrical Code, Volume 1*

Table 310.16. Allowable Ampacities of Insulated Conductors
Based On Not More Than Three Current-Carrying Conductors and Ambient Temperature of 30°C (86°F)

Size	Temperature Rating of Conductor, See Table 310.13						Size
	60°C (40°F)	75°C (167°F)	90°C (194°F)	60°C (40°F)	75°C (167°F)	90°C (194°F)	
AWG kcmil	TW UF	THHW THW THWN XHHW Wet Location	THHN THHW XHHW Dry Location	TW UF	THHW THW THWN XHHW Wet Location	THHN THHW XHHW Dry Location	AWG kcmil
	Copper			Aluminum/Copper-Clad Aluminum			
14*	20	20	25				
12*	25	25	30	20	20	25	12*
10*	30	35	40	25	30	35	10*
8	40	50	55	30	40	45	8
6	55	65	75	40	50	60	6
4	70	85	95	55	65	75	4
3	85	100	110	65	75	85	3
2	95	115	130	75	90	100	2
1	110	130	150	85	100	115	1
1/0	125	150	170	100	120	135	1/0
2/0	145	175	195	115	135	150	2/0
3/0	165	200	225	130	155	175	3/0
4/0	195	230	260	150	180	205	4/0
250	215	255	290	170	205	230	250
300	240	285	320	190	230	255	300
350	260	310	350	210	250	280	350
400	280	335	380	225	270	305	400
500	320	380	430	260	310	350	500

*See 240.4(D)

Article 310 Questions

(• Indicates that 75% or fewer exam takers get the question correct.)

1. Conductors must be insulated except where specifically allowed by the *NEC* to be bare, such as for equipment grounding or bonding purposes.

   (a) True   (b) False

2. Where _____ conductors are run in separate raceways or cables, the same number of conductors must be used in each raceway or cable.

   (a) parallel   (b) control   (c) communications   (d) aluminum

3. Insulated conductors and cables exposed to the direct rays of the sun must be _____.

   (a) covered with insulating material that is listed or listed and marked sunlight resistant
   (b) listed and marked sunlight resistant
   (c) listed for sunlight resistance
   (d) any of these

4. Which conductor has an insulation temperature rating of 90°C?

   (a) RH   (b) RHW   (c) THHN   (d) TW

5. The ampacity of a conductor can be different along the length of the conductor. The higher ampacity is permitted to be used beyond the point of transition for a distance of no more than _____ ft or, no more than _____ percent of the circuit length figured at the higher ampacity, whichever is less.

   (a) 10, 20   (b) 20, 10   (c) 10, 10   (d) 15, 15

6. •When bare grounding conductors are allowed, their ampacities are limited to _____.

   (a) 60°C   (b) 75°C
   (c) 90°C   (d) those permitted for the insulated conductors of the same size

Notes

312 Cabinets, Cutout Boxes, and Meter Socket Enclosures

Introduction

This article addresses the installation and construction specifications for the items mentioned in its title. In Article 310, we observed that you need different ampacities for the same conductor, depending on conditions of use. The same thing applies to these items—just in a different way. For example, you can't use just any enclosure in a wet area or in a hazardous (classified) location. The conditions of use impose special requirements for these situations.

For all such enclosures, certain requirements apply—regardless of the use. For example, you must cover any openings, protect conductors from abrasion, and allow sufficient bending room for conductors.

Part I is where you'll find the requirements most useful to the electrician in the field. Part II applies to manufacturers. If you use name brand components that are listed or labeled, you do not need to be concerned with Part II. However, if you are specifying custom enclosures, you need to be familiar with these requirements to ensure the authority having jurisdiction approves the enclosures.

312.1 Scope. Article 312 covers the installation and construction specifications for cabinets, cutout boxes, and meter socket enclosures. Figure 312–1

> **Author's Comment:** A cabinet is an enclosure for either surface mounting or flush mounting provided with a frame in which a door may be hung. A cutout box is designed for surface mounting with a swinging door [Article 100]. The industry name for a meter socket enclosure is "meter can."

PART I. INSTALLATION

312.2 Damp, Wet, or Hazardous (Classified) Locations.

(A) Damp and Wet Locations. Enclosures in damp or wet locations must prevent moisture or water from entering or accumulating within the enclosure, and must be weatherproof. When the enclosure is surface mounted in a wet location, the enclosure must be mounted with not less than a ¼ in. air space between it and the mounting surface. See 300.6(C).

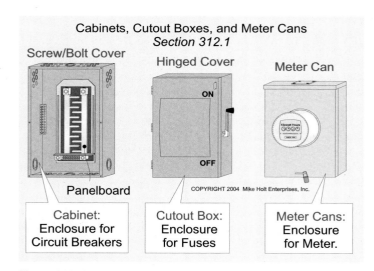

Cabinets, Cutout Boxes, and Meter Cans
Section 312.1

Screw/Bolt Cover — Hinged Cover — Meter Can

Panelboard

COPYRIGHT 2004 Mike Holt Enterprises, Inc.

Cabinet: Enclosure for Circuit Breakers

Cutout Box: Enclosure for Fuses

Meter Cans: Enclosure for Meter.

Figure 312–1

Where raceways or cables enter above the level of uninsulated live parts of an enclosure in a wet location, a fitting listed for wet locations must be used for termination.

Author's Comment: A fitting listed for use in a wet location with a sealing locknut would be suitable for this application.

Exception: The ¼ in. air space isn't required for nonmetallic equipment, raceways, or cables.

(B) Hazardous (Classified) Locations. Cabinets, cutout boxes, and meter socket enclosures installed in hazardous (classified) locations must comply with 501.10, 502.10, and 503.10(A)(1).

312.3 Installed in Walls.
Cabinets or cutout boxes installed in walls of concrete, tile, or other noncombustible material must be installed so that the front edge of the enclosure is set back no more than ¼ in. from the finished surface. In walls constructed of wood or other combustible material, cabinets or cutout boxes must be flush with the finished surface or project outward.

312.4 Repairing Gaps Around Plaster, Drywall, or Plasterboard Edges.
Gaps around cabinets and cutout boxes recessed in plaster, drywall, or plasterboard having a flush-type cover, must be repaired so there will be no gap greater than ⅛ in. at the edge of the cabinet or cutout box. **Figure 312–2**

312.5 Enclosures.

(A) Unused Openings. Openings intended to provide entry for conductors must be adequately closed. **Figure 312–3**

> **Author's Comment:** Unused openings for circuit breakers must be closed by means that provide protection substantially equivalent to the wall of the enclosure [408.7]. **Figure 312–4**

(C) Cable Termination. Cables must be secured to the enclosure with fittings designed and listed for the cable. See 300.12 and 300.15. **Figure 312–5**

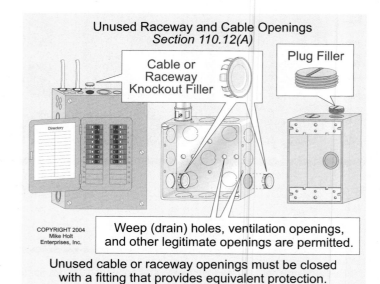

Unused Raceway and Cable Openings
Section 110.12(A)

Cable or Raceway Knockout Filler

Plug Filler

Weep (drain) holes, ventilation openings, and other legitimate openings are permitted.

Unused cable or raceway openings must be closed with a fitting that provides equivalent protection.

Figure 312–3

> **Author's Comment:** Cable clamps or cable connectors must be used with only one cable per fitting, unless that clamp or fitting is identified for more than one cable. Some NM cable clamps are listed to permit termination of two NM cables within a single fitting.

Exception: Cables with nonmetallic sheaths aren't required to be secured to the enclosure if the cables enter the top of a surface-mounted enclosure through a nonflexible raceway not less than 18 in. or more than 10 ft long, if all of the following conditions are met: **Figure 312–6**

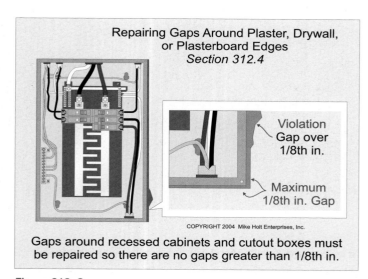

Repairing Gaps Around Plaster, Drywall, or Plasterboard Edges
Section 312.4

Violation Gap over 1/8th in.

Maximum 1/8th in. Gap

COPYRIGHT 2004 Mike Holt Enterprises, Inc.

Gaps around recessed cabinets and cutout boxes must be repaired so there are no gaps greater than 1/8th in.

Figure 312–2

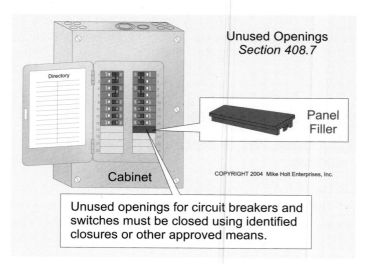

Unused Openings
Section 408.7

Directory

Panel Filler

Cabinet

COPYRIGHT 2004 Mike Holt Enterprises, Inc.

Unused openings for circuit breakers and switches must be closed using identified closures or other approved means.

Figure 312–4

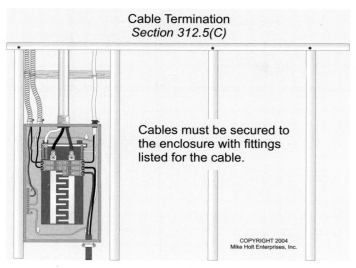

Cable Termination
Section 312.5(C)

Cables must be secured to the enclosure with fittings listed for the cable.

COPYRIGHT 2004
Mike Holt Enterprises, Inc.

Figure 312–5

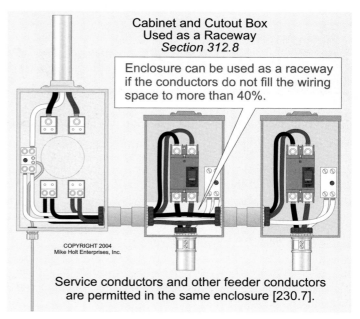

Cabinet and Cutout Box Used as a Raceway
Section 312.8

Enclosure can be used as a raceway if the conductors do not fill the wiring space to more than 40%.

COPYRIGHT 2004
Mike Holt Enterprises, Inc.

Service conductors and other feeder conductors are permitted in the same enclosure [230.7].

Figure 312–7

(a) Each cable is fastened within 1 ft from the raceway.

(b) The raceway doesn't penetrate a structural ceiling.

(c) Fittings are provided on the raceway to protect the cables from abrasion.

(d) The raceway is sealed.

(e) Each cable sheath extends not less than ¼ in. into the panelboard.

(f) The raceway is properly secured.

(g) Conductor fill is limited to Chapter 9 Table 1 percentages.

312.8 Used for Raceway and Splices. Cabinets, cutout boxes, and meter socket enclosures can be used as a raceway for conductors that feed through if the conductors do not fill the wiring space at any cross section to more than 40 percent. **Figure 312–7**

> **Author's Comment:** Service conductors and other conductors are permitted to be installed in the same enclosure [230.7].

Splices and taps can be installed in cabinets, cutout boxes, or meter socket enclosures if the splices or taps do not fill the wiring space at any cross section to more than 75 percent. **Figure 312–8**

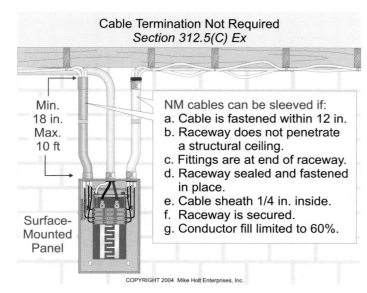

Cable Termination Not Required
Section 312.5(C) Ex

Min. 18 in. Max. 10 ft

Surface-Mounted Panel

NM cables can be sleeved if:
a. Cable is fastened within 12 in.
b. Raceway does not penetrate a structural ceiling.
c. Fittings are at end of raceway.
d. Raceway sealed and fastened in place.
e. Cable sheath 1/4 in. inside.
f. Raceway is secured.
g. Conductor fill limited to 60%.

COPYRIGHT 2004 Mike Holt Enterprises, Inc.

Figure 312–6

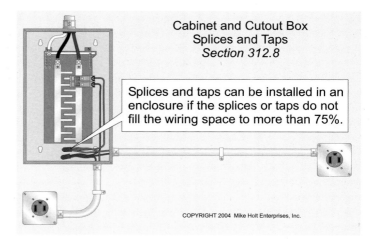

Cabinet and Cutout Box Splices and Taps
Section 312.8

Splices and taps can be installed in an enclosure if the splices or taps do not fill the wiring space to more than 75%.

COPYRIGHT 2004 Mike Holt Enterprises, Inc.

Figure 312–8

1. Cabinets or cutout boxes installed in wet locations must be _____.

    (a) waterproof   (b) raintight   (c) weatherproof   (d) watertight

2. Where raceways or cables enter above the level of uninsulated live parts of an enclosure in a wet location, a(n) _____ must be used.

    (a) fitting listed for wet locations   (b) explosion proof seal-off
    (c) fitting listed for damp locations   (d) insulated fitting

3. Plaster, drywall, or plasterboard surfaces that are broken or incomplete must be repaired so there will be no gaps or open spaces greater than _____ at the edge of a cabinet or cutout box employing a flush-type cover.

    (a) ¼ in.   (b) ½ in.   (c) ⅛ in.   (d) ¹⁄₁₆ in.

4. Each cable entering a cutout box _____.

    (a) must be secured to the cutout box   (b) can be sleeved through a chase
    (c) must have a maximum of two cables per connector   (d) all of these

5. A switch enclosure (cabinet) must not be used as a junction box, except where adequate space is provided so that the conductors don't fill the wiring space at any cross-section to more than 40 percent of the cross-sectional area of the space, and so that _____ don't fill the wiring space at any cross-section to more than 75 percent of the cross-sectional area of the space.

    (a) splices   (b) taps   (c) conductors   (d) all of these

Outlet, Device, Pull and Junction Boxes, Conduit Bodies, and Handhole Enclosures

Introduction

Article 314 contains installation requirements for outlet boxes, pull and junction boxes, conduit bodies, and handhole enclosures.

As with Article 312, conditions of use apply. If you're running a raceway in a hazardous (classified) location, for example, you must use the correct fittings and the proper installation methods. But consider something as simple as a splice. It makes sense that you wouldn't put a splice in the middle of a raceway—doing so means you cannot get to it. But if you put a splice in a conduit body, you're fine, right? Not necessarily. Suppose that conduit body is a "short radius" version (think of it as an elbow with the bend chopped off). You do not have much room inside such an enclosure and for that reason, you cannot put a splice inside a short radius conduit body.

Properly applying Article 314 means you will need to account for the internal volume of all boxes and fittings, and then determine the maximum wire fill. You'll also need to understand many other requirements, which we'll cover. If you start to get confused, take a break. Look carefully at the illustrations, and you'll learn more quickly and with greater retention.

PART I. SCOPE AND GENERAL

314.1 Scope. Article 314 contains the installation requirements for outlet boxes, conduit bodies, pull and junction boxes, and handhole enclosures. Figure 314–1

314.3 Nonmetallic Boxes. Nonmetallic boxes can only be used with nonmetallic cables and raceways.

Exception 1: Metal raceways and metal cables can be used with nonmetallic boxes, but only if an internal bonding means is provided in the box between all metal entries.

314.4 Metal Boxes. All metal boxes must be grounded (bonded) to an effective ground-fault current path in accordance with Article 250 [250.4(A)(3)].

> **Author's Comment:** Metal raceways and metal enclosures containing circuit conductors from a power supply that operates at 50V or less, aren't required to be bonded to an effective ground-fault current path [250.86 and 250.112(l)]. For example, metal boxes used with power-limited fire alarm circuits operating at a maximum of 50V are not required to be grounded (bonded) to an effective ground-fault current path.

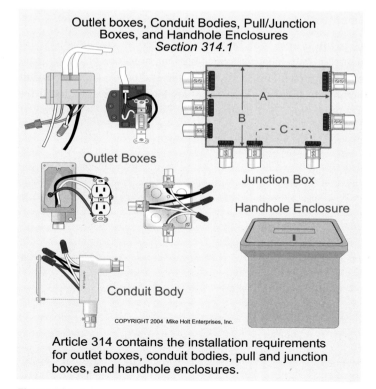

Outlet boxes, Conduit Bodies, Pull/Junction Boxes, and Handhole Enclosures
Section 314.1

Outlet Boxes

Junction Box

Handhole Enclosure

Conduit Body

COPYRIGHT 2004 Mike Holt Enterprises, Inc.

Article 314 contains the installation requirements for outlet boxes, conduit bodies, pull and junction boxes, and handhole enclosures.

Figure 314–1

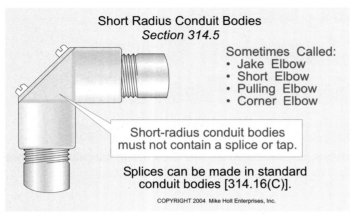

Figure 314–2

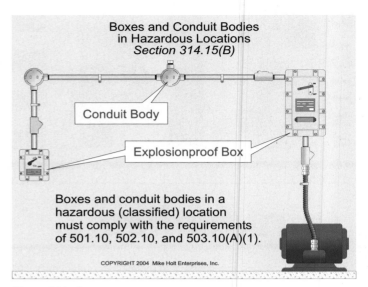

Figure 314–4

314.5 Short-Radius Conduit Bodies.

Short-radius conduit bodies, such as capped elbows, handy ells, and service-entrance elbows must not contain any splices or taps. Figure 314–2

Author's Comment: Splices and taps can be made in standard conduit bodies. See 314.16(C) for specific requirements.

PART II. INSTALLATION

314.15 Damp, Wet, or Hazardous (Classified) Locations.

(A) Damp and Wet Locations. Boxes and conduit bodies in damp or wet locations must prevent moisture or water from entering or accumulating within the enclosure. Boxes, conduit bodies, and fittings installed in wet locations must be listed for use in wet locations. Figure 314–3

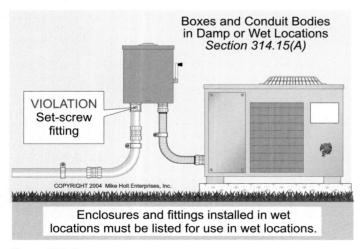

Figure 314–3

Author's Comment: Where handhole enclosures without bottoms are installed, all enclosed conductors and any splices or terminations must be listed as suitable for wet locations [314.30(C)].

(B) Hazardous (Classified) Locations. Boxes and conduit bodies installed in hazardous (classified) locations must comply with 501.10, 502.10, and 503.10(A)(1). Figure 314–4

314.16 Number of 6 AWG and Smaller Conductors in Boxes and Conduit Bodies.

Boxes containing 6 AWG and smaller conductors must be sized to provide sufficient free space for all conductors, devices, and fittings. In no case can the volume of the box, as calculated in 314.16(A), be less than the volume requirement as calculated in 314.16(B).

Conduit bodies must be sized in accordance with 314.16(C).

Author's Comment: The requirements for sizing boxes and conduit bodies containing conductors 4 AWG and larger are contained in 314.28.

(A) Box Volume Calculations. The volume of a box includes the total volume of its assembled parts, including plaster rings, extension rings, and domed covers that are either marked with their volume in cubic inches (cu in.) or are made from boxes listed in Table 314.16(A). Figure 314–5

(B) Box Fill Calculations. The calculated conductor volume determined by (1) through (5) and Table 314.16(B) are added together to determine the total volume of the conductors, devices, and fittings. Raceway and cable fittings, including locknuts and bushings, are not counted for box fill calculations. Figure 314–6

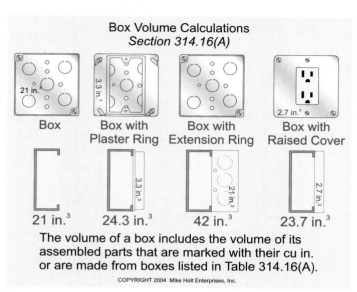

Box Volume Calculations
Section 314.16(A)

21 in.³ — Box

3.3 in.³ / 3.3 in.³ — Box with Plaster Ring — 24.3 in.³

21 in.³ — Box with Extension Ring — 42 in.³

2.7 in.³ / 2.7 in.³ — Box with Raised Cover — 23.7 in.³

21 in.³

The volume of a box includes the volume of its assembled parts that are marked with their cu in. or are made from boxes listed in Table 314.16(A).

COPYRIGHT 2004 Mike Holt Enterprises, Inc.

Figure 314–5

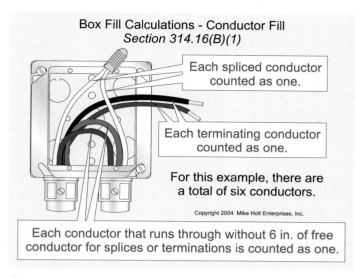

Box Fill Calculations - Conductor Fill
Section 314.16(B)(1)

Each spliced conductor counted as one.

Each terminating conductor counted as one.

For this example, there are a total of six conductors.

Copyright 2004 Mike Holt Enterprises, Inc.

Each conductor that runs through without 6 in. of free conductor for splices or terminations is counted as one.

Figure 314–7

(1) Conductor Fill. Each conductor that runs through a box and does not have 6 in. of free conductor for splices or terminations in accordance 300.14, and each conductor that terminates in a box is counted as a single conductor volume in accordance with Table 310.16(B). Conductors that originate and terminate within the box, such as pigtails, aren't counted at all. **Figure 314–7**

Author's Comments:

- According to 300.14, at least 6 in. of free conductor, measured from the point in the box where the conductors enter the enclosure, must be left at each outlet, junction, and switch point for splices or terminations of luminaires or devices.

- Conductor loops occupy space and a box can be excessively filled if we do not take into consideration the increased conductor volume. This can create a serious fire hazard, especially when an electronic device is installed in an outlet box without adequate room for heat dissipation.

Exception: Equipment grounding (bonding) conductors, and not more than four 16 AWG and smaller fixture wires, can be omitted from box fill calculations if they enter the box from a domed luminaire or similar canopy, such as a ceiling paddle fan canopy. **Figure 314–8**

(2) Cable Clamp Fill. One or more internal cable clamps count as a single conductor volume in accordance with Table 310.16(B), based on the largest conductor that enters the box. Cable connectors that have their clamping mechanism outside the box aren't counted.

(3) Support Fitting Fill. Each luminaire stud or luminaire hickey counts as a single conductor volume in accordance with Table 310.16(B), based on the largest conductor that enters the box. **Figure 314–9**

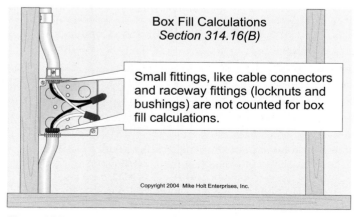

Box Fill Calculations
Section 314.16(B)

Small fittings, like cable connectors and raceway fittings (locknuts and bushings) are not counted for box fill calculations.

Copyright 2004 Mike Holt Enterprises, Inc.

Figure 314–6

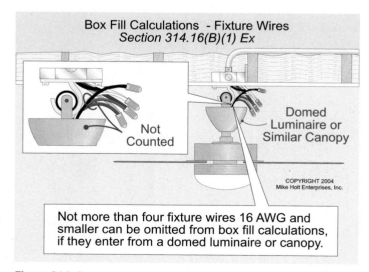

Box Fill Calculations - Fixture Wires
Section 314.16(B)(1) Ex

Not Counted

Domed Luminaire or Similar Canopy

COPYRIGHT 2004 Mike Holt Enterprises, Inc.

Not more than four fixture wires 16 AWG and smaller can be omitted from box fill calculations, if they enter from a domed luminaire or canopy.

Figure 314–8

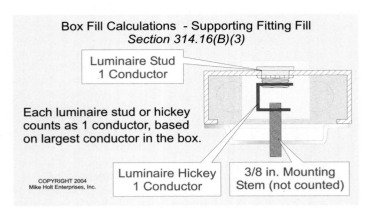

Box Fill Calculations - Supporting Fitting Fill
Section 314.16(B)(3)

Luminaire Stud
1 Conductor

Each luminaire stud or hickey
counts as 1 conductor, based
on largest conductor in the box.

COPYRIGHT 2004
Mike Holt Enterprises, Inc.

Luminaire Hickey
1 Conductor

3/8 in. Mounting
Stem (not counted)

Figure 314–9

(4) Device Yoke Fill. Each device yoke (regardless of the ampere rating of the device) counts as two conductor volumes in accordance with Table 310.16(B), based on the largest conductor that terminates on the device. Figure 314–10

Table 314.16(B)

Conductor AWG	Volume cu in.
18	1.50
16	1.75
14	2.00
12	2.25
10	2.50
8	3.00
6	5.00

(5) Equipment Grounding (Bonding) Conductor Fill. All equipment grounding (bonding) conductors in a box count as a single conductor volume in accordance with Table 310.16(B), based on the largest equipment grounding (bonding) conductor that enters the box. Equipment grounding (bonding) conductors for isolated ground circuits count as a single conductor volume in accordance with Table 310.16(B). Figure 314–11

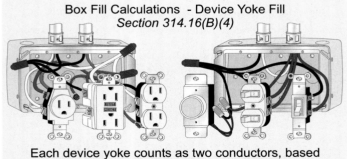

Box Fill Calculations - Device Yoke Fill
Section 314.16(B)(4)

Each device yoke counts as two conductors, based
on the largest conductor terminating on the device.

COPYRIGHT 2004 Mike Holt Enterprises, Inc.

Figure 314–10

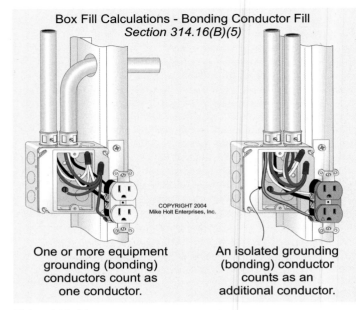

Box Fill Calculations - Bonding Conductor Fill
Section 314.16(B)(5)

COPYRIGHT 2004
Mike Holt Enterprises, Inc.

One or more equipment
grounding (bonding)
conductors count as
one conductor.

An isolated grounding
(bonding) conductor
counts as an
additional conductor.

Figure 314–11

Author's Comment: The conductor insulation is not a factor for box volume calculations.

Question: *How many 14 THHN conductors can be pulled through a 4 in. square x 2¼ in. deep box with a plaster ring with a marking of 3.6 cu in.? The box contains two receptacles, five 12 AWG conductors, and two 12 THHN equipment grounding (bonding) conductors.* Figure 314–12

(a) 3   (b) 5   (c) 7   (d) 9

Answer: *(b) 5*

Step 1. Volume of the box assembly [314.16(A)].
Box 30.3 cu in. + 3.6 cu in. plaster ring = 33.9 cu in.

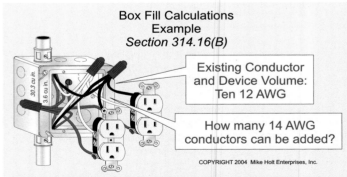

**Box Fill Calculations
Example
*Section 314.16(B)***

Existing Conductor
and Device Volume:
Ten 12 AWG

How many 14 AWG
conductors can be added?

COPYRIGHT 2004 Mike Holt Enterprises, Inc.

Step 1. Volume of box/ring: 30.3 + 3.6 cu in. = 33.9 cu in.
Step 2. Volume of existing conductors/devices = 22.5 cu in.
Step 3. Space remaining: 33.9 - 22.5 = 11.4 cu in.
Step 4. Number of 14 AWG added: 11.4/2.0 cu in. = 5

Figure 314–12

Author's Comment: A 4 x 4 x 2 ¼ in. box would have a gross volume of 34 cu in., but the interior volume is 30.3 cu in., as listed in Table 314.16(A).

Step 2. Determine the volume of the devices and conductors in the box.

Two—receptacles	4—12 AWG
Five—12 THHN	5—12 AWG
Two—12 AWG Grounds	1—12 AWG
Total 10—12 AWG x 2.25 cu in. = 22.5 cu in.	

Step 3. Determine the remaining volume permitted for the 14 AWG conductors.

33.9 cu in. − 22.5 cu in. = 11.4 cu in.

Step 4. Determine the number of 14 AWG conductors permitted in the remaining volume.

11.4 cu in./2.0 cu in. = 5 conductors

(C) Conduit Bodies.

(2) Splices. Splices are only permitted in conduit bodies that are legibly marked, by the manufacturer, with their volume. The maximum number of conductors permitted in a conduit body is limited in accordance with 314.16(B).

Question: *How many 12 AWG conductors can be spliced in a 15 cu in. conduit body?* **Figure 314–13**

(a) 4   (b) 6   (c) 8   (d) 10

Answer: *(b) 6 conductors (15 cu in./2.25 cu in.)*

314.17 Conductors That Enter Boxes or Conduit Bodies.

(A) Openings to Be Closed. Openings through which cables or raceways enter must be adequately closed.

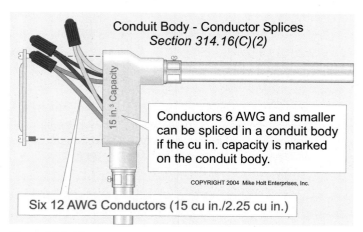

Conduit Body - Conductor Splices
Section 314.16(C)(2)

15 in.³ Capacity

Conductors 6 AWG and smaller can be spliced in a conduit body if the cu in. capacity is marked on the conduit body.

COPYRIGHT 2004 Mike Holt Enterprises, Inc.

Six 12 AWG Conductors (15 cu in./2.25 cu in.)

Figure 314–13

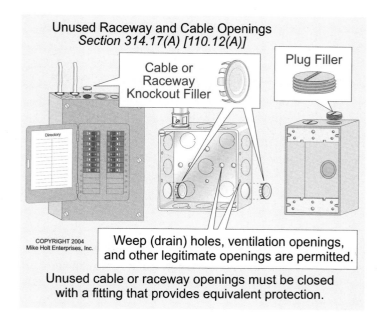

Unused Raceway and Cable Openings
Section 314.17(A) [110.12(A)]

Cable or Raceway Knockout Filler

Plug Filler

Directory

COPYRIGHT 2004 Mike Holt Enterprises, Inc.

Weep (drain) holes, ventilation openings, and other legitimate openings are permitted.

Unused cable or raceway openings must be closed with a fitting that provides equivalent protection.

Figure 314–14

Author's Comment: Unused cable or raceway openings in electrical equipment must be effectively closed by fittings that provide protection substantially equivalent to the wall of the equipment [110.12(A)]. **Figure 314–14**

(B) Metal Boxes and Conduit Bodies. Raceways and cables must be mechanically fastened to metal boxes or conduit bodies by fittings designed for the wiring method. See 300.12 and 300.15.

(C) Nonmetallic Boxes and Conduit Bodies. Raceways and cables must be securely fastened to nonmetallic boxes or conduit bodies by fittings designed for the wiring method [300.12 and 300.15]. Type NM cable must extend not less than ¼ in. into the nonmetallic box.

Author's Comment: Two NM cables are permitted to terminate in a single cable clamp, if the clamp is listed for this purpose.

Exception: Type NM cable terminating to a single-gang (2 ¼ x 4 in.) device box isn't required to be secured to the box if the cable is securely fastened within 8 in. of the box. **Figure 314–15**

314.20 Boxes Recessed in Walls or Ceilings.

Boxes having flush-type covers recessed in walls or ceilings of noncombustible material must have the front edge of the box, plaster ring, extension ring, or listed extender set back no more than ¼ in. from the finished surface. **Figure 314–16**

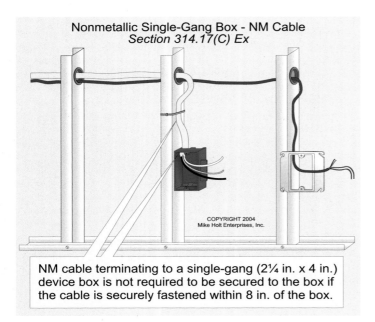

Nonmetallic Single-Gang Box - NM Cable
Section 314.17(C) Ex

NM cable terminating to a single-gang (2¼ in. x 4 in.) device box is not required to be secured to the box if the cable is securely fastened within 8 in. of the box.

Figure 314–15

In walls or ceilings constructed of wood or other combustible material, boxes must be installed so the front edge of the enclosure, plaster ring, extension ring, or listed extender is flush with, or projects out from, the finished surface. **Figure 314–17**

> **Author's Comment:** Plaster rings are available in a variety of depths to meet the above requirements.

314.21 Repairing Gaps Around Boxes.
Gaps around boxes recessed in plaster, drywall, or plasterboard having a flush-type cover, must be repaired so there will be no gap greater than ⅛ in. at the edge of the box. **Figure 314–18**

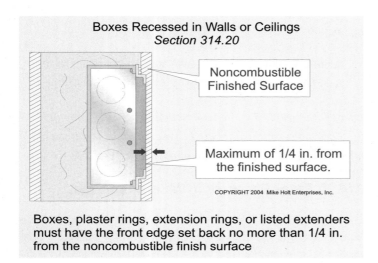

Boxes Recessed in Walls or Ceilings
Section 314.20

Noncombustible Finished Surface

Maximum of 1/4 in. from the finished surface.

Boxes, plaster rings, extension rings, or listed extenders must have the front edge set back no more than 1/4 in. from the noncombustible finish surface

Figure 314–16

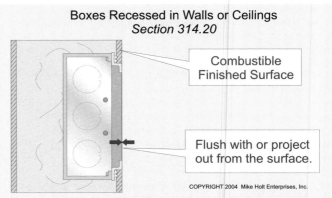

Boxes Recessed in Walls or Ceilings
Section 314.20

Combustible Finished Surface

Flush with or project out from the surface.

Boxes, plaster rings, extension rings, or listed extenders must have the front edge flush with, or project out from, the combustible finish surface

Figure 314–17

314.22 Surface Extensions.
Surface extensions can only be made from an extension ring mounted over a flush-mounted box. **Figure 314–19**

Exception: A surface extension can be made from the cover of a flush-mounted box if the cover is designed so it's unlikely to fall off if the mounting screws become loose. The surface extension wiring method must be flexible to permit the removal of the cover and provide access to the box interior, and bonding continuity must be independent of the connection between the box and the cover. **Figure 314–20**

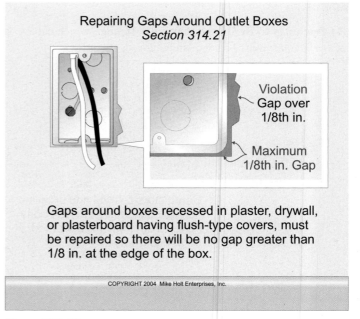

Repairing Gaps Around Outlet Boxes
Section 314.21

Violation
Gap over
1/8th in.

Maximum
1/8th in. Gap

Gaps around boxes recessed in plaster, drywall, or plasterboard having flush-type covers, must be repaired so there will be no gap greater than 1/8 in. at the edge of the box.

Figure 314–18

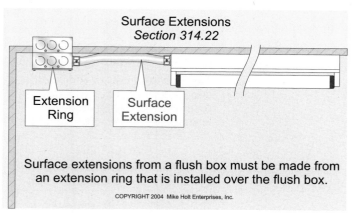

Surface Extensions
Section 314.22

Extension Ring

Surface Extension

Surface extensions from a flush box must be made from an extension ring that is installed over the flush box.

COPYRIGHT 2004 Mike Holt Enterprises, Inc.

Figure 314-19

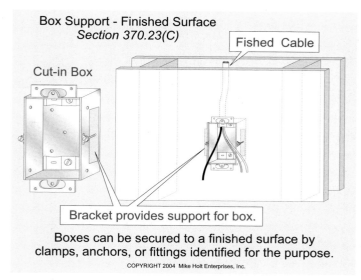

Box Support - Finished Surface
Section 370.23(C)

Cut-in Box

Fished Cable

Bracket provides support for box.

Boxes can be secured to a finished surface by clamps, anchors, or fittings identified for the purpose.

COPYRIGHT 2004 Mike Holt Enterprises, Inc.

Figure 314-21

314.23 Support of Boxes and Conduit Bodies. Boxes must be securely supported by one of the following methods:

(A) Surface. Boxes can be fastened to any surface that provides adequate support.

(B) Structural Mounting. Boxes can be supported from a structural member of a building or from grade by a metal, plastic, or wood brace.

(1) Nails and Screws. Nails or screws can be used to fasten boxes, provided the exposed threads of screws are protected to prevent abrasion of conductor insulation.

(2) Braces. Metal braces no less than 0.020 in. thick and wood braces not less than a nominal 1 x 2 in. can support a box.

(C) Finished Surface Support. Boxes can be secured to a finished surface (drywall or plaster walls or ceilings) by clamps, anchors, or fittings identified for the purpose. **Figure 314-21**

(D) Suspended-Ceiling Support. Outlet boxes can be supported to the structural or supporting elements of a suspended ceiling, if securely fastened by one of the following methods:

(1) Ceiling-Framing Members. An outlet box can be secured to suspended-ceiling framing members by bolts, screws, rivets, clips, or other means identified for the suspended-ceiling framing member(s). **Figure 314-22**

> **Author's Comment:** Luminaires can be supported to ceiling framing members as well [410.16(C)].

(2) Independent Support Wires. Outlet boxes can be secured, with fittings identified for the purpose, to independent support wires that are taut and secured at both ends [300.11(A)]. **Figure 314-23**

> **Author's Comment:** See 300.11(A) on the use of independent support wires to support raceways and cables.

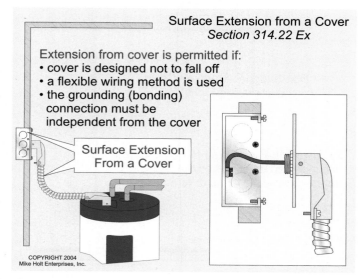

Surface Extension from a Cover
Section 314.22 Ex

Extension from cover is permitted if:
- cover is designed not to fall off
- a flexible wiring method is used
- the grounding (bonding) connection must be independent from the cover

Surface Extension From a Cover

COPYRIGHT 2004
Mike Holt Enterprises, Inc.

Figure 314-20

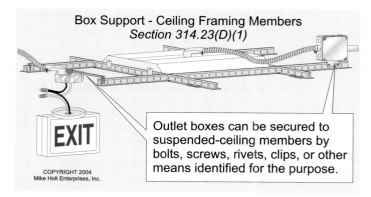

Box Support - Ceiling Framing Members
Section 314.23(D)(1)

EXIT

Outlet boxes can be secured to suspended-ceiling members by bolts, screws, rivets, clips, or other means identified for the purpose.

COPYRIGHT 2004
Mike Holt Enterprises, Inc.

Figure 314-22

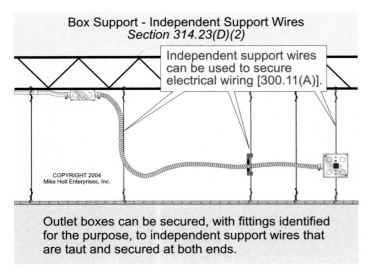

Figure 314–23

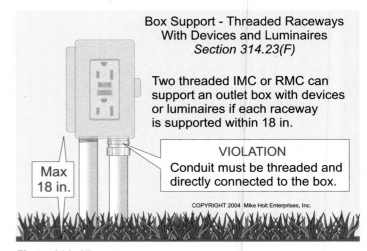

Figure 314–25

(E) Raceway Support—Boxes and Conduit Bodies Without Devices or Luminaires. Two intermediate metal or rigid metal conduits threaded wrenchtight can be used to support an outlet box that does not contain a device or luminaire, if each raceway is supported within 36 in. of the box, or within 18 in. if all conduit entries are on the same side. **Figure 314–24**

(F) Raceway Support—Boxes and Conduit Bodies with Devices or Luminaires. Two intermediate metal or rigid metal conduits threaded wrenchtight can be used to support an outlet box containing devices or luminaires, if each raceway is supported within 18 in. of the box. **Figure 314–25**

(H) Pendant Boxes.

(1) Flexible Cord. Boxes can be supported from a cord that is connected to fittings that prevent tension from being transmitted to joints or terminals [400.10]. **Figure 314–26**

314.25 Covers and Canopies. When the installation is complete, each outlet box must be provided with a cover or faceplate, unless covered by a fixture canopy, lampholder, or similar device. See 410.12. **Figure 314–27**

(A) Nonmetallic or Metallic. Nonmetallic covers or plates are permitted on any box, but metallic faceplates, where used, must be bonded to an effective ground-fault current path in accordance with 250.110 [250.4(A)(3)]. See 314.28(C).

> **Author's Comment:** Switch faceplates must be grounded (bonded) to an effective ground-fault current path [404.9(B)], and receptacle faceplates must be grounded (bonded) to an effective ground-fault current path [406.5(A)].

314.27 Outlet Box.

(A) Boxes at Luminaires. Lighting outlet boxes for luminaires must be designed for the purpose and must be installed for every

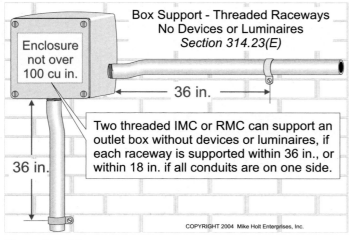

Figure 314–24

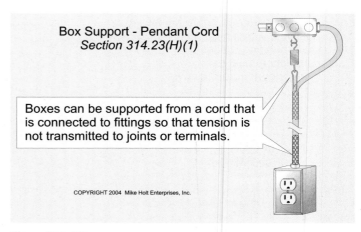

Figure 314–26

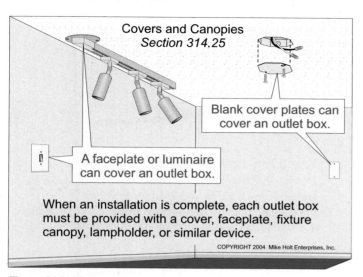

Figure 314–27

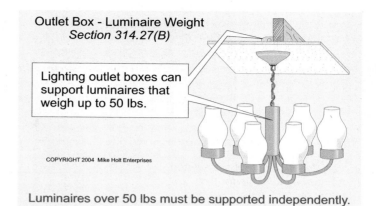

Luminaires over 50 lbs must be supported independently.

Figure 314–29

luminaire. Device outlet boxes (that use 6-32 screws) are only to be used for the support of switches or receptacles, not luminaires or lampholders.

Exception: A wall-mounted luminaire weighing no more than 6 lbs. can be supported to a device box or plaster ring secured to a box. Figure 314–28

(B) Luminaire Weight. Lighting outlet boxes (that use 8-32 and larger screws) can support luminaires that weigh up to 50 lbs. Luminaires weighing more than 50 lbs. must be supported independently of the lighting outlet box, unless the lighting outlet box is listed for the weight of the luminaire. Figure 314–29

(C) Floor Box. Floor boxes must be specifically listed for the purpose. Figure 314–30

(D) Ceiling Paddle Fan Box. Outlet boxes for a ceiling paddle fan must be listed and marked as suitable for the purpose, and must not support a fan weighing more than 70 lbs. Outlet boxes for a ceiling paddle fan that weighs more than 35 lbs. must include the maximum weight to be supported in the required marking. Figure 314–31

> **Author's Comment:** Where the maximum weight isn't marked on the box, and the fan weighs over 35 lbs., the fan must be supported independently of the outlet box. All ceiling paddle fans over 70 lbs. must be supported independently of the outlet box.

314.28 Boxes and Conduit Bodies for Conductors 4 AWG and Larger.
Boxes and conduit bodies containing conductors 4 AWG and larger must be sized so the conductor insulation will not be damaged.

> **Author's Comments:**
> * The requirements for sizing boxes and conduit bodies containing conductors 6 AWG and smaller are contained in 314.16.

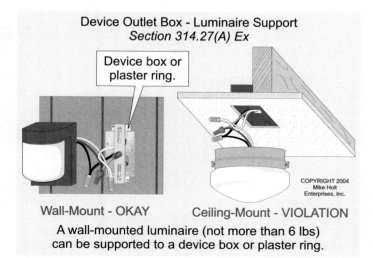

Figure 314–28

Figure 314–30

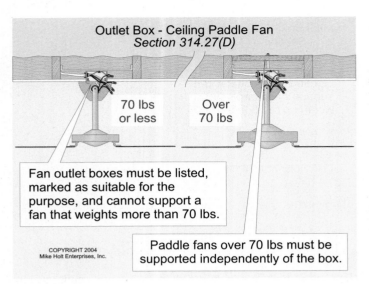

Figure 314–31

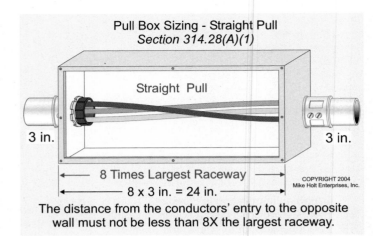

Figure 314–33

- Where conductors 4 AWG or larger enter a box or other enclosure, a fitting that provides a smooth, rounded, insulating surface, such as a bushing or adapter, is required to protect the conductors from abrasion during and after installation [300.4(F)]. **Figure 314–32**

(A) Minimum Size. For raceways containing conductors 4 AWG or larger, the minimum dimensions of boxes and conduit bodies must comply with the following:

(1) Straight Pulls. The minimum distance from where the conductors enter to the opposite wall cannot be less than eight times the trade size of the largest raceway. **Figure 314–33**

(2) Angle and U Pulls.

- Angle Pulls. The distance from the raceway entry to the opposite wall cannot be less than six times the trade size of the largest raceway, plus the sum of the trade sizes of the remaining raceways on the same wall and row. **Figure 314–34**

- U Pulls. When a conductor enters and leaves from the same wall, the distance from where the raceways enter to the opposite wall cannot be less than six times the trade size of the largest raceway, plus the sum of the trade sizes of the remaining raceways on the same wall and row. **Figure 314–35**

- Rows. Where there are multiple rows of raceway entries, each row is calculated individually and the row with the largest distance must be used.

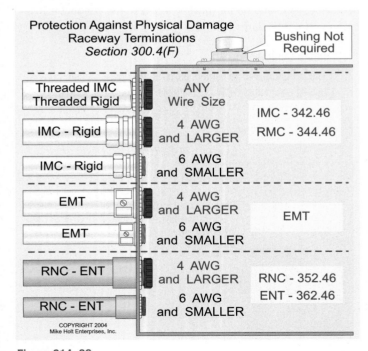

Figure 314–32

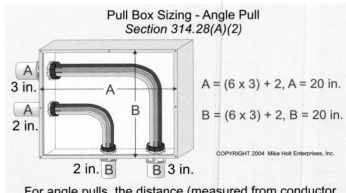

Figure 314–34

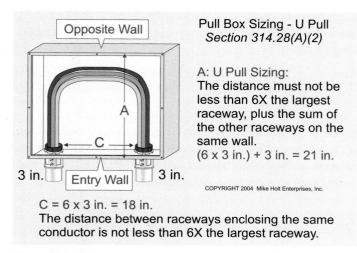

Pull Box Sizing - U Pull
Section 314.28(A)(2)

A: U Pull Sizing:
The distance must not be less than 6X the largest raceway, plus the sum of the other raceways on the same wall.
(6 x 3 in.) + 3 in. = 21 in.

COPYRIGHT 2004 Mike Holt Enterprises, Inc.

C = 6 x 3 in. = 18 in.
The distance between raceways enclosing the same conductor is not less than 6X the largest raceway.

Figure 314–35

- Distance Between Raceways. The distance between raceways enclosing the same conductor cannot be less than six times the trade size of the largest raceway, measured from the raceway's nearest edge-to-nearest edge. **See Figure314–35.**

Exception: When conductors enter an enclosure with a removable cover, such as a conduit body or wireway, the distance from where the conductors enter to the removable cover cannot be less than the bending distance as listed in Table 312.6(A) for one wire per terminal. **Figure 314–36**

(3) Smaller Dimensions. Boxes or conduit bodies of dimensions less than those required in 314.28(A)(1) and 314.28(A)(2) are permitted, if the enclosure is permanently marked with the maximum number and maximum size of conductors permitted.

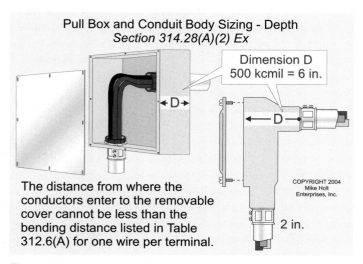

Pull Box and Conduit Body Sizing - Depth
Section 314.28(A)(2) Ex

Dimension D
500 kcmil = 6 in.

The distance from where the conductors enter to the removable cover cannot be less than the bending distance listed in Table 312.6(A) for one wire per terminal.

2 in.

COPYRIGHT 2004 Mike Holt Enterprises, Inc.

Figure 314–36

(C) Covers. All pull boxes, junction boxes, and conduit bodies must have a cover that is suitable for the conditions. Nonmetallic covers or plates are permitted on any box, but metallic face-plates, where used, must be grounded (bonded) to an effective ground-fault current path in accordance with 250.110 [250.4(A)(3)]. See 314.25(A).

314.29 Wiring to be Accessible. Boxes, conduit bodies, and handhole enclosures must be installed so that the wiring is accessible without removing any part of the building, sidewalks, paving, or earth. **Figure 314–37**

Exception: Listed boxes and handhole enclosures can be buried if covered by gravel, light aggregate, or noncohesive granulated soil, and their location is effectively identified and accessible for excavation.

314.30 Handhole Enclosures. Handhole enclosures must be designed and installed to withstand all loads likely to be imposed.

(A) Size. Handhole enclosures must be sized in accordance with 314.28(A). For handhole enclosures without bottoms, the measurement to the removable cover is taken from the end of the conduit or cable assembly.

(B) Mechanical Raceway and Cable Connection. Underground raceways and cables entering a handhole enclosure aren't required to be mechanically connected to the handhole enclosure. **Figure 314–38**

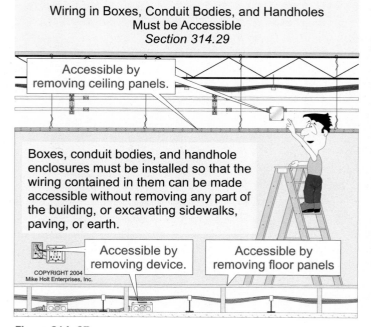

Wiring in Boxes, Conduit Bodies, and Handholes Must be Accessible
Section 314.29

Accessible by removing ceiling panels.

Boxes, conduit bodies, and handhole enclosures must be installed so that the wiring contained in them can be made accessible without removing any part of the building, or excavating sidewalks, paving, or earth.

Accessible by removing device.

Accessible by removing floor panels

COPYRIGHT 2004 Mike Holt Enterprises, Inc.

Figure 314–37

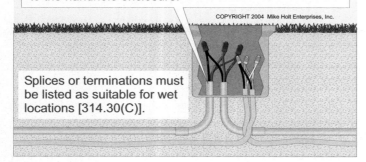

Handhole Enclosures
Mechanical Raceway and Cable Connection
Section 314.30(B)

Underground raceways and cable entering a handhole enclosure aren't required to be mechanically connected to the handhole enclosure.

COPYRIGHT 2004 Mike Holt Enterprises, Inc.

Splices or terminations must be listed as suitable for wet locations [314.30(C)].

Figure 314–38

(C) Handhole Enclosures Without Bottoms. All splices or terminations must be listed as suitable for wet locations [110.14(B)].

(D) Covers. Handhole enclosure covers must have an identifying mark or logo that prominently identifies the function of the enclosure, such as "electric."

Handhole enclosure covers must require the use of tools to open, or they must weigh over 100 lbs. Metal covers and other exposed conductive surfaces must be bonded to an effective ground-fault current path in accordance with 250.96(A) [250.4(A)(3)].

Article 314 Questions

1. •Nonmetallic boxes are permitted for use with _____.

    (a) flexible nonmetallic conduit   (b) liquidtight nonmetallic conduit
    (c) nonmetallic cables and raceways   (d) all of these

2. When counting the number of conductors in a box, a conductor running through the box with no loop in it is counted as _____ conductor(s).

    (a) one   (b) two   (c) zero   (d) none of these

3. Each yoke or strap containing one or more devices or equipment counts as _____ conductor(s), based on the largest conductor that terminates on that device.

    (a) 1   (b) 2   (c) 3   (d) none

4. •In noncombustible walls or ceilings, the front edge of a box, plaster ring, extension ring, or listed extender may be set back not more than _____ from the finished surface.

    (a) ⅜ in.   (b) ⅛ in.   (c) ½ in.   (d) ¼ in.

5. Surface mounted enclosures (boxes) must be _____ the building surface.

    (a) rigidly and securely fastened to
    (b) supported by cables that protrude from
    (c) supported by cable entries from the top and allowed to rest against
    (d) none of these

6. Outlet boxes can be secured to independent support wires, which are taut and secured at both ends, if the box is supported to the independent support wires using methods identified for the purpose.

    (a) True   (b) False

7. Outlet boxes used at luminaire or lampholder outlets must be _____.

    (a) designed for the purpose   (b) metal only
    (c) plastic only   (d) mounted using bar hangers only

8. Where a box is used as the sole support of a ceiling-suspended (paddle) fan, the box must be listed for the application and must be marked with the weight of the fan to be supported if over 35 lbs.

    (a) True   (b) False

9. Handhole enclosure covers must require the use of tools to open, or they must weigh over _____. Metal covers and other exposed conductive surfaces must be bonded to an effective ground-fault current path.

    (a) 45 lbs   (b) 100 lbs   (c) 70 lbs   (d) 200 lbs

320 Armored Cable (Type AC)

Introduction

Armored cable is an assembly of insulated conductors, 14 AWG through 1 AWG, that are individually wrapped within waxed paper and contained within a flexible spiral metal sheath. Armored cable looks like flexible metal conduit.

PART I. GENERAL

320.1 Scope. This article covers the use, installation, and construction specifications of armored cable, Type AC.

320.2 Definition.

Armored Cable (Type AC). A fabricated assembly of conductors in a flexible metal sheath with an internal bonding strip in intimate contact with the armor for its entire length. See 320.100. Figure 320–1

Author's Comment: The conductors are contained within a flexible metal (steel or aluminum) sheath that interlocks at the edges, giving Type AC cable an outside appearance similar to that of flexible metal conduit. Many electricians call this metal cable BX®. The advantages of any flexible cable system are that there is no limit to the number of bends between terminations and the cable can be quickly installed.

PART II. INSTALLATION

320.10 Uses Permitted. Type AC cable is permitted only where not subject to physical damage in the following locations, and in other locations and conditions not prohibited by 320.12 or elsewhere in the *Code*:

(1) Exposed and concealed,

(2) Cable trays,

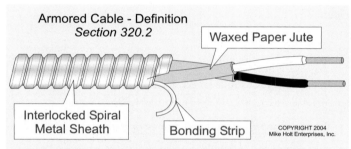

Armored Cable - Definition
Section 320.2

Waxed Paper Jute

Interlocked Spiral Metal Sheath

Bonding Strip

COPYRIGHT 2004
Mike Holt Enterprises, Inc.

Armored Cable: A fabricated assembly of conductors in a flexible metal sheath with an internal bonding strip in intimate contact with the armor for its entire length.

Figure 320–1

(3) Dry locations,

(4) Embedded in plaster or brick, except in damp or wet locations, and

(5) Air voids where not exposed to excessive moisture or dampness.

Author's Comment: Type AC cable is permitted in other environmental air spaces [300.22(C)(1)].

320.12 Uses Not Permitted. Type AC cable cannot be installed in the following locations:

(1) In theaters and similar locations, except as permitted by 518.4(A),

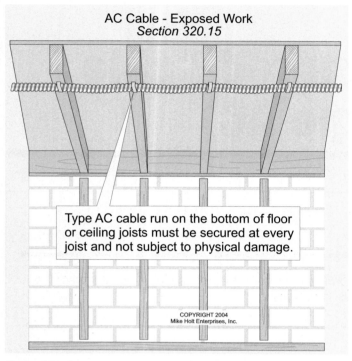

AC Cable - Exposed Work
Section 320.15

Type AC cable run on the bottom of floor or ceiling joists must be secured at every joist and not subject to physical damage.

COPYRIGHT 2004
Mike Holt Enterprises, Inc.

Figure 320–2

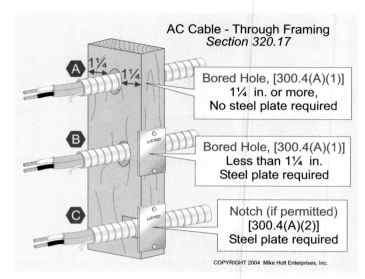

AC Cable - Through Framing
Section 320.17

Ⓐ 1¼ 1¼

Bored Hole, [300.4(A)(1)]
1¼ in. or more,
No steel plate required

Ⓑ LISTED

Bored Hole, [300.4(A)(1)]
Less than 1¼ in.
Steel plate required

Ⓒ LISTED

Notch (if permitted)
[300.4(A)(2)]
Steel plate required

COPYRIGHT 2004 Mike Holt Enterprises, Inc.

Figure 320–3

(2) In motion picture studios,

(3) In any hazardous (classified) location, except as permitted by 501.10(B)(3), 502.10(B)(3), and 504.20,

(4) Where exposed to corrosive fumes or vapors,

(5) Embedded in plaster finish on brick or other masonry in damp or wet locations.

320.15 Exposed Work. Exposed Type AC cable must closely follow the surface of the building finish or running boards. Type AC cable run on the bottom of floor or ceiling joists must be secured at every joist and not be subject to physical damage. **Figure 320–2**

320.17 Through or Parallel to Framing Members.
Type AC cable installed through, or parallel to, framing members or furring strips must be protected against physical damage from penetration by screws or nails by maintaining 1¼ in. of separation, or by installing a suitable metal plate in accordance with 300.4(A) and (D):

Author's Comments:
- 300.4(A)(1) Drilling Holes in Wood Members. When drilling holes through wood framing members for cables, the edge of the holes must be not less than 1 ¼ in. from the edge of the wood member. **Figure 320–3A**

If the edge of the hole is less than 1 ¼ in. from the edge, a ¹⁄₁₆ in. thick steel plate of sufficient length and width must be installed to protect the wiring method from screws and nails. **Figure 320–3B**

- 300.4(A)(2) Notching Wood Members. Where notching of wood framing members for cables is permitted by the building code, a ¹⁄₁₆ in. thick steel plate of sufficient length and width must be installed to protect the cables and raceways laid in these wood notches from screws and nails. **Figure 320–3C**

- 300.4(D) Cables Parallel to Framing Members and Furring Strips. Cables run parallel to framing members or furring strips must be protected where likely to be penetrated by nails or screws. The wiring method must be installed so it is at least 1 ¼ in. from the nearest edge of the framing members or furring strips, or a ¹⁄₁₆ in. thick steel plate must protect the wiring method. **Figure 320–4**

320.23 In Accessible Attics or Roof Spaces.

(A) On the Surface of Floor Joists, Rafters, or Studs. In attics and roof spaces that are accessible, substantial guards must protect the cables run across the top of floor joists, or across the face of rafters or studding within 7 ft of floor or floor joists. Where this space isn't accessible by permanent stairs or ladders, protection is required only within 6 ft of the nearest edge of the scuttle hole or attic entrance.

(B) Along the Side of Framing Members. When Type AC cable is run on the side of rafters, studs, or floor joists, no protection is required if the cable is installed and supported so the nearest outside surface of the cable or raceway is at least 1¼ in. from the nearest edge of the framing member where nails or screws are likely to penetrate [300.4(D)].

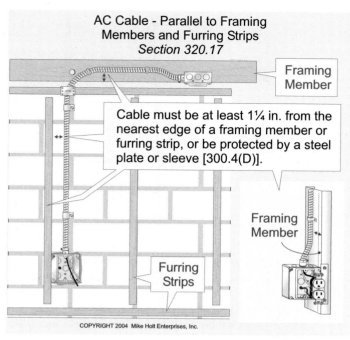

AC Cable - Parallel to Framing Members and Furring Strips
Section 320.17

Framing Member

Cable must be at least 1¼ in. from the nearest edge of a framing member or furring strip, or be protected by a steel plate or sleeve [300.4(D)].

Framing Member

Furring Strips

COPYRIGHT 2004 Mike Holt Enterprises, Inc.

Figure 320–4

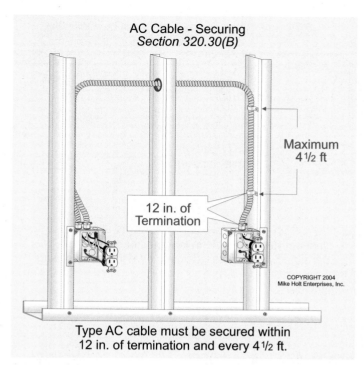

AC Cable - Securing
Section 320.30(B)

Maximum 4½ ft

12 in. of Termination

COPYRIGHT 2004 Mike Holt Enterprises, Inc.

Type AC cable must be secured within 12 in. of termination and every 4½ ft.

Figure 320–5

320.24 Bends. Type AC cable cannot be bent in a manner that will damage the cable. This is accomplished by limiting bending of the inner edge of the cable to a radius not less than five times the internal diameter of the cable.

320.30 Secured and Supported.

(A) General. Type AC cable must be supported and secured by staples, cable ties, straps, hangers, or similar fittings, designed and installed not to damage the cable.

(B) Securing. Type AC cable must be secured within 12 in. of every outlet box, junction box, cabinet, or fitting and at intervals not exceeding 4½ ft. **Figure 320–5**

> **Author's Comment:** Type AC cable is considered secured when installed horizontally through wooden or metal framing members [320.30(C)].

(C) Supporting. Type AC cable must be supported at intervals not exceeding 4½ ft.

Cables installed horizontally through wooden or metal framing members are considered secured and supported where support doesn't exceed 4½ ft. **Figure 320–6**

(D) Unsupported Cables. Type AC cable can be unsupported where the cable is:

(1) Fished through concealed spaces in finished buildings or structures, where support is impractical; or

(2) Not more than 2 ft long at terminals where flexibility is necessary; or

(3) Not more than 6 ft long from the last point of cable support to the point of connection to a luminaire or other piece of electrical equipment within an accessible ceiling.

For the purposes of this section, Type AC cable fittings are permitted as a means of cable support. **Figure 320–7**

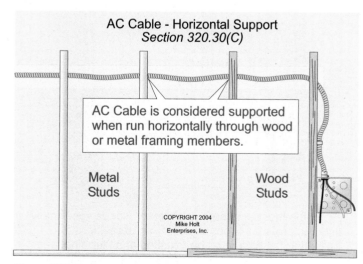

AC Cable - Horizontal Support
Section 320.30(C)

AC Cable is considered supported when run horizontally through wood or metal framing members.

Metal Studs

Wood Studs

COPYRIGHT 2004 Mike Holt Enterprises, Inc.

Figure 320–6

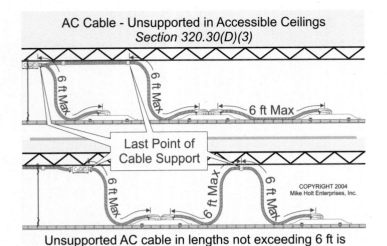

Unsupported AC cable in lengths not exceeding 6 ft is
permitted when installed within an accessible ceiling.

Figure 320–7

320.40 Boxes and Fittings. Type AC cable must terminate
in boxes or fittings specifically listed for Type AC cable to protect the conductors from abrasion [300.15]. An insulating anti-
short bushing, sometimes called a "redhead," must be installed at
all Type AC cable terminations. The termination fitting must
permit the visual inspection of the anti-short bushing once the
cable has been installed. **Figure 320–8**

Author's Comments:

- The internal aluminum-bonding strip within the cable serves
  no electrical purpose once outside the cable, but many electricians use it to secure the anti-short bushing to the cable.

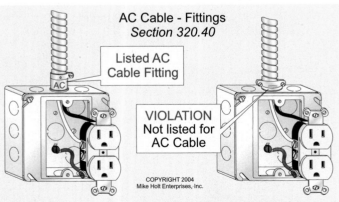

AC cable must terminate in boxes or fittings specifically
listed to protect the conductors from abrasion. An
insulating bushing must be installed at all terminations.

Figure 320–8

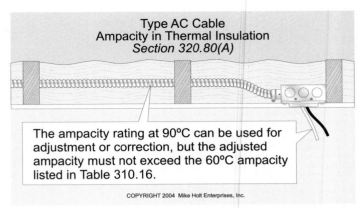

Figure 320–9

- Conductors 4 AWG and larger that enter an enclosure must
  be protected from abrasion during and after installation by a
  fitting that provides a smooth, rounded, insulating surface,
  such as an insulating bushing unless the design of the box,
  fitting, or enclosure provides equivalent protection in accordance with 300.4(F).

320.80 Conductor Ampacities. Conductor ampacity is
calculated in accordance with 310.15 based on 90°C insulation
rating of the conductors, provided the adjusted or corrected
ampacity doesn't exceed the temperature ratings of terminations
and equipment [110.14(C)].

(A) Thermal Insulation. Type AC cable installed in thermal
insulation must have conductors rated 90°C. The ampacity rating
at 90°C can be used for conductor ampacity adjustment and/or
correction, but the adjusted ampacity must not exceed the 60°C
ampacity listed in Table 310.16. **Figure 320–9**

> **Question:** *What is the ampacity of four 12 THHN current-car-
> rying conductors installed in Type AC cable?*
>
> *(a) 18A*   *(b) 24A*   *(c) 27A*   *(d) 30A*
>
> **Answer:** *(b) 24A*
>
> *Ampacity of 12 THHN at 90°C = 30A*
> *Ampacity Adjustment for Conductor Bundle*
>   *[Table 310.15(B)(2)(a)] = 0.8*
> *Conductor Adjusted Ampacity = 30A x 0.8*
> *Conductor Adjusted Ampacity = 24A*

PART III. CONSTRUCTION SPECIFICATIONS

320.100 Construction. Type AC cable has an armor of flexible metal tape with an internal aluminum-bonding strip in intimate contact with the armor for its entire length.

> **Author's Comment:** When cutting Type AC cable with a hacksaw, be sure to cut only one spiral of the cable and be careful not to nick the conductors; this is done by cutting the cable at an angle. Breaking the cable spiral (bending the cable very sharply), then cutting the cable with a pair of dikes isn't a good practice. The best method to separate the cable is to use a tool specially designed for the purpose, such as a rotary armor cutter.

320.108 Equipment Grounding (Bonding). The combination of the armor and 18 AWG aluminum-bonding strip [320.100] makes the cable suitable to serve as the effective ground-fault current path required by 250.4(A)(5). The effective ground-fault current path must be maintained by the use of fittings that are specifically listed for Type AC cable [320.40]. See 300.12, 300.15, and 300.100. **Figure 320–10**

> **Author's Comment:** The internal aluminum-bonding strip isn't an equipment grounding (bonding) conductor, but it serves to reduce the inductive reactance of the armored spirals to ensure that a short circuit will be cleared. Once the bonding strip exits the cable, it can be cut off because it no longer serves any purpose.

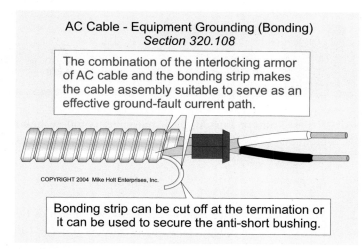

AC Cable - Equipment Grounding (Bonding)
Section 320.108

The combination of the interlocking armor of AC cable and the bonding strip makes the cable assembly suitable to serve as an effective ground-fault current path.

COPYRIGHT 2004 Mike Holt Enterprises, Inc.

Bonding strip can be cut off at the termination or it can be used to secure the anti-short bushing.

Figure 320–10

1. Exposed runs of Type AC cable must closely follow the surface of the building finish or of running boards. Exposed runs are also permitted to be installed on the underside of joists where supported at each joist and located so as not to be subject to physical damage.

   (a) True   (b) False

2. Where run across the top of floor joists, within 7 ft of floor or floor joists, or across the face of rafters or studding in attics and roof spaces that are accessible by permanent stairs or ladders, Type AC cable must be protected by substantial guard strips that are _____.

   (a) at least as high as the cable   (b) constructed of metal   (c) made for the cable   (d) none of these

3. When armored cable is run parallel to the sides of rafters, studs, or floor joists in an accessible attic, the cable must be protected with running boards.

   (a) True   (b) False

4. Type AC cable must be supported and secured at intervals not exceeding 4½ ft and the cable must be secured within _____ of every outlet box, cabinet, conduit body, or other armored cable termination.

   (a) 4 in.   (b) 8 in.   (c) 9 in.   (d) 12 in.

5. At all Type AC cable terminations, a(n) _____ must be provided.

   (a) fitting (or box design) that protects the wires from abrasion
   (b) insulating bushing between the conductors and the cable armor
   (c) both a and b
   (d) none of these

ARTICLE 330

Metal-Clad Cable (Type MC)

Introduction

Metal-clad cable encloses one or more insulated conductors in a metal sheath of either corrugated or smooth copper or aluminum tubing, or spiral interlocked steel or aluminum. The physical characteristics of MC cable make it a versatile wiring method you can use in almost any location and for almost any application. The most common type of MC cable is the interlocking type, which looks similar to armored cable or flexible metal conduit. Because the outer sheath of interlocking type cable isn't suitable as an effective ground-fault current path, it contains a separate equipment grounding (bonding) conductor.

PART I. GENERAL

330.1 Scope. Article 330 covers the use, installation, and construction specifications of metal-clad cable.

330.2 Definition.

Metal-Clad Cable (Type MC). A factory assembly of one or more insulated circuit conductors, with or without optical fiber members, enclosed in an armor of interlocking metal tape or a smooth or corrugated metallic sheath. **Figure 330–1**

> **Author's Comment:** Type MC cable encloses one or more insulated conductors in a metal sheath of either corrugated or smooth copper or aluminum tubing, or spiral-interlocked steel or aluminum. Because the outer sheath of interlocked Type MC cable isn't listed as an equipment grounding (bonding) conductor, it contains an equipment grounding (bonding) conductor [330.108]. The most common type of Type MC cable is the interlocking type, which looks similar to armored cable or flexible metal conduit.

PART II. INSTALLATION

330.10 Uses Permitted.

(A) General Uses. Type MC cable is permitted only where not subject to physical damage, and in other locations and conditions not prohibited by 330.12, or elsewhere in the *Code*:

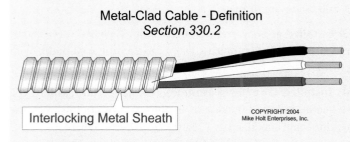

Metal-Clad Cable - Definition
Section 330.2

Interlocking Metal Sheath

COPYRIGHT 2004
Mike Holt Enterprises, Inc.

Metal-Clad Cable: A factory assembly of one or more insulated circuit conductors enclosed in an armor of interlocking metal tape, or a smooth or corrugated metallic sheath.

Figure 330–1

(1) In branch circuits, feeders and services.

(2) In power, lighting, control, and signal circuits.

(3) Indoors or outdoors.

(4) Exposed or concealed.

(5) Directly buried (if identified for the purpose).

(6) In a cable tray.

(7) In a raceway.

(8) As aerial cable on a messenger.

(9) In hazardous (classified) locations as permitted in 501.10(B), 502.10(B), and 503.10.

(10) Embedded in plaster or brick.

(11) In wet locations, if one or more of the following are met:

a. The metallic covering is impervious to moisture.

b. A lead sheath or moisture-impervious jacket is provided under the metal covering.

c. The insulated conductors under the metallic covering are listed for use in wet locations.

(12) Where single-conductor cables are used, all ungrounded conductors and, where used, the grounded neutral conductor must be grouped together to minimize induced voltage on the sheath [300.3(B)].

(B) Specific Uses. Type MC cable can be installed in compliance with Parts II and III of Article 725 and 770.133 as applicable, and in accordance with (1) through (4).

(1) Cable Tray. Type MC cable installed in cable tray must comply with 392.3, 392.4, 392.6, and 392.8 through 392.13.

(2) Direct Buried. Direct-buried cable must comply with 300.5 or 300.50, as appropriate.

(3) Installed as Service-Entrance Cable. Type MC cable is permitted for service entrances, when installed in accordance with 230.43.

(4) Installed Outside of Buildings or as Aerial Cable. Type MC cable installed outside of buildings must comply with the requirements contained in 225.10, 396.10, and 396.12.

> **Author's Comment:** Type MC cable is permitted in other environmental air spaces [300.22(C)(1)].

330.12 Uses Not Permitted. Type MC cable cannot be used where exposed to the following destructive or corrosive conditions, unless the metallic sheath is suitable for the conditions or is protected by material suitable for the conditions:

(1) Subject to physical damage.

(2) Direct burial in the earth, if not identified for this purpose [330.10(A)(5)].

(3) In concrete.

> **FPN:** Type MC cable identified for direct burial is suitable for installation in concrete.

(4) Where subject to cinder fills, strong chlorides, caustic alkalis, or vapors of chlorine or of hydrochloric acids.

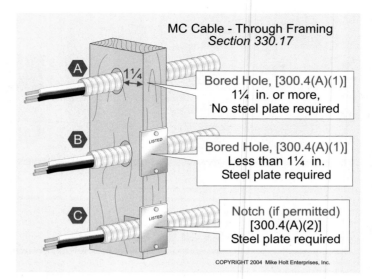

Figure 330–2

330.17 Through or Parallel to Framing Members. Type MC cable installed through or parallel to framing members or furring strips must be protected against physical damage from penetration from screws or nails by maintaining a 1¼ in. separation or by installing a suitable metal plate in accordance with 300.4(A) and (D).

Author's Comments:

- 300.4(A)(1) Drilling Holes in Wood Members. When drilling holes through wood framing members for cables, the edge of the holes must be not less than 1¼ in. from the edge of the wood member. **Figure 330–2A**

  If the edge of the hole is less than 1¼ in. from the edge, a ¹⁄₁₆ in. thick steel plate of sufficient length and width must be installed to protect the wiring method from screws and nails. **Figure 330–2B**

- 300.4(A)(2) Notching Wood Members. Where notching of wood framing members for cables is permitted by the building code, a ¹⁄₁₆ in. thick steel plate of sufficient length and width must be installed to protect the cables and raceways laid in these wood notches from screws and nails. **Figure 330–2C**

- 300.4(D) Cables Parallel to Framing Members and Furring Strips. Cables run parallel to framing members or furring strips must be protected where likely to be penetrated by nails or screws. The wiring method must be installed so it is at least 1¼ in. from the nearest edge of the framing member or furring strips, or a ¹⁄₁₆ in. thick steel plate must protect it. **Figure 330–3**

330.23 In Accessible Attics or Roof Spaces. Type MC cable installed in accessible attics or roof spaces must comply with 320.23.

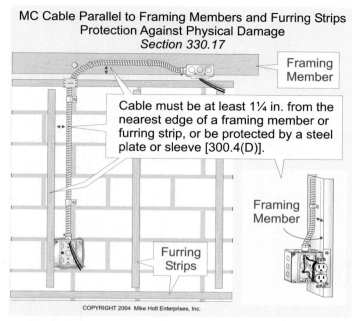

MC Cable Parallel to Framing Members and Furring Strips Protection Against Physical Damage
Section 330.17

Cable must be at least 1¼ in. from the nearest edge of a framing member or furring strip, or be protected by a steel plate or sleeve [300.4(D)].

Framing Member

Framing Member

Furring Strips

COPYRIGHT 2004 Mike Holt Enterprises, Inc.

Figure 330–3

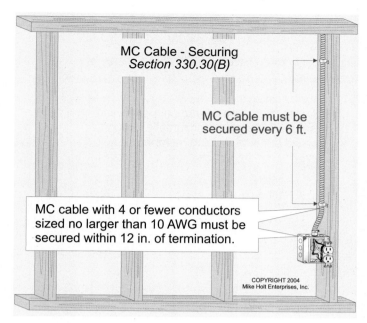

MC Cable - Securing
Section 330.30(B)

MC Cable must be secured every 6 ft.

MC cable with 4 or fewer conductors sized no larger than 10 AWG must be secured within 12 in. of termination.

COPYRIGHT 2004 Mike Holt Enterprises, Inc.

Figure 330–4

Author's Comments:

- On the Surface of Floor Joists [320.23(A)], Rafters, or Studs. In attics and roof spaces that are accessible, substantial guards must protect cables run across the top of floor joists, or across the face of rafters or studding within 7 ft of floor or floor joists. Where this space isn't accessible by permanent stairs or ladders, protection is required only within 6 ft of the nearest edge of the scuttle hole or attic entrance [320.23(A)].

- Along the Side of Framing Members [320.23(B)]. When Type MC cable is run on the side of rafters, studs, or floor joists, no protection is required if the cable is installed and supported so the nearest outside surface of the cable or raceway is at least 1¼ in. from the nearest edge of the framing member where nails or screws are likely to penetrate [300.4(D)].

330.24 Bends. All bends must be made so the cable will not be damaged, and the radius of the curve of any bend at the inner edge of any cable must not be less than what is dictated in each of the following instances:

(A) Smooth-Sheath Cables. Smooth-sheath cables must not be bent so that the bending radius of the inner edge of the cable is less than ten times the external diameter of the metallic sheath for cable up to ¾ in. in external diameter.

(B) Interlocked or Corrugated Sheath. Interlocked- or corrugated-sheath cables must not be bent where the radius of bend for the inner edge of the cable is less than seven times the external diameter of the cable.

PART II. INSTALLATION

330.30 Secured and Supported.

(A) General. Type MC cable must be supported and secured by staples, cable ties, straps, hangers, or similar fittings, designed and installed not to damage the cable.

(B) Securing. Where installed on or across framing members, Type MC cable with four or less conductors sized no larger than 10 AWG, must be secured within 12 in. of every outlet box, junction box, cabinet, or fitting and at intervals not exceeding 6 ft. **Figure 330–4**

(C) Supporting. Type MC cable must be supported at intervals not exceeding 6 ft. Cables installed horizontally through wooden or metal framing members are considered secured and supported where such support doesn't exceed 6 ft intervals. **Figure 330–5**

(D) Unsupported Cables. Type MC cable can be unsupported where the cable is:

(1) Fished through concealed spaces in finished buildings or structures, where support is impracticable; or

(2) Not more than 6 ft long from the last point of cable support to the point of connection to a luminaire or other piece of electrical equipment within an accessible ceiling.

For the purpose of this section, Type MC cable fittings are permitted as a means of cable support. **Figure 330–6**

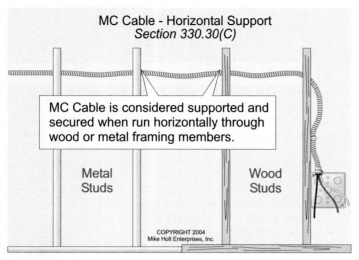

Figure 330–5

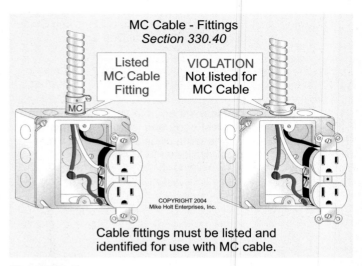

Figure 330–7

330.40 Fittings. Fittings used with Type MC cable must be listed and identified for use with the cable [300.15]. **Figure 330–7**

Author's Comments:

- The *NEC* doesn't require anti-short bushings (red heads) for termination of Type MC cable. If they are supplied with the cable, it is a good practice to use them at terminations, but they aren't required.

- Conductors 4 AWG and larger that enter an enclosure must be protected from abrasion during and after installation by a fitting that provides a smooth, rounded, insulating surface, such as an insulating bushing unless the design of the box, fitting, or enclosure provides equivalent protection in accordance with 300.4(F).

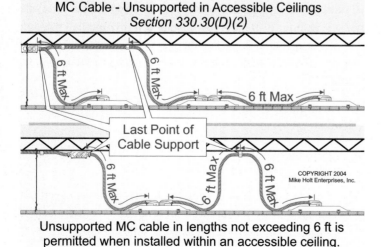

Unsupported MC cable in lengths not exceeding 6 ft is permitted when installed within an accessible ceiling.

Figure 330–6

330.80 Conductor Ampacities. Conductor ampacity is calculated in accordance with 310.15 based on 90°C insulation rating of the conductors, provided the adjusted or corrected ampacity doesn't exceed the temperature ratings of terminations and equipment [110.14(C)].

PART III. CONSTRUCTION SPECIFICATIONS

330.108 Equipment Grounding (Bonding). Where MC cable is used for equipment grounding (bonding), it must comply with 250.118(10) and 250.122.

Author's Comments:

- The armor on interlocked Type MC cable isn't listed as an effective ground-fault current path; therefore, an equipment grounding (bonding) conductor must be installed within the metal armor wrap [250.118(10)]. **Figure 330–8**

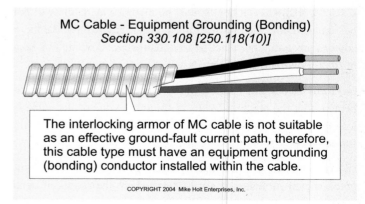

The interlocking armor of MC cable is not suitable as an effective ground-fault current path, therefore, this cable type must have an equipment grounding (bonding) conductor installed within the cable.

Figure 330–8

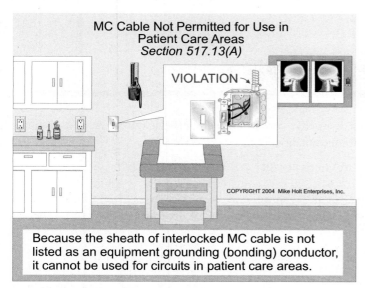

MC Cable Not Permitted for Use in
Patient Care Areas
Section 517.13(A)

VIOLATION

COPYRIGHT 2004 Mike Holt Enterprises, Inc.

Because the sheath of interlocked MC cable is not
listed as an equipment grounding (bonding) conductor,
it cannot be used for circuits in patient care areas.

Figure 330–9

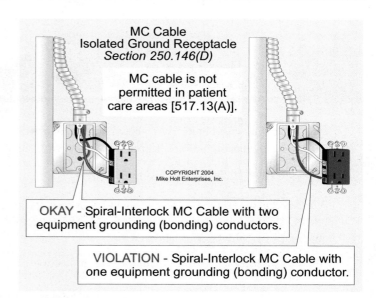

MC Cable
Isolated Ground Receptacle
Section 250.146(D)

MC cable is not
permitted in patient
care areas [517.13(A)].

COPYRIGHT 2004
Mike Holt Enterprises, Inc.

OKAY - Spiral-Interlock MC Cable with two
equipment grounding (bonding) conductors.

VIOLATION - Spiral-Interlock MC Cable with
one equipment grounding (bonding) conductor.

Figure 330–10

- Health Care Facilities. Because the outer sheath of interlocked Type MC cable is not listed as an equipment
  grounding (bonding) conductor [250.118(10)], it cannot be
  used for circuits in patient care areas [517.13(A)].
  Figure 330–9

- Isolated Ground Circuits. Interlocked Type MC cable containing one equipment grounding (bonding) conductor cannot
  be used for isolated circuits [250.110(1) and 250.146(D)].
  Figure 330–10

1. Type MC cable must not be used where exposed to the following destructive corrosive condition(s), unless the metallic sheath is suitable for the condition(s) or is protected by material suitable for the condition(s):

   (a) Direct burial in the earth   (b) In concrete   (c) In cinder fill   (d) all of these

2. Type MC cable installed in accessible attics or roof spaces must comply with the same requirements as given for AC cable in 320.24. This includes the installation of _____ to protect the cable when run across the top of floor joists if the space is accessible by permanent stairs or ladders.

   (a) GFCI protection   (b) arc fault protection   (c) rigid metal conduit   (d) guard strips

3. Smooth-sheath Type MC cable with an external diameter of not greater than 1 in. must have a bending radius of not more than _____ times the cable external diameter.

   (a) 5   (b) 10   (c) 12   (d) 13

4. Type MC cable must be supported and secured at intervals not exceeding _____.

   (a) 3 ft   (b) 6 ft   (c) 4 ft   (d) 2 ft

5. Type MC cable can be unsupported where it is:

   (a) Fished between concealed access points in finished buildings or structures and support is impracticable.
   (b) Not more than 2 ft in length at terminals where flexibility is necessary.
   (c) Not more than 6 ft from the last point of support within an accessible ceiling for the connection of luminaires.
   (d) a or c

334

Nonmetallic-Sheathed Cable (Types NM and NMC)

Introduction

Nonmetallic-sheathed cable is flexible, inexpensive, and easily installed. It provides very limited physical protection of the conductors, so the installation restrictions are strict. However, its low cost and relative ease of installation make it a common wiring method used for residential and commercial branch circuits.

PART I. GENERAL

334.1 Scope. Article 334 covers the use, installation, and construction specifications of nonmetallic-sheathed cable.

334.2 Definition.

Nonmetallic-Sheathed Cable (Type NM). Nonmetallic-sheathed cable is a wiring method that encloses two or more insulated conductors, 14 AWG through 2 AWG, within a nonmetallic jacket.

- Type NM has insulated conductors enclosed within an overall nonmetallic jacket.

- Type NMC has insulated conductors enclosed within an overall, corrosion resistant, nonmetallic jacket. **Figure 334–1**

Author's Comments:

- NM cable is normally marked with a "B" suffix, such as NM-B or NMC-B. The "B" indicates that the NM cable is comprised of conductors having 90°C insulation.

- Because this cable is nonmetallic, it contains a separate equipment grounding (bonding) conductor. Type NM cable is a common wiring method used for residential and small commercial branch circuits. It's the generally accepted practice in the electrical industry to call Type NM cable "Romex," a registered trademark of the Southwire Company.

334.6 Listed. NM cable must be listed [334.116].

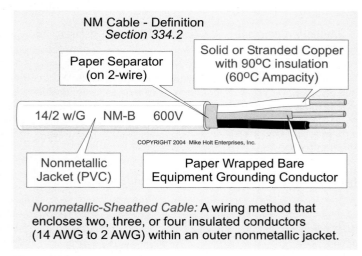

NM Cable - Definition
Section 334.2

Paper Separator (on 2-wire)

Solid or Stranded Copper with 90°C insulation (60°C Ampacity)

14/2 w/G   NM-B   600V

COPYRIGHT 2004 Mike Holt Enterprises, Inc.

Nonmetallic Jacket (PVC)

Paper Wrapped Bare Equipment Grounding Conductor

Nonmetallic-Sheathed Cable: A wiring method that encloses two, three, or four insulated conductors (14 AWG to 2 AWG) within an outer nonmetallic jacket.

Figure 334–1

PART II. INSTALLATION

334.10 Uses Permitted.

Type NM and Type NMC cables can be used in the following:

(1) One- and two-family dwellings. **Figure 334–2**

(2) Multifamily dwellings permitted to be of Types III, IV, and V construction. **Figure 334–3**

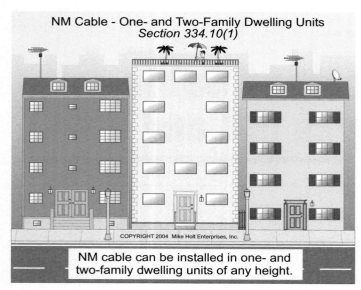

NM Cable - One- and Two-Family Dwelling Units
Section 334.10(1)

NM cable can be installed in one- and two-family dwelling units of any height.

Figure 334–2

(3) Other structures permitted to be of Types III, IV, and V construction, except as prohibited in 334.12. Cables must be concealed within walls, floors, or ceilings that provide a thermal barrier of material with at least a 15-minute finish rating as identified in listings of fire-rated assemblies. Figure 334–4

Author's Comment: See Article 100 for the definition of "Concealed."

FPN No. 1: Building constructions are defined in NFPA 220, *Standard on Types of Building Construction*, the applicable building code, or both.

FPN No. 2: See Annex E for determination of building types [NFPA 220, Table 3-1].

(4) Cable trays, where the cables are identified for this use.

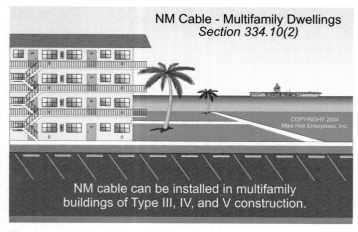

NM Cable - Multifamily Dwellings
Section 334.10(2)

NM cable can be installed in multifamily buildings of Type III, IV, and V construction.

Figure 334–3

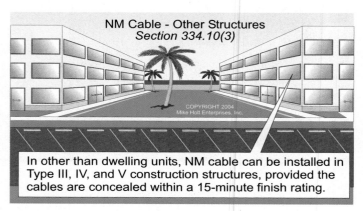

NM Cable - Other Structures
Section 334.10(3)

In other than dwelling units, NM cable can be installed in Type III, IV, and V construction structures, provided the cables are concealed within a 15-minute finish rating.

Figure 334–4

FPN: See 310.10 for temperature limitation of conductors.

334.12 Uses Not Permitted.

(A) Types NM and NMC.

(1) In any dwelling or structure not specifically permitted in 334.10(1), (2), and (3).

(2) Exposed in dropped or suspended ceilings in other than one- and two-family and multifamily dwellings. Figure 334–5

(3) As service-entrance cable.

(4) In commercial garages having hazardous (classified) locations as defined in 511.3.

(5) In theaters and similar locations, except where permitted in 518.4(B).

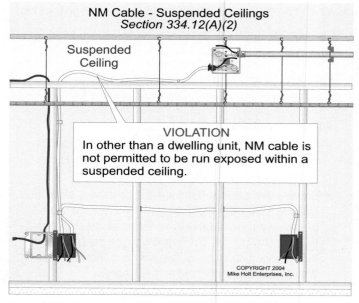

NM Cable - Suspended Ceilings
Section 334.12(A)(2)

Suspended Ceiling

VIOLATION
In other than a dwelling unit, NM cable is not permitted to be run exposed within a suspended ceiling.

Figure 334–5

(6) In motion picture studios.

(7) In storage battery rooms.

(8) In hoistways or on elevators or escalators.

(9) Embedded in poured cement, concrete, or aggregate.

(10) In hazardous (classified) locations, except where permitted by the following:

   a. 501.10(B)(3)

   b. 502.10(B)(3)

   c. 504.20

(B) Type NM. Type NM cables cannot be used under the following conditions or in the following locations:

(1) Where exposed to corrosive fumes or vapors.

(2) Where embedded in masonry, concrete, adobe, fill, or plaster.

(3) In a shallow chase in masonry, concrete, or adobe and covered with plaster, adobe, or similar finish.

(4) Where subject or exposed to excessive moisture or dampness.

Author's Comment: Type NM cable isn't permitted in ducts, plenums, or other environmental air spaces [300.22] or for wiring in patient care areas [517.13].

334.15 Exposed.

(A) Surface of the Building. Exposed Type NM cable must closely follow the surface of the building.

(B) Protected from Physical Damage. Nonmetallic-sheathed cable must be protected from physical damage by rigid metal conduit, intermediate metal conduit, Schedule 80 rigid nonmetallic conduit [352.10(F)], electrical metallic tubing, guard strips, or other means approved by the authority having jurisdiction.

Author's Comment: When installed in a raceway, the cable must be protected from abrasion by a fitting installed on the end of the raceway [300.15(C)].

Where NM cable passes through a floor, it must be protected by rigid metal conduit, intermediate metal conduit, Schedule 80 rigid nonmetallic conduit [352.10(F)], electrical metallic tubing, or other approved means for at least 6 in. above the floor.

Author's Comment: The above 6 in. protection rule "when passing through a floor," only applies when the cable is exposed and subject to physical damage. Honestly, I can't think of an example where this rule would be used!

Where Type NMC cable is installed in shallow chases in masonry, concrete, or adobe, the cable must be protected against nails or screws by a steel plate not less than 1/16 in. thick and covered with plaster, adobe, or similar finish.

Author's Comment: Where Type NM cable is installed in a metal raceway, the raceway isn't required to be grounded (bonded) to an effective ground-fault current path [250.86 Ex. 2 and 300.12 Ex.].

(C) In Unfinished Basements. Where the cable is run at angles with joists in unfinished basements, it is permissible to secure cables not smaller than two 6 AWG or three 8 AWG conductors directly to the lower edges of the joists. Smaller cables must be run through bored holes in joists or on running boards. **Figure 334–6**

NM cable on a wall of an unfinished basement can be installed in a listed raceway. A nonmetallic bushing or adapter must be installed at the point where the cable enters the raceway. The raceway and metal outlet boxes must be grounded (bonded) to an effective ground-fault current path in accordance with 250.148 [250.4(A)(3)].

334.17 Through or Parallel to Framing Members.

Type NM cable installed through or parallel to framing members or furring strips must be protected against physical damage from penetration by screws or nails by 1¼ in. of separation or by a suitable metal plate [300.4(A) and (D)]. **Figures 334–7 and 334–8**

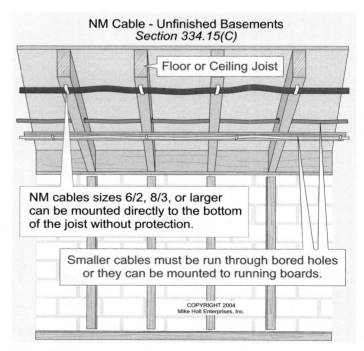

NM Cable - Unfinished Basements
Section 334.15(C)

Floor or Ceiling Joist

NM cables sizes 6/2, 8/3, or larger can be mounted directly to the bottom of the joist without protection.

Smaller cables must be run through bored holes or they can be mounted to running boards.

COPYRIGHT 2004
Mike Holt Enterprises, Inc.

Figure 334–6

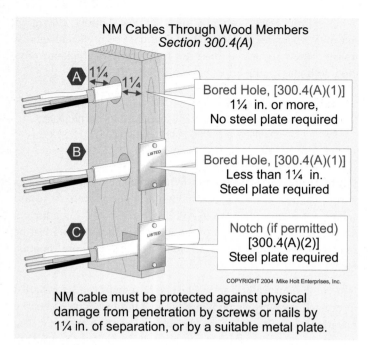

NM Cables Through Wood Members
Section 300.4(A)

A 1¼ ← → 1¼
Bored Hole, [300.4(A)(1)]
1¼ in. or more,
No steel plate required

B
Bored Hole, [300.4(A)(1)]
Less than 1¼ in.
Steel plate required

C
Notch (if permitted)
[300.4(A)(2)]
Steel plate required

COPYRIGHT 2004 Mike Holt Enterprises, Inc.

NM cable must be protected against physical
damage from penetration by screws or nails by
1¼ in. of separation, or by a suitable metal plate.

Figure 334–7

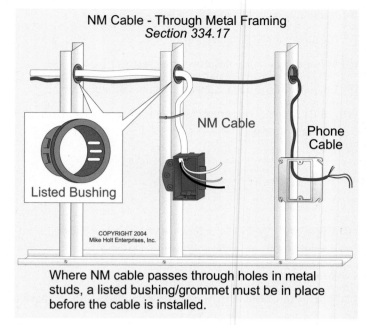

NM Cable - Through Metal Framing
Section 334.17

NM Cable

Phone
Cable

Listed Bushing

COPYRIGHT 2004
Mike Holt Enterprises, Inc.

Where NM cable passes through holes in metal
studs, a listed bushing/grommet must be in place
before the cable is installed.

Figure 334–9

Where Type NM cable passes through holes in metal studs, a
listed bushing or listed grommet is required [300.4(B)(1)] to be
in place before the cable is installed. Figure 334–9

334.23 Attics and Roof Spaces. Type NM cable installed
in accessible attics or roof spaces must comply with 320.23:

Author's Comments:

- On the Surface of Floor Joists [320.23(A)], Rafters, or Studs.
  In attics and roof spaces that are accessible, substantial
  guards must protect cables run across the top of floor joists,
  or across the face of rafters or studding within 7 ft of floor or
  floor joists. Where this space isn't accessible by permanent
  stairs or ladders, protection is required only within 6 ft of the
  nearest edge of the scuttle hole or attic entrance [320.23(A)].

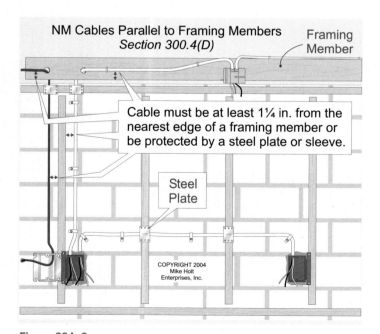

NM Cables Parallel to Framing Members
Section 300.4(D)   Framing
                            Member

Cable must be at least 1¼ in. from the
nearest edge of a framing member or
be protected by a steel plate or sleeve.

Steel
Plate

COPYRIGHT 2004
Mike Holt
Enterprises, Inc.

Figure 334–8

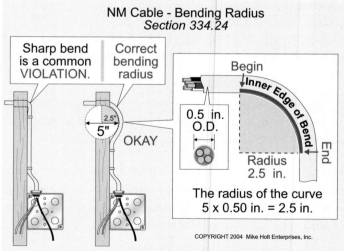

NM Cable - Bending Radius
Section 334.24

Sharp bend
is a common
VIOLATION.

Correct
bending
radius

Begin

Inner Edge of Bend

2.5"
5"
OKAY

0.5 in.
O.D.

Radius
2.5 in.

End

The radius of the curve
5 x 0.50 in. = 2.5 in.

COPYRIGHT 2004 Mike Holt Enterprises, Inc.

The radius of the inner edge of the curve must not
be less than 5 times the diameter of the cable.

Figure 334–10

• Along the Side of Framing Members [320.23(B)]. When Type NM cable is run on the side of rafters, studs, or floor joists, no protection is required if the cable is installed and supported so the nearest outside surface of the cable or raceway is at least 1¼ in. from the nearest edge of the framing member where nails or screws are likely to penetrate [300.4(D)].

334.24 Bends. When the cable is bent, it must not be damaged. The radius of the curve of the inner edge of any bend must not be less than five times the diameter of the cable. **Figure 334–10**

334.30 Secured or Supported. Staples, straps, cable ties, hangers, or similar fittings must secure Type NM so that the cable will not be damaged. Type NM cable must be secured within 12 in. of every box, cabinet, enclosure, or termination fitting, except as permitted by 314.17(C) Ex. or 312.5(C) Ex., and at intervals not exceeding 4½ ft. Two-wire (flat) NM cable is not permitted to be stapled on edge. **Figure 334–11**

Type NM cable installed in a raceway isn't required to be secured within the raceway. **Figure 334–12**

Author's Comment: Kind of like a no-brainer to me, but a few inspectors won't allow the sleeving of NM cables in a raceway because the cable can't be secured within 12 in. of the outlet box.

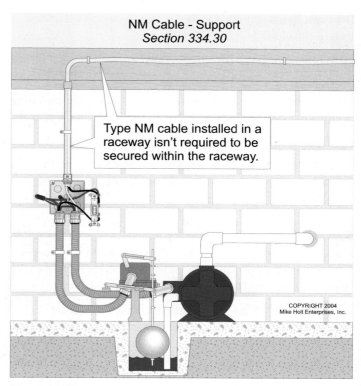

NM Cable - Support
Section 334.30

Type NM cable installed in a raceway isn't required to be secured within the raceway.

COPYRIGHT 2004
Mike Holt Enterprises, Inc.

Figure 334–12

(A) Horizontal Runs. Type NM cable installed horizontally in bored or punched holes in wood or metal framing members, or notches in wooden members is considered secured and supported, but the cable must be secured within 1 ft of termination. **Figure 334–13**

FPN: See 314.17(C) for support where nonmetallic boxes are used.

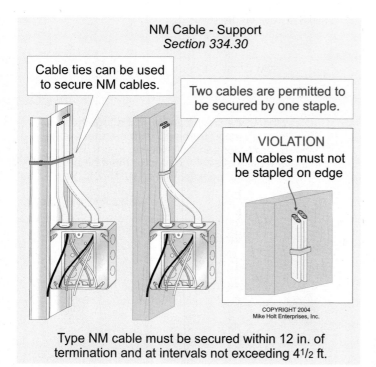

NM Cable - Support
Section 334.30

Cable ties can be used to secure NM cables.

Two cables are permitted to be secured by one staple.

VIOLATION
NM cables must not be stapled on edge

COPYRIGHT 2004
Mike Holt Enterprises, Inc.

Type NM cable must be secured within 12 in. of termination and at intervals not exceeding 4½ ft.

Figure 334–11

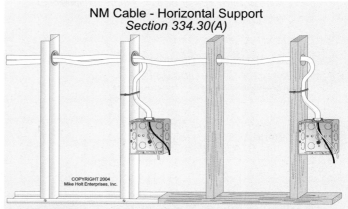

NM Cable - Horizontal Support
Section 334.30(A)

COPYRIGHT 2004
Mike Holt Enterprises, Inc.

NM cable run horizontally through framing members is considered supported if secured within 1 ft of termination.

Figure 334–13

(B) Unsupported. Type NM cable can be unsupported in the following situations:

(1) Where it's fished between concealed access points in finished buildings or structures and support is impracticable.

(2) Not more than 4½ ft of unsupported cable is permitted from the last point of support within an accessible ceiling for the connection of luminaires or equipment.

> **Author's Comment:** Type NM cable isn't permitted as a wiring method above accessible ceilings except in dwellings [334.12(A)(2)].

334.80 Conductor Ampacity. Conductor ampacity is calculated in accordance with 310.15 based on 90°C conductor insulation rating, provided the adjusted or corrected ampacity doesn't exceed that for a 60°C rated conductor.

> **Question:** What size Type NM cable is required to supply a 10 kW, 240V single-phase fixed space heater with a 3A blower motor? The terminals are rated 75°C. **Figure 334–14**
>
> (a) 2 AWG   (b) 4 AWG   (c) 6 AWG   (d) 8 AWG
>
> **Answer:** (b) 4 AWG
>
> Step 1. Determine the Total Load in Amperes
>
> > I = VA/E
> > I = 10,000W/240V + 3A
> > I = 44.67A
>
> Step 2. Size Conductor and Protection.
>
> > According to 424.3(B), the ungrounded conductors and overcurrent protection device for electric space-heating equipment must be sized no less than 125 percent of the total heating load. 44.67A x 1.25 = 56A
>
> > 4 AWG, rated 70A at 60°C would be required with a 60A protection device

> **Author's Comment:** AC, MC, and SE cables do not have a 60°C ampacity limitation; therefore, 6/2 AC, MC, or SE cable (rated 65A at 75°C) could be used. I know this doesn't make sense, but it's the *Code.*

Where NM cables containing two or more current-carrying conductors are bundled together and pass through wood framing that is to be fire- or draft-stopped using thermal insulation or sealing foam, the allowable ampacity of each conductor must be adjusted in accordance with Table 310.15(B)(2)(a). **Figure 334–15**

> **Author's Comments:**
>
> • This new requirement only applies to wood framing members, and has no effect unless you bundle more than nine current-carrying conductors together.
>
> • 15A Circuit. If we bundle three 14/2 and one 14/3 cable (nine current-carrying 14 THHN conductors), the ampacity for each conductor (25A at 90°C, Table 310.16) is adjusted by a 70 percent adjustment factor [Table 310.15(B)(2)(a)].
>   Adjusted Conductor Ampacity = 25A x 0.70
>   Adjusted Conductor Ampacity = 17.5A
>
> • 20A Circuit. If we bundle three 12/2 and one 12/3 cable (nine current-carrying 12 THHN conductors), the ampacity for each conductor (30A at 90°C, Table 310.16) is adjusted by a 70 percent adjustment factor [Table 310.15(B)(2)(a)].
>   Adjusted Conductor Ampacity = 30A x 0.70
>   Adjusted Conductor Ampacity = 21A
>
> • For more information on sizing conductors, see 110.14(C) and 310.10 in this textbook.

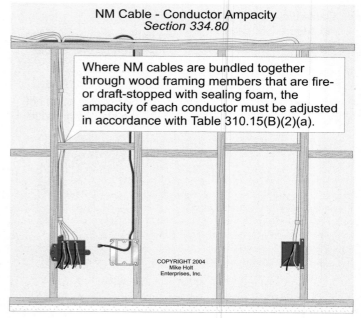

NM Cable - Conductor Ampacity
Section 334.80

Where NM cables are bundled together through wood framing members that are fire- or draft-stopped with sealing foam, the ampacity of each conductor must be adjusted in accordance with Table 310.15(B)(2)(a).

COPYRIGHT 2004
Mike Holt
Enterprises, Inc.

Figure 334–15

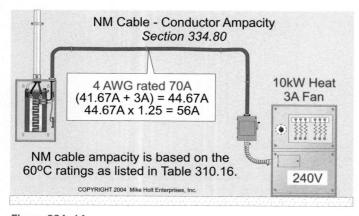

NM Cable - Conductor Ampacity
Section 334.80

4 AWG rated 70A
(41.67A + 3A) = 44.67A
44.67A x 1.25 = 56A

10kW Heat
3A Fan

240V

NM cable ampacity is based on the 60°C ratings as listed in Table 310.16.

COPYRIGHT 2004 Mike Holt Enterprises, Inc.

Figure 334–14

PART III. CONSTRUCTION SPECIFICATIONS

334.100 Construction. The outer cable sheath of non-metallic-sheathed cable must be a nonmetallic material.

334.104 Conductors. The conductors must be 14 AWG through 2 AWG copper, or 12 AWG through 2 AWG aluminum or copper-clad aluminum.

334.108 Equipment Grounding. The cable must have an insulated or bare conductor for equipment grounding (bonding) only.

334.112 Insulation. Conductor insulation must be rated 90°C (194°F).

1. Type NM cable must be _____.

   (a) marked   (b) approved   (c) identified   (d) listed

2. Type NM cable must not be used _____.

   (a) in commercial buildings   (b) in the air void of masonry block not subject to excessive moisture
   (c) for exposed work   (d) embedded in poured cement, concrete, or aggregate

3. Type NM cable installed through, or parallel to, framing members must be protected against physical damage in accordance with 300.4. Grommets or bushings for the protection of Type NM cable as required in 300.4(B)(1) must be _____ for the purpose, and they must remain in place.

   (a) marked   (b) approved   (c) identified   (d) listed

4. Flat two-conductor Type NM cables cannot be stapled on edge.

   (a) True   (b) False

5. Where more than two NM cables containing two or more current-carrying conductors are bundled together and pass through wood framing that is to be fire- or draft-stopped using thermal insulation or sealing foam, the allowable ampacity of each conductor is _____.

   (a) no more than 20A   (b) adjusted in accordance with 310.15(B)(2)(a)
   (c) limited to 30A   (d) calculated by an engineer

ARTICLE 336

Power and Control Tray Cable (Type TC)

Introduction

Tray Cable is primarily used for installation in cable trays and associated raceways. It can also be used where supported by a messenger wire. It has an outer sheath that is much stiffer than Type NM cable, particularly when cold, and can be difficult to remove. It is usually labeled as "sunlight resistant," although this is not part of the *Code* requirement for this cable type.

PART I. GENERAL

336.1 Scope.

Article 336 covers the use, installation, and construction specifications of power and control tray cable, Type TC.

336.2. Definition.

Tray Cable (Type TC). This is a factory assembly of two or more insulated conductors under a nonmetallic sheath for installation in cable trays, raceways, or where supported by a messenger wire.

PART II. INSTALLATION

336.10 Uses Permitted. Type TC cable can be used:

(1) For power, lighting, control, and signal circuits.

(2) In cable trays.

(3) In raceways.

(4) In outdoor locations supported by a messenger wire.

(5) For Class I circuits as permitted in Parts II and III of Article 725.

(6) For nonpower-limited fire alarm circuits if conductors comply with 760.27.

(7) In industrial establishments, where the conditions of maintenance and supervision ensure that only qualified persons service the installation, and where the cable is continuously supported and protected against physical damage using mechanical protection such as struts, angles, or channel. Type TC cable that complies with the crush and impact requirements of Type MC cable and is identified for such use with the marking Type TC-ER is permitted between a cable tray and the utilization equipment or device.

(8) Where installed in wet locations, Type TC cable must also be resistant to moisture and corrosive agents.

> **FPN:** See 310.10 for temperature limitation of conductors.

336.12 Uses Not Permitted. Type TC cable cannot be:

(1) Installed where it will be exposed to physical damage.

(2) Installed outside a raceway or cable tray system, except as permitted in 336.10(6).

(3) Used where exposed to direct rays of the sun, unless identified as sunlight resistant.

(4) Direct buried, unless identified for such use.

1. Type TC cable can be used _____.

   (a) for power and lighting circuits

   (b) in cable trays in hazardous locations

   (c) in Class 1 control circuits

   (d) all of these

2. Type TC cable must not be installed _____.

   (a) where it will be exposed to physical damage

   (b) outside of a raceway or cable tray system

   (c) direct buried unless identified for such use

   (d) all of these

338 Service-Entrance Cable (Types SE and USE)

Introduction

Service-entrance cable is a single conductor or multiconductor assembly with or without an overall moisture-resistant covering. This cable is used primarily for services not over 600V, but can also be used for feeders and branch circuits.

PART I. GENERAL

338.1 Scope. Article 338 covers the use, installation, and construction specifications of service-entrance cable, Types SE and USE.

338.2 Definition.

Service-Entrance Cable. Service-entrance cable is a single or multiconductor assembly with or without an overall covering used primarily for services not over 600V. **Figure 338–1**

Type SE. SE and SER cable has a flame-retardant, moisture-resistant covering and is permitted only in aboveground installations. This cable is permitted for branch circuits or feeders when installed according to 338.10(B).

> **Author's Comment:** SER cable is simply SE cable with an insulated neutral, resulting in three insulated conductors with an uninsulated ground. SER cable is round, whereas 2-wire SE cable is flat.

Type USE. USE cable is identified for underground use. Its covering is moisture resistant but not flame retardant and it isn't suitable for use within a premises [338.10(B)].

PART II. INSTALLATION

338.10 Uses Permitted.

(A) Service-Entrance Conductors. Service-entrance cable used as service-entrance conductors must be installed in accordance

Service-Entrance Cable
Section 338.2 Definition

Service-Entrance Cable. A single or multiconductor assembly with or without an overall covering used primarily for services not over 600V.

Aboveground

SE cable is permitted only in aboveground installations and is permitted for branch circuits or feeders when installed according to 338.10(B).

Underground Only

USE cable is identified for underground use. Its covering is moisture resistant but not flame retardant, and it isn't suitable for use within a premises.

COPYRIGHT 2004 Mike Holt Enterprises, Inc.

Figure 338–1

with Article 230. Type USE used for underground service laterals can emerge from the ground if protected in accordance with 300.5(D).

(B) Branch Circuits or Feeders.

(1) Insulated Conductors. Type SE cable is permitted in wiring systems where all of the circuit conductors of the cable are of the rubber-covered or thermoplastic type.

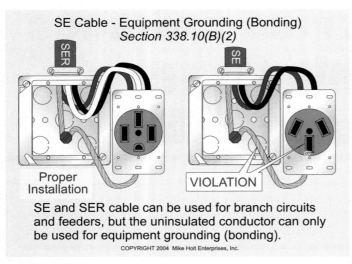

Figure 338–2

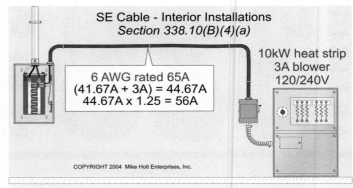

Figure 338–3

(2) Uninsulated Conductor. Type SE cable is permitted for use where the insulated conductors are used for circuit wiring and the uninsulated conductor is only used for equipment grounding (bonding). **Figure 338–2**

Exception: The uninsulated conductor can be used for the grounded neutral conductors when in compliance with 250.32, 250.140, and 225.30 through 225.40.

> **Author's Comment:** USE cable cannot be used for interior wiring because it doesn't have flame-retardant insulation.

(3) Temperature Limitations. Type SE cable used to supply appliances cannot be subjected to conductor temperatures exceeding its insulation rating.

(4) Installation Methods for Branch Circuits and Feeders. Type SE cable used for branch circuits or feeders must comply with (a) and (b).

(a) Interior Installations. Type SE cable used for interior branch circuit of feeder wiring must be installed in accordance with Parts I and II of Article 334, excluding 334.80.

> **Author's Comment:** This means that when Type SE cables are used for interior branch circuits or feeders, the 60°C ampacity limitation of Type NM cable contained in 334.80 doesn't apply and the conductors can be sized to the terminal ratings in accordance with 110.14(C). **Figure 338–3**

> *Question: What size Type SE cable is required to supply a 10 kW, 240V single-phase fixed space heater with a 3A blower motor? The terminals are rated 75°C.*
>
> *(a) 2 AWG   (b) 4 AWG   (c) 6 AWG   (d) 8 AWG*
>
> **Answer:** *(b) 4 AWG*

Step 1. Determine the Total Load in Amperes

$I = VA/E$

$I = 10,000W/240V + 3A$

$I = 44.67A$

Step 2. Size Conductor and Protection.

> *According to 424.3(B), the ungrounded conductors and overcurrent protection device for electric space-heating equipment must be sized no less than 125 percent of the total heating load. 44.67A x 1.25 = 56A*
>
> *6 AWG, rated 65A at 75°C would be required with a 60A protection device*

FPN: See 310.10 for temperature limitation of conductors.

(b) Exterior Installations. Service-entrance cable used for exterior branch circuits or feeders must be installed in accordance with Article 225.

> **Author's Comment:** Type SE cable isn't permitted in ducts, plenums, other environmental air spaces [300.22], or in patient care areas [517.13].

338.24 Bends. Bends in cable must be made so the protective coverings of the cable are not damaged, and the radius of the curve of the inner edge is at least five times the diameter of the cable.

(• Indicates that 75% or fewer exam takers get the question correct.)

1. Type _____ is a type of multiconductor cable permitted for use as an underground service-entrance cable.

   (a) SE   (b) NMC   (c) UF   (d) USE

2. Type SE cable is permitted for use as _____ in wiring systems where all of the circuit conductors of the cable are of the rubber-covered or thermoplastic type.

   (a) branch circuits   (b) feeders   (c) either a or b   (d) neither a nor b

3. Type SE cables are permitted for use for branch circuits or feeders where the insulated conductors are used for circuit wiring and the uninsulated conductor is used only for _____ purposes.

   (a) grounded neutral connection   (b) equipment grounding
   (c) remote control and signaling   (d) none of these

4. Bends made in USE and SE cable must be made so that the cable is not damaged. The radius of the curve of the inner edge of any bend during or after installation must not be less than _____ the diameter of the cable.

   (a) 5 times   (b) 7 times   (c) 10 times   (d) 125% of

5. •Type USE or SE cable must have a minimum of _____ conductors (including the uninsulated one) in order for one of the conductors to be uninsulated.

   (a) one   (b) two   (c) three   (d) four

Notes

ARTICLE 340

Underground Feeder and Branch-Circuit Cable (Type UF)

Introduction

UF cable is a moisture-, fungus-, and corrosion-resistant cable system suitable for direct burial in the earth.

PART I. GENERAL

340.1 Scope. Article 340 covers the use, installation, and construction specifications of underground feeder and branch-circuit cable, Type UF.

340.2 Definition.

Underground Feeder and Branch-Circuit Cable (Type UF). A factory assembly of one or more insulated conductors with an integral or an overall covering of nonmetallic material suitable for direct burial in the earth.

Author's Comments:

- UF cable is a moisture-, fungus-, and corrosion-resistant cable system suitable for direct burial in the earth. It comes in sizes 14 AWG through 4/0 AWG [340.104]. The covering of multiconductor UF cable is molded plastic that encapsulates the insulated conductors.

- Because the covering of UF cable encapsulates the insulated conductors, it's difficult to strip off the outer jacket to gain access to the conductors, but this encapsulated cover provides excellent corrosion protection. Be careful not to damage the conductor insulation or cut yourself when you remove the encapsulated outer cover.

340.6 Listing Requirements. Type UF cable must be listed.

PART II. INSTALLATION

340.10 Uses Permitted.

(1) Underground, in accordance with 300.5.

(2) As a single conductor in the same trench or raceway with the other circuit conductors.

(3) As interior or exterior wiring in wet, dry, or corrosive locations.

(4) As Type NM cable when installed in accordance with Article 334.

(5) For solar photovoltaic systems in accordance with 690.31.

(6) As single-conductor cables for nonheating leads for heating cables as provided in 424.43.

(7) Supported by cable trays.

340.12 Uses Not Permitted.

(1) Services [230.43].

(2) Commercial garages [511.3].

(3) Theaters [520.5].

(4) Motion picture studios [530.11].

(5) Storage battery rooms [Article 480].

(6) Hoistways [Article 620].

(7) Hazardous (classified) locations.

(8) Embedded in concrete.

(9) Exposed to direct sunlight unless identified.

(10) Where subject to physical damage. **Figure 340–1**

(11) Overhead messenger supported wiring.

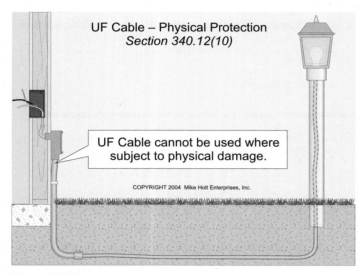

Figure 340–1

Author's Comment: Type UF cable isn't permitted in ducts, plenums, other environmental air spaces [300.22], or in patient care areas [517.13].

340.24 Bends. Bends in cables must be made so the protective coverings of the cable are not damaged, and the radius of the curve of the inner edge must not be less than five times the diameter of the cable.

340.80 Ampacity. The ampacity of conductors contained in UF cable is based on 60°C insulation rating listed in Table 310.16.

340.112 Insulation. The conductors of Type UF cable must be one of the moisture-resistant types listed in Table 310.13 that is suitable for branch-circuit wiring. Where installed as a substitute wiring method for NM cable, the conductor insulation must be rated 90°C (194°F).

1. Underground feeder and branch circuit (Type UF) cable is allowed to be used as service entrance cable.

   (a) True   (b) False

2. Type UF cable must not be used _____.

   (a) in any hazardous (classified) location
   (b) embedded in poured cement, concrete, or aggregate
   (c) where exposed to direct rays of the sun, unless identified as sunlight-resistant
   (d) all of these

3. Bends made in UF cable must be made so that the cable is not damaged. The radius of the curve of the inner edge of any bend during or after installation must not be less than _____ the diameter of the cable.

   (a) 5 times   (b) 7 times   (c) 10 times   (d) 125% of

4. The ampacity of Type UF cable must be that of _____ in accordance with 310.15.

   (a) 90°C conductors   (b) 75°C conductors   (c) 60°C conductors   (d) none of these

5. The overall covering of Type UF cable must be _____.

   (a) flame retardant   (b) moisture, fungus, and corrosion resistant
   (c) suitable for direct burial in the earth   (d) all of these

Notes

342 Intermediate Metal Conduit (Type IMC)

Introduction

Intermediate metal conduit (IMC) is a circular metal raceway with an outside diameter equal to that of rigid metal conduit. The wall thickness of intermediate metal conduit is less than that of rigid metal conduit (RMC), so it has a greater interior cross-sectional area. Intermediate metal conduit is lighter and less expensive than rigid metal conduit, but it can be used in all of the same locations as rigid metal conduit. Intermediate metal conduit also uses a different steel alloy that makes it stronger than rigid metal conduit, even though the walls are thinner. IMC is manufactured in both galvanized steel and aluminum; the steel type is much more common.

PART I. GENERAL

342.1 Scope. Article 342 covers the use, installation, and construction specifications of intermediate metal conduit and associated fittings.

342.2 Definition.

Intermediate Metal Conduit (Type IMC). This is a listed steel raceway of circular cross section that can be threaded with integral or associated couplings. It is listed for the installation of electrical conductors, and is used with listed fittings to provide electrical continuity

> **Author's Comment:** The wall thickness of intermediate metal conduit is less than rigid metal conduit, so it has a greater interior cross-sectional area and can hold more conductors. In addition, it's lighter and is permitted in all of the same locations as rigid metal conduit. The type of steel from which intermediate metal conduit is manufactured, the process by which it's made, and the corrosion protection applied are equal, or superior, to that of rigid metal conduit. Impact, crush, and hydrostatic tests on both intermediate metal conduit and rigid metal conduit show intermediate metal conduit to be equal to, or better than, rigid conduit.

342.6 Listing Requirements. Intermediate metal conduit and its associated fitting such as elbows and couplings must be listed.

PART II. INSTALLATION

342.10 Uses Permitted.

(A) All Atmospheric Conditions and Occupancies. Intermediate metal conduit is permitted in all atmospheric conditions and occupancies.

(B) Corrosion Environments. Intermediate metal conduit, elbows, couplings, and fittings can be installed in concrete, in direct contact with the earth, or in areas subject to severe corrosive influences where provided with corrosion protection and judged suitable for the condition in accordance with 300.6.

(D) Wet Locations. All support fittings, such as screws, straps, etc., installed in a wet location must be made of corrosion-resistant material or be protected by corrosion-resistant coatings in accordance with 300.6.

> **CAUTION:** *Supplementary coatings for corrosion protection have not been investigated by a product testing and listing agency, and these coatings are known to cause cancer in laboratory animals. I know of a case where an electrician was taken to the hospital for lead poisoning after using a supplemental coating product (asphaltic paint) in a poorly ventilated area. As with all products, be sure to read and follow all product instructions, particularly when petroleum-based chemicals (volatile organic compounds) may be in the material.*

342.14 Dissimilar Metals.

Where practical, contact with dissimilar metals should be avoided to prevent the deterioration of the metal because of galvanic action. However, aluminum fittings and enclosures are permitted with steel intermediate metal conduit.

342.20 Trade Size.

(A) Minimum. Intermediate metal conduit smaller than trade size ½ cannot be used.

(B) Maximum. Intermediate metal conduit larger than trade size 4 cannot be used.

342.22 Number of Conductors.

Raceways must be large enough to permit the installation and removal of conductors without damaging the conductor insulation. When all conductors in a raceway are the same size and insulation, the number of conductors permitted can be found in Annex C for the raceway type.

> *Question: How many 10 THHN conductors can be installed in trade size 1 IMC?*
>
> *(a) 12   (b) 14   (c) 16   (d) 18*
>
> *Answer: (d) 18 conductors, Annex C, Table C4*

> **Author's Comment:** See 300.17 for additional examples on how to size raceways when conductors aren't all the same size.

Cables can be installed in intermediate metal conduit where not prohibited by the cable article. The number of cables cannot exceed the allowable percentage fill specified in Table 1, Chapter 9.

342.24 Bends.

Raceway bends cannot be made in any manner that would damage the raceway or significantly change its internal diameter (no kinks). The radius of the curve of the inner edge of any field bend must not be less than shown in Table 2, Chapter 9.

> **Author's Comment:** This is usually not a problem because benders are made to comply with this table. However, when using a hickey bender (short-radius bender), be careful not to over-bend the raceway.

342.26 Number of Bends (360°).

To reduce the stress and friction on conductor insulation, the maximum number of bends (including offsets) between pull points cannot exceed 360°. Figure 342–1

> **Author's Comment:** There is no maximum distance between pull boxes because this is a design issue, not a safety issue.

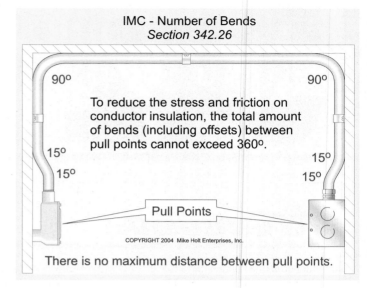

IMC - Number of Bends
Section 342.26

90°   90°

To reduce the stress and friction on conductor insulation, the total amount of bends (including offsets) between pull points cannot exceed 360°.

15°   15°
15°   15°

Pull Points

COPYRIGHT 2004 Mike Holt Enterprises, Inc.

There is no maximum distance between pull points.

Figure 342–1

342.28 Reaming.

When the raceway is cut in the field, reaming is required to remove the burrs and rough edges.

> **Author's Comment:** It's an accepted practice to ream small raceways with a screwdriver or the backside of pliers. However, when the raceway is cut with a three-wheel pipe cutter, a reaming tool is required to remove the sharp edge of the indented raceway. In addition, when conduit is threaded in the field, the threads must be coated with an electrically conductive, corrosion-resistant compound (cold zinc) in accordance with 300.6(A).

342.30 Secured and Supported.

Intermediate metal conduit must be installed as a complete system in accordance with 300.18 [300.10 and 300.12], and it must be securely fastened in place and supported in accordance with (A) and (B).

(A) Securely Fastened. Intermediate metal conduit must be securely fastened within 3 ft of every box, cabinet, or termination fitting. Figure 342–2

> **Author's Comment:** Support is required within 3 ft of terminations, not within 3 ft of each coupling!

When structural members do not permit the raceway to be secured within 3 ft of a box or termination fitting, the raceway must be secured within 5 ft of the termination. Figure 342–3

(B) Supports.

(1) General. Intermediate metal conduit must be supported at intervals not exceeding 10 ft.

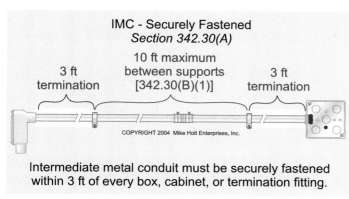

Figure 342-2

(2) Straight Horizontal Runs. Straight horizontal runs made with threaded couplings can be supported in accordance with the distances listed in Table 344.30(B)(2).

Table 344.30(B)(2)

Trade Size	Support Spacing
½ - ¾	10 ft
1	12 ft*
1¼ - 1½	14 ft
2 - 2½	16 ft
3 and larger	20 ft

*Figure 342-4

(3) Vertical Risers. Exposed vertical risers for fixed equipment can be supported at intervals not exceeding 20 ft, if the conduit is made up with threaded couplings, firmly supported and securely fastened at the top and bottom of the riser, and no other means of support is available. **Figure 342-5**

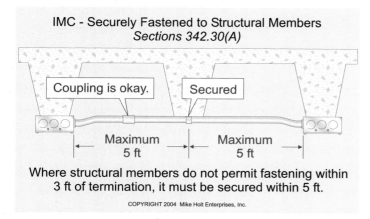

Figure 342-3

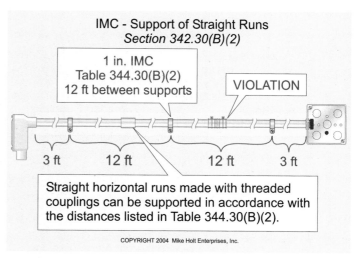

Figure 342-4

(4) Horizontal Runs. Conduits installed horizontally in bored or punched holes in wood or metal framing members, or notches in wooden members are considered supported, but the raceway must be secured within 3 ft of termination.

> **Author's Comment:** IMC must be provided with expansion fittings where necessary to compensate for thermal expansion and contraction [300.7(B)]. The expansion characteristics for metal raceways are determined by multiplying the values from Table 352.44(A) by 0.20.

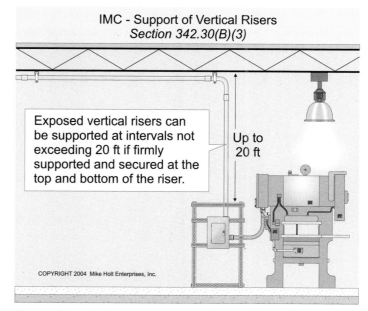

Figure 342-5

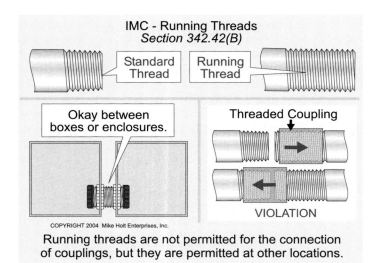

Running threads are not permitted for the connection of couplings, but they are permitted at other locations.

Figure 342–6

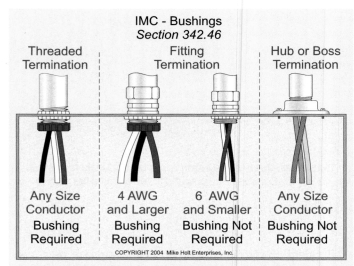

Conductors 4 AWG and larger must be protected by a fitting that provides a smooth, rounded, insulating surface, such as an insulating bushing [300.4(F)].

Figure 342–7

342.42 Couplings and Connectors.

(A) Installation. Threadless couplings and connectors must be made up tight to maintain an effective ground-fault current path to safely conduct fault current in accordance with 250.4(A)(5), 250.96(A), and 300.10.

> **Author's Comment:** Loose locknuts have been found to burn clear before a fault was cleared because loose termination fittings increase the impedance of the fault-current path.

Where buried in masonry or concrete, threadless fittings must be the concrete-tight type. Where installed in wet locations, fittings must be listed for use in wet locations in accordance with 314.15(A).

Threadless couplings and connectors cannot be used on threaded conduit ends unless listed for the purpose.

(B) Running Threads. Running threads are not permitted for the connection of couplings, but they are permitted at other locations. Figure 342–6

342.46 Bushings. To protect conductors of any size from abrasion, a metal or plastic bushing must be installed on conduit termination threads, unless the design of the box, fitting, or enclosure is such as to afford equivalent protection in accordance with 300.4(F). Figure 342–7

> **Author's Comment:** Conductors 4 AWG and larger that enter an enclosure must be protected from abrasion during and after installation by a fitting that provides a smooth, rounded insulating surface, such as an insulating bushing, unless the design of the box, fitting, or enclosure provides equivalent protection in accordance with 300.4(F).

1. Materials such as straps, bolts, screws, etc. that are associated with the installation of IMC in wet locations are required to be _____.

    (a) weatherproof   (b) weathertight   (c) corrosion-resistant   (d) none of these

2. One-inch IMC raceway containing three or more conductors must not exceed _____ percent conductor fill.

    (a) 53   (b) 31   (c) 40   (d) 60

3. IMC must be firmly fastened within _____ of each outlet box, junction box, device box, fitting, cabinet, or other conduit termination.

    (a) 12 in.   (b) 18 in.   (c) 2 ft   (d) 3 ft

4. Exposed vertical risers of IMC for industrial machinery or fixed equipment can be supported at intervals not exceeding _____ if the conduit is made up with threaded couplings, firmly supported at the top and bottom of the riser, and no other means of support is available.

    (a) 10 ft   (b) 12 ft   (c) 15 ft   (d) 20 ft

5. Threadless couplings approved for use with IMC in wet locations must be _____.

    (a) rainproof   (b) listed for wet locations   (c) moistureproof   (d) concrete-tight

Notes

ARTICLE 344

Rigid Metal Conduit (Type RMC)

Introduction

Rigid metal conduit (RMC), commonly called "rigid," has long been the standard raceway for providing protection from physical impact and from difficult environments. The outside diameter of rigid metal conduit is the same as intermediate metal conduit. However, the wall thickness of rigid metal conduit is greater than intermediate metal conduit, therefore it has a smaller interior cross-sectional area. Rigid metal conduit is heavier and more expensive than intermediate metal conduit, and it can be used in any location. RMC is manufactured in both galvanized steel and aluminum; the steel type is much more common.

PART I. GENERAL

344.1 Scope. Article 344 covers the use, installation, and construction specifications of rigid metal conduit and associated fittings.

344.2 Definition.

Rigid Metal Conduit (Type RMC). This is a listed metal raceway of circular cross section with integral or associated couplings, listed for the installation of electrical conductors and used with listed fittings to provide electrical continuity.

Author's Comment: When the mechanical and physical characteristics of rigid metal conduit are desired and a corrosive environment is anticipated, a PVC-coated raceway system is commonly used. This type of raceway is frequently used in the petrochemical industry; the common trade name is "Plasti-bond®." The benefits of the improved corrosion protection can be achieved only when the system is properly installed. All joints must be sealed in accordance with the manufacturer's instructions, and the coating must not be damaged with tools such as benders, pliers, and pipe wrenches. Couplings are available with an extended skirt that can be properly sealed after installation.

344.6 Listing Requirements. Rigid metal conduit, elbows, couplings, and associated fittings must be listed.

PART II. INSTALLATION

344.10 Uses Permitted.

(A) All Atmospheric Conditions and Occupancies. Rigid metal conduit is permitted in all atmospheric conditions and occupancies.

(B) Corrosive Environments. Rigid metal conduit, elbows, couplings, and fittings can be installed in concrete, in direct contact with the earth, or in areas subject to severe corrosive influences where protected by corrosion protection and judged suitable for the condition in accordance with 300.6.

(C) Wet Locations. All support fittings, such as screws, straps, etc., installed in a wet location must be made of corrosion-resistant material or protected by corrosion-resistant coatings in accordance with 300.6.

> **CAUTION:** *Supplementary coatings (asphaltic paint) for corrosion protection have not been investigated by a product testing and listing agency, and these coatings are known to cause cancer in laboratory animals.*

344.14 Dissimilar Metals. Where practical, contact with dissimilar metals should be avoided to prevent the deterioration of the metal because of galvanic action. However, aluminum fittings and enclosures are permitted with rigid metal conduit.

344.20 Trade Size.

(A) Minimum. Rigid metal conduit smaller than trade size ½ cannot be used.

(B) Maximum. Rigid metal conduit larger than trade size 6 cannot be used.

344.22 Number of Conductors.
Raceways must be large enough to permit the installation and removal of conductors without damaging the conductors' insulation. When all conductors in a raceway are the same size and insulation, the number of conductors permitted can be found in Annex C for the raceway type.

> *Question: How many 8 THHN conductors can be installed in trade size 1½ RMC?*
>
> *(a) 16   (b) 18   (c) 20   (d) 22*
>
> *Answer: 22 conductors, Annex C, Table C8*

Author's Comment: See 300.17 for additional examples on how to size raceways when conductors aren't all the same size.

Cables can be installed in rigid metal conduit where not prohibited by the cable article. The number of cables cannot exceed the allowable percentage fill specified in Table 1, Chapter 9.

344.24 Bends.
Raceway bends cannot be made in any manner that would damage the raceway or significantly change its internal diameter (no kinks). The radius of the curve of the inner edge of any field bend must not be less than shown in Table 2, Chapter 9.

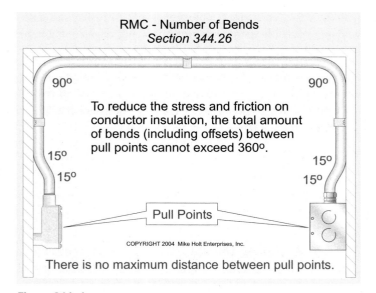

RMC - Number of Bends
Section 344.26

90°   90°

To reduce the stress and friction on conductor insulation, the total amount of bends (including offsets) between pull points cannot exceed 360°.

15°   15°
15°   15°

Pull Points

COPYRIGHT 2004 Mike Holt Enterprises, Inc.

There is no maximum distance between pull points.

Figure 344–1

Author's Comment: This is usually not a problem because benders are made to comply with this table. However, when using a hickey bender (short-radius bender), be careful not to over-bend the raceway.

344.26 Number of Bends (360°).
To reduce the stress and friction on the conductors' insulation, the maximum number of bends (including offsets) between pull points cannot exceed 360°. **Figure 344–1**

Author's Comment: The *NEC* doesn't specify a maximum distance between pull points.

344.28 Reaming.
When the raceway is cut in the field, reaming is required to remove the burrs and rough edges.

Author's Comment: It's an accepted practice to ream small raceways with a screwdriver or the backside of pliers. However, when the raceway is cut with a three-wheel pipe cutter, a reaming tool is required to remove the sharp edge of the indented raceway. In addition, when conduit is threaded in the field, the threads must be coated with an electrically conductive, corrosion-resistant compound (cold zinc) in accordance with 300.6(A).

344.30 Secured and Supported.
Rigid metal conduit must be installed as a complete system in accordance with 300.18 [300.10 and 300.12], and it must be securely fastened in place and supported in accordance with (A) or (B).

(A) Securely Fastened. Rigid metal conduit must be securely fastened within 3 ft of every box, cabinet, or termination fitting. **Figure 344–2**

When structural members do not permit the raceway to be secured within 3 ft of a box or termination fitting, the raceway must be secured within 5 ft of termination. **Figure 344–3**

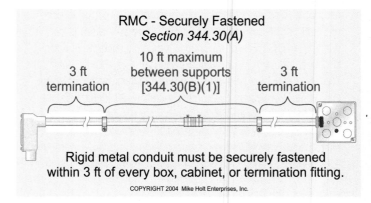

RMC - Securely Fastened
Section 344.30(A)

10 ft maximum
3 ft   between supports   3 ft
termination   [344.30(B)(1)]   termination

Rigid metal conduit must be securely fastened within 3 ft of every box, cabinet, or termination fitting.

COPYRIGHT 2004 Mike Holt Enterprises, Inc.

Figure 344–2

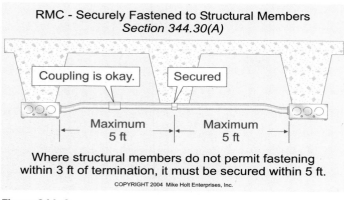

Figure 344–3

Author's Comment: Support is required within 3 ft of terminations, not within 3 ft of each coupling!

(B) Supports.

(1) General. Rigid metal conduit must be supported at intervals not exceeding 10 ft.

(2) Straight Horizontal Runs. Straight horizontal runs made with threaded couplings can be supported in accordance with the distances listed in Table 344.30(B)(2).

Table 344.30(B)(2)

Trade Size	Support Spacing
½ - ¾	10 ft
1	12 ft*
1¼ - 1½	14 ft
2 - 2½	16 ft
3 and larger	20 ft

*Figure 344–4

(3) Vertical Risers. Exposed vertical risers for fixed equipment can be supported at intervals not exceeding 20 ft, if the conduit is made up with threaded couplings, firmly supported and securely fastened at the top and bottom of the riser, and no other means of support is available. Figure 344–5

(4) Horizontal Runs. Conduits installed horizontally in bored or punched holes in wood or metal framing members, or notches in wooden members are considered supported, but the raceway must be secured within 3 ft of termination.

Author's Comment: RMC must be provided with expansion fittings where necessary to compensate for thermal expansion and contraction [300.7(B)]. The expansion characteristics for metal raceways are determined by multiplying the values from Table 352.44(A) by 0.20.

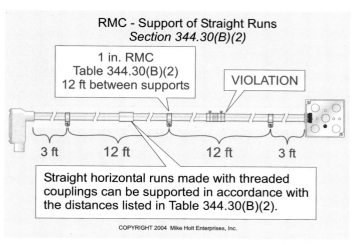

Figure 344–4

344.42 Couplings and Connectors.

(A) Installation. Threadless couplings and connectors must be made up tight to maintain an effective ground-fault current path to safely conduct fault current in accordance with 250.4(A)(5), 250.96(A), and 300.10.

Author's Comment: Loose locknuts have been found to burn clear before a fault was cleared because loose connections increase the impedance of the fault-current path.

Where buried in masonry or concrete, threadless fittings must be the concrete-tight type. Where installed in wet locations, fittings must be listed for use in wet locations in accordance with 314.15(A).

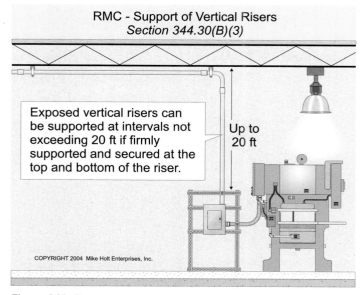

Figure 344–5

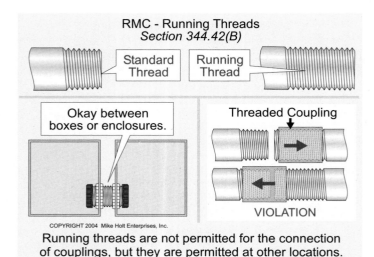

RMC - Running Threads
Section 344.42(B)

Standard Thread | Running Thread

Okay between boxes or enclosures.

Threaded Coupling

VIOLATION

COPYRIGHT 2004 Mike Holt Enterprises, Inc.

Running threads are not permitted for the connection of couplings, but they are permitted at other locations.

Figure 344–6

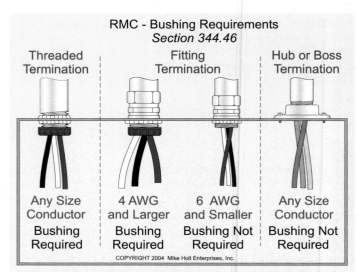

RMC - Bushing Requirements
Section 344.46

Threaded Termination | Fitting Termination | Hub or Boss Termination

Any Size Conductor | 4 AWG and Larger | 6 AWG and Smaller | Any Size Conductor
Bushing Required | Bushing Required | Bushing Not Required | Bushing Not Required

COPYRIGHT 2004 Mike Holt Enterprises, Inc.

Conductors 4 AWG and larger must be protected by a fitting that provides a smooth, rounded, insulating surface, such as an insulating bushing [300.4(F)].

Figure 344–7

Threadless couplings and connectors cannot be used on threaded conduit ends unless listed for the purpose.

(B) Running Threads. Running threads are not permitted for the connection of couplings, but they are permitted at other locations. Figure 344–6

344.46 Bushings. To protect conductors of any size from abrasion, a metal or plastic bushing must be installed on conduit threads at terminations in accordance with 300.4(F), unless the design of the box, fitting, or enclosure is such as to afford equivalent protection. Figure 344–7

> **Author's Comment:** Conductors 4 AWG and larger that enter an enclosure must be protected from abrasion during and after installation by a fitting that provides a smooth, rounded insulating surface, such as an insulating bushing, unless the design of the box, fitting, or enclosure provides equivalent protection in accordance with 300.4(F).

Article 344 Questions

1. RMC can be installed in or under cinder fill subject to permanent moisture, when protected on all sides by a layer of noncinder concrete not less than _____ thick.

    (a) 2 in.   (b) 4 in.   (c) 6 in.   (d) 18 in.

2. The minimum radius of a field bend on 1¼ in. RMC is _____.

    (a) 7 in.   (b) 8 in.   (c) 14 in.   (d) 10 in.

3. Straight runs of 1 in. RMC using threaded couplings may be secured at intervals not exceeding _____.

    (a) 5 ft   (b) 10 ft   (c) 12 ft   (d) 14 ft

4. Threadless couplings and connectors used with RMC and installed in wet locations must be _____.

    (a) listed for wet locations   (b) listed for damp locations   (c) nonabsorbent   (d) weatherproof

5. Where rigid metal conduit enters a box, fitting, or other enclosure, a bushing must be provided to protect the wire from abrasion unless the design of the box, fitting, or enclosure is such as to afford equivalent protection.

    (a) True   (b) False

Notes

Mike Holt Enterprises, Inc. • www.NECcode.com • 1.888.NEC.Code

348 Flexible Metal Conduit (Type FMC)

Introduction

Flexible metal conduit (FMC), commonly called Greenfield or "flex," is a raceway of an interlocked metal strip of either steel or aluminum. It's primarily used for the final 6 ft or less of raceways between a more rigid raceway system and equipment that moves, shakes, or vibrates. Examples of such equipment include pump motors and industrial machinery.

PART I. GENERAL

348.1 Scope. Article 348 covers the use, installation, and construction specifications for flexible metal conduit and associated fittings.

348.2 Definition.

Flexible Metal Conduit (Type FMC). This is a raceway of circular cross section made of a helically wound, formed, interlocked metal strip of either steel or aluminum.

348.6 Listing Requirements. Flexible metal conduit with associated fittings must be listed.

PART II. INSTALLATION

348.10 Uses Permitted. Flexible metal conduit is permitted exposed or concealed.

348.12 Uses Not Permitted.

(1) In wet locations.

(2) In hoistways, other than as permitted in 620.21(A)(1).

(3) In storage battery rooms.

(4) In hazardous (classified) locations, except as permitted by 501.10(B).

(5) Exposed to material having a deteriorating effect on the installed conductors.

(6) Underground or embedded in poured concrete.

(7) Where subject to physical damage.

348.20 Trade Size.

(A) Minimum. Flexible metal conduit smaller than trade size ½ cannot be used, except that trade size ⅜ can be used for the following applications: Figure 348–1

(1) For enclosing the leads of motors,

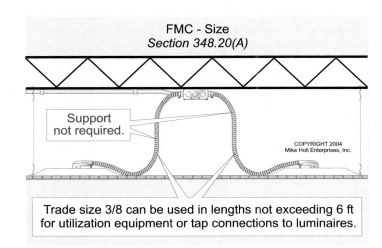

FMC - Size
Section 348.20(A)

Support not required.

COPYRIGHT 2004
Mike Holt Enterprises, Inc.

Trade size 3/8 can be used in lengths not exceeding 6 ft for utilization equipment or tap connections to luminaires.

Figure 348–1

(2) Not exceeding 6 ft in length:

    a. For utilization equipment,

    b. As part of a listed assembly, or

    c. For tap connections to luminaires as permitted by 410.67(C).

(3) In manufactured wiring systems, 604.6(A).

(4) In hoistways, 620.21(A)(1).

(5) As part of a listed luminaire assembly, 410.77(C).

(B) Maximum. Flexible metal conduit larger than trade size 4 cannot be used.

348.22 Number of Conductors.

Trade Size ½ and Larger. Flexible metal conduit must be large enough to permit the installation and removal of conductors without damaging the conductors' insulation. When all conductors in a raceway are the same size and insulation, the number of conductors permitted can be found in Annex C for the raceway type.

> **Question:** *How many 6 THHN conductors can be installed in trade size 1 FMC?*
>
> *(a) 2   (b) 4   (c) 6   (d) 8*
>
> **Answer:** *(c) 6 conductors*
>   *Annex C, Table C3*

> **Author's Comment:** See 300.17 for additional examples on how to size raceways when conductors aren't all the same size.

Trade Size ⅜. The number and size of conductors in trade size ⅜ flexible metal conduit must comply with Table 348.22.

> **Question:** *How many 12 THHN conductors can be installed in trade size ⅜ FMC that uses outside fitting?*
>
> *(a) 1   (b) 3   (c) 5   (d) 7*
>
> **Answer:** *(b) 3 conductors, Table 348.22*

One covered or bare equipment grounding (bonding) conductor of the same size is permitted with the circuit conductors. See the * note at the bottom of Table 348.22.

Cables can be installed in flexible metal conduit where not prohibited by the cable's article. The number of cables cannot exceed the allowable percentage fill specified in Table 1, Chapter 9.

348.24 Bends.
Bends must be made so the conduit will not be damaged and its internal diameter will not be effectively reduced. The radius of the curve of the inner edge of any field bend must not be less than shown in Table 2, Chapter 9 using the column "Other Bends."

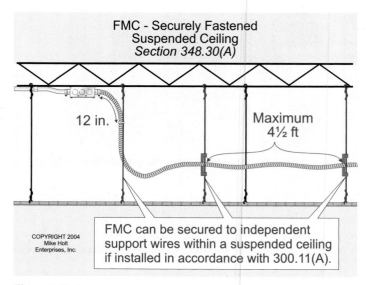

FMC - Securely Fastened Suspended Ceiling
Section 348.30(A)

12 in.

Maximum 4½ ft

COPYRIGHT 2004 Mike Holt Enterprises, Inc.

FMC can be secured to independent support wires within a suspended ceiling if installed in accordance with 300.11(A).

Figure 348–2

348.26 Number of Bends (360°).
Bends between pull points such as boxes and conduit bodies cannot exceed 360°, and the raceway must be installed so the conductors can be pulled or removed without damaging the conductor insulation.

348.28 Trimming.
The cut ends of flexible metal conduit must be trimmed to remove the rough edges, but this isn't necessary where fittings are threaded into the convolutions.

348.30 Secured and Supported.
Flexible metal conduit must be installed as a complete system [300.10, 300.12, and 300.18(A)], and it must be securely fastened in place and supported in accordance with (A) or (B).

(A) Securely Fastened. Flexible metal conduit must be securely fastened by a means approved by the authority having jurisdiction within 1 ft of termination and at intervals not exceeding 4½ ft. **Figure 348–2**

Exception 1: Where fished, flexible metal conduit is not required to be secured.

Exception 2: At terminals where flexibility is required, unsecured lengths cannot exceed: **Figure 348–3**

    (1) 3 ft for trade sizes ½ through 1¼

    (2) 4 ft for trade sizes 1½ through 2

    (3) 5 ft for trade size 2½ and larger

Exception 4: FMC to a luminaire or electrical equipment within an accessible ceiling is permitted to be unsupported for not more than 6 ft from the last point where the raceway is securely fastened. **Figure 348–4**

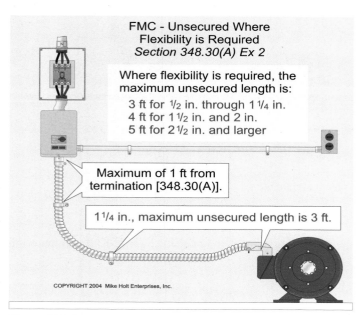

FMC - Unsecured Where
Flexibility is Required
Section 348.30(A) Ex 2

Where flexibility is required, the
maximum unsecured length is:

3 ft for ½ in. through 1 ¼ in.
4 ft for 1 ½ in. and 2 in.
5 ft for 2 ½ in. and larger

Maximum of 1 ft from
termination [348.30(A)].

1 ¼ in., maximum unsecured length is 3 ft.

COPYRIGHT 2004 Mike Holt Enterprises, Inc.

Figure 348–3

(B) Horizontal Runs. Flexible metal conduit installed horizontally in bored or punched holes in wood or metal framing members, or notches in wooden members is considered supported, but the raceway must be secured within 1 ft of termination. **Figure 348–5**

348.42 Fittings. Angle connectors cannot be installed in concealed locations.

Author's Comment: Conductors 4 AWG and larger that enter an enclosure must be protected from abrasion during and after installation by a fitting that provides a smooth, rounded insulating surface, such as an insulating bushing, unless the design of the box, fitting, or enclosure provides equivalent protection in accordance with 300.4(F).

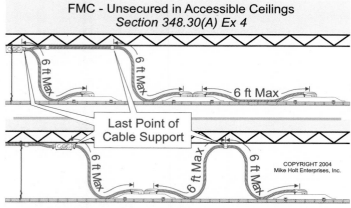

FMC - Unsecured in Accessible Ceilings
Section 348.30(A) Ex 4

6 ft Max
6 ft Max
6 ft Max
Last Point of
Cable Support
6 ft Max
6 ft Max
6 ft Max

COPYRIGHT 2004
Mike Holt Enterprises, Inc.

Lengths not exceeding 6 ft can be unsecured within an
accessible ceiling for luminaire(s) or other equipment.

Figure 348–4

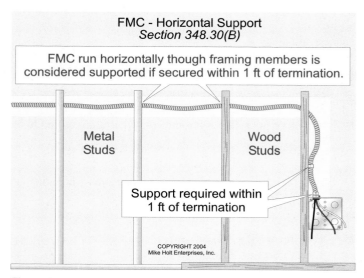

FMC - Horizontal Support
Section 348.30(B)

FMC run horizontally though framing members is
considered supported if secured within 1 ft of termination.

Metal
Studs

Wood
Studs

Support required within
1 ft of termination

COPYRIGHT 2004
Mike Holt Enterprises, Inc.

Figure 348–5

348.60 Bonding. Where flexibility is required, an equipment grounding (bonding) conductor must be installed with the circuit conductors [300.3(B)].

Where flexibility is not required, the metal armor of flexible metal conduit is permitted to serve as an equipment grounding (bonding) conductor if the circuit conductors contained in the raceway are protected by overcurrent devices rated 20A or less and the combined length of the flexible metal raceway in the same ground return path doesn't exceed 6 ft [250.118(5)].

Where an equipment grounding (bonding) conductor is required, it must be installed with the circuit conductors inside or outside the raceway. Where an equipment bonding jumper is installed outside a raceway, the length of the equipment bonding jumper cannot exceed 6 ft and it must be routed with the raceway or enclosure [250.102(E), 250.134(B), and 300.3(B)].

1. Which of the following conductor types are required to be used when FMC is installed in a wet location?

    (a) THWN   (b) XHHW   (c) THW   (d) any of these

2. FMC can be installed exposed or concealed where not subject to physical damage.

    (a) True   (b) False

3. How many 12 AWG XHHW conductors, not counting a bare ground wire, are allowed in trade size ⅜ in. FMC (maximum of 6 ft) with outside fittings?

    (a) 4   (b) 3   (c) 2   (d) 5

4. Bends in flexible metal conduit _____ between pull points.

    (a) must not be made   (b) nned not be limited (in degrees)
    (c) must not exceed more than 360 degrees   (d) must not exceed 180 degrees

5. FMC must be supported and secured _____.

    (a) at intervals not exceeding 4½ ft   (b) within 8 in. on each side of a box where fished
    (c) where fished   (d) at intervals not exceeding 6 ft at motor terminals

350

Liquidtight Flexible Metal Conduit (Type LFMC)

Introduction

Liquidtight flexible metal conduit (LFMC), with its associated connectors and fittings, is a flexible raceway system commonly used for connections to equipment that vibrate or are required to move occasionally. Liquidtight flexible metal conduit is commonly called Sealtight® or "liquidtight." Liquidtight flexible metal conduit is of similar construction to flexible metal conduit, but has an outer liquidtight thermoplastic covering. It has the same primary purpose as flexible metal conduit, but it also provides protection from moisture and corrosive effects.

PART I. GENERAL

350.1 Scope. Article 350 covers the use, installation, and construction specifications of liquidtight flexible metal conduit and associated fittings.

350.2 Definition.

Liquidtight Flexible Metal Conduit (Type LFMC). This is a raceway of circular cross section having an outer liquidtight, nonmetallic, sunlight-resistant jacket over an inner flexible metal core with associated connectors and fittings, and listed for the installation of electric conductors.

350.6 Listing Requirement. Liquidtight flexible metal conduit and its associated fittings must be listed. Figure 350–1

PART II. INSTALLATION

350.10 Uses Permitted.

(A) Permitted Use. Listed liquidtight flexible metal conduit is permitted either exposed or concealed at any of the following locations:

(1) Where flexibility or protection from liquids, vapors, or solids is required.

LFMC - Listing Required
Section 350.6

Listed Liquidtight

LFMC and associated fittings must be listed for the purpose.

COPYRIGHT 2002
Mike Holt Enterprises, Inc.

Figure 350–1

(2) In hazardous (classified) locations, as permitted in 501.10(B), 502.10(A)(2), 502.10(B)(2), or 503.10(A)(2).

(3) For direct burial, if listed and marked for this purpose.

350.12 Uses Not Permitted.

(1) Where subject to physical damage.

(2) Where the combination of the ambient and conductor operating temperatures exceeds the rating of the raceway.

350.20 Trade Size.

(A) Minimum. Liquidtight flexible metal conduit smaller than trade size ½ cannot be used.

Exception: Liquidtight flexible metal conduit can be smaller than trade size ½ if installed in accordance with 348.20(A).

(1) For enclosing the leads of motors.

(2) Not exceeding 6 ft in length:

a. For utilization equipment,

b. As part of a listed assembly, or

c. For tap connections to luminaires as permitted by 410.67(C).

(3) In manufactured wiring systems, 604.6(A).

(4) In hoistways, 620.21(A)(1).

(5) As part of a listed assembly to connect wired luminaire sections, 410.77(C).

(B) Maximum. Liquidtight flexible metal conduit larger than trade size 4 cannot be used.

350.22 Number of Conductors.

(A) Raceway Trade Size ½ and Larger. Raceways must be large enough to permit the installation and removal of conductors without damaging the insulation. When all conductors in a raceway are the same size and insulation, the number of conductors permitted can be found in Annex C for the raceway type.

> **Question:** How many 6 THHN conductors can be installed in trade size 1 LFMC? **Figure 350–2**
>
> (a) 3   (b) 5   (c) 7   (d) 9
>
> **Answer:** (c) 7 conductors, Annex C, Table C.7

> **Author's Comment:** See 300.17 for additional examples on how to size raceways when conductors aren't all the same size.

Cables can be installed in liquidtight flexible metal conduit where not prohibited by the cable's *Code* article. The number of cables cannot exceed the allowable percentage fill specified in Table 1, Chapter 9.

(B) Raceway Trade Size ⅜. The number and size of conductors in a trade size ⅜ liquidtight flexible metal conduit must comply with Table 348.22.

> **Question:** How many 12 THHN conductors can be installed in trade size ⅜ LFMC that uses outside fittings?
>
> (a) 1   (b) 3   (c) 5   (d) 7
>
> **Answer:** (b) 3 conductors, Table 348.22

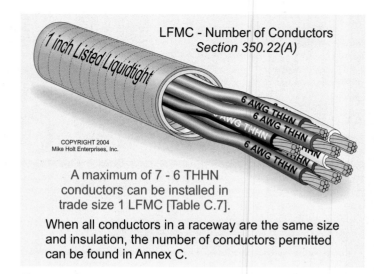

A maximum of 7 - 6 THHN conductors can be installed in trade size 1 LFMC [Table C.7].

When all conductors in a raceway are the same size and insulation, the number of conductors permitted can be found in Annex C.

Figure 350–2

One covered or bare equipment grounding (bonding) conductor of the same size is permitted with the circuit conductors. See the * note at the bottom of Table 348.22

350.24 Bends. Bends must be made so the conduit will not be damaged and the internal diameter of the conduit will not be effectively reduced. The radius of the curve of the inner edge of any field bend cannot be less than shown in Table 2, Chapter 9 using the column "Other Bends."

350.26 Number of Bends (360°). Bends between pull points, such as boxes and conduit bodies, cannot exceed 360°, and the raceway must be installed so the conductors can be pulled or removed without damaging the conductor insulation.

350.30 Secured and Supported. Liquidtight flexible metal conduit must be installed as a complete system [300.10, 300.12, and 300.18(A)], and it must be securely fastened in place and supported in accordance with (A) or (B).

(A) Securely Fastened. Liquidtight flexible metal conduit must be securely fastened by a means approved by the authority having jurisdiction within 1 ft of termination and at intervals not exceeding 4½ ft.

Exception 1: Where fished, liquidtight flexible metal conduit isn't required to be secured.

Exception 2: Securing the raceway isn't required for lengths up to 3 ft where flexibility is required.

Exception 4: Lengths not exceeding 6 ft can be unsecured within an accessible ceiling for luminaire(s) or other equipment.

(B) Horizontal Runs. Liquidtight flexible metal conduit installed horizontally in bored or punched holes in wood or metal framing members, or notches in wooden members is considered supported, but the raceway must be secured within 1 ft of termination.

350.42 Fittings. Angle connector fittings cannot be installed in concealed locations.

> **Author's Comment:** Conductors 4 AWG and larger that enter an enclosure must be protected from abrasion during and after installation by a fitting that provides a smooth, rounded insulating surface, such as an insulating bushing, unless the design of the box, fitting, or enclosure provides equivalent protection in accordance with 300.4(F).

350.60 Bonding. Where flexibility is required, an equipment grounding (bonding) conductor must be installed with the circuit conductors [300.3(B)].

Where flexibility is not required, the metal armor of liquidtight flexible metal conduit is permitted to serve as an equipment grounding (bonding) conductor if the circuit conductors are protected by overcurrent devices rated 20A or less for trade size ½ or smaller liquidtight flexible metal conduit, or 60A or less for trade size ¾ through 1¼, and the combined length of flexible raceway in the same ground return path doesn't exceed 6 ft [250.118(6)].

Where an equipment grounding (bonding) conductor is required, it must be installed with the circuit conductors.

Where an equipment bonding jumper is installed outside a raceway, the length of the equipment bonding jumper cannot exceed 6 ft and it must be routed with the raceway or enclosure [250.102(E), 250.134(B), and 300.3(B)].

(• Indicates that 75% or fewer exam takers get the question correct.)

1. •LFMC smaller than trade size _____ must not be used, except as permitted in 348.20(A).

   (a) ⅜   (b) ½   (c) 1½   (d) 1¼

2. Where flexibility is necessary, securing LFMC is not required for lengths not exceeding _____ at terminals.

   (a) 2 ft   (b) 3 ft   (c) 4 ft   (d) 6 ft

3. Liquidtight flexible metal conduit is not required to be fastened when used for tap conductors to luminaires up to _____ in length.

   (a) 4½ ft   (b) 18 in.   (c) 6 ft   (d) no limit on length

4. _____ connectors must not be used for concealed installations of liquidtight flexible metal conduit.

   (a) Straight   (b) Angle   (c) Grounding-type   (d) none of these

5. Where flexibility _____ liquidtight flexible metal conduit is permitted to be used as an equipment grounding (bonding) conductor when installed in accordance with 250.118(6).

   (a) is required   (b) is not required   (c) either a or d   (d) is optional

352 Rigid Nonmetallic Conduit (Type RNC)

Introduction

Rigid nonmetallic conduit (RNC) is commonly called "PVC." This type of conduit gives you many of the advantages of rigid conduit, while allowing installation in areas that are wet or corrosive. It is an inexpensive raceway, and easily installed. It is lightweight, easily cut, glued together, and relatively strong. However, RNC is brittle when cold, and it sags when hot. It is commonly used as an underground raceway because of its low cost, ease of installation, and resistance to corrosion and decay.

PART I. GENERAL

352.1 Scope. Article 352 covers the use, installation, and construction specifications of rigid nonmetallic conduit and associated fittings.

352.2 Definition.

Rigid Nonmetallic Conduit (Type RNC). This is a listed nonmetallic raceway of circular cross section with integral or associated couplings, listed for the installation of electrical conductors and cables [350.22]. **Figure 352–1**

PART II. INSTALLATION

352.10 Uses Permitted.

> **FPN:** In extreme cold, rigid nonmetallic conduit can become brittle and is more susceptible to physical damage.

(A) Concealed. Rigid nonmetallic conduit can be concealed within walls, floors, or ceilings, directly buried or embedded in concrete in buildings of any height.

(B) Corrosive Influences. Rigid nonmetallic conduit is permitted in areas subject to severe corrosion for which the material is specifically approved.

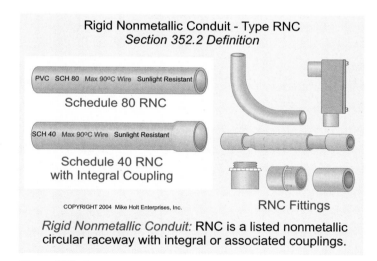

Rigid Nonmetallic Conduit - Type RNC
Section 352.2 Definition

PVC SCH 80 Max 90ºC Wire Sunlight Resistant

Schedule 80 RNC

SCH 40 Max 90ºC Wire Sunlight Resistant

Schedule 40 RNC
with Integral Coupling

COPYRIGHT 2004 Mike Holt Enterprises, Inc.

RNC Fittings

Rigid Nonmetallic Conduit: RNC is a listed nonmetallic circular raceway with integral or associated couplings.

Figure 352–1

Author's Comment: Where subject to exposure to chemical solvents, vapors, splashing, or immersion, materials or coatings must either be inherently resistant to chemicals based upon their listing or be identified for the specific chemical reagent [300.6(C)(2)].

(D) Wet Locations. Rigid nonmetallic conduit is permitted in wet locations such as dairies, laundries, canneries, car washes, and other areas that are frequently washed or outdoors.

Supporting fittings such as straps, screws, and bolts must be made of corrosion-resistant materials, or must be protected with a corrosion-resistant coating, in accordance with 300.6(A).

(E) Dry and Damp Locations. Rigid nonmetallic conduit is permitted in dry and damp locations, except where limited in 352.12.

(F) Exposed. Schedule 40 rigid nonmetallic conduit is permitted for exposed locations where not subject to physical damage [352.12(C)]. In areas exposed to physical damage, Schedule 80 rigid nonmetallic conduit, intermediate metal conduit, or rigid metal conduit must be used.

(G) Underground. Rigid nonmetallic conduit installed underground must comply with the burial requirements of 300.5.

(H) Support of Conduit Bodies. Rigid nonmetallic conduit is permitted to support nonmetallic conduit bodies not larger than the largest trade size of an entering raceway. These conduit bodies cannot support luminaires or other equipment and cannot contain devices other than splicing devices as permitted by 110.14(B) and 314.16(C)(2).

352.12 Uses Not Permitted.

(A) Hazardous (Classified) Locations.

(1) In hazardous (classified) locations, except as permitted in 503.10(A), 504.20, 514.8 Ex. 2, and 515.8. **Figure 352–2**

(2) In Class I, Division 2 locations, except as permitted in 501.10(B)(3).

(B) Support of Luminaires. Rigid nonmetallic conduit cannot be used for the support of luminaires or other equipment not described in 352.10(H).

> **Author's Comment:** Rigid nonmetallic conduit is permitted to support conduit bodies in accordance with 314.23(E) Ex.

(C) Physical Damage. Schedule 40 rigid nonmetallic conduit cannot be installed where subject to physical damage, but heavier and thicker Schedule 80 rigid nonmetallic conduit can be used where subject to physical damage. See 300.5(D).

(D) Ambient Temperature. Rigid nonmetallic conduit cannot be installed where the ambient temperature exceeds 50°C (122°F).

(E) Operating Temperature. Rigid nonmetallic conduit cannot contain conductors or cables whose operating temperature would exceed the conduit temperature rating. **Figure 352–3**

Exception: Conductors rated above the rigid nonmetallic conduit temperature rating can be installed if the conductors do not operate at a temperature above the listed raceway temperature rating.

(F) Places of Assembly. Because of toxicity, rigid nonmetallic conduit must not be installed in assembly occupancies and theaters, unless encased in not less than 2 in. of concrete, as permitted in 518.4(B) and 520.5(C).

> **Author's Comment:** Rigid nonmetallic conduit is prohibited as a wiring method for patient care areas in health care facilities [517.13(A)] or in ducts, plenums, and other environmental air spaces [300.22].

352.20 Trade Size.

(A) Minimum. Rigid nonmetallic conduit smaller than trade size ½ cannot be used.

(B) Maximum. Rigid nonmetallic conduit larger than trade size 6 cannot be used.

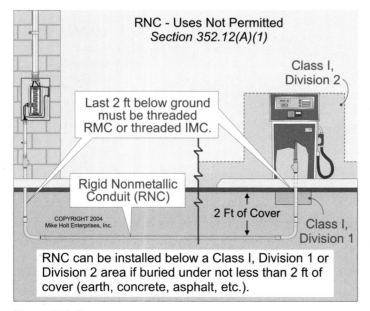

RNC - Uses Not Permitted
Section 352.12(A)(1)

Class I, Division 2

Last 2 ft below ground must be threaded RMC or threaded IMC.

Rigid Nonmetallic Conduit (RNC)

COPYRIGHT 2004 Mike Holt Enterprises, Inc.

2 Ft of Cover

Class I, Division 1

RNC can be installed below a Class I, Division 1 or Division 2 area if buried under not less than 2 ft of cover (earth, concrete, asphalt, etc.).

Figure 352–2

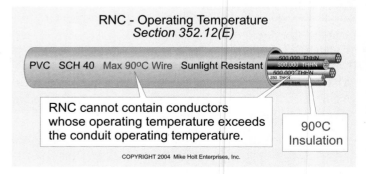

RNC - Operating Temperature
Section 352.12(E)

PVC   SCH 40   Max 90°C Wire   Sunlight Resistant

RNC cannot contain conductors whose operating temperature exceeds the conduit operating temperature.

90°C Insulation

COPYRIGHT 2004 Mike Holt Enterprises, Inc.

Figure 352–3

352.22 Number of Conductors.

Raceways must be large enough to permit the installation and removal of conductors without damaging the conductors' insulation, and the number of conductors cannot exceed that permitted by the percentage fill specified in Table 1, Chapter 9.

When all conductors in a raceway are the same size and insulation, the number of conductors permitted can be found in Annex C for the raceway type.

Question: *How many 4/0 THHN conductors can be installed in trade size 2 Schedule 40 RNC?*

(a) 2   (b) 4   (c) 6   (d) 8

Answer: *(b) 4 conductors, Annex C, Table C10.*

Author's Comment: Schedule 80 rigid nonmetallic conduit has the same outside diameter as Schedule 40 rigid nonmetallic conduit, but the wall thickness of Schedule 80 is greater, which results in a reduced interior area for conductor fill.

Question: *How many 4/0 THHN conductors can be installed in trade size 2 Schedule 80 RNC?*

(a) 3   (b) 5   (c) 7   (d) 9

Answer: *(a) 3 conductors, Annex C, Table C9.*

Author's Comment: See 300.17 for additional examples on how to size raceways when conductors aren't all the same size.

Cables can be installed in rigid nonmetallic conduit where not prohibited by the cable's article. The number of cables cannot exceed the allowable percentage fill specified in Table 1, Chapter 9.

352.24 Bends.

Raceway bends cannot be made in any manner that would damage the raceway or significantly change its internal diameter (no kinks). The radius of the curve of the inner edge of any field bend cannot be less than shown in Table 2, Chapter 9.

352.26 Number of Bends (360°).

To reduce the stress and friction on the conductor insulation, the number of bends (including offsets) between pull points cannot exceed 360°. Figure 352–4

Author's Comment: Be sure to use equipment designed for heating the raceway so it is pliable for bending (for example, a "hot box"). Do not use open-flame torches.

352.28 Trimming.

The cut ends of rigid nonmetallic conduit must be trimmed (inside and out) to remove the burrs and rough

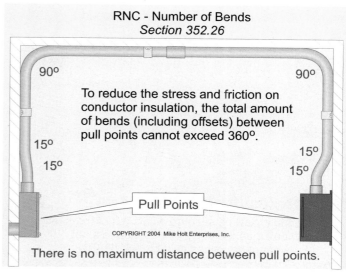

RNC - Number of Bends
Section 352.26

90°   90°

To reduce the stress and friction on conductor insulation, the total amount of bends (including offsets) between pull points cannot exceed 360°.

15°   15°
15°   15°

Pull Points

COPYRIGHT 2004 Mike Holt Enterprises, Inc.

There is no maximum distance between pull points.

Figure 352–4

edges. Trimming nonmetallic raceway is very easy; most of the burrs rub off with fingers, and a knife will smooth the rough edges.

352.30 Secured and Supported.

Rigid nonmetallic conduit must be installed as a complete system [300.10, 300.12, and 300.18(A)], and it must be securely fastened in place, and supported in accordance with (A) or (B).

(A) Secured. Rigid nonmetallic conduit must be secured within 3 ft of every box, cabinet, or termination fitting, such as a conduit body. Figure 352–5

(B) Supports. Rigid nonmetallic conduit must be supported at intervals not exceeding the values in the following table, and the raceway must be fastened in a manner that permits movement from thermal expansion or contraction. See Figure 352–5.

Table 352.30(B)	
Trade Size	**Support Spacing**
½ – 1	3 ft
1 ¼ – 2	5 ft
2 ½ – 3	6 ft
3 ½ – 5	7 ft
6	8 ft

Rigid nonmetallic conduit installed horizontally in bored or punched holes in wood or metal framing members, or notches in wooden members, is considered supported, but the raceway must be secured within 3 ft of termination.

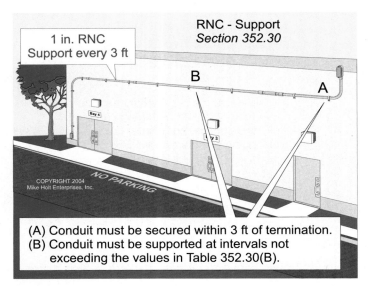

Figure 352–5

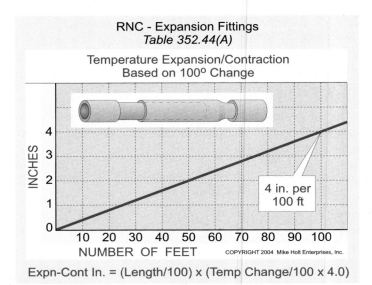

Expn-Cont In. = (Length/100) x (Temp Change/100 x 4.0)

Figure 352–7

352.44 Expansion Fittings. Where rigid nonmetallic conduit is installed in a straight run between securely mounted items such as boxes, cabinets, elbows, or other conduit terminations, expansion fittings must be provided to compensate for thermal expansion and contraction of the raceway in accordance with Tables 352.44(A) or (B), if the length change is determined to be ¼ in. or greater. **Figure 352–6**

> **Author's Comment:** Table 352.44(A) was created based on the following formula:
>
> Expansion-Contraction Inches = Raceway Length/100 x ((Temp Change/100) x 4.0) **Figure 352–7**

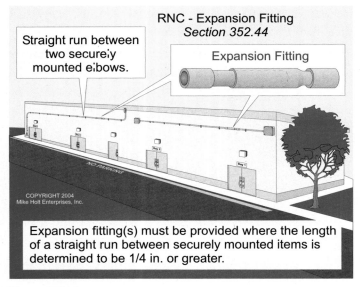

Figure 352–6

Example: What is the temperature change for a 25 ft run of RNC located in an ambient temperature change of 25°F?
(a) 1 in.   (b) 2 in.   (c) 3 in. (d) 4 in.
Answer: (a) 1 in.

Expansion/Contraction Inches = Raceway Length/100 x ((Temp °F Change/100) x 4.0)
Expansion/Contraction Inches = (25/100) x ((25/100) x 4.0)
Expansion/Contraction Inches = 1 in.

352.46 Bushings. Conductors 4 AWG and larger that enter an enclosure must be protected from abrasion during and after installation by a fitting that provides a smooth, rounded insulating surface, such as an insulating bushing, unless the design of the box, fitting, or enclosure provides equivalent protection in accordance with 300.4(F). Some rigid nonmetallic conduit adapters (connectors) and all bell-ends provide the conductor protection required in this section. **Figure 352–8**

> **Author's Comment:** When rigid nonmetallic conduit is stubbed into an open-bottom switchboard or other apparatus, the raceway, including the end fitting (bell-end), cannot rise more than 3 in. above the bottom of the switchboard enclosure [300.16(B) and 408.5].

352.48 Joints. Joints, such as couplings and connectors, must be made in a manner approved by the authority having jurisdiction.

> **Author's Comment:** Follow the manufacturer's instructions for the raceway, fitting, and glue. Some glues require the raceway surface to be cleaned with a solvent before the application of the glue. After applying glue to both surfaces, a quarter turn of the fitting is required.

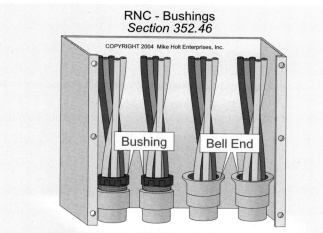

RNC - Bushings
Section 352.46

Bushing   Bell End

Conductors 4 AWG and larger require a fitting that provides a smooth, rounded, insulating surface to protect the wire during and after installation, see 300.4(F).

Figure 352–8

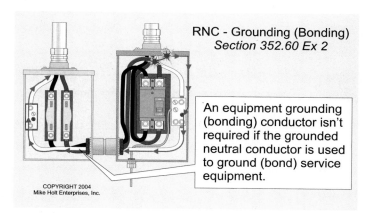

RNC - Grounding (Bonding)
Section 352.60 Ex 2

An equipment grounding (bonding) conductor isn't required if the grounded neutral conductor is used to ground (bond) service equipment.

Figure 352–9

352.60 Grounding (Bonding). Where equipment grounding (bonding) is required, a separate equipment grounding (bonding) conductor must be installed within the conduit [300.2(B)].

Exception 2: An equipment grounding (bonding) conductor isn't required in rigid nonmetallic conduit if the grounded neutral conductor is used to ground (bond) service equipment, as permitted in 250.142(A) [250.24(C)]. **Figure 352–9**

1. Extreme _____ may cause rigid nonmetallic conduit to become brittle, and therefore more susceptible to damage from physical contact.

   (a) sunlight   (b) corrosive conditions   (c) heat   (d) cold

2. Rigid nonmetallic conduit can be used to support nonmetallic conduit bodies not larger than the largest raceway, but the conduit bodies must not contain devices, luminaires, or other equipment.

   (a) True   (b) False

3. When installing rigid nonmetallic conduit, _____.

   (a) all cut ends must be trimmed inside and outside to remove rough edges
   (b) there must be a support within 2 ft of each box and cabinet
   (c) all joints must be made by an approved method
   (d) a and c

4. Expansion fittings for rigid nonmetallic conduit must be provided to compensate for thermal expansion and contraction when the length change in a straight run between securely mounted boxes, cabinets, elbows, or other conduit terminations is expected to be _____ or greater.

   (a) ¼ in.   (b) ½ in.   (c) 1 in.   (d) none of these

5. An equipment grounding (bonding) conductor is not required in rigid nonmetallic conduit if the grounded neutral conductor is used to ground equipment as permitted in 250.142.

   (a) True   (b) False

ARTICLE 353

High-Density Polyethylene Conduit (Type HDPE)

Introduction

This is a new article in the 2005 *NEC*; HDPE was previously included in Article 352. It's lightweight and durable. It resists decomposition, oxidation, and hostile elements that cause damage to other materials. HDPE is mechanically and chemically resistant to a host of environmental conditions. Uses include communication, data, cable television, and general-purpose raceways.

PART I. GENERAL

353.1 Scope. This article covers the use, installation, and construction specifications for high-density polyethylene conduit and associated fittings.

353.2 Definition.

High Density Polyethylene Conduit (Type HDPE). A nonmetallic raceway of circular cross section, with couplings, connectors, and fittings for the installation of electrical conductors. Figure 353–1

> **Author's Comment:** High-density polyethylene conduit is a flexible nonmetallic raceway used in concrete and for underground or innerduct applications. A typical application can be parking lot lights or street lighting. It can be installed for direct boring, plowing, and open trench wiring. It can also be pulled through an existing conduit installation. When HDPE conduit comes with pre-installed conductors, it's called Nonmetallic Underground Conduit with Conductors, and must be installed in accordance with Article 354.

353.6 Listing Requirements. High-density polyethylene conduit with associated fittings must be listed.

PART II. INSTALLATION

353.10 Uses Permitted. High-density polyethylene conduit is permitted as follows:

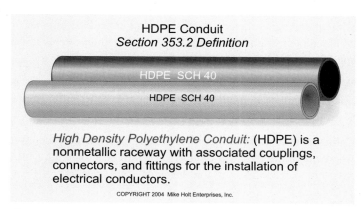

HDPE Conduit
Section 353.2 Definition

HDPE SCH 40

HDPE SCH 40

High Density Polyethylene Conduit: (HDPE) is a nonmetallic raceway with associated couplings, connectors, and fittings for the installation of electrical conductors.

COPYRIGHT 2004 Mike Holt Enterprises, Inc.

Figure 353–1

(1) In specific lengths or continuous lengths from a reel.

(2) In locations subject to severe corrosive influences as covered in 300.6(C) and where subject to chemicals for which the conduit is listed [300.6(C)(2)].

(3) In cinder fill.

(4) In direct burial installations in earth or concrete.

> **FPN:** See 300.5 for specific burial depth requirements.

353.12 Uses Not Permitted. High-density polyethylene conduit is not permitted:

(1) Where exposed.

(2) Within a building.

(3) In hazardous (classified) locations, except as permitted in 504.20.

(4) Where subject to ambient temperatures that exceed 50°C, unless listed otherwise.

(5) For conductors operating at a temperature above the rating of the raceway.

Exception: Conductors or cables rated at a temperature above the high-density polyethylene conduit temperature rating are permitted, provided the conductors do not operate at a temperature above the conduit listed temperature rating.

353.20 Trade Size.

(A) Minimum. High-density polyethylene conduit smaller than trade size ½ is not permitted.

(B) Maximum. High-density polyethylene conduit larger than trade size 4 is not permitted.

353.22 Number of Conductors. Raceways must be large enough to permit the installation and removal of conductors without damaging the conductors' insulation, and the number of conductors must not exceed that permitted by the percentage fill specified in Table 1, Chapter 9.

When all conductors in a raceway are the same size and insulation, the number of conductors permitted can be found in Annex C for the raceway type.

Cables can be installed in high-density polyethylene conduit where not prohibited by the cable's article. The number of cables must not exceed the allowable percentage fill specified in Table 1, Chapter 9.

353.24 Bends—How Made. Raceway bends must not be made in any manner that would damage the raceway or significantly change its internal diameter (no kinks). The radius of the curve of the inner edge of any field bend must not be less than shown in Table 354.24.

353.26 Bends—Number in One Run. To reduce the stress and friction on the conductor insulation, the number of bends (including offsets) between pull points must not exceed 360°.

> **Author's Comment:** Be sure to use equipment designed to heat the raceway so it's pliable for bending (for example, a "hot box"). Do not use open-flame torches.

353.28 Trimming. The cut ends of high-density polyethylene conduit must be trimmed (inside and out) to remove the burrs and rough edges. Trimming nonmetallic raceways is very easy; most of the burrs rub off with fingers, and a knife will smooth the rough edges.

353.46 Bushings. Conductors 4 AWG and larger that enter an enclosure must be protected from abrasion during and after installation by a fitting that provides a smooth, rounded insulating surface, such as an insulating bushing, unless the design of the box, fitting, or enclosure provides equivalent protection in accordance with 300.4(F).

> **Author's Comment:** When high-density polyethylene conduit is stubbed into an open-bottom switchboard or other apparatus, the raceway, including the end fitting (bell-end), must not rise more than 3 in. above the bottom of the switchboard enclosure. See 300.16(B) and 408.5.

353.48 Joints. Joints, such as couplings and connectors, must be made in a manner approved by the authority having jurisdiction.

> **Author's Comment:** This means you must follow the manufacturer's instructions for the raceway, fitting, and glue. Some glues require the raceway surface to be cleaned with a solvent before the application of the glue. After the application of the glue to both surfaces, a quarter turn of the fitting is required.

353.56 Splices and Taps. Splices and taps must be made in accordance with 300.15.

353.60 Grounding (Bonding). Where equipment grounding (bonding) is required, a separate equipment grounding (bonding) conductor must be installed within the conduit [300.2(B)].

Exception 2: An equipment grounding (bonding) conductor isn't required in HDPE conduit, if the grounded neutral conductor is used to ground (bond) service equipment as permitted in 250.142(A) [250.24(C)].

Article 353 Questions

1. High-Density Polyethylene Conduit (HDPE) can be manufactured _____.

    (a) in discrete lengths   (b) in continuous lengths from a reel
    (c) only in 20 ft. lengths   (d) either a or b

2. HDPE is not permitted where it will be subject to ambient temperatures in excess of _____.

    (a) 50°C   (b) 60°C   (c) 75°C   (d) 90°C

3. The cut ends of HDPE must be _____ to avoid rough edges.

    (a) filed on the inside   (b) trimmed inside and outside
    (c) cut only with a hack saw   (d) all of these

4. HDPE must be resistant to _____.

    (a) moisture   (b) corrosive chemical atmospheres
    (c) impact and crushing   (d) all of these

5. Each length of HDPE must be clearly and durably marked not less than every _____ ft, as required in 110.21.

    (a) 10   (b) 3   (c) 5   (d) 20

Get Ready Now!
2005 NEC

Order Mike's Code Change Library and SAVE OVER $450

354 Nonmetallic Underground Conduit with Conductors (Type NUCC)

Introduction

Nonmetallic underground conduit with conductors (NUCC) is a factory assembly of conductors or cables inside a nonmetallic, smooth-wall conduit with a circular cross section. The nonmetallic conduit is manufactured from a material called HDPE that is resistant to moisture and corrosive agents. It can also be supplied on reels without damage or distortion and is of sufficient strength to withstand abuse, such as impact or crushing, in handling and during installation without damage to conduit or conductors.

PART I. GENERAL

354.1 Scope. Article 354 covers the use, installation, and construction specifications of nonmetallic underground conduit with conductors (NUCC).

354.2 Definition.

Nonmetallic Underground Conduit Conductors (Type NUCC). This is a factory assembly of conductors or cables inside a nonmetallic, smooth-wall conduit with a circular cross section.

Author's Comment: The nonmetallic conduit for this wiring method is composed of a material that is resistant to moisture and corrosive agents. It's capable of being supplied on reels without damage or distortion and must be of sufficient strength to withstand abuse in handling and during installation without damage to conduit or conductors.

354.6 Listing Requirement. NUCC and its associated fittings must be listed. Figure 354–1

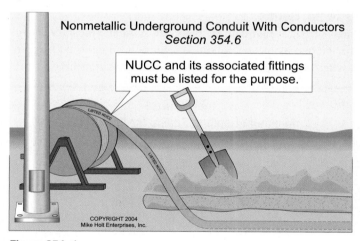

Nonmetallic Underground Conduit With Conductors
Section 354.6

NUCC and its associated fittings must be listed for the purpose.

COPYRIGHT 2004
Mike Holt Enterprises, Inc.

Figure 354–1

(3) In cinder fill.

(4) In underground locations subject to severe corrosive influences.

354.12 Uses Not Permitted.

(1) In exposed locations.

(2) Inside buildings. Figure 354–2

(3) In hazardous (classified) locations, except as permitted by 503.10(A), 504.20, 514.8, and 501.10(B).

PART II. INSTALLATION

354.10 Uses Permitted.

(1) For direct burial underground installation. See 300.5.

(2) Encased or embedded in concrete.

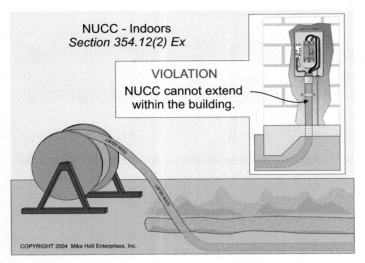

NUCC - Indoors
Section 354.12(2) Ex

VIOLATION
NUCC cannot extend
within the building.

LISTED NUCC

COPYRIGHT 2004 Mike Holt Enterprises, Inc.

Figure 354–2

354.20 Trade Size.

(A) Minimum. NUCC smaller than trade size ½ is not permitted.

(B) Maximum. NUCC larger than trade size 4 is not permitted.

354.24 Bends. Raceway bends must not be made in any manner that would damage the raceway or significantly change its internal diameter (no kinks). This is accomplished by complying with the following bending radius requirements:

Table 354.24
Minimum Bending Radius for Type NUCC

Trade Size	Minimum Bending Radius
½	10 in.
¾	12 in.
1	14 in.
1¼	18 in.
1½	20 in.
2	26 in.
2½	36 in.
3	48 in.
4	60 in.

354.26 Number of Bends (360°). To reduce stress and friction on the conductor insulation, the number of bends (including offsets) between pull points must not exceed 360°.

354.28 Trimming. At terminations, the conduit must be trimmed inside and out to remove rough edges using a method approved by the authority having jurisdiction that will not damage the conductor or cable insulation or jacket.

354.46 Bushings. Conductors 4 AWG and larger that enter an enclosure must be protected from abrasion during and after installation by a fitting that provides a smooth, rounded insulating surface, such as an insulating bushing, unless the design of the box, fitting, or enclosure provides equivalent protection in accordance with 300.4(F).

354.48 Joints. All joints between the conduit, fittings, and boxes must be made by a method approved by the authority having jurisdiction.

354.50 Conductor Terminations. All terminations between the conductors or cables and equipment must be made by a method approved by the authority having jurisdiction.

1. NUCC and its associated fittings must be _____.

   (a) listed   (b) approved   (c) identified   (d) none of these

2. NUCC must not be used _____.

   (a) in exposed locations   (b) inside buildings   (c) in hazardous (classified) locations   (d) all of these

3. Bends in nonmetallic underground conduit with conductors (NUCC) must be _____ so that the conduit will not be damaged and the internal diameter of the conduit will not be effectively reduced.

   (a) manually made   (b) made only with approved benders
   (c) made with rigid metal conduit bending shoes   (d) made using an open flame torch

4. In order to _____ NUCC, the conduit must be trimmed away from the conductors or cables using an approved method that will not damage the conductor or cable insulation or jacket.

   (a) facilitate installing   (b) enhance the appearance of the installation of
   (c) terminate   (d) provide safety to the persons installing

5. All joints between nonmetallic underground conduit with conductors (NUCC), fittings, and boxes must be made by _____.

   (a) a qualified person   (b) set screw fittings   (c) an approved method   (d) exothermic welding

Notes

Introduction

Liquidtight flexible nonmetallic conduit (LFNC) is a listed raceway of circular cross section having an outer liquidtight, nonmetallic, sunlight-resistant jacket over an inner flexible core with associated couplings, connectors, and fittings.

PART I. GENERAL

356.1 Scope. Article 356 covers the use, installation, and construction specifications of liquidtight flexible nonmetallic conduit and associated fittings.

356.2 Definition.

Liquidtight Flexible Nonmetallic Conduit (Type LFNC). This is a listed raceway of circular cross section having an outer liquidtight, nonmetallic, sunlight-resistant jacket over a flexible inner core with associated couplings, connectors, and fittings and listed for the installation of electrical conductors.

(1) Type LFNC-A (orange color). A smooth seamless inner core and cover having one or more reinforcement layers between the core and cover.

(2) Type LFNC-B (gray color). A smooth inner surface with integral reinforcement within the conduit wall.

(3) Type LFNC-C (black color). A corrugated internal and external surface without integral reinforcement.

356.6 Listing Requirement. Liquidtight flexible nonmetallic conduit and its associated fittings must be listed.

PART II. INSTALLATION

356.10 Uses Permitted. Listed liquidtight flexible nonmetallic conduit is permitted either exposed or concealed at any of the following locations:

(1) Where flexibility is required.

(2) Where protection from liquids, vapors, or solids is required.

(3) Outdoors, if listed and marked for this purpose.

(4) Directly buried in the earth, if listed and marked for this purpose.

(5) LFNC-B (gray color) is permitted in lengths over 6 ft secured according to 356.30.

(6) LFNC-B (black color) as a listed manufactured prewired assembly.

356.12 Uses Not Permitted.

(1) Where subject to physical damage.

(2) Where the combination of ambient and conductor temperature will produce an operating temperature above the rating of the raceway.

(3) Longer than 6 ft, except if approved by the authority having jurisdiction as essential for a required degree of flexibility.

(4) Where the operating voltage of the contained conductors exceeds 600 volts, nominal.

(5) In a hazardous (classified) location, except as permitted by 501.10(B), 502.10(A) and (B), and 504.20.

356.20 Trade Size.

(A) Minimum. Liquidtight flexible nonmetallic conduit smaller than trade size ½ is not permitted, except as permitted in the following:

(1) Enclosing the leads of motors, 430.245(B).

(2) Tap connections to lighting fixtures as permitted by 410.67(C).

(3) Electric sign connections, 600.32(A).

(B) Maximum. Liquidtight flexible nonmetallic conduit larger than trade size 4 is not permitted.

356.22 Number of Conductors.

Raceways must be large enough to permit the installation and removal of conductors without damaging the insulation. When all conductors in a raceway are the same size and insulation, the number of conductors permitted can be found in Annex C for the raceway type.

> *Question:* How many 8 THHN conductors can be installed in trade size ¾ LFNC-B?

> *Answer:* Six conductors, Annex C, Table C5.

Cables can be installed in liquidtight flexible nonmetallic conduit where not prohibited by the cable's article. The number of cables must not exceed the allowable percentage fill specified in Table 1, Chapter 9.

356.24 Bends.
Raceway bends must not be made in any manner that would damage the raceway or significantly change its internal diameter (no kinks). The radius of the curve of the inner edge of any field bend must not be less than shown in Table 2, Chapter 9 using the column "Other Bends."

356.26 Number of Bends (360°).
Bends between pull points such as boxes and conduit bodies must not exceed 360°, and the raceway must be installed so the conductors can be pulled or removed without damaging the conductor insulation.

356.30 Secured and Supported.
Type LFNC-B (gray color) must be securely fastened and supported in accordance with one of the following:

(1) The conduit must be securely fastened at intervals not exceeding 3 ft and within 1 ft of termination when installed longer than 6 ft.

(2) Securing or supporting isn't required where it's fished, installed in lengths not exceeding 3 ft at terminals where flexibility is required, or installed in lengths not exceeding 6 ft for tap conductors to luminaires, as permitted in 410.67(C).

(3) Horizontal runs of liquidtight flexible nonmetallic conduit installed horizontally in bored or punched holes in wood or metal framing members, or notches in wooden members, is considered supported, but the raceway must be secured within 1 ft of termination.

(4) Securing or supporting of LFNC-B isn't required where installed in lengths not exceeding 6 ft from the last point where the raceway is securely fastened for connections within an accessible ceiling to luminaire(s) or other equipment.

356.42 Fittings.
Only fittings listed for use with liquidtight flexible nonmetallic conduit must be used [300.15]. Angle connector fittings must not be installed in concealed raceway installations. Straight liquidtight flexible nonmetallic conduit fittings are permitted for direct burial or encasement in concrete.

> **Author's Comment:** Conductors 4 AWG and larger that enter an enclosure must be protected from abrasion during and after installation by a fitting that provides a smooth, rounded insulating surface, such as an insulating bushing, unless the design of the box, fitting, or enclosure provides equivalent protection in accordance with 300.4(F).

356.60 Grounding (Bonding).
Where flexibility is required, an equipment grounding (bonding) conductor must be installed inside or outside the raceway. Where the equipment bonding jumper is installed outside the raceway, the length of the equipment bonding jumper must not exceed 6 ft and must be routed with the raceway or enclosure [250.102(E), 250.134(B), and 300.3(B)].

Article 356 Questions

1. Type LFNC-B can be installed in lengths longer than _____ where secured in accordance with 356.30.

   (a) 2 ft   (b) 3 ft   (c) 6 ft   (d) 10 ft

2. The number of conductors allowed in LFNC must not exceed that permitted by the percentage fill specified in _____.

   (a) Table 1, Chapter 9   (b) Table 250.66   (c) Table 310.16   (d) 240.6

3. Bends in LFNC must _____ between pull points.

   (a) not be made   (b) not be limited in degrees
   (c) be limited to not more than 360 degrees   (d) be limited to 180 degrees

4. Where flexibility is necessary, securing LFNC is not required for lengths less than _____ at terminals.

   (a) 2 ft   (b) 3 ft   (c) 4 ft   (d) 6 ft

5. When LFNC is used to connect equipment requiring flexibility, a separate _____ must be installed.

   (a) equipment grounding (bonding) conductor   (b) expansion fitting
   (c) flexible nonmetallic connector   (d) none of these

Notes

358 Electrical Metallic Tubing (Type EMT)

Introduction

Electrical metallic tubing (EMT) is a lightweight raceway that is relatively easy to bend, cut, and ream. Because it isn't threaded, all connectors and couplings are of the threadless type and provide quick, easy, and inexpensive installation when compared to other metallic conduit systems. This makes it very popular. EMT is manufactured in both galvanized steel and aluminum; the steel type is much more common.

PART I. GENERAL

358.1 Scope. Article 358 covers the use, installation, and construction specifications of electrical metallic tubing.

358.2 Definition.

Electrical Metallic Tubing (Type EMT). This is a listed thinwall, metallic tubing of circular cross section used for the installation and physical protection of electrical conductors when joined together with listed fittings. It also can be used as an equipment grounding (bonding) conductor [250.118] when installed utilizing appropriate fittings.

358.6 Listing Requirement. Electrical metallic tubing, elbows, and associated fittings must be listed.

PART II. INSTALLATION

358.10 Uses Permitted.

(A) Exposed and Concealed. Electrical metallic tubing is permitted exposed or concealed.

(B) Corrosion Protection. Electrical metallic tubing, elbows, couplings, and fittings can be installed in concrete, in direct contact with the earth, or in areas subject to severe corrosive influences where protected by corrosion protection and judged suitable for the condition. **Figure 358–1**

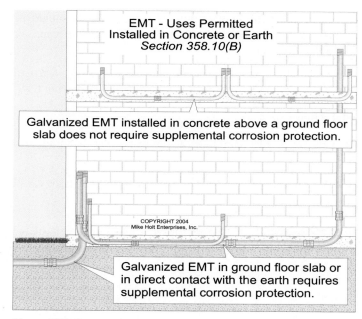

EMT - Uses Permitted
Installed in Concrete or Earth
Section 358.10(B)

Galvanized EMT installed in concrete above a ground floor slab does not require supplemental corrosion protection.

COPYRIGHT 2004
Mike Holt Enterprises, Inc.

Galvanized EMT in ground floor slab or in direct contact with the earth requires supplemental corrosion protection.

Figure 358–1

CAUTION: *Supplementary coatings for corrosion protection (asphaltic paint) have not been investigated by a product testing and listing agency, and these coatings are known to cause cancer in laboratory animals.*

(C) Wet Locations. All support fittings, such as screws, straps, etc., installed in a wet location must be made of corrosion-resistant material, or a corrosion-resistant coating must protect them in accordance with 300.6.

Author's Comment: Fittings used in wet locations must be listed for the application (wet location). For more information, visit http://www.etpfittings.com/.

358.12 Uses Not Permitted. EMT must not be used under the following conditions:

(1) Where, during installation or afterward, it will be subject to severe physical damage.

(2) Where protected from corrosion solely by enamel.

(3) In cinder concrete or cinder fill where subject to permanent moisture, unless protected on all sides by a layer of noncinder concrete not less than 2 in., unless the tubing is at least 18 in. under the fill.

(4) In any hazardous (classified) location, except as permitted by 502.10, 503.10, and 504.20.

(5) For the support of luminaires or other equipment (like boxes), except conduit bodies no larger than the largest trade size of the tubing can be supported by the raceway. Figure 358–2

(6) Where practical, contact with dissimilar metals should be avoided to prevent the deterioration of the metal because of galvanic action.

Exception: Aluminum fittings are permitted on steel electrical metallic tubing and steel fittings are permitted on aluminum EMT.

358.20 Trade Size.

(A) Minimum. Electrical metallic tubing smaller than trade size ½ is not permitted.

(B) Maximum. Electrical metallic tubing larger than trade size 4 is not permitted.

358.22 Number of Conductors. Raceways must be large enough to permit the installation and removal of conductors

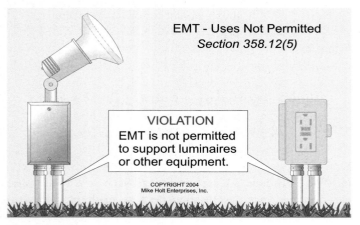

Figure 358–2

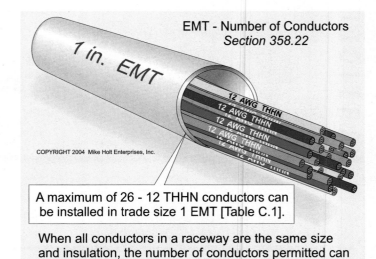

A maximum of 26 - 12 THHN conductors can be installed in trade size 1 EMT [Table C.1].

When all conductors in a raceway are the same size and insulation, the number of conductors permitted can be found in Annex C.

Figure 358–3

without damaging the conductors' insulation. When all conductors in a raceway are the same size and insulation, the number of conductors permitted can be found in Annex C for the raceway type.

> **Question:** *How many 12 THHN conductors can be installed in trade size 1 EMT?* Figure 358–3
>
> *(a) 26   (b) 28   (c) 30   (d) 32*
>
> **Answer:** *(a) 26 conductors, Annex C, Table C.1.*

Author's Comment: See 300.17 for additional examples on how to size raceways when conductors aren't all the same size.

Cables can be installed in electrical metallic tubing where not prohibited by the cable's article. The number of cables must not exceed the allowable percentage fill specified in Table 1, Chapter 9.

358.24 Bends. Raceway bends must not be made in any manner that would damage the raceway or significantly change its internal diameter (no kinks). The radius of the curve of the inner edge of any field bend must not be less than shown in Chapter 9, Table 2 for one-shot and full shoe benders.

Author's Comment: This isn't a problem, because most benders are made to comply with this table.

358.26 Number of Bends (360°). To reduce the stress and friction on the conductor insulation, the number of bends (including offsets) between pull points must not exceed 360°. Figure 358–4

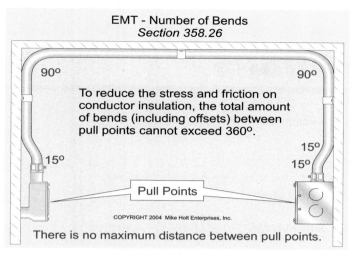

EMT - Number of Bends
Section 358.26

90°   90°

To reduce the stress and friction on conductor insulation, the total amount of bends (including offsets) between pull points cannot exceed 360°.

15°   15°
15°

Pull Points

COPYRIGHT 2004 Mike Holt Enterprises, Inc.

There is no maximum distance between pull points.

Figure 358–4

358.28 Reaming and Threading.

(A) Reaming. Reaming to remove the burrs and rough edges is required when the raceway is cut.

> **Author's Comment:** It's considered an accepted practice to ream small raceways with a screwdriver or the backside of pliers.

(B) Threading. Electrical metallic tubing must not be threaded.

358.30 Secured and Supported. Electrical metallic tubing must be installed as a complete system in accordance with 300.18 [300.10 and 300.12], and it must be securely fastened in place and supported in accordance with (A) or (B).

(A) Securely Fastened. Electrical metallic tubing must be securely fastened within 3 ft of every box, cabinet, or termination fitting, and at intervals not exceeding 10 ft. **Figure 358–5**

> **Author's Comment:** Support is required within 3 ft of termination, not within 3 ft of a coupling!

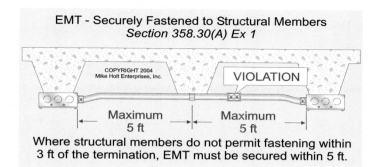

EMT - Securely Fastened to Structural Members
Section 358.30(A) Ex 1

COPYRIGHT 2004
Mike Holt Enterprises, Inc.   VIOLATION

Maximum   Maximum
5 ft   5 ft

Where structural members do not permit fastening within 3 ft of the termination, EMT must be secured within 5 ft.

Figure 358–6

Exception 1: When structural members do not permit the raceway to be secured within 3 ft of a box or termination fitting, an unbroken raceway can be secured within 5 ft of a box or termination fitting. **Figure 358–6**

(B) Horizontal Runs. Electrical metallic tubing installed horizontally in bored or punched holes in wood or metal framing members, or notches in wooden members, is considered supported, but the raceway must be secured within 3 ft of termination.

358.42 Coupling and Connectors. Threadless couplings and connectors must be made up tight to maintain an effective ground-fault current path to conduct fault current in accordance with 250.4(A)(5) [250.96(A) and 300.10].

Where buried in masonry or concrete, threadless electrical metallic tubing fittings must be of the concrete-tight type. Where installed in wet locations, fittings must be listed for use in wet locations in accordance with 314.15(A).

> **Author's Comment:** Conductors 4 AWG and larger that enter an enclosure must be protected from abrasion during and after installation by a fitting that provides a smooth, rounded insulating surface, such as an insulating bushing, unless the design of the box, fitting, or enclosure provides equivalent protection in accordance with 300.4(F). **Figure 358–7**

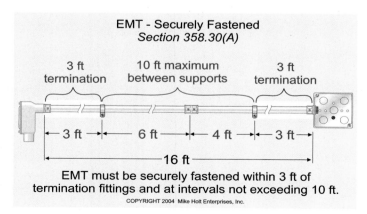

EMT - Securely Fastened
Section 358.30(A)

3 ft   10 ft maximum   3 ft
termination   between supports   termination

← 3 ft →←   6 ft   →← 4 ft →← 3 ft →
←————————— 16 ft —————————→

EMT must be securely fastened within 3 ft of termination fittings and at intervals not exceeding 10 ft.

COPYRIGHT 2004 Mike Holt Enterprises, Inc.

Figure 358–5

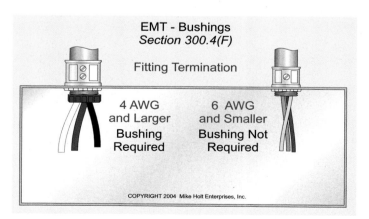

EMT - Bushings
Section 300.4(F)

Fitting Termination

4 AWG   6 AWG
and Larger   and Smaller
Bushing   Bushing Not
Required   Required

COPYRIGHT 2004 Mike Holt Enterprises, Inc.

Figure 358–7

(• Indicates that 75% or fewer exam takers get the question correct.)

1. EMT must not be used where _____.

   (a) subject to severe physical damage   (b) protected from corrosion only by enamel
   (c) used for the support of luminaires   (d) any of these

2. A run of EMT between outlet boxes must not exceed _____ offsets close to boxes.

   (a) 360° plus   (b) 360° total including   (c) four quarter bends plus   (d) 180° total including

3. •EMT must be supported within 3 ft of each coupling.

   (a) True   (b) False

4. Horizontal runs of EMT supported by openings through framing members at intervals not greater than _____, and securely fastened within 3 ft of termination points, are permitted.

   (a) 1.4 ft   (b) 12 in.   (c) 4½ ft   (d) 10 ft

5. Couplings and connectors used with EMT must be made up _____.

   (a) of metal   (b) in accordance with industry standards
   (c) tight   (d) none of these

ARTICLE 362

Electrical Nometallic Tubing (Type ENT)

Introduction

Electrical nonmetallic tubing (ENT) is a pliable, corrugated, circular raceway made of polyvinyl chloride (PVC).

PART I. GENERAL

362.1 Scope.
Article 362 covers the use, installation, and construction specifications of electrical nonmetallic tubing and associated fittings.

362.2 Definition.

Electrical Nonmetallic Tubing (Type ENT). This is a pliable corrugated raceway of circular cross section with integral or associated couplings, connectors, and fittings listed for the installation of electrical conductors.

> **Author's Comment:** In some parts of the country, the field name for electrical nonmetallic tubing is "Smurf Pipe" or "Smurf Tube," because it was available only in blue when it originally came out at the height of popularity of the children's cartoon characters, the "Smurfs." Today, the raceway is available in a rainbow of colors such as white, yellow, red, green, and orange, and is sold in both fixed lengths and on reels.

Electrical nonmetallic tubing is a pliable raceway that can be bent by hand with a reasonable force, but without other assistance.

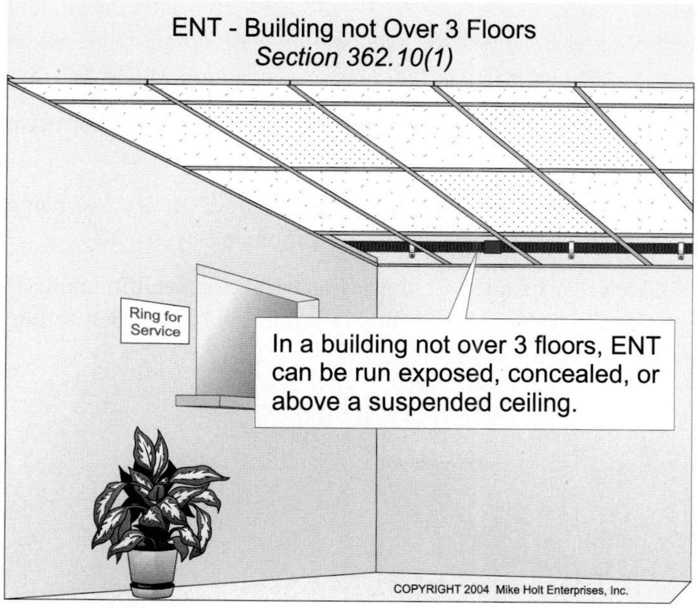

ENT - Building not Over 3 Floors
Section 362.10(1)

Ring for Service

In a building not over 3 floors, ENT can be run exposed, concealed, or above a suspended ceiling.

COPYRIGHT 2004 Mike Holt Enterprises, Inc.

Figure 362–1

PART II. INSTALLATION

362.10 Uses Permitted.

Definition of First Floor: The first floor of a building is the floor with 50 percent or more of the exterior wall surface area level with or above finished grade. If one additional level not designed for human habitation and used only for vehicle parking, storage, or similar use is at ground level, then the first of the three permissible floors can be the next higher floor.

(1) In buildings not exceeding three floors. Figure 362–1

    a. Exposed, where not prohibited by 362.12.

    b. Concealed within walls, floors, and ceilings.

(2) In buildings exceeding three floors, electrical nonmetallic tubing can be installed concealed in walls, floors, or ceilings that provide a thermal barrier having a 15-minute finish rating as identified in listings of fire-rated assemblies. Figure 362–2

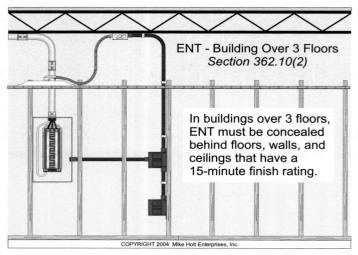

Figure 362–2

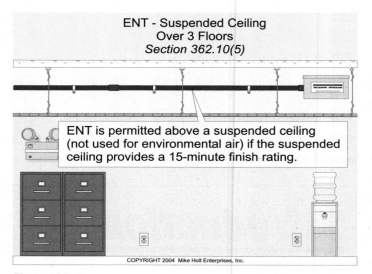

Figure 362–4

Exception to (2): Where a fire sprinkler system is installed on all floors, in accordance with NFPA 13, Standard for the Installation of Sprinkler Systems, electrical nonmetallic tubing is permitted exposed or concealed in buildings of any height. Figure 362–3

(3) Electrical nonmetallic tubing is permitted in severe corrosive and chemical locations when identified for this use.

(4) Electrical nonmetallic tubing is permitted in dry and damp concealed locations, where not prohibited by 362.12.

(5) Above Suspended Ceilings. Electrical nonmetallic tubing is permitted above a suspended ceiling if the suspended ceiling

provides a thermal barrier having a 15-minute finish rating, as identified in listings of fire-rated assemblies. Figure 362–4

Exception: Where a fire sprinkler system is installed on all floors, in accordance with NFPA 13, Standard for the Installation of Sprinkler Systems, electrical nonmetallic tubing is permitted above a suspended ceiling that doesn't have a 15-minute finish rated thermal barrier material. Figure 362–5

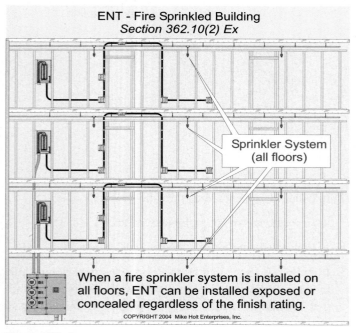

Figure 362–3

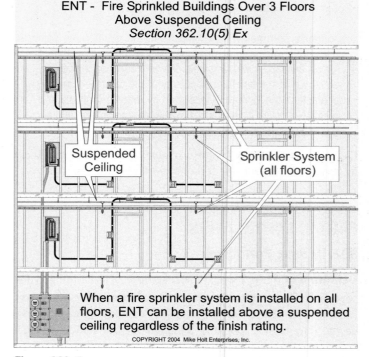

Figure 362–5

(6) Electrical nonmetallic tubing can be encased or embedded in a concrete slab on grade where the tubing is placed on sand or screenings approved by the authority having jurisdiction, provided fittings identified for the purpose are used.

Author's Comment: Electrical nonmetallic tubing is not permitted in the earth [362.12(5)].

(7) Electrical nonmetallic tubing is permitted in wet locations indoors or in a concrete slab on or below grade, with fittings listed for the purpose.

(8) Listed prewired electrical nonmetallic tubing is permitted in trade sizes ½, ¾, and 1.

362.12 Uses Not Permitted.

(1) In hazardous (classified) locations, except as permitted by 504.20 and 505.15(A)(1).

(2) For the support of luminaires or equipment. See 314.2.

(3) Where the ambient temperature exceeds 50°C (122°F).

(4) To contain conductors that operate at a temperature above the temperature rating of the raceway. **Figure 362–6**

Exception: Conductors rated at a temperature above the electrical nonmetallic tubing temperature rating are permitted, provided the conductors do not operate at a temperature above the electrical nonmetallic tubing listed temperature rating.

(5) For direct earth burial.

Author's Comment: Electrical nonmetallic tubing can be encased in concrete [362.10(6)].

(6) As a wiring method for systems over 600V.

(7) Exposed in buildings over three floors, except as permitted by 362.10(2) and (5) Ex.

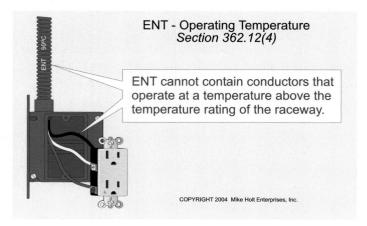

Figure 362–6

(8) In assembly occupancies or theaters, except as permitted by 518.4 and 520.5.

(9) Exposed to the direct rays of the sun for an extended period, unless listed as sunlight resistant.

Author's Comment: Exposing electrical nonmetallic tubing to direct rays of the sun for an extended time may result in the product becoming brittle, unless listed to resist the effects of ultraviolet (UV) radiation. **Figure 362–7**

(10) Where subject to physical damage.

Author's Comment: Electrical nonmetallic tubing is prohibited as a wiring method in ducts, plenums, and other environmental air handling spaces [300.22] and for patient care area circuits in health care facilities [517.13(A)].

362.20 Trade Sizes.

(A) Minimum. Electrical nonmetallic tubing smaller than trade size ½ is not permitted.

(B) Maximum. Electrical nonmetallic tubing larger than trade size 4 is not permitted.

362.22 Number of Conductors.
Raceways must be large enough to permit the installation and removal of conductors without damaging the conductors' insulation, and the number of conductors must not exceed that permitted by the percentage fill specified in Table 1, Chapter 9.

When all conductors in a raceway are the same size and insulation, the number of conductors permitted can be found in Annex C for the raceway type.

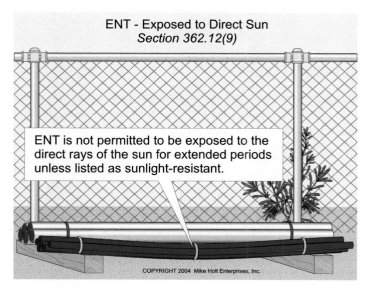

Figure 362–7

Question: *How many 12 THHN conductors can be installed in trade size ½ ENT?*

(a) 5   (b) 7   (c) 9   (d) 11

Answer: *(b) 7 conductors, Annex C, Table C2*

Author's Comment: See 300.17 for additional examples on how to size raceways when conductors aren't all the same size.

Cables can be installed in electrical nonmetallic tubing where not prohibited by the cable's article. The number of cables must not exceed the allowable percentage fill specified in Table 1, Chapter 9.

362.24 Bends. Raceway bends must not be made in any manner that would damage the raceway or significantly change its internal diameter (no kinks). The radius of the curve of the inner edge of any field bend must not be less than shown in Chapter 9, Table 2, using the column "Other Bends."

362.26 Number of Bends (360°). To reduce the stress and friction on the conductor insulation, the number of bends (including offsets) between pull points must not exceed 360°.

362.28 Trimming. The cut ends of electrical nonmetallic tubing must be trimmed (inside and out) to remove the burrs and rough edges. Trimming nonmetallic raceways is very easy; most of the burrs rub off with fingers, and a knife can be used to smooth the rough edges.

362.30 Secured and Supported. Electrical nonmetallic tubing must be installed as a complete system in accordance with 300.18 [300.10 and 300.12], and it must be securely fastened in place and supported in accordance with (A) or (B).

(A) Securely Fastened. Electrical nonmetallic tubing must be secured within 3 ft of every box, cabinet, or termination fitting, such as a conduit body, and at intervals not exceeding 3 ft. Figure 362–8

Exception 2: Lengths not exceeding 6 ft from the last point where the raceway is securely fastened within an accessible ceiling to luminaire(s) or other equipment.

(B) Horizontal Runs. Electrical nonmetallic tubing installed horizontally in bored or punched holes in wood or metal framing members, or notches in wooden members, is considered supported, but the raceway must be secured within 3 ft of terminations.

362.46 Bushings. Conductors 4 AWG and larger that enter an enclosure from a fitting must be protected from abrasion during and after installation by a fitting that provides a smooth,

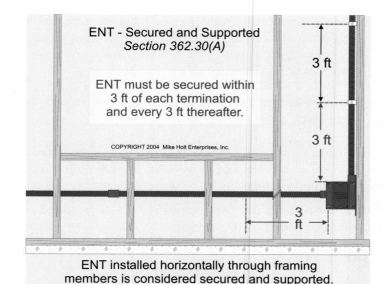

ENT - Secured and Supported
Section 362.30(A)

ENT must be secured within 3 ft of each termination and every 3 ft thereafter.

3 ft

3 ft

3 ft

COPYRIGHT 2004 Mike Holt Enterprises, Inc.

ENT installed horizontally through framing members is considered secured and supported.

Figure 362–8

rounded insulating surface, such as an insulating bushing, unless the design of the box, fitting, or enclosure provides equivalent protection in accordance with 300.4(F).

362.48 Joints. Joints, such as couplings and connectors, must be made in a manner approved by the authority having jurisdiction.

Author's Comment: Follow the manufacturer's instructions for the raceway, fittings, and glue. According to product listings, rigid nonmetallic conduit fittings are permitted with electrical nonmetallic tubing.

CAUTION: *Glue used with electrical nonmetallic tubing must be listed for ENT. Glue for rigid nonmetallic conduit made from PVC generally cannot be used with electrical nonmetallic tubing, because it damages the plastic from which electrical nonmetallic tubing is manufactured.*

362.60 Grounding (Bonding). Where equipment grounding (bonding) is required, a separate equipment grounding (bonding) conductor must be installed in the raceway. Figure 362–9

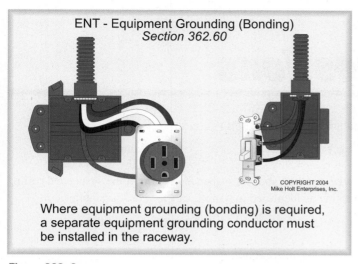

ENT - Equipment Grounding (Bonding)
Section 362.60

COPYRIGHT 2004
Mike Holt Enterprises, Inc.

Where equipment grounding (bonding) is required, a separate equipment grounding conductor must be installed in the raceway.

Figure 362–9

1. ENT is composed of a material that is resistant to moisture, chemical atmospheres, and is _____.

   (a) flexible   (b) flame-retardant   (c) fireproof   (d) flammable

2. In a building without a fire sprinkler system, ENT is permitted to be installed above a suspended ceiling if the suspended ceiling provides a thermal barrier having at least a _____-minute finish rating as identified in listings of fire-rated assemblies.

   (a) 5   (b) 10   (c) 15   (d) none of these

3. ENT is permitted for direct earth burial when used with fittings listed for this purpose.

   (a) True   (b) False

4. The number of conductors allowed in ENT must not exceed that permitted by the percentage fill specified in _____.

   (a) Table 1, Chapter 9   (b) Table 250.66   (c) Table 310.16   (d) 240.6

5. Bushings or adapters are required at ENT terminations to protect the conductors from abrasion, unless the box, fitting, or enclosure design provides equivalent protection.

   (a) True   (b) False

ARTICLE 376 Metal Wireways

Introduction

A metal wireway or "trough" is commonly used where access to the conductors within the raceway is required to make terminations, splices, or taps to several devices at a single location. Its cost precludes its use for other than short distances, except in some commercial or industrial occupancies where the wiring is frequently revised.

PART I. GENERAL

376.1 Scope.
Article 376 covers the use, installation, and construction specifications of metallic wireways and associated fittings.

376.2 Definition.

Metal Wireway. This is a sheet metal raceway with hinged or removable covers for housing and protecting electric wires and cable, and in which conductors are placed after the wireway has been installed. Figure 376–1

> **Author's Comment:** Some people improperly call this wiring method an auxiliary gutter, which it's not. See Article 366 of the *NEC*.

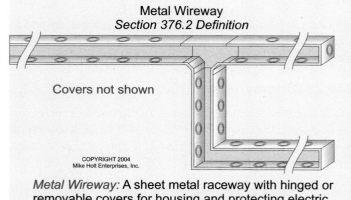

Metal Wireway
Section 376.2 Definition

Covers not shown

COPYRIGHT 2004
Mike Holt Enterprises, Inc.

Metal Wireway: A sheet metal raceway with hinged or removable covers for housing and protecting electric wires and cable, and in which conductors are placed after the wireway has been installed.

Figure 376–1

PART II. INSTALLATION

376.10 Uses Permitted.

(1) Exposed.

(2) Concealed, as permitted by 376.10(4).

(3) In hazardous (classified) locations as permitted by 501.10(B), 502.10(B), or 504.20.

(4) Unbroken through walls, partitions, and floors.

376.12 Uses Not Permitted.

(1) Subject to severe physical damage.

(2) Subject to corrosive environments.

376.21 Conductor—Maximum Size.
The maximum size conductor permitted in a wireway must not be larger than that for which the wireway is designed.

376.22 Number of Conductors.
The maximum number of conductors permitted in a wireway is limited to 20 percent of the cross-sectional area of the wireway. Figure 376–2

> **Author's Comment:** Splices and taps must not fill more than 75 percent of the wiring space at any cross section [376.56].

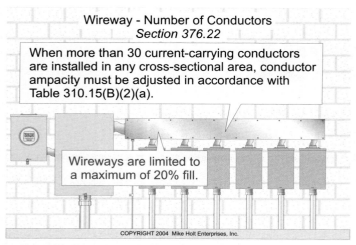

Figure 376–2

Ampacity Adjustment. When more than 30 current-carrying conductors are installed in any cross-sectional area of the wireway, the conductor ampacity, as listed in Table 310.16, must be reduced according to the adjustment factors listed in Table 310.15(B)(2)(a).

Signaling and motor control conductors between a motor and its starter and used only for starting duty aren't considered current carrying.

376.23 Wireway Sizing.

(A) Sizing for Conductor Bending Radius. Where conductors are bent within a wireway, the wireway must be sized to meet the bending radius requirements contained in Table 312.6(A), based on one wire per terminal. Figure 376–3

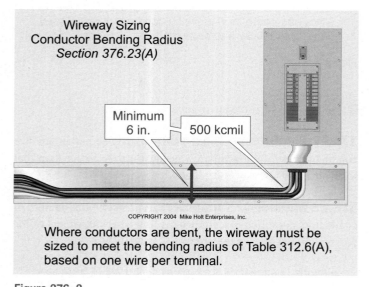

Where conductors are bent, the wireway must be sized to meet the bending radius of Table 312.6(A), based on one wire per terminal.

Figure 376–3

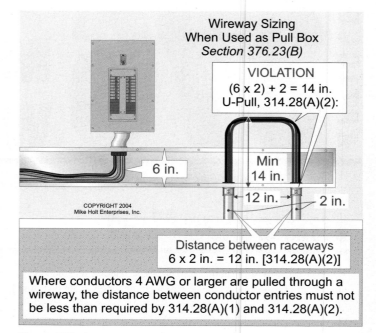

Where conductors 4 AWG or larger are pulled through a wireway, the distance between conductor entries must not be less than required by 314.28(A)(1) and 314.28(A)(2).

Figure 376–4

(B) Wireway Used as Pull Box. Where conductors 4 AWG or larger are pulled through a wireway, the distance between raceway and cable entries enclosing the same conductor must not be less than required by 314.28(A)(1) and 314.28(A)(2). Figure 376–4

Author's Comments:

- Straight Pulls. The minimum distance from where the conductors enter to the opposite wall must not be less than eight times the trade size of the largest raceway [314.28(A)(1)].

- Angle Pulls. The distance from the raceway entry to the opposite wall must not be less than six times the trade diameter of the largest raceway, plus the sum of the trade sizes of the remaining raceways on the same wall [314.28(A)(2)].

- U Pulls. When a conductor enters and leaves from the same wall, the distance from where the raceways enter to the opposite wall must not be less than six times the trade size of the largest raceway, plus the sum of the trade sizes of the remaining raceways on the same wall [314.28(A)(2)].

- The distance between raceways enclosing the same conductor must not be less than six times the trade size of the largest raceway, measured from raceway edge-to-nearest edge [314.28(A)(2)].

376.30 Supports. Wireways must be supported in accordance with (A) or (B).

(A) Horizontal Support. Where run horizontally, wireways must be supported at each end and at intervals not exceeding 5 ft.

(B) Vertical Support. Where run vertically, wireways must be securely supported at intervals not exceeding 15 ft, with no more than one joint between supports.

376.56 Splices and Taps.

(A) Splices and Taps. Splices and taps in wireways must be accessible, and they must not fill the wireway to more than 75 percent of its cross-sectional area. **Figure 376–5**

> **Author's Comment:** The maximum number of conductors permitted in a wireway is limited to 20 percent of its cross-sectional area at any point [376.22].

(B) Power Distribution Blocks.

(1) Installation. Power distribution blocks installed in wireways must be listed.

(2) Size of Enclosure. In addition to the wiring space requirement [356.56(A)], the power distribution block must be installed in a wireway not smaller than that specified in the installation instructions of the power distribution block.

(3) Wire Bending Space. Wire bending space at the terminals of power distribution blocks must comply with 312.6(B).

(4) Live Parts. Power distribution blocks must not have exposed live parts in the wireway after installation.

> **Author's Comment:** Power distribution blocks make for a cleaner installation than the use of multiple split-bolts.

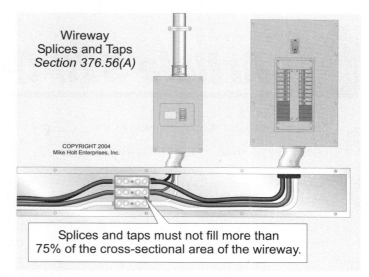

Wireway
Splices and Taps
Section 376.56(A)

COPYRIGHT 2004
Mike Holt Enterprises, Inc.

Splices and taps must not fill more than 75% of the cross-sectional area of the wireway.

Figure 376–5

1. Metal wireways are sheet metal troughs with _____ for housing and protecting electric wires and cable.

    (a) removable covers   (b) hinged covers   (c) a or b   (d) none of these

2. Wireways are permitted to pass transversely through a wall _____. Access to the conductors must be maintained on both sides of the wall.

    (a) if the length passing through the wall is unbroken   (b) if the wall is not fire rated
    (c) in hazardous locations   (d) if the wall is fire rated

3. The derating factors in 310.15(B)(2)(a) must be applied to a metal wireway only where the number of current-carrying conductors in the wireway exceeds _____.

    (a) 30   (b) 20   (c) 80   (d) 3

4. Wireways must be supported where run horizontally at each end and at intervals not to exceed _____, or for individual lengths longer than _____ at each end or joint, unless listed for other support intervals.

    (a) 5 ft   (b) 10 ft   (c) 3 ft   (d) 6 ft

5. Power distribution blocks installed in metal wireways must be listed.

    (a) True   (b) False

ARTICLE 378 Nonmetallic Wireways

Introduction

A nonmetallic wireway is a "trough" made from PVC, fiber reinforced plastic (FRP), fiberglass, or similar materials. It is used in harsh environments, but has severe limitations on its use.

PART I. GENERAL

378.1 Scope. Article 378 covers the use, installation, and construction specifications of nonmetallic wireways and associated fittings.

> **CAUTION:** *Nonmetallic wireways do not have the same heat-transfer characteristics as metal wireways. Therefore, the ampacity adjustment factors of Table 310.15(B)(2)(a) apply any time there are four or more current-carrying conductors in any cross-sectional area of a nonmetallic wireway [378.22]! Effectively, this severely limits the use of nonmetallic wireways.*

378.2 Definition.

Nonmetallic Wireway. This is a flame-retardant raceway with hinged or removable covers for housing and protecting electric wires and cable, and in which conductors are placed after the wireway has been installed as a complete system.

PART II. INSTALLATION

378.10 Uses Permitted.

(1) Exposed, except as permitted in 378.10(4).

(2) Subject to corrosive environments, where identified for the use.

(3) In wet locations, where listed for the purpose.

(4) In unbroken extensions passing transversely through walls.

378.12 Uses Not Permitted.

(1) Where subject to severe physical damage.

(2) In hazardous (classified) locations, except as permitted by 504.20.

(3) Where exposed to sunlight, unless listed and marked as suitable for this purpose.

(4) Where the ambient temperatures exceed the wireway listing.

(5) Where the conductors operate at a temperature above the raceway temperature rating.

378.21 Conductor—Maximum Size.
The maximum size conductor permitted in a wireway must not be larger than that for which the wireway is designed.

378.22 Number of Conductors.

The maximum number of conductors permitted in a wireway is limited to 20 percent of the cross-sectional area of the wireway, except that splices and taps must not fill the nonmetallic wireway to more than 75 percent of its area at any point [378.56].

The ampacity adjustment factors of Table 310.15(B)(2)(a) apply any time there are four or more current-carrying conductors in any cross-sectional area of the wireway, but signaling and motor control conductors aren't considered current carrying.

378.23 Nonmetallic Wireway Sizing.

(A) Bending Radius. Where conductors are bent within a wireway, the nonmetallic wireway must be sized to meet the bending radius requirements in Table 312.6(A), based on one wire per terminal. For example, a nonmetallic wireway must not be smaller than 6 in. to permit sufficient bending radius for 500 kcmil wire.

(B) Wireway Used as Pull Box. Where insulated conductors 4 AWG or larger are pulled through a wireway, the distance between raceway and cable entries enclosing the same conductor must not be less than required by 314.28(A)(1) and 314.28(A)(2).

Author's Comments:

- Straight Pulls. The minimum distance from where the conductors enter to the opposite wall must not be less than eight times the trade size of the largest raceway [314.28(A)(1)].

- Angle Pulls. The distance from the raceway entry to the opposite wall must not be less than six times the trade size of the largest raceway, plus the sum of the trade sizes of the remaining raceways on the same wall [314.28(A)(2)].

- U Pulls. When a conductor enters and leaves from the same wall, the distance from where the raceways enter to the opposite wall must not be less than six times the trade size of the largest raceway, plus the sum of the trade sizes of the remaining raceways on the same wall [314.28(A)(2)].

- The distance between raceways enclosing the same conductor must not be less than six times the trade size of the largest raceway, measured from raceway edge-to-nearest edge [314.28(A)(2)].

378.30 Supports.
Nonmetallic wireways must be supported in accordance with (A) or (B).

(A) Horizontal Support. Where run horizontally, nonmetallic wireways must be supported at each end and at intervals not to exceed 3 ft.

(B) Vertical Support. Where run vertically, nonmetallic wireways must be securely supported at intervals not exceeding 4 ft, with no more than one joint between supports.

378.44 Expansion Fittings.
Expansion fittings for nonmetallic wireways are required to compensate for thermal expansion and contraction where the length change is expected to be ¼ in. or greater in a straight run. The expansion characteristics of PVC nonmetallic wireway are identical to those for PVC rigid nonmetallic conduit in Table 352.44(A).

378.56 Splices and Taps.
Splices and taps in nonmetallic wireways must be accessible and must not fill the wireway to more than 75 percent of its cross-sectional area.

> **Author's Comment:** The maximum number of conductors permitted in a wireway is limited to 20 percent of its cross-sectional area [378.22].

378.60 Grounding (Bonding).
Where an equipment grounding (bonding) conductor is required, it must be installed within the raceway [250.102(E) and 300.3(B)].

An equipment grounding (bonding) conductor isn't required if the grounded neutral conductor is used to bond service equipment to the power source [250.24(C)], as permitted in 250.142(A).

1. Nonmetallic wireways can pass transversely through a wall _____.

   (a) if the length through the wall is unbroken
   (b) if the wall is not fire rated
   (c) in hazardous (classified) locations
   (d) if the wall is fire rated

2. The derating factors in 310.15(B)(2)(a) apply to a nonmetallic wireway.

   (a) True   (b) False

3. Nonmetallic wireways must be supported where run horizontally at each end and at intervals not to exceed _____ and at each end or joint, unless listed for other support intervals.

   (a) 5 ft   (b) 10 ft   (c) 3 ft   (d) 6 ft

4. Expansion fittings for nonmetallic wireways must be provided to compensate for thermal expansion and contraction, where the length change is expected to be _____ or greater in a straight run.

   (a) ¼ in.   (b) ½ in.   (c) 6 in.   (d) 1/16 in.

5. Where equipment grounding is required by Article 250 for nonmetallic wireway installations, a separate equipment grounding (bonding) conductor must _____.

   (a) be run outside the raceway using solid copper wire
   (b) be installed in the raceway
   (c) be obtained using a separate driven ground rod
   (d) not be required

Electrical Estimating

Estimate Like the Experts!

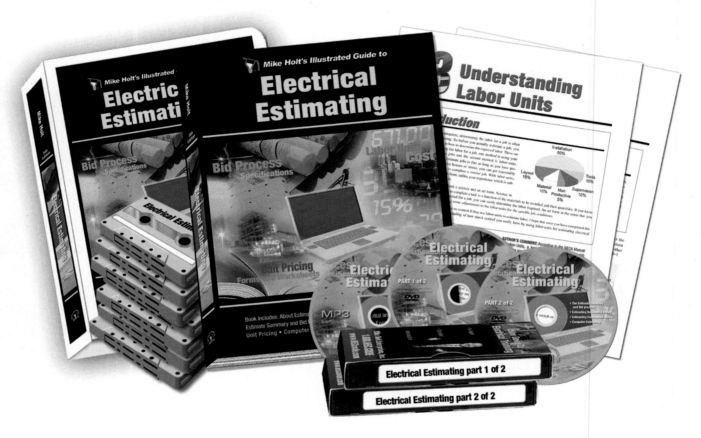

Running a business requires an understanding of how to estimate a job. This program will show you how to take-off a job, how to determine the material cost, how to accurately estimate labor cost, and how to determine your overhead, profit, and break-even point. This course also gives tips on selling and marketing your business. You'll learn how to estimate and project manage residential, commercial, and industrial projects; and how to make more money on every job by proper estimating and project management. Watch the videos or DVDs at home or office and listen to the cassettes or MP3s while driving. You'll quickly learn how to make more money on every job.

Call us today at 1.888.NEC.Code, or visit us online at www.NECcode.com, for the latest information and pricing.

380 Multioutlet Assemblies

Introduction

A multioutlet assembly is a surface, flush, or freestanding raceway designed to hold conductors and receptacles, and is assembled in the field or at the factory [Article 100]. It is not limited to systems commonly referred to by trade names of "plugtrak" or "plugmold." Figure 380–1

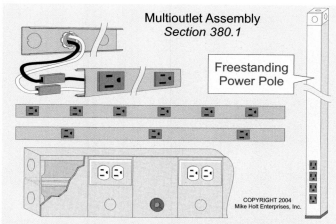

Multioutlet Assembly
Section 380.1

Freestanding Power Pole

COPYRIGHT 2004
Mike Holt Enterprises, Inc.

Multioutlet Assembly: A surface, flush, or freestanding raceway designed to hold conductors and receptacles assembled in the field or at the factory.

Figure 380–1

380.1 Scope. Article 380 covers the use, installation, and construction specifications of multioutlet assemblies.

380.2 Uses.

(A) Permitted. Dry locations only.

(B) Not Permitted.

(1) Concealed.

(2) Where subject to severe physical damage.

(3) Where the voltage is 300V or more between conductors, unless the metal has a thickness not less than 0.040 in.

(4) Where subject to corrosive vapors.

(5) In hoistways.

(6) In hazardous (classified) locations, except as permitted by 501.10(B).

380.3 Through Partitions. Metal multioutlet assemblies can pass through a dry partition provided no receptacle is concealed in the wall and the cover of the exposed portion of the system can be removed.

1. A multioutlet assembly can be installed in _____.

   (a) dry locations   (b) wet locations   (c) a and b   (d) none of these

2. A multioutlet assembly cannot be installed _____.

   (a) in concealed locations   (b) where subject to severe physical damage

   (c) where subject to corrosive vapors   (d) all of these

3. Metal multioutlet assemblies can pass through a dry partition, provided no receptacle is concealed in the partition and the cover of the exposed portion of the system can be removed.

   (a) True   (b) False

384 Strut-Type Channel Raceways

Introduction

Strut-type channel raceway is a metallic raceway formed by installing a cover onto strut. Strut is usually utilized as a supporting assembly, but when paired with a matching cover can be used as a raceway.

PART I. GENERAL

384.1 Scope. Article 384 covers the use, installation, and construction specifications of strut-type channel raceways and associated fittings.

384.2 Definition.

Strut-type Channel Raceway. This is a metallic raceway in which conductors and cables are placed after the raceway has been installed as a complete system intended to be mounted to a surface or suspended with associated accessories [300.18(A)]. Figure 384–1

PART II. INSTALLATION

384.10 Uses Permitted.

(1) Exposed.

(2) In dry locations.

(3) Where subject to corrosive environments, where protected by finishes suitable for the condition.

(4) Where the voltage is 600V or less.

(5) As power poles.

(6) In hazardous (classified) locations as permitted by 501.10(B)(3).

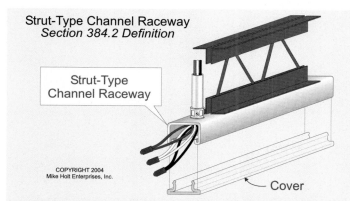

Strut-Type Channel Raceway
Section 384.2 Definition

Strut-Type
Channel Raceway

COPYRIGHT 2004
Mike Holt Enterprises, Inc.

Cover

Strut-Type Channel Raceway: A metallic raceway mounted to the surface or suspended, in which conductors or cables are laid in place after the raceway has been installed.

Figure 384–1

(7) Unbroken through walls, partitions, and floors.

(8) Indoors only, if made from a ferrous material and protected by enamel.

384.12 Uses Not Permitted. Strut-type channel raceways are not permitted as follows:

(1) Where concealed.

(2) Ferrous raceways and fittings protected from corrosion solely by enamel cannot be installed where subject to severe corrosive influences.

384.21 Conductor—Maximum Size. The maximum size conductor must not be larger than that for which the raceway is designed.

384.22 Number of Conductors. The number of conductors permitted in strut-type channel raceways must not exceed the percentage fill values in Table 384.22.

Raceways with external joiners must use 40 percent wire fill to calculate the number of conductors permitted, and raceways with internal joiners must use 25 percent wire fill to calculate the number of conductors permitted.

The ampacity adjustment factors of 310.15(B)(2)(a) do not apply to conductors installed in strut-type channel raceways where all of the following conditions are met:

(1) The cross-sectional area of the raceway is at least 4 square inches.

(2) The number of current-carrying conductors doesn't exceed 30.

(3) The sum of the cross-sectional areas of all contained conductors doesn't exceed 20 percent of the interior cross-sectional area of the strut-type channel raceways.

384.30 Securing and Supporting.

(A) Surface Mounted. A surface mount strut-type channel raceway must be secured to the mounting surface with retention straps external to the channel at intervals not exceeding 10 ft and within 3 ft of each outlet box, cabinet, junction box, or other channel raceway termination.

(B) Suspension Mount. Strut-type channel raceways can be suspension mounted with methods approved by the authority having jurisdiction designed for the purpose, at intervals not to exceed 10 ft.

384.56 Splices and Taps. Splices and taps must be accessible and must not fill the wireway to more than 75 percent of its cross-sectional area at any point.

384.60 Grounding (Bonding). Strut-type channel raceway enclosures must have a means for connecting an equipment grounding (bonding) conductor, and the strut-type channel raceways are permitted as an equipment grounding (bonding) conductor in accordance with 250.118(14).

1. A strut-type channel raceway is a metallic raceway intended to be mounted to the surface of, or suspended from, a structure with associated accessories for the installation of electrical conductors.

   (a) True   (b) False

2. A strut-type channel raceway can be installed _____.

   (a) where exposed   (b) as a power pole
   (c) unbroken through walls, partitions, and floors   (d) all of these

3. A strut-type channel raceway cannot be installed _____.

   (a) in concealed locations   (b) where subject to corrosive vapors if protected solely by enamel
   (c) a or b   (d) none of these

4. A surface mount strut-type channel raceway must be secured to the mounting surface with retention straps external to the channel at intervals not exceeding _____ and within 3 ft of each outlet box, cabinet, junction box, or other channel raceway termination.

   (a) 3 ft   (b) 5 ft   (c) 6 ft   (d) 10 ft

5. Splices and taps are permitted within a strut-type channel raceway provided they are accessible. The conductors, including splices and taps, must not fill the raceway to more than _____ percent of its area at that point.

   (a) 25   (b) 80   (c) 125   (d) 75

Notes

386 Surface Metal Raceways

Introduction

A surface metal raceway is a common method of adding a raceway when exposed conduit systems are not acceptable and concealing the raceway is not economically feasible. It comes in several colors, and is now available with colored or real wood inserts designed to make it look like molding rather than a raceway.

PART I. GENERAL

386.1 Scope.
This article covers the use, installation, and construction specifications of surface metal raceways and associated fittings.

386.2 Definition.

Surface Metal Raceway. A surface metal raceway is a metallic raceway intended to be mounted to the surface, with associated accessories, in which conductors are placed after the raceway has been installed as a complete system [300.18(A)]. **Figure 386–1**

> **Author's Comment:** Surface raceways are available in different shapes and sizes and can be mounted on walls, ceilings, or floors. They have removable covers that eliminate the need for wire pulling. Some surface raceways have two or more separate compartments, which permit the separation of power and lighting conductors from low-voltage or limited-energy conductors or cables (control, signal, and communications cables and conductors) [386.70].

386.6 Listing Requirements.
Surface metal raceways with associated fittings must be listed.

> **Author's Comment:** Enclosures for switches, receptacles, luminaires, and other devices are identified by the markings on their packaging, which identify the type of surface metal raceway with which the enclosure can be used.

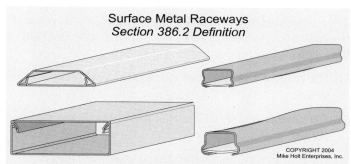

Surface Metal Raceways
Section 386.2 Definition

COPYRIGHT 2004
Mike Holt Enterprises, Inc.

Surface Raceway: A raceway intended to be mounted to the surface, in which conductors are placed after the raceway has been installed as a complete system.

Figure 386–1

PART II. INSTALLATION

386.10 Uses Permitted.

(1) In dry locations.

(2) In Class I, Division 2 hazardous (classified) locations as permitted in 501.10(B)(3).

(3) Under raised floors, as permitted in 645.5(D)(2).

(4) Extension through walls and floors, if the length passing through is unbroken and access to the conductors is maintained on both sides of the wall, partition, or floor.

386.12 Uses Not Permitted.

(1) Where subject to severe physical damage, unless otherwise approved by the authority having jurisdiction.

(2) Where the voltage is 300V or more between conductors, unless the metal has a thickness not less than 0.040 in.

(3) Where subject to corrosive vapors.

(4) In hoistways.

(5) Where concealed, except as permitted in 386.10.

386.21 Size of Conductors. The maximum size conductor permitted in a wireway must not be larger than that for which the wireway is designed.

> **Author's Comment:** Because partial packages are often purchased, you may not always get this information.

386.22 Number of Conductors. The number of conductors or cables installed in a surface metal raceway must not be greater than the number for which the raceway is designed. Cables can be installed in surface metal raceways where not prohibited by the cable's article.

The ampacity adjustment factors of 310.15(B)(2)(a) do not apply to conductors installed in surface metal raceways where all of the following conditions are met: **Figure 386–2**

(1) The cross-sectional area of the raceway is at least 4 square inches,

(2) The number of current-carrying conductors doesn't exceed 30, and

(3) The sum of the cross-sectional areas of all contained conductors doesn't exceed 20 percent of the interior cross-sectional area of the raceways.

386.30 Securing and Supporting. Surface metal raceways must be secured and supported at intervals in accordance with the manufacturer's installation instructions.

386.56 Splices and Taps. Splices and taps must be accessible and must not fill the raceway to more than 75 percent of its cross-sectional area.

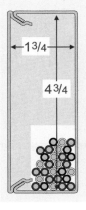

Surface Raceway - Conductor Fill
Section 386.22

← 1 3/4 →

4 3/4

The ampacity adjustment factors of 310.15(B)(2)(a) do not apply where all of the following conditions are met:

(1) Cross-sectional area exceeds 4 sq in.,
(2) Number of current-carrying conductors does not exceed 30, and
(3) Conductor fill does not exceed 20% of the cross-sectional area of the raceway.

COPYRIGHT 2004 Mike Holt Enterprises, Inc.

Figure 386–2

386.60 Grounding (Bonding). Surface metal raceway fittings must be mechanically and electrically joined together in a manner that doesn't subject the conductors to abrasion. Surface metal raceways that allow a transition to another wiring method, such as knockouts for connecting conduits, must have a means for the termination of an equipment grounding (bonding) conductor. The surface metal raceway is considered suitable as an equipment grounding (bonding) conductor in accordance with 250.118(14).

386.70 Separate Compartments. Where surface metal raceways have separate compartments within a single raceway, power and lighting conductors can occupy one compartment, and the other compartment may contain control, signaling, or communications wiring. Stamping, imprinting, or color coding of the interior finish must identify the separate compartments, and the same relative position of compartments must be maintained throughout the premises.

> **Author's Comments:**
> - Separation from power conductors is required by the *NEC* for the following low-voltage and limited-energy systems:
>   - CATV, 820.44(F)(1)
>   - Communications, 800.133(A)(1)
>   - Control and Signaling, 725.55(B)
>   - Fire Alarm, 760.55(B)
>   - Intrinsically Safe Systems, 504.30(A)(2)
>   - Instrument Tray Cable, 727.5
>   - Radio and Television, 810.18(C)
>   - Sound Systems, 640.9(C)
> - Nonconductive optical fiber cables are permitted to occupy the same cable tray or raceway as conductors for electric light, power, Class 1, or nonpower-limited fire alarm circuits [770.133(A)].

Article 386 Questions

1. It is permissible to run unbroken lengths of surface metal raceways through dry _____.

    (a) walls   (b) partitions   (c) floors   (d) all of these

2. The adjustment factors of 310.15(B)(2)(a), (Notes to Ampacity Tables of 0 through 2,000V), do not apply to conductors installed in surface metal raceways where _____.

    (a) the cross-sectional area exceeds 4 sq in
    (b) the current-carrying conductors do not exceed 30 in number
    (c) the total cross-sectional area of all conductors does not exceed 20 percent of the interior cross-sectional area of the raceway
    (d) all of these

3. Surface metal raceways must be secured and supported at intervals _____.

    (a) in accordance with the manufacturer's installation instructions
    (b) appropriate for the building design
    (c) not exceeding 8 ft
    (d) not exceeding 4 ft

4. Surface metal raceway enclosures providing a transition from other wiring methods must have a means for connecting a(n) _____.

    (a) grounded neutral conductor   (b) ungrounded conductor
    (c) equipment grounding (bonding) conductor   (d) all of these

5. Surface metal raceways and their fittings must be so designed that the sections can be _____.

    (a) electrically coupled together   (b) mechanically coupled together
    (c) installed without subjecting the wires to abrasion   (d) all of these

Notes

ARTICLE 388 — Surface Nonmetallic Raceways

Introduction

A surface nonmetallic raceway is a common method of adding a raceway when exposed conduit systems are not acceptable and concealing the raceway is not economically feasible. Surface nonmetallic raceway is less expensive than a comparable surface metallic raceway and more easily installed, but may not be as impact resistant. However, it does not dent, deform, or lose paint like the metallic versions, so it may retain its appearance longer.

PART I. GENERAL

388.1 Scope. Article 388 covers the use, installation, and construction specifications of surface nonmetallic raceways and associated fittings.

388.2 Definition.

Surface Nonmetallic Raceway. This is a nonmetallic raceway intended to be mounted to a surface, with associated accessories, in which conductors are placed after the raceway has been installed as a complete system.

388.6 Listing Requirements. Surface nonmetallic raceways with associated fittings must be listed.

PART II. INSTALLATION

388.10 Uses Permitted.

(1) In dry locations.

(2) Unbroken through walls, partitions, and floors.

388.12 Uses Not Permitted.

(1) Concealed, except as permitted by 388.10(2).

(2) Where subject to severe physical damage.

(3) Where the voltage is 300V or more between conductors, unless listed for higher voltage.

(4) In hoistways.

(5) In hazardous (classified) locations, except as permitted by 501.10(B)(3).

(6) Where subject to ambient temperatures that exceed the wireway listing.

(7) Where the conductors operate at a temperature above the raceway temperature rating.

388.21 Size of Conductors. The maximum size conductor permitted in a raceway must not be larger than that for which the raceway is designed.

388.22 Number of Conductors. The number of conductors permitted in a surface nonmetallic raceway must not exceed the number for which the raceway has been listed.

> **CAUTION:** *Surface nonmetallic raceways do not have the same heat-transfer characteristics as surface metal raceways. Therefore, the ampacity adjustment factors of Table 310.15(B)(2)(a) apply any time there are four or more current-carrying conductors in any cross-sectional area of a nonmetallic surface raceway! Effectively, this severely limits the use of surface nonmetallic raceways in these applications.*

Cables can be installed in surface nonmetallic raceways where not prohibited by the cable's article.

388.56 Splices and Taps. Splices and taps must be accessible and must not fill the surface nonmetallic raceway to more than 75 percent of its cross-sectional area.

388.60 Grounding (Bonding). Where an equipment grounding (bonding) conductor is required, it must be installed within the raceway [250.102(E) and 300.3(B)].

388.70 Separate Compartments. Where surface nonmetallic raceways have separate compartments within a single raceway, power and lighting conductors can occupy one compartment, and the other compartment may contain control, signaling, or communications wiring. Stamping, imprinting, or color coding of the interior finish must identify the separate compartments.

Author's Comments:

- Separation from power conductors is required for the following low-voltage and limited-energy systems:
  - CATV, 820.44(F)(1)
  - Communications, 800.133(A)(1)
  - Control and Signaling, 725.55(B)
  - Fire Alarm, 760.55(B)
  - Intrinsically Safe Systems, 504.30(A)(2)
  - Instrument Tray Cable, 727.5
  - Radio and Television, 810.18(C)
  - Sound Systems, 640.9(C)

- Nonconductive optical fiber cables are permitted to occupy the same cable tray or raceway with conductors for electric light, power, Class 1, or nonpower-limited fire alarm circuits [770.133(A)].

Article 388 Questions

1. It is permissible to run unbroken lengths of surface nonmetallic raceways through dry _____.

    (a) walls   (b) partitions   (c) floors   (d) all of these

2. The maximum number of conductors permitted in any surface nonmetallic raceway must be _____.

    (a) no more than 30 percent of the inside diameter   (b) no greater than the number for which it was designed
    (c) no more than 75 percent of the cross-sectional area   (d) that which is permitted in Table 312.6(A)

3. The conductors, including splices and taps, in a nonmetallic surface raceway having a removable cover, must not fill the raceway to more than _____ percent of its cross-sectional area at that point.

    (a) 75   (b) 40   (c) 38   (d) 53

4. Where combination surface nonmetallic raceways are used for both signaling conductors and for lighting and power circuits, the different systems must be run in separate compartments identified by _____ of the interior finish.

    (a) stamping   (b) imprinting   (c) color coding   (d) any of these

Notes

392 Cable Trays

Introduction

A cable tray system is a unit or assembly of units or sections with associated fittings that forms a structural system used to securely fasten or support cables and raceways. Cable systems include ladder, ventilated trough, ventilated channel, solid bottom, and other similar structures.

392.1 Scope. Article 392 covers cable tray systems, including ladder, ventilated trough, ventilated channel, solid bottom, and other similar structures.

392.2 Definition.

Cable Tray System. A unit or assembly of units or sections with associated fittings forming a rigid structural system used to securely fasten or support cables and raceways.

> **Author's Comment:** A cable tray isn't a raceway. It's a support system for cables and raceways.

392.3 Uses Permitted. Cable trays can be used as a support system for service, feeder, or branch-circuit conductors, as well as communications circuits, control circuits, and signaling circuits. Figure 392-1

Author's Comments:

- Cable trays used to support service-entrance conductors must contain only service-entrance conductors unless a solid fixed barrier separates the service-entrance conductors from other conductors [230.44].

  Cable tray installations aren't limited to industrial establishments. Where exposed to direct rays of the sun, insulated conductors and jacketed cables must be identified as being sunlight resistant. The manufacturer must identify cable trays and associated fittings for their intended use.

- Cable trays are manufactured in many forms, from a simple hanger or wire mesh to a substantial, rigid, steel support system. Cable trays are designed and manufactured to support specific wiring methods.

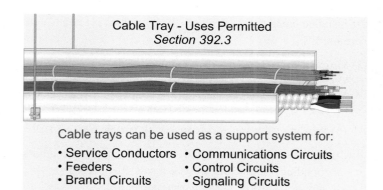

Cable Tray - Uses Permitted
Section 392.3

Cable trays can be used as a support system for:

- Service Conductors
- Feeders
- Branch Circuits
- Communications Circuits
- Control Circuits
- Signaling Circuits

COPYRIGHT 2004 Mike Holt Enterprises, Inc.

Figure 392–1

(A) Wiring Methods. Any of the following wiring methods can be installed in a cable tray.

Armored cable	320
Communications raceways	800
Electrical metallic tubing	358
Electrical nonmetallic tubing	362
Fire alarm cables	760
Flexible metal conduit	348
Instrumentation tray cable	727
Intermediate metal conduit	342
Liquidtight flexible metal conduit	350
Liquidtight flexible nonmetallic conduit	356
Metal-clad cable	330
Multipurpose and communications cables	800
Nonmetallic-sheathed cable	334

Power and control tray cable	336
Power-limited tray cable	725.61(C)
and	725.71(F)
Optical fiber cables and raceways	770
Rigid metal conduit	344
Rigid nonmetallic conduit	352
Service-entrance cable	338
Underground feeder and branch-circuit cable	340

Author's Comment: Control, signal, and communications cables must be separated from the power conductors by a barrier or maintain a 2 in. separation.

- Coaxial Cables, 820.133(A)(2) Ex. 1
- Class 2 and 3 Cables, 725.55(H)
- Communications Cables, 800.133(A)(2) Ex. 1
- Fire Alarm Cables, 760.55(G)
- Intrinsically Safe Systems Cables, 504.30(A)(2) Ex. 1
- Radio and Television Cables, 810.18(B) Ex. 1

(B) In Industrial Establishments. Where conditions of maintenance and supervision ensure that only qualified persons service the installed cable tray system, any of the cables (1) and (2) are permitted.

(1) Single Conductors. Single-conductor cables:

(a) 1/0 AWG or larger listed and marked for use in cable trays.

(b) Welding cables in accordance with Article 630, Part IV.

(c) Single conductors used as equipment grounding (bonding) conductors must be insulated, covered, or bare, and they must be 4 AWG or larger.

(C) Equipment Grounding (Bonding) Conductors. Metallic cable trays can serve as equipment grounding (bonding) conductors where qualified persons service the cable tray system, and the cable tray complies with 392.7.

(D) Hazardous (Classified) Locations. Cable trays in hazardous (classified) locations must contain only the cable types permitted in 501.10, 502.10, 503.10, 504.20, and 505.15.

(E) Nonmetallic Cable Tray. In addition to the uses permitted elsewhere in Article 392, nonmetallic cable trays can be installed in corrosive areas and in areas requiring voltage isolation.

392.4 Uses Not Permitted. Cable tray systems are not permitted:

- In hoistways.
- Where subject to severe physical damage.
- In ducts, plenums, and other air-handling spaces, except as permitted by 300.22 to support wiring methods recognized for use in such spaces.

392.6 Installation.

(A) Complete System. Cable trays must be installed as a complete system, except that mechanically discontinuous segments between cable tray runs, or between cable tray runs and equipment are permitted. A bonding jumper sized in accordance with 250.102 and installed in accordance with 250.96 must bond the sections of cable tray, or the cable tray and the raceway or equipment.

(B) Completed Before Installation. Each run of cable tray must be completed before the installation of cables or conductors.

(C) Support. Supports for cable trays must be provided to prevent stress on cables where they enter raceways or other enclosures from cable tray systems. Cable trays must be supported in accordance with the manufacturer's installation instructions.

(G) Through Partitions and Walls. Cable trays can extend through partitions and walls, or vertically through platforms and floors where the installation is made in accordance with the fire seal requirements of 300.21.

(H) Exposed and Accessible. Cable trays must be exposed and accessible, except as permitted by 392.6(G).

(I) Adequate Access. Sufficient space must be provided and maintained about cable trays to permit adequate access for installing and maintaining the cables.

(J) Raceways, Cables, and Boxes Supported from Cable Trays. In industrial facilities where conditions of maintenance and supervision ensure that only qualified persons will service the installation, and where the cable tray system is designed and installed to support the load, cable tray systems are permitted to support raceways, cables, boxes, and conduit bodies. Figure 392–2

For raceways terminating at the tray, a listed cable tray clamp or adapter must be used to securely fasten the raceway to the cable tray system. The raceway must be supported in accordance with the appropriate raceway article.

Raceways or cables running parallel to the cable tray system can be attached to the bottom or side of a cable tray system. The raceway or cable must be fastened and supported in accordance with the appropriate raceway or cable's article.

Boxes and conduit bodies attached to the bottom or side of a cable tray system must be fastened and supported in accordance with 314.23.

392.7 Grounding (Bonding).

(A) Metallic Cable Trays. Metallic cable trays supporting electrical conductors must be grounded (bonded) to an effective ground-fault current path in accordance with 250.96.

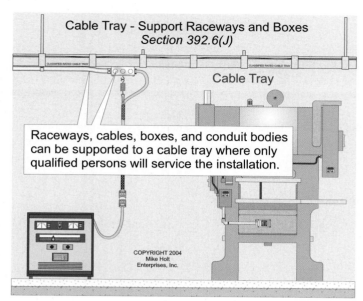

Cable Tray - Support Raceways and Boxes
Section 392.6(J)

Cable Tray

Raceways, cables, boxes, and conduit bodies can be supported to a cable tray where only qualified persons will service the installation.

COPYRIGHT 2004
Mike Holt
Enterprises, Inc.

Figure 392–2

(B) Steel or Aluminum Cable Tray Systems. Steel or aluminum cable tray systems are permitted to function as an equipment grounding (bonding) conductor if all the following requirements are met:

(1) Cable tray and fittings are identified for bonding purposes.

(2) The minimum cross-sectional area of cable tray conforms to Table 392.7(B).

(3) Cable tray and fittings are legibly and durably marked to show the metallic cross-sectional area of channel or one-piece construction trays, or total metallic cross-sectional area of both side rails for ladder or trough cable trays.

(4) Cable tray, fittings, and raceways are bonded in accordance with 250.96 using bolted mechanical connectors or bonding jumpers sized in accordance with 250.102.

392.8 Cable Installation.

(A) Cable Splices. Accessible splices are permitted if the splice is insulated by a method approved by the authority having jurisdiction, and they do not project above the side rails of the cable tray.

(B) Fastened Securely. Cables run vertically must be securely fastened to the cable tray.

(C) Bushed Conduit and Tubing. A box isn't required where cables or conductors exit a bushed raceway that is used for the support or protection of the conductors.

(D) Connected in Parallel. To prevent unbalanced current in the parallel conductors due to inductive reactance, all circuit conductors of a parallel set (A, B, C, N) [310.4] must be bundled together and secured to prevent excessive movement due to fault-current magnetic forces.

(E) Single Conductors. Single conductors sized 1/0 through 4/0 AWG must be installed in a single layer, unless the conductors are part of a circuit group as permitted in 392.8(D) for parallel sets.

1. A cable tray is a unit or assembly of units or sections and associated fittings forming a _____ system used to securely fasten or support cables and raceways.

    (a) structural   (b) flexible   (c) movable   (d) secure

2. Cable trays and their associated fittings must be _____ for the intended use.

    (a) listed   (b) approved   (c) identified   (d) none of these

3. Supports for cable trays must be provided in accordance with _____.

    (a) installation instructions   (b) the *NEC*   (c) a or b   (d) none of these

4. For raceways terminating at the tray, a(n) _____ cable tray clamp or adapter must be used to securely fasten the raceway to the cable tray system.

    (a) listed   (b) approved   (c) identified   (d) none of these

5. One of the requirements that must be met to use steel or aluminum cable tray systems as equipment grounding (bonding) conductors, is that the cable tray sections and fittings have been _____ marked to show the cross-sectional area of metal in channel cable trays, or cable trays of one-piece construction and total cross sectional area of both side rails for ladder or trough cable trays.

    (a) legibly   (b) durably   (c) a or b   (d) a and b

CHAPTER 4
Equipment for General Use

Introduction

With the first three chapters behind you, the final chapter in the *NEC* for building a solid foundation in general work is Chapter 4. This chapter helps you apply the first three chapters to general equipment. These first four chapters follow a natural sequential progression. Each of the next four Chapters—5, 6, 7, and 8—builds upon the first four, but in no particular order. You do not need to understand any of the other "next chapters" to work with any one of them. But you do need to understand all of the first four chapters to properly apply any of the next four.

Chapter 4 has some logical arrangements of its own. Here are the groupings:

- Flexible cords and cables, fixture wires, switches, receptacles
- Switchboards and panelboards
- Lamps, lighting, appliances, and space heaters
- Motors, refrigeration equipment, generators, and transformers
- Capacitors and other components

These groupings make sense. For example, motors, refrigeration equipment, generators, and transformers are all inductive equipment.

This logical arrangement of the *NEC* is something to keep in mind when you're searching for a particular item. You know, for example, that transformers are general equipment. So you will find the *Code* requirements in Chapter 4. You know they are wound devices. So you will find transformer requirements located somewhere near motor requirements. Refrigeration equipment in the *NEC* means hermetically sealed motors. So the requirements should logically be located right next to motor requirements. And that's exactly where they are.

Article 400. Flexible Cords and Cables. This article covers the general requirements, applications, and construction specifications for flexible cords and flexible cables.

Article 402. Fixture Wires. This article covers the general requirements and construction specifications for fixture wires.

Article 404. Switches. The requirements of Article 404 apply to switches of all types. These include snap (toggle) switches, dimmers, fan switches, knife switches, circuit breakers used as switches, and automatic switches such as time clocks, timers, and switches and circuit breakers used for disconnecting means.

Article 406. Receptacles, Cord Connectors, and Attachment Plugs (Caps). This article covers the rating, type, and installation of receptacles, cord connectors, and attachment plugs (cord caps).

Article 408. Switchboards and Panelboards. Article 408 covers specific requirements for switchboards, panelboards, and distribution boards that control light and power circuits.

Article 410. Luminaires, Lampholders, and Lamps. Article 410 contains the requirements for luminaires, lampholders, and lamps. Because of the many types and applications of luminaires, manufacturer's instructions are very important and helpful for proper installation. UL produces a pamphlet called the *Luminaire Marking Guide*, which provides information for properly installing common types of incandescent, fluorescent, and high-intensity discharge (HID) luminaires.

Article 411. Lighting Systems Operating at 30V or Less. This article covers lighting systems and their associated components operating at 30V or less.

Article 422. Appliances. Article 422 covers electric appliances used in any occupancy.

Article 424. Fixed Electric Space-Heating Equipment. This article covers fixed electrical equipment used for space heating. For the purpose of this article, heating equipment includes heating cable, unit heaters, boilers, central systems, and other fixed electric space-heating equipment. This article does not apply to process heating and room air conditioning.

Article 430. Motors, Motor Circuits, and Controllers. Article 430 contains the specific requirements for conductor sizing, overcurrent protection, control circuit conductors, motor controllers, and disconnecting means. The installation requirements for motor control centers are covered in Article 430, Part VIII.

Article 440. Air-Conditioning and Refrigeration Equipment. This article applies to electrically driven air-conditioning and refrigeration equipment with a motorized hermetic refrigerant compressor. The requirements in this article are in addition to, or amend, the requirements in Article 430 and other articles.

Article 445. Generators. This article contains the electrical installation requirements for generators, such as where they can be installed, nameplate markings, conductor ampacity, and disconnecting means.

Article 450. Transformers and Transformer Vaults. This article covers the installation of all transformers.

Article 460. Capacitors. This article covers the installation of capacitors, including those in hazardous (classified) locations as modified by Articles 501 through 503.

Chapter 4 doesn't end with Article 460. The remaining articles are also important, but they do not address topics the typical electrician deals with. These are:

- **Article 480. Batteries.**

- **Article 490. Equipment, Over 600 Volts, Nominal.**

You may want to read through these, just to see what's there. After you finish Article 460 in this book, you will have completed your study of the first four Chapters of the *NEC*, and will have a solid foundation for applying it to your work.

400 Flexible Cords and Flexible Cables

Introduction

This article covers the general requirements, applications, and construction specifications for flexible cords and flexible cables. The *NEC* doesn't consider flexible cords to be wiring methods like those defined in Chapter 3.

Always use a cord (and fittings) identified for the application. For example, use cords listed for a wet location if you're using the cord outdoors in a wet location. The jacket material of any cord is tested to maintain its insulation properties and other characteristics only in the environments for which it has been listed.

This textbook doesn't go into detail about Table 400.4, but you should take a few moments to review it. It's not limited to extension cords—three of the entries are for elevator cables. You do not need to memorize this table, but do become aware of the types of cords and cables it covers.

400.1 Scope. Article 400 covers the general requirements, applications, and construction specifications for flexible cords and flexible cables as contained in Table 400.4.

400.3 Suitability. Flexible cords and flexible cables, as well as their fittings must be suitable for the use and location.

> **Author's Comment:** Other *NEC* Articles might have specific requirements. **Figure 400–1**

400.4 Types of Flexible Cords and Flexible Cables. The use of flexible cords and flexible cables must conform to the description contained in Table 400.4.

400.5 Ampacity of Flexible Cords and Flexible Cables. Tables 400.5(A) and 400.5(B) list the allowable ampacity for copper conductors in flexible cords and flexible cables with not more than three current-carrying conductors at an ambient temperature of 86°F.

Where the number of current-carrying conductors in a cable or raceway exceeds three, the allowable ampacity of each conductor must be adjusted in accordance with the following multipliers:

Table 400.5 Adjustment Factor

Current–Carrying	Ampacity Multiplier
4–6 Conductors	0.80
7–9 Conductors	0.70
10–20 Conductors	0.50

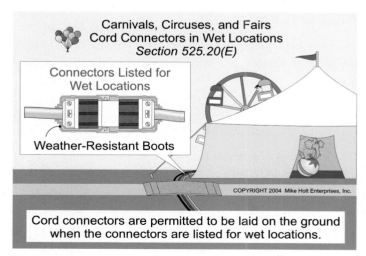

Carnivals, Circuses, and Fairs
Cord Connectors in Wet Locations
Section 525.20(E)

Connectors Listed for Wet Locations

Weather-Resistant Boots

Cord connectors are permitted to be laid on the ground when the connectors are listed for wet locations.

Figure 400–1

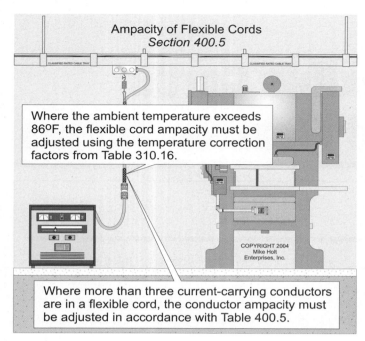

Ampacity of Flexible Cords
Section 400.5

Where the ambient temperature exceeds 86ºF, the flexible cord ampacity must be adjusted using the temperature correction factors from Table 310.16.

Where more than three current-carrying conductors are in a flexible cord, the conductor ampacity must be adjusted in accordance with Table 400.5.

COPYRIGHT 2004
Mike Holt
Enterprises, Inc.

Figure 400–2

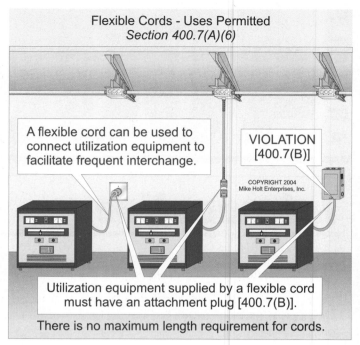

Flexible Cords - Uses Permitted
Section 400.7(A)(6)

A flexible cord can be used to connect utilization equipment to facilitate frequent interchange.

VIOLATION
[400.7(B)]

COPYRIGHT 2004
Mike Holt Enterprises, Inc.

Utilization equipment supplied by a flexible cord must have an attachment plug [400.7(B)].

There is no maximum length requirement for cords.

Figure 400–3

Where the ambient temperature exceeds 86°F, the flexible cord or flexible cable ampacity, as listed in Table 400.5(A) or 400.5(B), must be adjusted by using the temperature correction factors listed in Table 310.16. **Figure 400–2**

Author's Comments:

- Temperature rating for flexible cords and flexible cables are not contained in the *NEC*, but UL listing standards state that flexible cords and flexible cables are rated for 60°C unless marked otherwise.

- See 400.13 for overcurrent protection requirements for flexible cords and flexible cables.

400.7 Uses Permitted.

(A) Uses Permitted. Flexible cords and flexible cables within the scope of this article can be used for the following applications:

(1) Pendants [210.50(A) and 314.23(H)].

(2) Wiring of luminaires [410.14 and 410.30(B)].

(3) Connection of portable lamps, portable and mobile signs, or appliances [422.16].

(4) Elevator cables.

(5) Wiring of cranes and hoists.

(6) Connection of utilization equipment to facilitate frequent interchange [422.16]. **Figure 400–3**

(7) Prevention of the transmission of noise or vibration [422.16].

(8) Appliances where the fastening means and mechanical connections are specifically designed to permit ready removal for maintenance and repair, and the appliance is intended or identified for flexible cord connections [422.16].

(9) Connection of moving parts.

(10) Where specifically permitted elsewhere in this *Code*.

Author's Comment: Flexible cords and flexible cables are permitted for fixed permanent wiring by 501.10(A)(2) and (B)(2), 501.140, 502.4(A)(1)(e), 502.4(B)(2), 503.3(A)(2), 550.10(B), 553.7(B), and 555.13(A)(2).

(B) Attachment Plugs. Attachment plugs are required for flexible cords used for any of the following applications: **Figure 400–4**

- Portable lamps, portable and mobile signs, or appliances [400.7(A)(3)].
- Stationary equipment to facilitate its frequent interchange [400.7(A)(6)]. See 422.16.
- Appliances specifically designed to permit ready removal for maintenance and repair, and identified for flexible cord connection [400.7(A)(8)].

Author's Comment: An attachment plug can serve as the disconnecting means for stationary appliances [422.33] and room air conditioners [440.63].

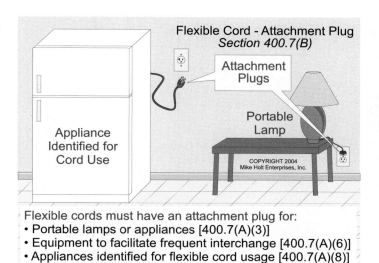

Flexible Cord - Attachment Plug
Section 400.7(B)

Attachment Plugs

Portable Lamp

Appliance Identified for Cord Use

COPYRIGHT 2004 Mike Holt Enterprises, Inc.

Flexible cords must have an attachment plug for:
• Portable lamps or appliances [400.7(A)(3)]
• Equipment to facilitate frequent interchange [400.7(A)(6)]
• Appliances identified for flexible cord usage [400.7(A)(8)]

Figure 400–4

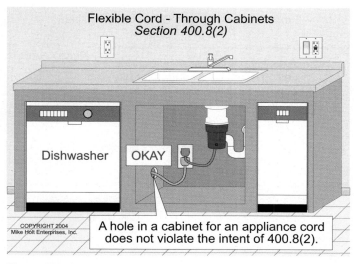

Flexible Cord - Through Cabinets
Section 400.8(2)

Dishwasher   OKAY

COPYRIGHT 2004 Mike Holt Enterprises, Inc.

A hole in a cabinet for an appliance cord does not violate the intent of 400.8(2).

Figure 400–6

400.8 Uses Not Permitted.
Unless specifically permitted in 400.7, flexible cords must not be:

(1) Used as a substitute for the fixed wiring of a structure.

(2) Run through holes in walls, structural ceilings, suspended/dropped ceilings, or floors. **Figure 400–5**

Author's Comment: According to an article in the International Association of Electrical Inspectors magazine (*IAEI News*), a flexible cord run through a cabinet for an appliance isn't considered as being run through a wall. **Figure 400–6**

(3) Run through doorways, windows, or similar openings.

(4) Attached to building surfaces.

(5) Where concealed by walls, floors, or ceilings, or located above suspended or dropped ceilings. **Figure 400–7**

Author's Comments:
• Flexible cords are permitted within a raised floor used for environmental air because this space isn't considered a concealed space. See Article 100 for the definition of "Exposed."
• Receptacle outlets are permitted above a suspended ceiling, but they cannot be used for flexible cords used as a wiring method. **Figure 400–8**

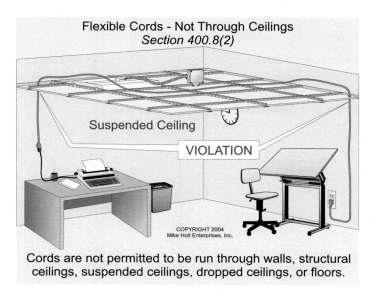

Flexible Cords - Not Through Ceilings
Section 400.8(2)

Suspended Ceiling

VIOLATION

COPYRIGHT 2004 Mike Holt Enterprises, Inc.

Cords are not permitted to be run through walls, structural ceilings, suspended ceilings, dropped ceilings, or floors.

Figure 400–5

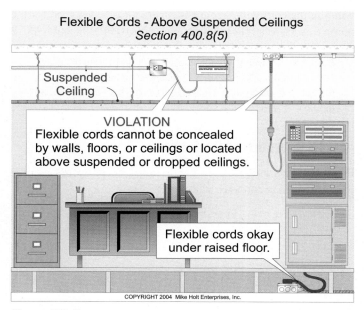

Flexible Cords - Above Suspended Ceilings
Section 400.8(5)

Suspended Ceiling

VIOLATION
Flexible cords cannot be concealed by walls, floors, or ceilings or located above suspended or dropped ceilings.

Flexible cords okay under raised floor.

COPYRIGHT 2004 Mike Holt Enterprises, Inc.

Figure 400–7

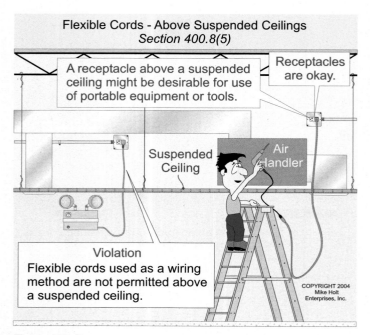

Flexible Cords - Above Suspended Ceilings
Section 400.8(5)

A receptacle above a suspended ceiling might be desirable for use of portable equipment or tools.

Receptacles are okay.

Suspended Ceiling

Air Handler

Violation
Flexible cords used as a wiring method are not permitted above a suspended ceiling.

COPYRIGHT 2004 Mike Holt Enterprises, Inc.

Figure 400–8

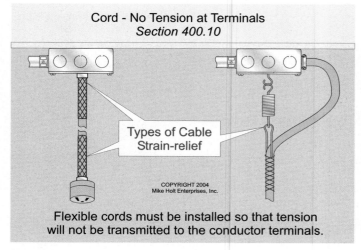

Cord - No Tension at Terminals
Section 400.10

Types of Cable Strain-relief

COPYRIGHT 2004 Mike Holt Enterprises, Inc.

Flexible cords must be installed so that tension will not be transmitted to the conductor terminals.

Figure 400–9

(6) Installed in raceways.

(7) Where subject to physical damage.

Author's Comment: You would think the *Code* doesn't have to specify everything, but it does!

400.10 Pull at Joints and Terminals.
Flexible cords must be installed so tension will not be transmitted to the conductor terminals. This can be accopmlished by knotting the cord, winding the cord with tape, or by using fittings designed for the purpose, such as strain-relief fittings [400.10 FPN]. **Figure 400–9**

Author's Comment: When critical health and economic activities are dependant on flexible cord supplied equipment, the best method is a factory-made, stress-relieving, listed device; not an old-timer's knot.

400.13 Overcurrent Protection.

Flexible cords and flexible cables must be protected against overcurrent in accordance with 240.5

Author's Comment: Section 240.5 contains the following requirements:

- Rated no higher than the cord's ampacity as specified in Table 400.5(A) and Table 400.5(B) [240.5(A)].
- Flexible cord for listed utilization equipment is considered protected when used in accordance with the equipment listing requirements [240.5(B)(1)].
- Extension cord sets are considered protected when used in accordance with the extension cord listing requirements [240.5(B)(3)].
- Flexible cord used in field-installed extension cords, made with separately listed and installed components, can be supplied by a 20A branch circuit for 16 AWG and larger conductors [240.5(B)(4)].

400.14 Protection from Damage.
Flexible cords must be protected by bushings or fittings where passing through holes in covers, outlet boxes, or similar enclosures.

In industrial establishments where the conditions of maintenance and supervision ensure that only qualified persons will service the installation, flexible cords or flexible cables not exceeding 50 ft can be installed in aboveground raceways.

1. The allowable ampacity of flexible cords and cables is found in _____.

   (a) Table 310.16   (b) Table 400.5(A) and (B)   (c) Table 1, Chapter 9   (d) Table 430.52

2. A 3-conductor 16 AWG, SJE cable (one conductor is used for grounding) has a maximum ampacity of _____ for each conductor.

   (a) 13A   (b) 12A   (c) 15A   (d) 8A

3. Flexible cords must not be used as a substitute for _____ wiring unless specifically permitted in 400.7.

   (a) temporary   (b) fixed   (c) overhead   (d) none of these

4. Flexible cords and cables must not be concealed behind building _____, or run through doorways, windows, or similar openings.

   (a) structural ceilings   (b) suspended or dropped ceilings
   (c) floors or walls   (d) all of these

5. In industrial establishments where conditions of maintenance and supervision ensure that only qualified persons service the installation, flexible cords and cables are permitted to be installed in aboveground raceways that are no longer than _____, to protect the flexible cord or cable from physical damage.

   (a) 25 ft   (b) 50 ft   (c) 100 ft   (d) no limit

Notes

Mike Holt Enterprises, Inc. • www.NECcode.com • 1.888.NEC.Code

Fixture Wires

Introduction

This article covers the general requirements and construction specifications for fixture wires. One such requirement is that no fixture wire can be smaller than 18 AWG. Another requirement is that fixture wires must be of a type listed in Table 402.3. That table makes up the bulk of Article 402.

402.1 Scope. Article 402 covers the general requirements and construction specifications for fixture wires.

402.3 Types. Fixture wires must be of a type contained in Table 402.3.

402.5 Allowable Ampacity of Fixture Wires. The allowable ampacity of fixture wires is as follows:

Table 402.5
Allowable Ampacity for Fixture Wires

Wire AWG	Wire Ampacity
18	6A
16	8A
14	17A
12	23A
10	28A

402.6 Minimum Size. Fixture wires must not be smaller than 18 AWG.

402.7 Raceway Size. Raceways must be large enough to permit the installation and removal of conductors without damaging the conductors' insulation. The number of fixture wires permitted in a single conduit or tubing must not exceed the percentage fill specified in Table 1, Chapter 9.

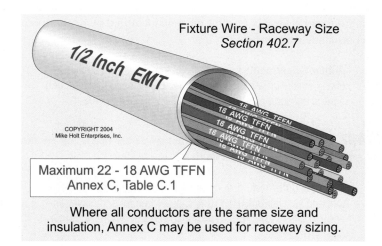

Fixture Wire - Raceway Size
Section 402.7

1/2 Inch EMT

COPYRIGHT 2004
Mike Holt Enterprises, Inc.

Maximum 22 - 18 AWG TFFN
Annex C, Table C.1

Where all conductors are the same size and insulation, Annex C may be used for raceway sizing.

Figure 402–1

Author's Comment: When all conductors in a raceway are the same size and insulation, the number of conductors permitted can be found in Annex C for the raceway type.

Question: How many 18 TFFN conductors can be installed in trade size ½ electrical metallic tubing? **Figure 402–1**

(a) 12   (b) 14   (c) 19   (d) 22

Answer: (d) 22 conductors, Annex C, Table C.1

Author's Comment: See 300.17 for additional examples on how to size raceways when conductors aren't all the same size.

Figure 402–2

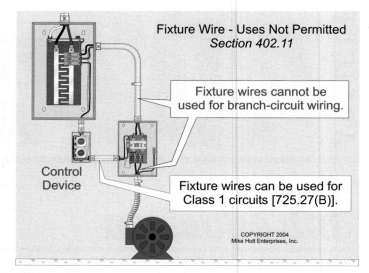

Figure 402–4

402.8 Grounded Neutral Conductor. Fixture wire used as a grounded neutral conductor must be identified by one or more continuous white stripes.

> **Author's Comment:** To prevent electric shock, the screw shell of a luminaire or lampholder must be connected to the grounded neutral conductor [200.10(C) and 410.23]. **Figure 402–2**

402.10 Uses Permitted. Fixture wires are permitted for the connection of luminaires. **Figure 402–3**

> **Author's Comment:** Fixture wires can also be used for elevators and escalators [620.11(C)], Class 1 control and power-limited circuits [725.27(B)], and nonpower-limited fire alarm circuits [760.27(B)].

402.11 Uses Not Permitted. Fixture wires cannot be used for branch-circuit wiring. **Figure 402–4**

> **Author's Comment:** Fixture wires can be used for Class 1 remote-control circuits [725.27(B)].

402.12 Overcurrent Protection. Fixture wires must be protected against overcurrent according to the requirements contained in 240.5.

> **Author's Comment:** Fixture wires used for motor control circuit taps must have overcurrent protection in accordance with 430.72(A), and Class 1 remote-control circuits must have overcurrent protection in accordance with 725.23.

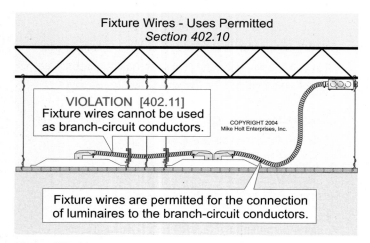

Figure 402–3

1.   18 TFFN has an ampacity of _____.

   (a) 14A   (b) 10A   (c) 8A   (d) 6A

2.   The smallest size fixture wire permitted in the *NEC* is _____ AWG.

   (a) 22   (b) 20   (c) 18   (d) 16

3.   Fixture wires are used to connect luminaires to the _____ conductors supplying the luminaires.

   (a) service   (b) branch-circuit   (c) feeder   (d) none of these

4.   Fixture wires are permitted for installation in luminaires and in similar equipment where enclosed or protected and not subject to _____ in use, or for connecting luminaires to the branch-circuit conductors supplying the luminaires.

   (a) bending or twisting   (b) knotting   (c) stretching or straining   (d) none of these

5.   Fixture wires cannot be used for branch-circuit wiring.

   (a) True   (b) False

Notes

404 Switches

Introduction

The requirements of Article 404 apply to switches of all types, such as snap (toggle) switches, dimmers, fan switches, knife switches, circuit breakers used as switches, and automatic switches. Automatic switches include those used as time clocks and timers, plus switches and circuit breakers used for disconnecting means. Here are a few key points to remember:

- Enclosures for switches or circuit breakers can contain splices if you meet certain conditions.
- Observe the wet location requirements. These include locations subject to saturation with water such as those near some showers, tubs, and pools.
- Observe switch grouping and accessibility requirements.
- Observe requirements for mounting, marking, grounding (bonding), orientation, rating, and labeling of various kinds of switches.

404.1 Scope. The requirements of Article 404 apply to all types of switches, such as snap (toggle) switches, safety (cutout) switches, knife switches, circuit breakers used as switches, and automatic switches such as time clocks, including switches and circuit breakers used for disconnecting means.

404.2 Switch Connections.

(A) Three-Way and Four-Way Switches. Wiring for three-way and four-way switching must be done so that only the ungrounded conductor is switched. **Figure 404–1**

> **Author's Comment:** In other words, the grounded neutral conductor must not be switched. The white insulated conductor within a cable assembly can be used for single-pole, three-way or four-way switch loops if it's permanently reidentified to indicate its use as an ungrounded conductor at each location where the conductor is visible and accessible [200.7(C)(2)].

Where metal raceway or metal-clad cable contains the ungrounded conductors for switches, the wiring must be arranged to avoid heating the surrounding metal by induction. This is accomplished by installing all circuit conductors in the same raceway in accordance with 300.3(B) and 300.20(A), or ensuring that they're all within the same cable.

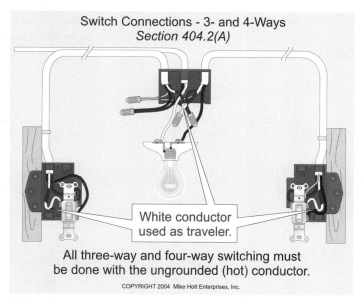

Switch Connections - 3- and 4-Ways
Section 404.2(A)

White conductor used as traveler.

All three-way and four-way switching must be done with the ungrounded (hot) conductor.

COPYRIGHT 2004 Mike Holt Enterprises, Inc.

Figure 404–1

Exception: A grounded neutral conductor isn't required in the same raceway or cable with travelers and switch leg (switch loops) conductors. **Figure 404–2**

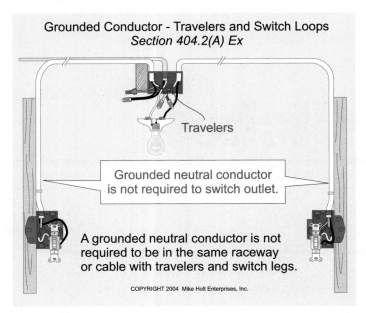

Figure 404–2

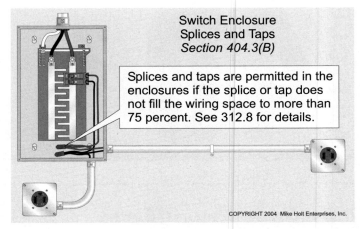

Figure 404–4

(B) Switching Grounded Neutral Conductors. Only the ungrounded conductor is permitted to be used for switching. Figure 404–3

404.3 Switch Enclosures.

(A) General. Switches and circuit breakers used as switches must be of the externally operable type mounted in an enclosure listed for the intended use.

(B) Used for Raceways or Splices. Switch or circuit breaker enclosures are permitted to contain splices and taps if the splices and/or taps do not fill the wiring space at any cross section to more than 75 percent. Figure 404–4

Switch or circuit breaker enclosures are permitted to have conductors feed through them if the wiring at any cross section doesn't exceed 40 percent [312.8].

404.4 Wet Locations. Switches and circuit breakers used as switches installed in wet locations must be installed in weatherproof enclosures. The enclosure must be installed so that not less than ¼ in. airspace is provided between the enclosure and the wall or other supporting surface [312.2(A)]. Figure 404–5

Switches can be located next to, but not within, a bathtub or shower space. Figure 404–6

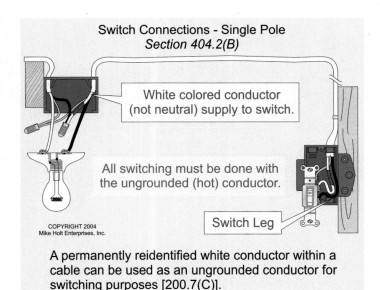

Figure 404–3

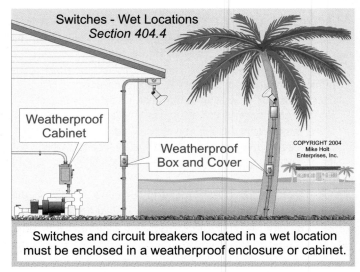

Figure 404–5

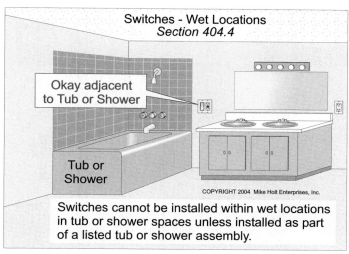

Switches - Wet Locations
Section 404.4

Okay adjacent
to Tub or Shower

Tub or
Shower

COPYRIGHT 2004 Mike Holt Enterprises, Inc.

Switches cannot be installed within wet locations
in tub or shower spaces unless installed as part
of a listed tub or shower assembly.

Figure 404–6

Author's Comment: Switches must be located not less than 5 ft from pools [680.22(C)], outdoor spas and hot tubs [680.41], and indoor spas or hot tubs [680.43(C)], **Figure 404–7**. This 5 ft rule doesn't apply to switches located adjacent to bathtubs, shower stalls, or hydromassage bathtubs [680.70 and 680.72].

404.6 Position of Knife Switches.

(A) Single-Throw Knife Switch. Single-throw knife switches must be installed so that gravity will not tend to close them.

404.7 Indicating. Switches, motor circuit switches, and circuit breakers used as switches must be marked to indicate whether they are in the "on" or "off" position. When the switch is operated vertically, it must be installed so the "up" position is the "on" position [240.81]. **Figure 404–8**

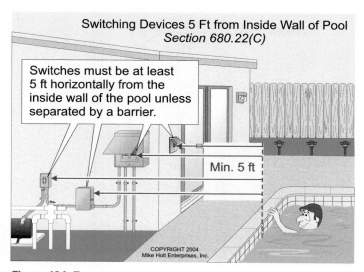

Switching Devices 5 Ft from Inside Wall of Pool
Section 680.22(C)

Switches must be at least
5 ft horizontally from the
inside wall of the pool unless
separated by a barrier.

Min. 5 ft

COPYRIGHT 2004
Mike Holt Enterprises, Inc.

Figure 404–7

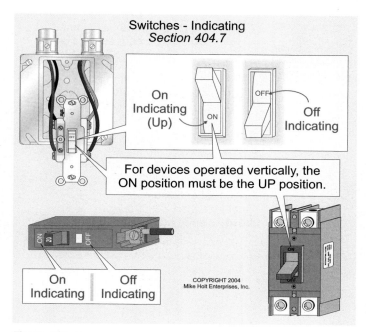

Switches - Indicating
Section 404.7

On
Indicating
(Up)

OFF

Off
Indicating

ON

For devices operated vertically, the
ON position must be the UP position.

On
Indicating

Off
Indicating

COPYRIGHT 2004
Mike Holt Enterprises, Inc.

Figure 404–8

Exception 1: Double-throw switches, such as three-way and four-way switches, aren't required to be marked "on" or "off."

Exception 2: On busway installations, tap switches employing a center-pivoting handle can be open or closed with either end of the handle in the up or down position. The switch position must be clearly indicated and must be visible from the floor or from the usual point of operation.

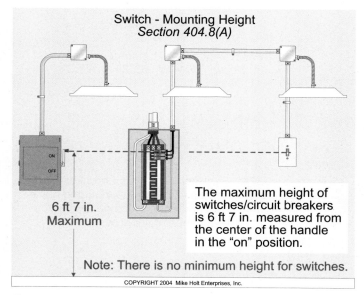

Switch - Mounting Height
Section 404.8(A)

ON

OFF

6 ft 7 in.
Maximum

The maximum height of
switches/circuit breakers
is 6 ft 7 in. measured from
the center of the handle
in the "on" position.

Note: There is no minimum height for switches.

COPYRIGHT 2004 Mike Holt Enterprises, Inc.

Figure 404–9

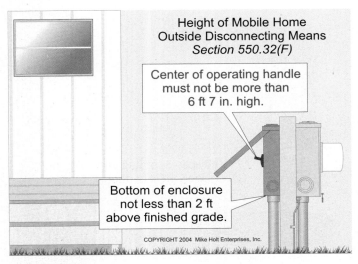

Height of Mobile Home
Outside Disconnecting Means
Section 550.32(F)

Center of operating handle
must not be more than
6 ft 7 in. high.

Bottom of enclosure
not less than 2 ft
above finished grade.

COPYRIGHT 2004 Mike Holt Enterprises, Inc.

Figure 404–10

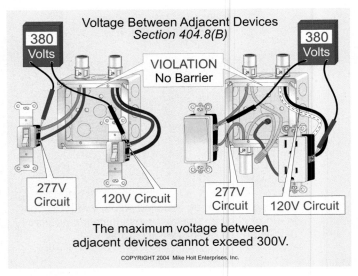

Voltage Between Adjacent Devices
Section 404.8(B)

380 Volts   380 Volts

VIOLATION
No Barrier

277V Circuit   120V Circuit   277V Circuit   120V Circuit

The maximum voltage between
adjacent devices cannot exceed 300V.

COPYRIGHT 2004 Mike Holt Enterprises, Inc.

Figure 404–12

404.8 Accessibility and Grouping.

(A) Location. All switches and circuit breakers used as switches must be capable of being operated from a readily accessible location. They must be installed so the center of the grip of the operating handle of the switch or circuit breaker, when in its highest position, isn't more than 6 ft 7 in. above the floor or working platform. **Figure 404–9**

> **Author's Comment:** There's no minimum height requirement for switches and circuit breakers used as switches. However, 550.32(F) requires the service disconnecting means enclosure for mobile and manufactured homes to be mounted a minimum of 2 ft above the finished grade. **Figure 404–10**

Exception 1: On busways, fusible switches, and circuit breakers where suitable means is provided to operate the handle of the device from the floor.

Exception 2: Switches and circuit breakers used as switches can be mounted above 6 ft 7 in. if they are next to the equipment they supply and are accessible by portable means [240.24(A)(4)]. **Figure 404–11**

(B) Voltage Between Devices. Snap switches must not be grouped or ganged in enclosures with other snap switches, receptacles, or similar devices if the voltage between devices exceeds 300V. **Figures 404–12** and **404–13**

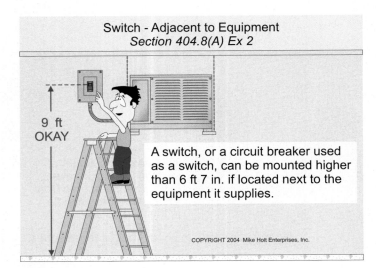

Switch - Adjacent to Equipment
Section 404.8(A) Ex 2

9 ft
OKAY

A switch, or a circuit breaker used
as a switch, can be mounted higher
than 6 ft 7 in. if located next to the
equipment it supplies.

COPYRIGHT 2004 Mike Holt Enterprises, Inc.

Figure 404–11

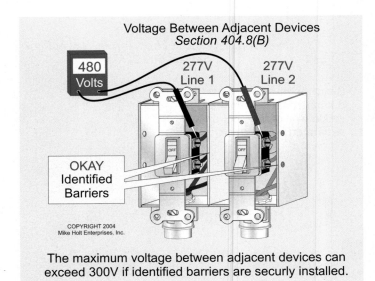

Voltage Between Adjacent Devices
Section 404.8(B)

480 Volts   277V Line 1   277V Line 2

OKAY
Identified
Barriers

COPYRIGHT 2004
Mike Holt Enterprises, Inc.

The maximum voltage between adjacent devices can
exceed 300V if identified barriers are securly installed.

Figure 404–13

Author's Comment: According to UL product standard requirements, a multipole general-use snap switch must not be fed from more than a single circuit, unless listed and marked as a two-circuit or three-circuit switch.

404.9 Switch Cover Plates (Faceplate).

(A) Mounting. Faceplates for switches must be installed so that they completely cover the outlet box opening, and where flush mounted, the faceplate must seat against the wall surface.

(B) Grounding (Bonding). The metal mounting yokes for switches, dimmers, and similar control switches, must be grounded (bonded) to an effective ground-fault current path, whether or not a metal faceplate is installed. The metal mounting yoke must be grounded (bonded) by one of the following means:

(1) Mounting Screw. The switch is mounted with metal screws to a metal box. **Figure 404–14**

> **Author's Comment:** Direct metal-to-metal contact between the device yoke of a switch and the box isn't required.

(2) Equipment Bonding Conductor. An equipment grounding (bonding) conductor, or equipment bonding jumper is connected to the grounding terminal of the metal mounting yoke. **Figure 404–15**

Exception: The metal mounting yoke of a replacement switch isn't required to be bonded to an effective ground-fault current path at an existing installation where no bonding means exists in the outlet box, but only if the switch faceplate is nonmetallic or the replacement switch is GFCI protected.

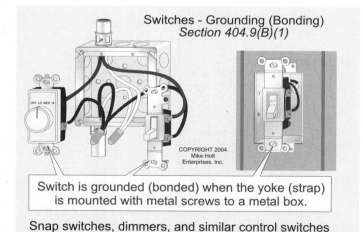

Switch is grounded (bonded) when the yoke (strap) is mounted with metal screws to a metal box.

Snap switches, dimmers, and similar control switches must be bonded to an effective ground-fault current path, whether or not a metal faceplate is installed [404.9(B)].

Figure 404–14

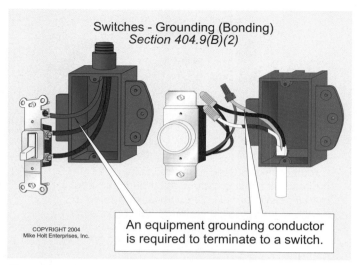

An equipment grounding conductor is required to terminate to a switch.

Figure 404–15

404.10 Mounting Snap Switches.

(A) Mounting of Snap Switches. Snap switches installed in recessed boxes must have the ears of the switch yoke seated firmly against the finished wall surface.

> **Author's Comment:** In walls or ceilings of noncombustible material, boxes must not be set back more than ¼ in. from the finished surface. In combustible walls or ceilings, boxes must be flush with, or project slightly from, the finished surface [314.20]. There cannot be any gaps greater than ⅛ in. at the edge of the box [314.21].

404.11 Circuit Breakers Used as Switches. A manu-

ally operable circuit breaker used as a switch must show when it's in the "on" (closed) or "off" (open) position [404.7].

> **Author's Comment:** Circuit breakers used to switch 120V or 277V fluorescent lighting circuits must be listed and marked "SWD" or "HID." Circuit breakers used to switch high-intensity discharge lighting circuits must be listed and must be marked as "HID" [240.83(D)]. **Figure 404–16**

404.12 Grounding (Bonding) Metal Enclosures. Metal

enclosures for switches and circuit breakers used as switches must be grounded (bonded) to an effective ground-fault current path, in accordance with Part VI of Article 250, with an equipment grounding (bonding) conductor of a type specified in 250.118 [250.4(A)(3)].

Nonmetallic enclosures containing switches are permitted if the wiring method includes an equipment grounding (bonding) conductor, as required by 404.9(B).

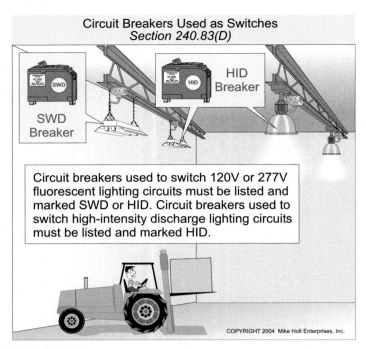

Figure 404–16

404.14 Rating and Use of Snap Switches.

(A) AC General-Use Snap Switch. Alternating-current general-use snap switches are permitted to control:

(1) Resistive and inductive loads, including electric-discharge lamps not exceeding the ampere rating of the switch, at the voltage involved.

(2) Tungsten-filament lamp loads not exceeding the ampere rating of the switch at 120V.

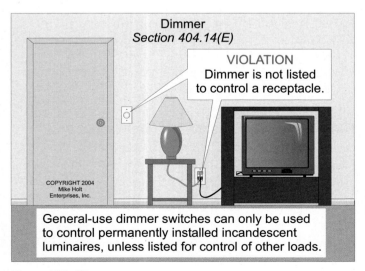

Figure 404–17

(3) Motor loads, 2 hp or less, that do not exceed 80 percent of the ampere rating of the switch. See 430.109(C).

(C) CO/ALR Snap Switches. Snap switches rated 15 or 20A that are connected to aluminum wire must be marked CO/ALR. See 406.2(C).

> **Author's Comment:** According to UL requirements, aluminum conductors cannot terminate onto screwless (push-in) terminals of a snap switch.

(E) Dimmer. General-use dimmer switches can only be used to control permanently installed incandescent luminaires.

> **Author's Comment:** Dimmers aren't listed to control a receptacle. Figure 404–17

404.15 Switch Marking.

(A) Markings. Switches must be marked with the current, voltage, and, if horsepower rated, the maximum rating for which they are designed.

(B) Off Indication. Where in the off position, a switching device with a marked "off" position must completely disconnect all ungrounded conductors of the load it controls.

> **Author's Comment:** Where an electronic occupancy sensor is used for switching, voltage will be present and a small current of 0.05 mA can flow through the circuit when the switch is in the "off" position. This small amount of current can startle a person, perhaps causing a fall. To solve this problem, manufacturers have simply removed the word "off" from the switch. Figure 404–18

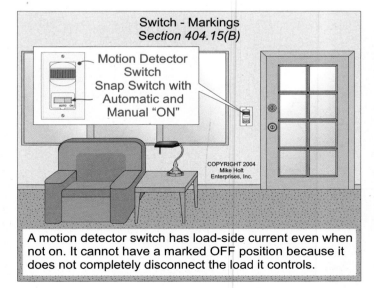

Figure 404–18

1. Three-way and four-way switches must be wired so that all switching is done only in the _____ circuit conductor.

    (a) ungrounded   (b) grounded   (c) equipment ground   (d) neutral

2. Switches must not be installed within wet locations in tub or shower spaces unless installed as part of a listed tub or shower assembly.

    (a) True   (b) False

3. Snap switches must not be grouped or ganged in enclosures with other _____ if the voltage between adjacent devices exceeds 300V, unless identified barriers are securely installed between adjacent devices.

    (a) snap switches   (b) receptacles   (c) similar devices   (d) all of these

4. A hand-operable circuit breaker equipped with a _____, or a power operated circuit breaker capable of being opened by hand in the event of a power failure, is permitted to serve as a switch if it has the required number of poles.

    (a) lever   (b) handle   (c) shunt trip   (d) a or b

5. Snap switches rated _____ or less directly connected to aluminum conductors must be listed and marked CO/ALR.

    (a) 15A   (b) 20A   (c) 25A   (d) 30A

Introduction

This article covers the rating, type, and installation of receptacles, cord connectors, and attachment plugs (cord caps). It also addresses their grounding (bonding) requirements. Some key points to remember include:

- Follow the grounding (bonding) requirements of the specific type of device you're using.
- Use GFCIs where specified by 406.3(D)(2), and install them according to manufacturer's instructions.
- Mount receptacles according to the requirements of 406.4. These are highly detailed.

406.1 Scope. Article 406 covers the rating, type, and installation of receptacles, cord connectors, and attachment plugs (cord caps).

406.2 Receptacle Rating and Type.

(B) Rating. Receptacles and cord connectors cannot be rated less than 15A.

> **FPN:** Single receptacles must have an ampere rating not less than the rating of the branch circuit [210.21(B)(1)], and multi-outlet receptacles (duplex receptacles) must have a rating in accordance with Table 210.21(B)(3).

Table 210.21(B)(3) Receptacle Ratings

Circuit Rating	Receptacle Rating
15A	15A
20A	15 or 20A*
30A	30A
40A	40 or 50A
50A	50A

*Figure 406–1

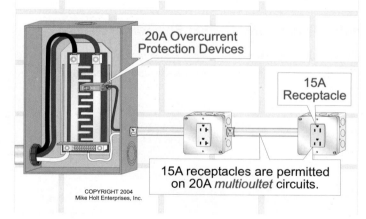

20A Circuit - Receptacles Permitted
Table 210.21(B)(3)

20A Overcurrent Protection Devices

15A Receptacle

15A receptacles are permitted on 20A *multioutlet* circuits.

COPYRIGHT 2004
Mike Holt Enterprises, Inc.

Figure 406–1

(C) Receptacles for Aluminum Conductors. Receptacles rated 20A or less for the connection to aluminum wire must be marked CO/ALR.

> **Author's Comment:** According to UL requirements, aluminum conductors cannot terminate onto screwless (push-in) terminals of a receptacle.

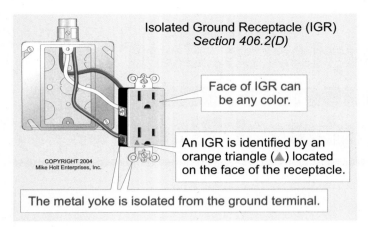

Figure 406–2

(D) Isolated Ground Receptacles. Isolated ground receptacles, used for the reduction of electric noise in accordance with 250.146(D), must be identified by an orange triangle marking on the face of the receptacle. **Figure 406–2**

(1) Isolated ground receptacles can only be used with an insulated equipment (bonding) conductor installed with the circuit conductors in accordance with 250.146(D).

(2) Isolated ground receptacles installed in nonmetallic boxes must be covered with a nonmetallic faceplate, because a metal faceplate secured to an isolated ground receptacle cannot be bonded to an effective ground-fault current path [250.4(A)(3)]. **Figure 406–3**

406.3 General Installation Requirements.

(A) Grounding (Bonding) Type. Receptacles installed on 15 and 20A branch circuits must be of the grounding (bonding) type.

Exception: Nongrounding-type receptacles are permitted for replacement in an existing outlet box where no bonding means exists, in accordance with 406.3(D).

(B) Receptacle Grounding (Bonding). Receptacles must have their grounding terminals bonded to an effective ground-fault current path in accordance with 250.146, 250.148, and 406.3(C).

Exception 2: Replacement receptacles aren't required to have their grounding contacts grounded (bonded) to an effective ground-fault current path if they are GFCI protected and installed in accordance with 406.3(D).

(C) Methods of Grounding (Bonding) Receptacles. The grounding terminals for receptacles must be connected to the branch-circuit equipment grounding (bonding) conductor in accordance with 250.146. **Figure 406–4**

> **FPN:** See 250.146(D) and 406.2(D) for the installation requirements for isolated ground receptacles.

(D) Receptacle Replacement.

(1) Where Grounding (Bonding) Means Exist. Where an equipment bonding means exists in the receptacle enclosure, grounding-type receptacles must replace nongrounding-type receptacles, and the receptacle's grounding terminal must be grounded (bonded) to an effective ground-fault current path in accordance with 406.3(C).

(2) GFCI Protection Required. When receptacles are replaced in locations where GFCI protection is required, the replacement receptacles must be GFCI protected. This includes the replacement of receptacles in dwelling unit bathrooms, garages, outdoors, crawl spaces, unfinished basements, kitchen countertops, rooftops, or within 6 ft of laundry, utility, and wet bar sinks. See 210.8 in this textbook for specific GFCI protection requirements.

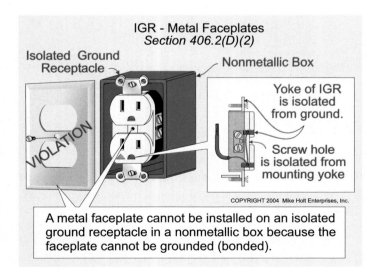

Figure 406–3

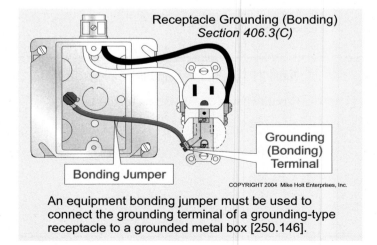

Figure 406–4

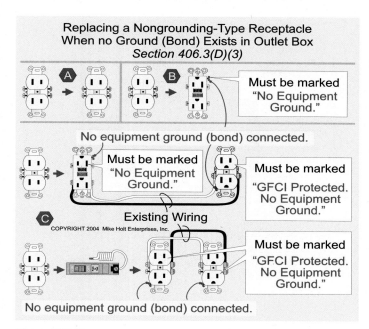

Figure 406–5

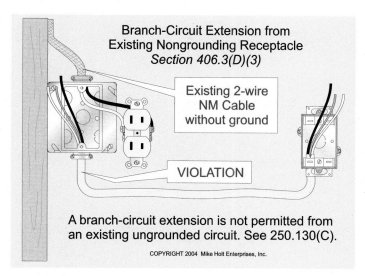

Figure 406–7

(3) Where No Ground Exists. Where no equipment bonding means exists in the outlet box, such as old 2-wire Type NM cable without a ground, nongrounding-type receptacles can be replaced with (a), (b), or (c): **Figure 406–5**

   (a) Another nongrounding-type receptacle.

   (b) A GFCI grounding-type receptacle marked "No Equipment Ground."

   (c) A grounding-type receptacle, if GFCI protected and marked "GFCI Protected" and "No Equipment Ground."

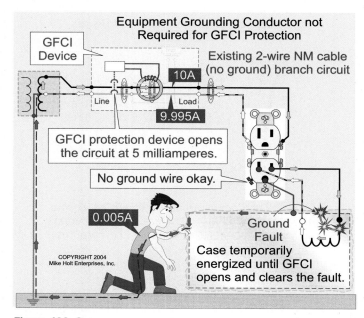

Figure 406–6

Author's Comment: GFCI protection functions properly on a 2-wire circuit without an equipment grounding (bonding) conductor, because the equipment grounding (bonding) conductor serves no role in the operation of the GFCI-protection device. See Article 100 for the definition of "Ground-Fault Circuit-Interrupter" for additional details. **Figure 406–6**

CAUTION: *The permission to replace nongrounding-type receptacles with GFCI-protected grounding-type receptacles doesn't apply to new receptacle outlets that extend from an existing ungrounded outlet box. Once you add a receptacle outlet (branch-circuit extension), the receptacle must be of the grounding (bonding) type and it must have its grounding terminal grounded (bonded) to an effective ground-fault current path in accordance with 250.130(C).* **Figure 406–7**

406.4 Receptacle Mounting. Receptacles must be installed in outlet boxes designed for the purpose, and the outlet box must be securely fastened in place, unless the outlet box for the receptacle is supported in accordance with 314.23.

Author's Comments:

- 314.23(H) permits a pendant cord to support an outlet box [314.23(H)], but fittings on the outlet box must ensure that tension will not be transmitted to the conductor joints or terminals [400.10]. **Figure 406–8**

- The orientation (position) of the ground terminal of a receptacle isn't specified in the *NEC*. The ground terminal can be up, down, or to the side. Proposals to specify the mounting position of the ground were all rejected. For more information on this subject, visit www.MikeHolt.com. **Figure 406–9**

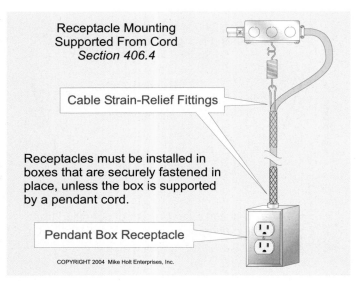

Receptacle Mounting
Supported From Cord
Section 406.4

Cable Strain-Relief Fittings

Receptacles must be installed in
boxes that are securely fastened in
place, unless the box is supported
by a pendant cord.

Pendant Box Receptacle

COPYRIGHT 2004 Mike Holt Enterprises, Inc.

Figure 406–8

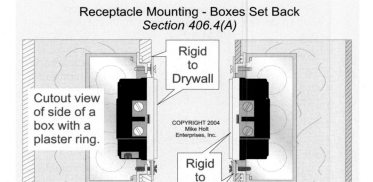

Receptacle Mounting - Boxes Set Back
Section 406.4(A)

Rigid
to
Drywall

Cutout view
of side of a
box with a
plaster ring.

COPYRIGHT 2004
Mike Holt
Enterprises, Inc.

Rigid
to
Box

Receptacles in outlet boxes must be installed so that
the mounting yoke (strap) of the receptacle is held
rigidly to the finished surface or to the outlet box.

Figure 406–10

(A) Boxes That Are Set Back. Receptacles in outlet boxes that are set back from the finished surface, as permitted by 314.20, must be installed so the mounting yoke of the receptacle is held rigidly to the finished surface or outlet box. **Figure 406–10**

> **Author's Comment:** In walls or ceilings of noncombustible material, outlet boxes must not be set back more than ¼ in. from the finished surface. In walls or ceilings of combustible material, outlet boxes must be flush with the finished surface [314.20]. There cannot be any gaps greater than ⅛ in. at the edge of the outlet box [314.21]. **Figure 406–11**

(B) Boxes that are Flush with the Surface. Receptacles mounted in outlet boxes that are flush with the finished surface must be installed so that the mounting yoke of the receptacle is held rigidly against the outlet box or raised box cover.

(C) Receptacles Mounted on Covers. Receptacles supported by a cover must be held rigidly to the cover with two screws. **Figure 406–12**

(D) Position of Receptacle Faces. Receptacles must be flush with, or project from, the faceplates.

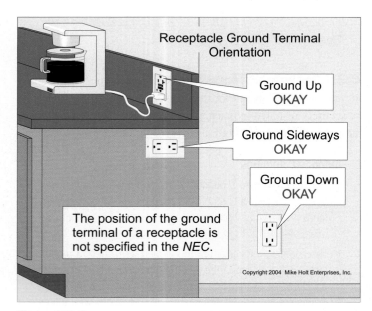

Receptacle Ground Terminal
Orientation

Ground Up
OKAY

Ground Sideways
OKAY

Ground Down
OKAY

The position of the ground
terminal of a receptacle is
not specified in the *NEC*.

Copyright 2004 Mike Holt Enterprises, Inc.

Figure 406–9

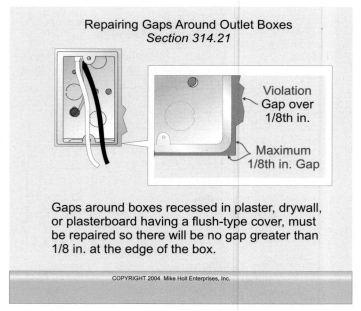

Repairing Gaps Around Outlet Boxes
Section 314.21

Violation
Gap over
1/8th in.

Maximum
1/8th in. Gap

Gaps around boxes recessed in plaster, drywall,
or plasterboard having a flush-type cover, must
be repaired so there will be no gap greater than
1/8 in. at the edge of the box.

COPYRIGHT 2004 Mike Holt Enterprises, Inc.

Figure 406–11

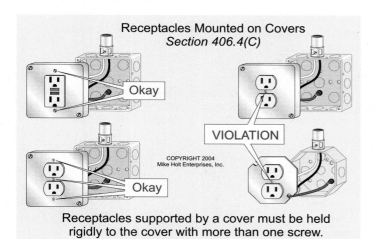

Receptacles Mounted on Covers
Section 406.4(C)

Okay

Okay

VIOLATION

COPYRIGHT 2004
Mike Holt Enterprises, Inc.

Receptacles supported by a cover must be held rigidly to the cover with more than one screw.

Figure 406–12

(E) Receptacles in Countertops and Similar Work Surfaces in Dwelling Units. Receptacles must not be installed in a face-up position in countertops or similar work surface areas. Figure 406–13

406.5 Receptacle Faceplates. Faceplates for receptacles must completely cover the outlet openings, and they must seat firmly against the mounting surface.

(B) Bonding. Metal faceplates for receptacles must be bonded to an effective ground-fault current path.

> **Author's Comment:** The *NEC* doesn't specify how this is to be accomplished, but 517.13(B) Exception 1 for health care facilities, permits the metal mounting screw(s) securing the faceplate to a metal outlet box or wiring device to accomplish the bonding requirement. **Figure 406–14**

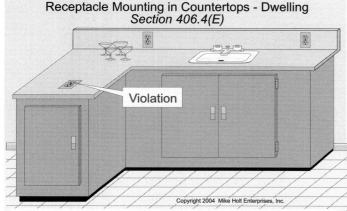

Receptacle Mounting in Countertops - Dwelling
Section 406.4(E)

Violation

Copyright 2004 Mike Holt Enterprises, Inc.

Receptacles in a dwelling unit cannot be installed in the face-up position in a countertop or similar work surface.

Figure 406–13

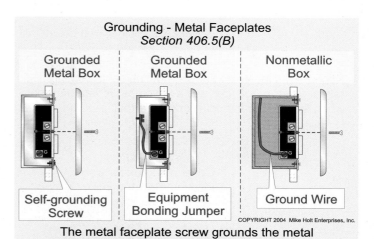

Grounding - Metal Faceplates
Section 406.5(B)

Grounded Metal Box | Grounded Metal Box | Nonmetallic Box

G | G | G

Self-grounding Screw | Equipment Bonding Jumper | Ground Wire

COPYRIGHT 2004 Mike Holt Enterprises, Inc.

The metal faceplate screw grounds the metal faceplate to the receptacle's grounding terminal.

Figure 406–14

406.6 Attachment Plugs, Cord Connectors, and Flanged Surface Devices. Attachment plugs and cord connectors must be listed for the purpose.

(A) Exposed Live Parts. Attachment plugs, cord connectors, and flanged surface devices must have no exposed current-carrying parts, except the prongs, blades, or pins.

(B) No Energized Parts. Attachment plugs must be installed so their prongs, blades, or pins aren't energized unless inserted into an energized receptacle. **Figure 406–15**

(D) Flanged Surface Inlet. A flanged surface inlet must be installed so that the prongs, blades, or pins aren't energized unless an energized cord connector is inserted into it.

> **Author's Comment:** The use of flanged "inlets," such as those on computers, transfer switches, etc., for detachable power cords is increasing.

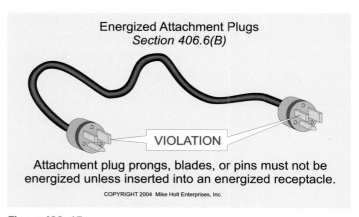

Energized Attachment Plugs
Section 406.6(B)

VIOLATION

Attachment plug prongs, blades, or pins must not be energized unless inserted into an energized receptacle.

COPYRIGHT 2004 Mike Holt Enterprises, Inc.

Figure 406–15

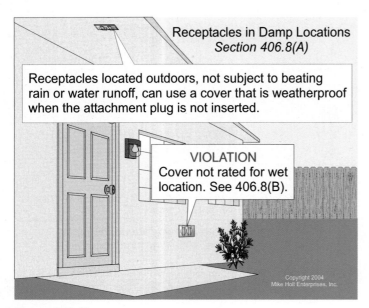

Figure 406–16

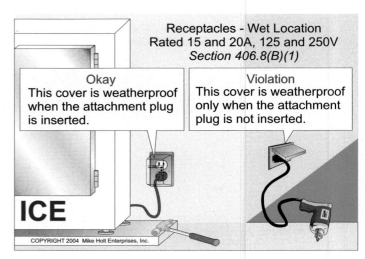

Figure 406–17

406.8 Receptacles in Damp or Wet Locations.

(A) Damp Locations. Receptacles installed outdoors under roofed open porches, canopies, marquees, and the like, and not subject to beating rain or water runoff or in other damp locations, must be installed in an enclosure that is weatherproof when the attachment plug cap isn't inserted and receptacle covers are closed. **Figure 406–16**

A receptacle installed within an enclosure that is weatherproof when an attachment plug is inserted is also suitable for a damp location.

> **Author's Comment:** See Article 100 for the definition of "Location, Damp."

(B) Receptacles in Wet Locations.

(1) 15 and 20A Receptacles. All 15 and 20A, 125V and 250V receptacles installed in a wet location must be within an enclosure that is weatherproof even when an attachment plug is inserted. **Figures 406–17 and 406–18**

(2) Other Receptacles. Receptacles rated other than 15 or 20A, 125V and 250V installed in a wet location must comply with (a) or (b):

(a) Wet Location Cover. A receptacle installed in a wet location, where the load isn't attended while in use, must have an enclosure that is weatherproof with the attachment plug cap inserted or removed.

(b) Damp Location Cover. A receptacle installed in a wet location for use with portable tools can have an enclosure that is weatherproof when the attachment plug is removed.

(C) Bathtub and Shower Space. Receptacles must not be installed within or directly over a bathtub or shower stall. **Figure 406–19**

> **Author's Comment:** Receptacles must be located no less than 5 ft from any spas or hot tubs [680.22(A)(1) and 680.43(A)(1)]. Hydromassage bathtubs are treated like bathtubs [680.70].

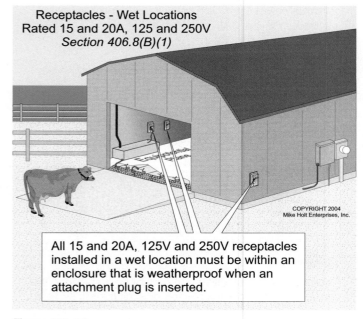

Figure 406–18

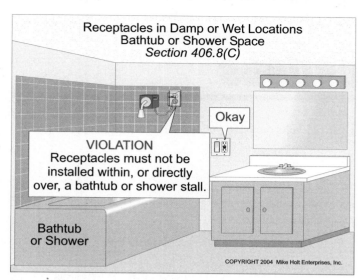

Figure 406–19

(E) Flush Mounting with Faceplate. The enclosure for a receptacle installed in an outlet box that is flush-mounted on a finished surface must be made weatherproof by a weatherproof faceplate that provides a watertight connection between the plate and the finished surface.

406.10 Connecting Receptacle Grounding Terminal to Box.
The grounding terminal of receptacles must be bonded to an effective ground-fault current path in accordance with 250.146.

1. Receptacles and cord connectors must be rated not less than _____ at 125 volts or at 250 volts, and must be of a type not suitable for use as lampholders.

   (a) 30A   (b) 20A   (c) 15A   (d) 10A

2. When replacing a receptacle, and the grounding (bonding) means exists in the receptacle enclosure, or a grounding (bonding) conductor is installed in accordance with 250.130(C), _____-type receptacles must be used.

   (a) isolated ground   (b) grounding   (c) GFCI   (d) two-wire

3. Receptacles mounted in boxes flush with the wall surface or projecting beyond it must be installed so that the mounting yoke or strap of the receptacle is _____.

   (a) held rigidly against the box or box cover   (b) mounted behind the wall surface
   (c) held rigidly at the finished surface   (d) none of these

4. Attachment plugs, cord connectors, and flanged-surface devices, must be listed with the manufacturer's name or identification and voltage and ampere ratings.

   (a) True   (b) False

5. An outdoor receptacle in a location protected from the weather, or another damp location, must be installed in an enclosure that is weatherproof when the receptacle is _____.

   (a) covered   (b) enclosed   (c) protected   (d) none of these

6. A receptacle must not be installed within, or directly over, a bathtub or shower space.

   (a) True   (b) False

408 Switchboards and Panelboards

Introduction

Article 408 covers the specific requirements for switchboards, panelboards, and distribution boards that control light and power circuits. Some key points to remember:

- One objective of Article 408 is that the installation prevents contact between current-carrying conductors and people or maintenance equipment.
- The circuit directory of a panelboard must clearly identify the purpose or use of each circuit that originates in that panelboard.
- You must know the difference between a "lighting and appliance panelboard" and a "power panelboard."
- You must understand the detailed grounding (bonding) requirements for panelboards.

PART I. GENERAL

408.1 Scope.

(1) Article 408 covers the specific requirements for switchboards, panelboards, and distribution boards that control light and power circuits. **Figure 408–1**

(2) Article 408 covers battery-charging panels supplied from light or power circuits.

Author's Comment: For the purposes of this textbook, we will only cover the requirements for panelboards.

408.3 Arrangement of Busbars and Conductors.

(C) Used as Service Equipment. Each panelboard used as service equipment must be provided with a main bonding jumper to connect the grounded neutral service conductor to the panelboard metal frame.

(E) Panelboard Phase Arrangement. Since 1975, panelboards supplied by a 4-wire three-phase delta-connected system must have the high-leg conductor (208V) terminate to the "B" (center) phase of the panelboard [408.3(E)]. **Figure 408–2**

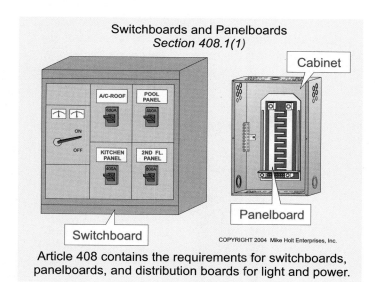

Switchboards and Panelboards
Section 408.1(1)

Cabinet

A/C-ROOF   POOL PANEL
600A   400A

ON
OFF

KITCHEN PANEL   2ND FL. PANEL
400A   600A

Switchboard

Panelboard

COPYRIGHT 2004 Mike Holt Enterprises, Inc.

Article 408 contains the requirements for switchboards, panelboards, and distribution boards for light and power.

Figure 408–1

Exception: The high-leg conductor can terminate to the "C" phase when the meter is located in the same section of a switchboard or panelboard.

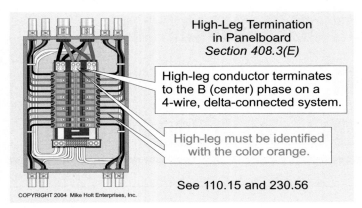

Figure 408–2

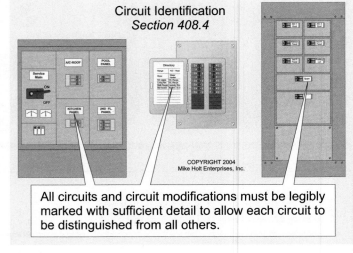

Figure 408–4

FPN: Orange identification, or some other effective means, is required for the high-leg conductor [110.15 and 230.56].

WARNING: *The ANSI standard for meter equipment requires the high-leg conductor (208V-to-neutral) to terminate on the "C" (right) phase of the meter enclosure. This is because the demand meter needs 120V and it gets this from the "B" phase.* Figure 408–3

WARNING: *When replacing equipment in existing facilities that contain a high-leg conductor, care must be taken to ensure that the high-leg conductor is replaced in the original location. Prior to 1975, the high-leg conductor was required to terminate on the "C" phase of panelboards and switchboards. Failure to re-terminate the high-leg in accordance with the existing installation can result in 120V circuits inadvertently connected to the 208V high-leg, with disastrous results.*

408.4 Circuit Directory or Circuit Identification. All circuits and circuit modifications must be legibly identified as to their clear, evident, and specific purpose. The identification must include sufficient detail to allow each circuit to be distinguished from all others, and the identification must be on a circuit directory located on the face or inside of the door of a panelboard. See 110.22. Figure 408–4

408.5 Clearance for Conductor Entering Bus Enclosures. Where conduits or other raceways enter a switchboard, floor-standing panelboard, or similar enclosure, the conduit or raceways, including their end fittings, must not rise more than 3 in. above the bottom of the enclosure.

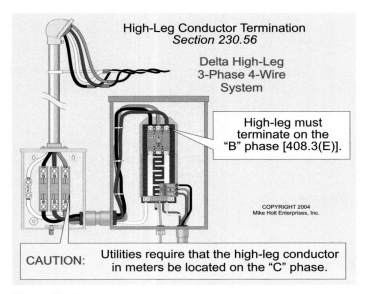

Figure 408–3

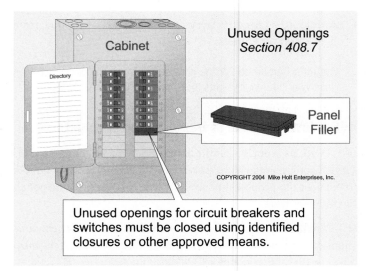

Figure 408–5

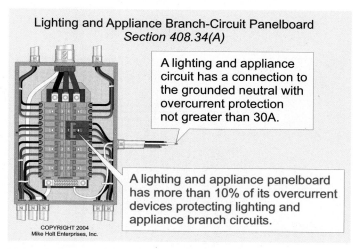

Lighting and Appliance Branch-Circuit Panelboard
Section 408.34(A)

A lighting and appliance circuit has a connection to the grounded neutral with overcurrent protection not greater than 30A.

A lighting and appliance panelboard has more than 10% of its overcurrent devices protecting lighting and appliance branch circuits.

COPYRIGHT 2004
Mike Holt Enterprises, Inc.

Figure 408–6

408.7 Unused Openings.
Unused openings for circuit breakers and switches must be closed using identified closures, or other means approved by the authority having jurisdiction, that provide protection substantially equivalent to the wall of the enclosure. **Figure 408–5**

PART III. PANELBOARDS

408.34 Classification of Panelboards.

(A) Lighting and Appliance Panelboard. A lighting and appliance branch-circuit panelboard is one with more than 10 percent of its overcurrent protection devices protecting "lighting and appliance branch circuits." **Figure 408–6**

A lighting and appliance branch circuit is rated 30A or less and has a connection to the grounded neutral conductor of the panelboard [408.34].

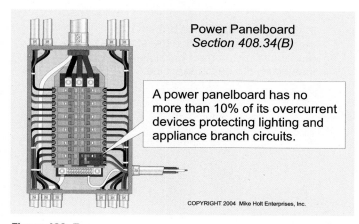

Power Panelboard
Section 408.34(B)

A power panelboard has no more than 10% of its overcurrent devices protecting lighting and appliance branch circuits.

COPYRIGHT 2004 Mike Holt Enterprises, Inc.

Figure 408–7

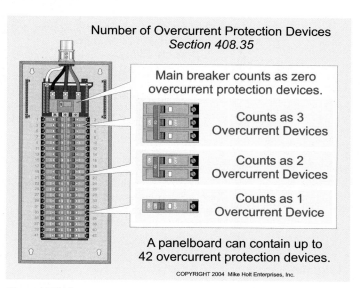

Number of Overcurrent Protection Devices
Section 408.35

Main breaker counts as zero overcurrent protection devices.

Counts as 3 Overcurrent Devices

Counts as 2 Overcurrent Devices

Counts as 1 Overcurrent Device

A panelboard can contain up to 42 overcurrent protection devices.

COPYRIGHT 2004 Mike Holt Enterprises, Inc.

Figure 408–8

(B) Power Panelboard. A power panelboard is one having 10 percent or fewer of its overcurrent protection devices protecting "lighting and appliance branch circuits." **Figure 408–7**

408.35 Number of Overcurrent Protection Devices.
Not counting the main breaker, the maximum number of overcurrent protection devices permitted in a lighting and appliance branch-circuit panelboard is 42. A 2-pole circuit breaker counts as two overcurrent protection devices, and a 3-pole circuit breaker counts as three overcurrent protection devices. **Figure 408–8**

> **Author's Comment:** The maximum number of circuit breakers in a panelboard is also limited by the listing instructions posted inside the panelboard [110.3(B)].

408.36 Overcurrent Protection of Panelboard.

(A) Lighting and Appliance Branch-Circuit Panelboard. Each lighting and appliance branch-circuit panelboard must be provided with overcurrent protection having a rating not greater than that of the panelboard. **Figure 408–9**

Exception 1: Individual protection for a lighting and appliance branch-circuit panelboard isn't required at the panelboard, if the panelboard feeder has overcurrent protection sized not greater than the rating of the panelboard. **Figure 408–10**

> **CAUTION:** *When tap conductors supply a lighting and appliance branch-circuit panelboard, as permitted in 240.21(B), overcurrent protection must be installed within or ahead of the panelboard.* **Figure 408–11**

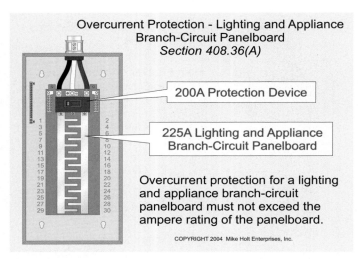

Overcurrent Protection - Lighting and Appliance
Branch-Circuit Panelboard
Section 408.36(A)

200A Protection Device

225A Lighting and Appliance
Branch-Circuit Panelboard

Overcurrent protection for a lighting
and appliance branch-circuit
panelboard must not exceed the
ampere rating of the panelboard.

COPYRIGHT 2004 Mike Holt Enterprises, Inc.

Figure 408–9

(D) Panelboards Supplied Through a Transformer. When a lighting and appliance branch-circuit panelboard is supplied from a transformer, as permitted in 240.21(C), the overcurrent protection for the panelboard must be on the secondary side of the transformer. The required overcurrent protection can be in a separate enclosure ahead of the panelboard, or it can be in the panelboard. **Figure 408–12**

Exception: A panelboard supplied by a 2-wire system or a 3-wire delta/delta connected transformer, is considered protected by the primary protection device when installed in accordance with 240.4(F) and 240.21(C)(1).

(F) Back-Fed Devices. Plug-in circuit breakers that are back-fed must be secured in place by an additional fastener that requires other than a pull to release the breaker from the panelboard.

Author's Comments:

- The purpose of the breaker fastener is to prevent the circuit breaker from being accidentally removed from the panel-

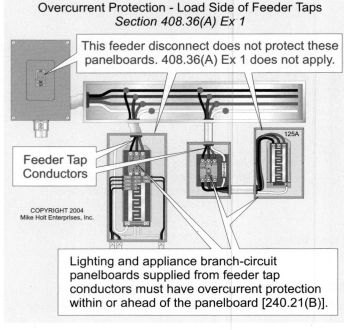

Overcurrent Protection - Load Side of Feeder Taps
Section 408.36(A) Ex 1

This feeder disconnect does not protect these
panelboards. 408.36(A) Ex 1 does not apply.

125A

Feeder Tap
Conductors

COPYRIGHT 2004
Mike Holt Enterprises, Inc.

Lighting and appliance branch-circuit
panelboards supplied from feeder tap
conductors must have overcurrent protection
within or ahead of the panelboard [240.21(B)].

Figure 408–11

board while energized, thereby exposing someone to dangerous voltage. **Figure 408–13**

- Circuit breakers are often back-fed to provide the overcurrent protection for lighting and appliance panelboards required by 408.36(A).

CAUTION: *Circuit breakers marked "Line" and "Load" must be installed in accordance with listing or labeling instructions [110.3(B)]; therefore, these types of devices must not be back-fed.* **Figure 408–14**

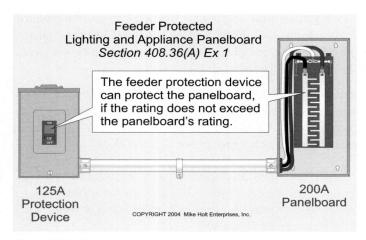

Feeder Protected
Lighting and Appliance Panelboard
Section 408.36(A) Ex 1

The feeder protection device
can protect the panelboard,
if the rating does not exceed
the panelboard's rating.

125A
Protection
Device

200A
Panelboard

COPYRIGHT 2004 Mike Holt Enterprises, Inc.

Figure 408–10

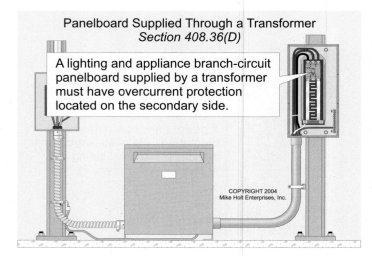

Panelboard Supplied Through a Transformer
Section 408.36(D)

A lighting and appliance branch-circuit
panelboard supplied by a transformer
must have overcurrent protection
located on the secondary side.

COPYRIGHT 2004
Mike Holt Enterprises, Inc.

Figure 408–12

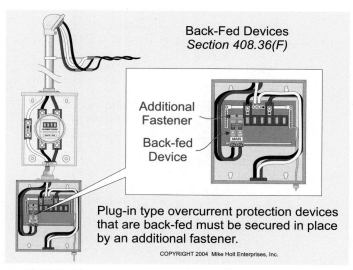

Back-Fed Devices
Section 408.36(F)

Additional Fastener

Back-fed Device

Plug-in type overcurrent protection devices that are back-fed must be secured in place by an additional fastener.

COPYRIGHT 2004 Mike Holt Enterprises, Inc.

Figure 408–13

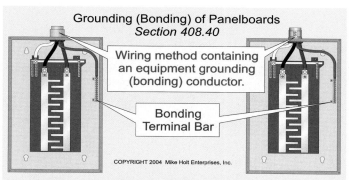

Grounding (Bonding) of Panelboards
Section 408.40

Wiring method containing an equipment grounding (bonding) conductor.

Bonding Terminal Bar

COPYRIGHT 2004 Mike Holt Enterprises, Inc.

Where a panelboard is used with a nonmetallic raceway or cable, or where separate equipment grounding conductors are provided, a terminal bar for the equipment grounding conductors must be bonded to the cabinet.

Figure 408–15

408.37 Panelboards in Damp or Wet Locations. The

enclosures for panelboards (cabinets) must prevent moisture or water from entering or accumulating within the enclosure, and they must be weatherproof when located in a wet location. When the enclosure is surface mounted in a wet location, the enclosure must be mounted with not less than ¼ in. air space between it and the mounting surface [312.2(A)].

408.40 Grounding (Bonding) of Panelboards. Metal

panelboard cabinets and frames must be grounded (bonded) to an effective ground-fault current path, in accordance with Part VI of Article 250, with an equipment grounding (bonding) conductor of a type specified in 250.118 [215.6 and 250.4(A)(3)].

Where the panelboard is used with nonmetallic raceways or cables, or where separate equipment grounding (bonding) con-

ductors are provided, a terminal bar for the equipment grounding (bonding) conductors must be bonded to the metal cabinet. **Figure 408–15**

Exception: Isolated equipment grounding (bonding) conductors, installed for the reduction of electrical noise as permitted in 250.146(D), are permitted to pass through the panelboard without terminating onto the equipment grounding terminal of the panelboard cabinet. **Figure 408–16**

Equipment grounding (bonding) conductors must not terminate on the grounded neutral terminal bar, and grounded neutral conductors must not terminate on the equipment grounding terminal bar, except as permitted by 250.142 for services, for separately derived systems, or for separate building disconnects. **Figure 408–17**

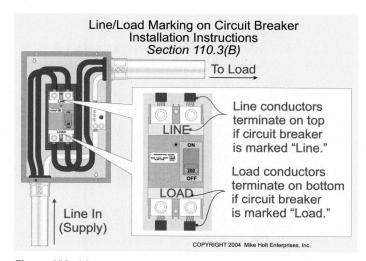

Line/Load Marking on Circuit Breaker
Installation Instructions
Section 110.3(B)

To Load

LINE

ON
200
OFF

LOAD

Line In (Supply)

Line conductors terminate on top if circuit breaker is marked "Line."

Load conductors terminate on bottom if circuit breaker is marked "Load."

COPYRIGHT 2004 Mike Holt Enterprises, Inc.

Figure 408–14

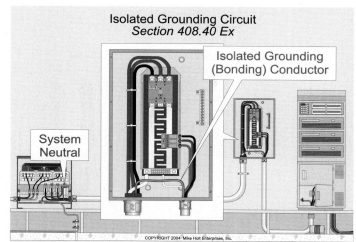

Isolated Grounding Circuit
Section 408.40 Ex

Isolated Grounding (Bonding) Conductor

System Neutral

COPYRIGHT 2004 Mike Holt Enterprises, Inc.

An isolated grounding conductor can pass through a metal enclosure, but it must terminate to the system neutral.

Figure 408–16

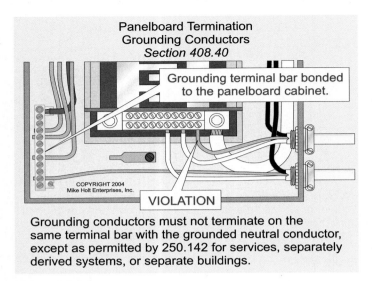

Figure 408–17

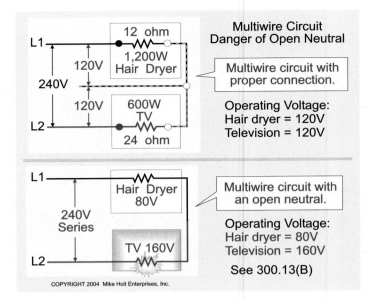

Figure 408–19

Grounding conductors must not terminate on the same terminal bar with the grounded neutral conductor, except as permitted by 250.142 for services, separately derived systems, or separate buildings.

Author's Comment: See Article 100 for the definition of "Separately Derived System."

CAUTION: *Most panelboards are rated as "suitable for use as service equipment," which means that they are supplied with a main bonding jumper [250.28]. This screw or strap must not be installed except when the panelboard is used for service equipment, for separately derived systems, or for remote building disconnects as permitted by 250.142(A) [250.24(A)(5)]. In addition, a panelboard marked "suitable only for use as service equipment" means that the neutral bar or terminal of the panelboard has been bonded to the case at the factory, and this panelboard is restricted in it's use for service equipment, for separately derived systems, or for remote building disconnects [250.142(B)].*

408.41 Grounded Neutral Conductor Terminations.

Each grounded neutral conductor within a panelboard must terminate to an individual terminal. Figure 408–18

Author's Comment: If two grounded neutral conductors were connected to the same terminal, and someone removed one of them, the other might unintentionally be removed as well. If that happened to the grounded neutral conductor of a multiwire circuit, it could result in excessive line-to-neutral voltage for one of the circuits. See 300.13(B) for details. Figure 408–19

This requirement doesn't apply to equipment grounding (bonding) conductors because the voltage of a circuit is not affected if an equipment grounding (bonding) conductor is accidentally removed. Figure 408–20

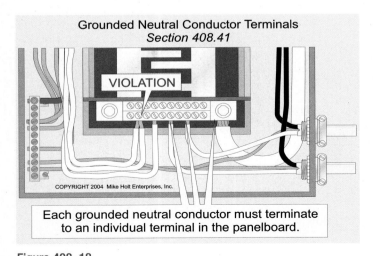

Figure 408–18

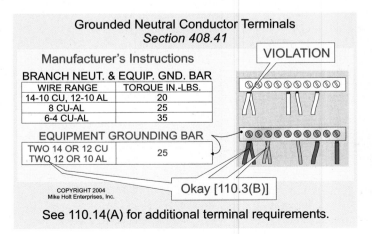

Figure 408–20

1. Panelboards supplied by a 3Ø, 4-wire, delta-connected system must have that phase with the higher voltage-to-ground (high-leg) connected to the _____ phase.

   (a) A   (b) B   (c) C   (d) any of these

2. Conduits and raceways, including end fittings, must not rise more than _____ above the bottom of a switchboard enclosure.

   (a) 3 in.   (b) 4 in.   (c) 5 in.   (d) 6 in.

3. To qualify as a lighting and appliance branch-circuit panelboard, the number of circuits rated at 30A or less and having a neutral conductor must be _____ of the total.

   (a) more than 10 percent   (b) 10 percent   (c) 20 percent   (d) 40 percent

4. A lighting and appliance branch-circuit panelboard is not required to be individually protected if the panelboard _____ conductor has protection not greater than the panelboard rating.

   (a) grounded neutral   (b) feeder   (c) branch circuit   (d) none of these

5. Plug-in-type circuit breakers that are back-fed (to supply a panelboard) must be _____ by an additional fastener that requires more than a pull to release.

   (a) grounded   (b) secured in place   (c) shunt tripped   (d) none of these

Notes

410 Luminaires, Lampholders, and Lamps

Introduction

This article covers luminaires, lampholders, pendants, and the wiring and equipment of such lamps and luminaires. The scope of Article 410 has been expanded to include decorative lighting products, lighting accessories for temporary seasonal and holiday use, and portable flexible lighting products.

This article is highly detailed, but it's broken down into sixteen parts. The first five are sequential, and apply to all luminaires, lampholders, and lamps—General, Location, Boxes and Covers, Supports, and Grounding (Bonding). This is mostly mechanical information, and it's not hard to follow or absorb. Part VI, Wiring, ends the sequence. The seventh, ninth, and tenth parts provide requirements for manufacturers to follow—use only equipment that conforms to these requirements. Part VIII provides requirements for Installing Lampholders. The rest of Article 410 addresses specific types of lighting.

> **Author's Comment:** Article 410 doesn't include "Lighting Systems Operating at 30 Volts or Less"; Article 411 addresses them.

PART I. GENERAL

410.1 Scope. This article covers luminaires, lampholders, lamps, decorative lighting products, lighting accessories for temporary seasonal and holiday use, portable flexible lighting products, and the wiring and equipment of such products and lighting installations. **Figure 410–1**

> **Author's Comment:** Because of the many types and applications of luminaires, manufacturers' instructions are very important and helpful for proper installation. UL produces a pamphlet called the "Luminaire Marking Guide," which provides information for properly installing common types of incandescent, fluorescent, and high-intensity discharge (HID) luminaires.

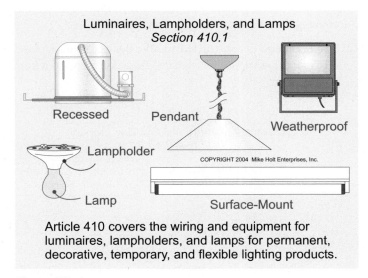

Luminaires, Lampholders, and Lamps
Section 410.1

Recessed　　Pendant　　Weatherproof

Lampholder

Lamp　　Surface-Mount

COPYRIGHT 2004 Mike Holt Enterprises, Inc.

Article 410 covers the wiring and equipment for luminaires, lampholders, and lamps for permanent, decorative, temporary, and flexible lighting products.

Figure 410–1

PART II. LUMINAIRE LOCATIONS

410.4 Luminaires in Specific Locations.

(A) Wet or Damp Locations. Luminaires in wet or damp locations must be installed in a manner that prevents water from accumulating in any part of the luminaire. Luminaires marked "Suitable for Dry Locations Only" must be installed only in a dry location; luminaires marked "Suitable for Damp Locations" can be installed in either a damp or dry location; and luminaires marked "Suitable for Wet Locations" can be installed in a dry, damp, or wet location. **Figure 410–2**

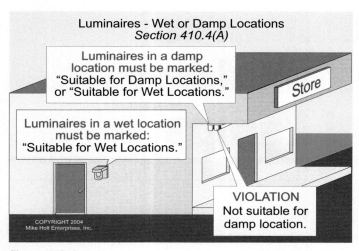

Figure 410–2

Author's Comment: A dry location can be subjected to occasional dampness or wetness. See Article 100 for the definition of "Location, Dry."

(B) Corrosive Locations. Luminaires installed in corrosive locations must be suitable for the location.

(C) In Ducts or Hoods. Luminaires can be installed in commercial cooking hoods where all of the following conditions are met: **Figure 410–3**

(1) The luminaire is identified for use within commercial cooking hoods.

(2) The luminaire is constructed so all exhaust vapors, grease, oil, or cooking vapors are excluded from the lamp and wiring compartment.

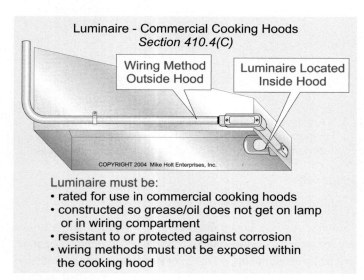

Figure 410–3

(3) The luminaire is corrosion resistant or protected against corrosion, and the surface must be smooth so as not to collect deposits and to facilitate cleaning.

(4) Wiring methods and materials supplying the luminaire must not be exposed within the cooking hood.

Author's Comment: Standard gasketed luminaires cannot be installed in a commercial cooking hood because accumulations of grease and oil can result in a fire caused by high temperatures on the glass globe.

(D) Bathtub and Shower Areas. No part of chain-, cable-, or cord-connected luminaires, track lighting, or ceiling paddle fans can be located within 3 ft horizontally and 8 ft vertically from the top of the bathtub rim or shower stall threshold (bathtub/shower luminaire zone). **Figure 410–4**

Author's Comment: The 3 x 8 ft bathtub/shower luminaire zone does not apply to recessed or surface mounted luminaires, switches, or receptacles. See 404.4 for switch requirements and 406.8(C) for receptacle requirements. **Figure 410–5**

Luminaires located in bathtub and shower zones must be listed for damp locations, or listed for wet locations where subject to shower spray.

(E) Luminaires in Indoor Sports, Mixed-Use, and All-Purpose Facilities. Luminaires subject to physical damage, using a mercury vapor or metal-halide lamp, installed in playing and spectator seating areas of indoor sports, mixed-use, or all-purpose facilities must be of the type that has a glass or plastic lamp shield. Such luminaires are permitted to have an additional guard. **Figure 410–6**

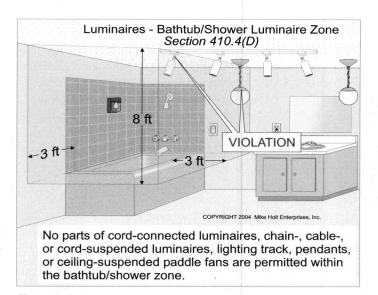

Figure 410–4

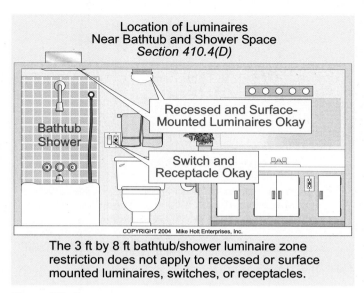

Location of Luminaires
Near Bathtub and Shower Space
Section 410.4(D)

Recessed and Surface-
Mounted Luminaires Okay

Bathtub
Shower

Switch and
Receptacle Okay

COPYRIGHT 2004 Mike Holt Enterprises, Inc.

The 3 ft by 8 ft bathtub/shower luminaire zone
restriction does not apply to recessed or surface
mounted luminaires, switches, or receptacles.

Figure 410–5

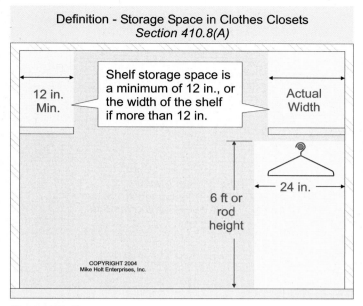

Definition - Storage Space in Clothes Closets
Section 410.8(A)

12 in.
Min.

Shelf storage space is
a minimum of 12 in., or
the width of the shelf
if more than 12 in.

Actual
Width

24 in.

6 ft or
rod
height

COPYRIGHT 2004
Mike Holt Enterprises, Inc.

Figure 410–7

410.8 Clothes Closets.

(A) Definition of Storage Space. Storage space is defined as a volume bounded by the sides and back closet walls extending from the closet floor vertically to a height of 6 ft or the highest clothes-hanging rod at a horizontal distance of 2 ft from the sides and back of the closet walls. Storage space continues vertically to the closet ceiling for a distance of 1 ft or the width of the shelf, whichever is greater. **Figure 410–7**

(B) Luminaire Types Permitted in Clothes Closets. The following types of luminaires can be installed in a clothes closet:

(1) A surface or recessed incandescent luminaire with enclosed lamp.

(2) A surface or recessed fluorescent luminaire.

(C) Luminaire Types not Permitted in Clothes Closets. Incandescent luminaires with open or partially open lamps and pendant-type luminaires cannot be installed in a clothes closet. **Figure 410–8**

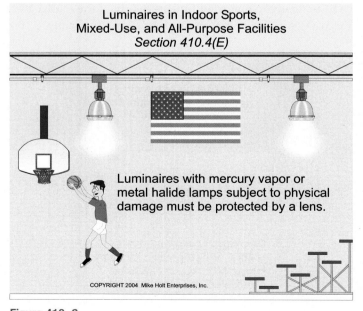

Luminaires in Indoor Sports,
Mixed-Use, and All-Purpose Facilities
Section 410.4(E)

Luminaires with mercury vapor or
metal halide lamps subject to physical
damage must be protected by a lens.

COPYRIGHT 2004 Mike Holt Enterprises, Inc.

Figure 410–6

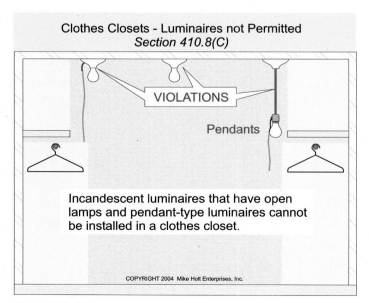

Clothes Closets - Luminaires not Permitted
Section 410.8(C)

VIOLATIONS

Pendants

Incandescent luminaires that have open
lamps and pendant-type luminaires cannot
be installed in a clothes closet.

COPYRIGHT 2004 Mike Holt Enterprises, Inc.

Figure 410–8

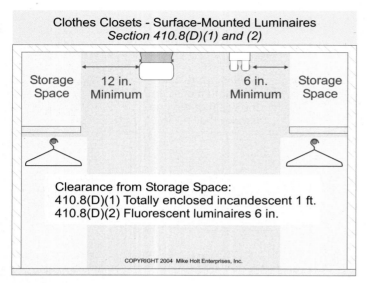

Clothes Closets - Surface-Mounted Luminaires
Section 410.8(D)(1) and (2)

Storage Space — 12 in. Minimum — 6 in. Minimum — Storage Space

Clearance from Storage Space:
410.8(D)(1) Totally enclosed incandescent 1 ft.
410.8(D)(2) Fluorescent luminaires 6 in.

COPYRIGHT 2004 Mike Holt Enterprises, Inc.

Figure 410–9

(D) Installation of Luminaires in Clothes Closets.

(1) Surface-mounted totally enclosed incandescent luminaires must maintain a minimum clearance of 12 in. from the storage space. Figure 410–9

(2) Surface-mounted fluorescent luminaires must maintain a minimum clearance of 6 in. from the storage space. See Figure 410–9.

(3) Recessed incandescent luminaires with a completely enclosed lamp must maintain a minimum clearance of 6 in. from the storage space. Figure 410–10

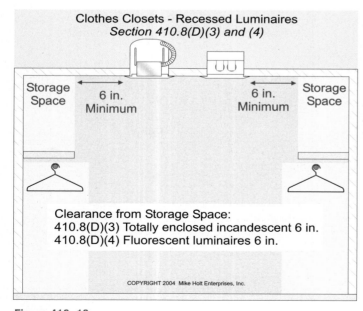

Clothes Closets - Recessed Luminaires
Section 410.8(D)(3) and (4)

Storage Space — 6 in. Minimum — 6 in. Minimum — Storage Space

Clearance from Storage Space:
410.8(D)(3) Totally enclosed incandescent 6 in.
410.8(D)(4) Fluorescent luminaires 6 in.

COPYRIGHT 2004 Mike Holt Enterprises, Inc.

Figure 410–10

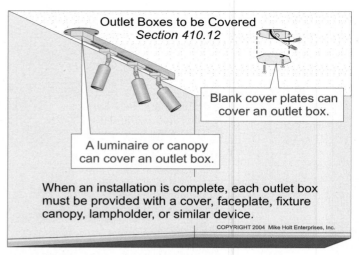

Outlet Boxes to be Covered
Section 410.12

Blank cover plates can cover an outlet box.

A luminaire or canopy can cover an outlet box.

When an installation is complete, each outlet box must be provided with a cover, faceplate, fixture canopy, lampholder, or similar device.

COPYRIGHT 2004 Mike Holt Enterprises, Inc.

Figure 410–11

(4) Recessed fluorescent luminaires must maintain a minimum clearance of 6 in. from the storage space. See Figure 410–10.

PART III. LUMINAIRE OUTLET BOXES AND COVERS

410.12 Outlet Boxes to be Covered. All outlet boxes for luminaires must be covered with a luminaire, lampholder, or blank faceplate. See 314.25. Figure 410–11

410.14 Connection of Electric-Discharge Luminaires.

(A) Luminaires Supported Independently of the Outlet Box. Electric-discharge luminaires supported independently of the

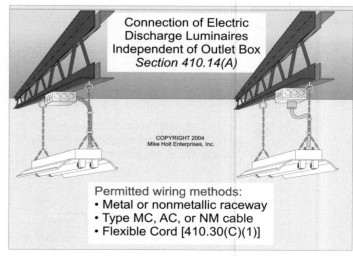

Connection of Electric Discharge Luminaires Independent of Outlet Box
Section 410.14(A)

COPYRIGHT 2004 Mike Holt Enterprises, Inc.

Permitted wiring methods:
• Metal or nonmetallic raceway
• Type MC, AC, or NM cable
• Flexible Cord [410.30(C)(1)]

Figure 410–12

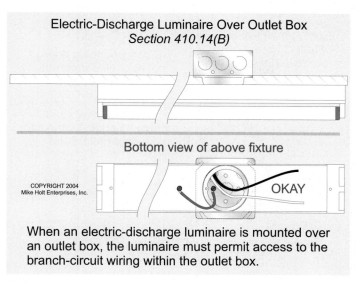

Electric-Discharge Luminaire Over Outlet Box
Section 410.14(B)

Bottom view of above fixture

COPYRIGHT 2004
Mike Holt Enterprises, Inc.

OKAY

When an electric-discharge luminaire is mounted over an outlet box, the luminaire must permit access to the branch-circuit wiring within the outlet box.

Figure 410–13

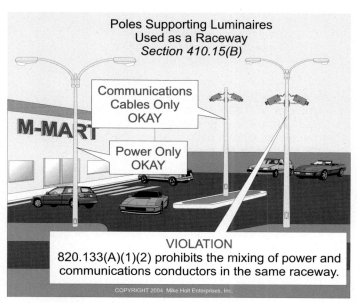

Poles Supporting Luminaires Used as a Raceway
Section 410.15(B)

Communications Cables Only OKAY

M-MART

Power Only OKAY

VIOLATION
820.133(A)(1)(2) prohibits the mixing of power and communications conductors in the same raceway.

COPYRIGHT 2004 Mike Holt Enterprises, Inc.

Figure 410–14

outlet box must be connected to the branch circuit with a raceway, or with MC, AC, or NM cable. **Figure 410–12**

Electric-discharge luminaires can be cord-connected if the luminaires are provided with internal adjustments to position the lamp [410.30(B)].

Electric-discharge luminaires can be cord-connected if the cord is visible for its entire length and is plugged into a receptacle, and the installation is in accordance with 410.30(C).

(B) Access to Outlet Box. When an electric-discharge luminaire is surface mounted over a concealed outlet box, and not supported by the outlet box, the luminaire must be provided with suitable openings to permit access to the branch-circuit wiring within the outlet box. **Figure 410–13**

PART IV. LUMINAIRE SUPPORTS

410.15 Supports.

(A) General Support Requirement. Luminaires and lampholders must be securely supported.

(B) Metal or Nonmetallic Poles. Metal or nonmetallic poles can be used to support luminaires and they can be used as a raceway.

> **Author's Comment:** With homeland security a high priority, many owners want to install security cameras on existing parking lot poles. However, the installation requirements of 820.133(A)(1)(2) prohibit the mixing of power and communications conductors in the same raceway. **Figure 410–14**

In addition, they must comply with the requirements of (1) through (6).

(1) The pole must have an accessible 2 x 4 in. handhole with a raintight cover that provides access to the supply conductors within the pole.

Exception 1: The handhole isn't required for a pole 8 ft or less in height, if the supply conductors for the luminaire are accessible by removing the luminaire. **Figure 410–15**

Exception 2: The handhole can be omitted on poles 20 ft or less in height, if the pole is provided with a hinged base.

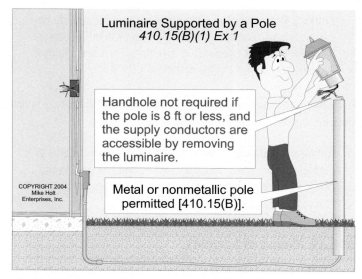

Luminaire Supported by a Pole
410.15(B)(1) Ex 1

Handhole not required if the pole is 8 ft or less, and the supply conductors are accessible by removing the luminaire.

COPYRIGHT 2004
Mike Holt
Enterprises, Inc.

Metal or nonmetallic pole permitted [410.15(B)].

Figure 410–15

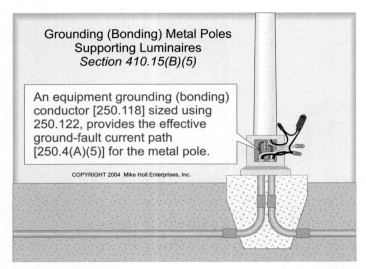

Figure 410-16

Figure 410-17

(2) When the supply raceway or cable doesn't enter the pole, a threaded fitting or nipple must be welded, brazed, or attached to the pole opposite the handhole opening for the supply conductors.

(3) An accessible bonding terminal must be installed inside the pole with a handhole or hinged base.

Exception: The bonding terminal can be omitted on metal poles 8 ft or less in height, if the supply conductors for the luminaire are accessible by removing the luminaire.

(5) The metal pole must be grounded (bonded) to an effective ground-fault current path in accordance with Part VI of Article 250 with an equipment grounding (bonding) conductor of a type specified in 250.118 [250.4(A)(5)]. **Figure 410–16**

Author's Comment: The earth cannot be used to provide the low-impedance fault-current path to the power source, because the earth is a poor conductor of electricity. For more information, see 250.4(A)(5). **Figure 410–17**

(6) Conductors in vertical metal poles must be supported when the vertical rise exceeds 100 ft [Table 300.19(A)].

Author's Comment: When provided by the manufacturer of roadway lighting poles, so-called J-hooks must be used to support conductors, as they are part of the listing instructions [110.3(B)].

410.16 Means of Support.

(A) Outlet Boxes. Outlet boxes designed for the support of luminaires must be supported as follows:

- Fastened to any surface that provides adequate support [314.23(A)].

- Supported from a structural member of a building or from grade by a metal, plastic, or wood brace [314.23(B)].

- Secured to a finished surface (drywall or plaster walls or ceilings) by clamps, anchors, or fittings identified for the application [314.23(C)].

- Secured to the structural or supporting elements of a suspended ceiling [314.23(D)]. **Figure 410–18**

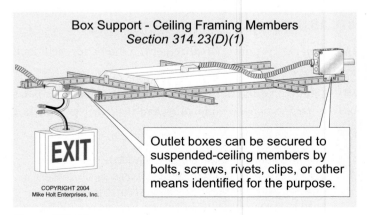

Figure 410-18

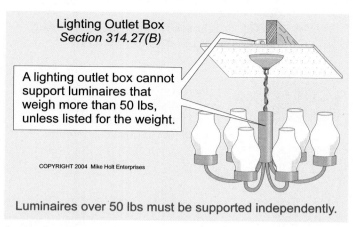

Lighting Outlet Box
Section 314.27(B)

A lighting outlet box cannot support luminaires that weigh more than 50 lbs, unless listed for the weight.

COPYRIGHT 2004 Mike Holt Enterprises

Luminaires over 50 lbs must be supported independently.

Figure 410–19

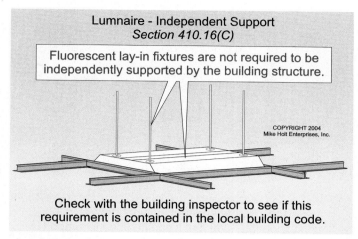

Lumnaire - Independent Support
Section 410.16(C)

Fluorescent lay-in fixtures are not required to be independently supported by the building structure.

COPYRIGHT 2004 Mike Holt Enterprises, Inc.

Check with the building inspector to see if this requirement is contained in the local building code.

Figure 410–21

- Supported by two intermediate metal conduits or rigid metal conduits threaded wrenchtight [314.23(E) and (F)].

- Embedded in concrete or masonry [314.23(G)].

Maximum Luminaire Weight. Outlet boxes for luminaires are permitted to support a luminaire that weighs up to 50 lbs, unless the box is listed for the luminaire's weight [314.27(B)]. **Figure 410–19**

(B) Inspection. Luminaires must be installed so that the connections between the fixture conductors and the branch-circuit conductors can be inspected without requiring the disconnection of any part of the wiring.

(C) Suspended-Ceiling Framing Members. To protect firefighters, luminaires must be attached to the suspended-ceiling framing with screws, bolts, rivets, or clips listed and identified for such use. **Figure 410–20**

Author's Comments:

- The *NEC* doesn't require independent support wires for suspended-ceiling luminaires that aren't installed in a fire-rated ceiling; however, building codes often do. **Figure 410–21**

- Raceways and cables within a suspended ceiling must be supported in accordance with 300.11(A). Outlet boxes can be secured to the ceiling-framing members by bolts, screws, rivets, clips, or independent support wires that are taut and secured at both ends [314.23(D)]. **Figure 410–22**

(H) Luminaires Supported by Trees. Trees can be used for the support of luminaires, but they cannot be used for the support of overhead conductor spans [225.26]. **Figure 410–23**

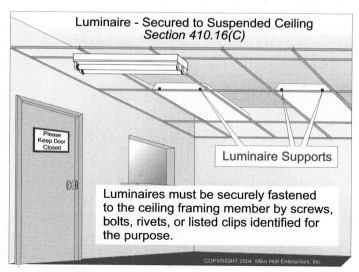

Luminaire - Secured to Suspended Ceiling
Section 410.16(C)

Please Keep Door Closed

Luminaire Supports

Luminaires must be securely fastened to the ceiling framing member by screws, bolts, rivets, or listed clips identified for the purpose.

COPYRIGHT 2004 Mike Holt Enterprises, Inc.

Figure 410–20

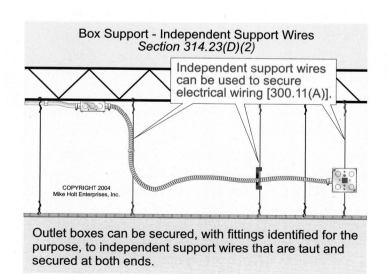

Box Support - Independent Support Wires
Section 314.23(D)(2)

Independent support wires can be used to secure electrical wiring [300.11(A)].

COPYRIGHT 2004 Mike Holt Enterprises, Inc.

Outlet boxes can be secured, with fittings identified for the purpose, to independent support wires that are taut and secured at both ends.

Figure 410–22

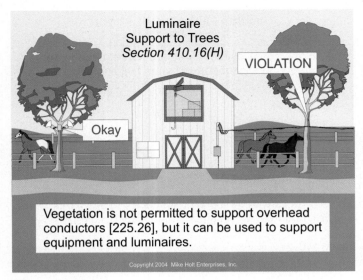

Vegetation is not permitted to support overhead conductors [225.26], but it can be used to support equipment and luminaires.

Figure 410–23

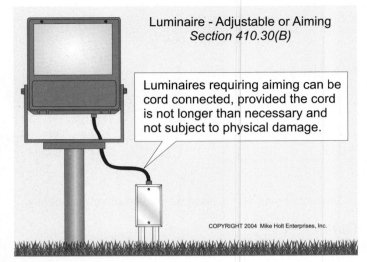

Figure 410–24

PART V. GROUNDING (BONDING)

410.18 Exposed Luminaire Parts.

(A) Exposed Conductive Parts. Exposed metal parts of luminaires must be grounded (bonded) to an effective ground-fault current path in accordance with Part VI of Article 250.

(B) Made of Insulating Material. Luminaires supplied by a wiring method that doesn't provide a ready means for bonding must be made of insulating material and must have no exposed conductive parts.

Exception 1: Replacement luminaires can be installed in an outlet box that doesn't contain an equipment grounding (bonding) conductor, if the luminaire is bonded to one of the following:

> *(1) Grounding electrode (earthing) system [250.50].*
>
> *(2) Grounding electrode conductor.*
>
> *(3) Panelboard equipment grounding terminal.*
>
> *(4) Grounded neutral service conductor within the service equipment enclosure.*

Exception 2: Replacement luminaires that are GFCI-protected aren't required to be connected to an equipment grounding (bonding) conductor if no equipment grounding (bonding) conductor exists at the outlet.

> **Author's Comment:** This is similar to the rule for receptacle replacements in locations where an equipment grounding (bonding) conductor isn't present in the outlet box [406.3(D)(3)].

PART VI. WIRING OF LUMINAIRES

410.23 Polarization of Luminaires.
Luminaires must have the grounded neutral conductor connected to the screw shell of the lampholder [200.10(C)], and the grounded neutral conductor must be properly identified in accordance with 200.6.

410.30 Cord-Connected Luminaires.

(B) Adjustable Luminaires. Luminaires that require adjusting or aiming after installation can be cord-connected with or without an attachment plug, provided the exposed cord is of the hard usage or extra-hard usage type. The cord must not be longer than necessary for luminaire adjustment, and it must not be subject to strain or physical damage [400.10]. **Figure 410–24**

(C) Electric-Discharge Luminaires. A listed luminaire or listed luminaire assembly can be cord-connected if: **Figure 410–25**

(1) The luminaire is mounted directly below the outlet box or busway, and

(2) The flexible cord:

> a. Is visible its entire length,
>
> b. Isn't subject to strain or physical damage [400.10], and
>
> c. Terminates in an attachment plug, busway plug, or canopy with strain relief, or is a part of a listed assembly incorporating a manufactured wiring system connector in accordance with 604.6(C).

> **Author's Comment:** The *Code* doesn't require twist-lock receptacles for this application.

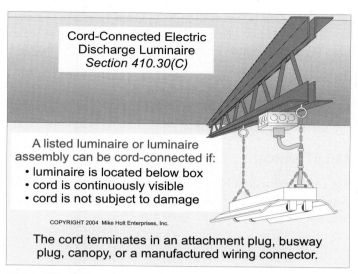

Cord-Connected Electric Discharge Luminaire
Section 410.30(C)

A listed luminaire or luminaire assembly can be cord-connected if:
- luminaire is located below box
- cord is continuously visible
- cord is not subject to damage

COPYRIGHT 2004 Mike Holt Enterprises, Inc.

The cord terminates in an attachment plug, busway plug, canopy, or a manufactured wiring connector.

Figure 410–25

410.31 Luminaires Used as a Raceway. Luminaires cannot be used as a raceway for circuit conductors unless the luminaire is listed and marked for use as a raceway. Figure 410–26

410.32 Wiring Luminaires Connected Together. Luminaires designed for end-to-end assembly, or luminaires connected together by recognized wiring methods, are permitted to contain a 2-wire branch circuit, or one multiwire branch circuit, supplying the connected luminaires. One additional 2-wire branch circuit supplying one or more of the connected luminaires, such as a nightlight, is also permitted. Figure 410–27

410.33 Branch-Circuit Conductors and Ballast. Conductors within 3 in. of fixture ballasts must have an insulation rating not less than a 90°C.

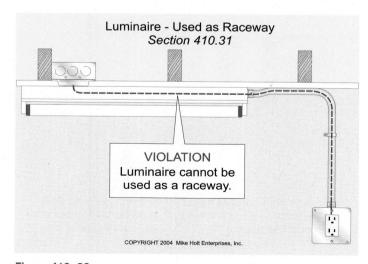

Luminaire - Used as Raceway
Section 410.31

VIOLATION
Luminaire cannot be used as a raceway.

COPYRIGHT 2004 Mike Holt Enterprises, Inc.

Figure 410–26

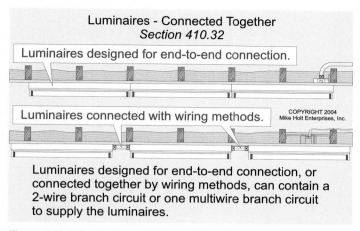

Luminaires - Connected Together
Section 410.32

Luminaires designed for end-to-end connection.

Luminaires connected with wiring methods.

COPYRIGHT 2004 Mike Holt Enterprises, Inc.

Luminaires designed for end-to-end connection, or connected together by wiring methods, can contain a 2-wire branch circuit or one multiwire branch circuit to supply the luminaires.

Figure 410–27

PART VIII. LAMPHOLDERS

410.47 Screw-Shell Lampholder. Lampholders of the screw-shell type must be installed for use as lampholders only.

Author's Comment: A receptacle adapter that screws into a lampholder is a violation of this section. Figure 410–28

PART XI. RECESSED LUMINAIRES

410.65 Thermally Protected.

(C) Recessed Incandescent Luminaires. Recessed incandescent luminaires must be identified as thermally protected.

Author's Comment: When higher-wattage lamps or improper trims are installed, the lampholder contained in a recessed luminaire can overheat, activating the thermal protection device and causing the luminaire to cycle on and off.

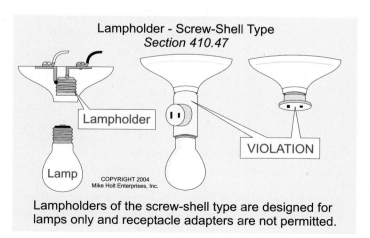

Lampholder - Screw-Shell Type
Section 410.47

Lampholder

VIOLATION

Lamp

COPYRIGHT 2004 Mike Holt Enterprises, Inc.

Lampholders of the screw-shell type are designed for lamps only and receptacle adapters are not permitted.

Figure 410–28

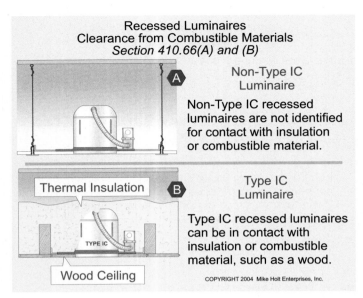

Figure 410–29

Exception 2: Thermal protection isn't required for recessed Type IC luminaires whose design, construction, and thermal performance characteristics are equivalent to a thermally protected fixture.

410.66 Recessed Luminaire Clearances.

(A) Clearances From Combustible Materials.

(1) Non-Type IC Luminaire. A recessed luminaire that isn't identified for contact with insulation must have all recessed parts, except the points of supports, spaced not less than ¼ in. from combustible materials. **Figure 410–29A**

(2) Type IC Luminaire. A Type IC luminaire (identified for contact with insulation) can be in contact with combustible materials. **Figure 410–29B**

(B) Installation. Thermal insulation must not be installed above a recessed luminaire or within 3 in. of the recessed luminaire's enclosure, wiring compartment, or ballast unless identified for contact with insulation; Type IC.

410.67 Wiring.

(C) Tap Conductors. Fixture wires installed in accordance with Article 402, protected in accordance with 240.5(B)(2), and not over 6 ft long are permitted. **Figure 410–30**

PART XIII. ELECTRIC-DISCHARGE LIGHTING

410.73 General.

(F) High-Intensity Discharge Luminaires.

(5) Metal-Halide Lamp Containment. Luminaires containing a metal-halide lamp, other than a thick-glass parabolic reflector lamp (PAR), must be provided with a containment barrier that encloses the lamp, or the luminaire must only allow the use of a Type "O" lamp that has an internal arc-tube shield. **Figure 410–31**

> **Author's Comment:** Fires have resulted from aging arc lamps that explode in the morning on startup, shattering the lamp globe and showering glass and hot quartz fragments from metal-halide lamp failures. The possibility of failure increases significantly as the lamp approaches and exceeds its rated life. It is projected that one violent rupture occurs in 100,000 failures.

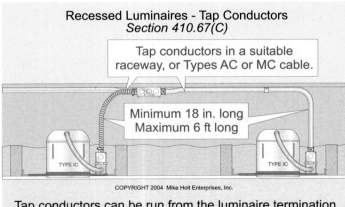

Figure 410–30

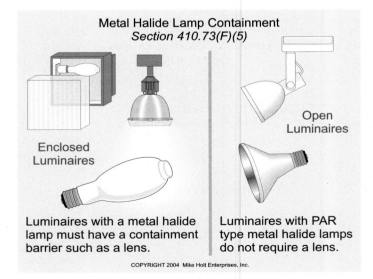

Figure 410–31

"O-rated" lamps, which have an internal arc-tube shield, have been designed to meet ANSI containment tests for the installation of a metal-halide lamp in an open fixture.

(G) Disconnecting Means. In indoor locations, other than dwellings and associated accessory structures, fluorescent luminaires that utilize double-ended lamps (typical fluorescent lamps) and contain ballast(s), or ballasted luminaires that are supplied from multiwire branch circuits and contain ballast(s) that can be serviced in place or re-ballasted, must have a disconnecting means to disconnect simultaneously all conductors of the ballast, including the grounded neutral conductor, if any. The disconnecting means must be accessible to qualified persons.

This requirement will become effective January 1, 2008.

> **Author's Comment:** Changing the ballast out while the circuit feeding the luminaire is energized has become a regular practice because a local disconnect isn't available. Also, ballasts are often serviced from a ladder, adding the possibility of increased injury from a fall. The rule requires the disconnecting means to open "all circuit conductors," including the grounded neutral conductor. If the grounded neutral conductor in a multiwire circuit isn't disconnected at the same time as the ungrounded conductor, a false sense of security could result in an unexpected shock, and its consequences, from the grounded neutral conductor.

Exception 1: A disconnecting means isn't required for luminaires installed in hazardous (classified) locations.

Exception 2: A disconnecting means isn't required for the emergency illumination required in 700.16.

Exception 3: For cord-and-plug connected luminaires, an accessible separable connector, or an accessible plug and receptacle, is permitted to serve as the disconnecting means.

Exception 4: A disconnecting means isn't required in industrial establishments with restricted public access where written procedures and conditions of maintenance and supervision ensure that only qualified persons service the installation.

Exception 5: Where more than one luminaire is installed and supplied by other than a multiwire branch circuit, a disconnecting means isn't required for every luminaire when the design of the installation includes locally accessible disconnects, such that the illuminated space won't be left in total darkness.

410.76 Luminaire Mounting.

(B) Surface-Mounted Luminaires with Ballasts. Surface-mounted luminaires containing ballasts must have a minimum of 1½ in. clearance from combustible low-density fiberboard. Figure 410–32

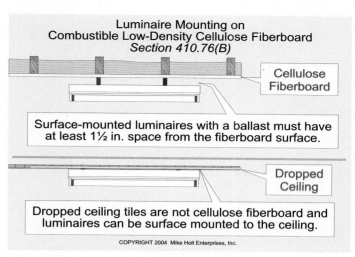

Figure 410–32

> **Author's Comment:** This rule doesn't apply to the mounting of surface-mounted fluorescent luminaires onto wood, plaster, concrete, or drywall. Figure 410–33

PART XV. TRACK LIGHTING

410.100 Definition.

Track Lighting. This is a manufactured assembly designed to support and energize luminaires that can be readily repositioned on track whose length may be altered by the addition or subtraction of sections of track.

410.101 Installation.

(A) Track Lighting. Track lighting must be permanently installed and permanently connected to the branch-circuit

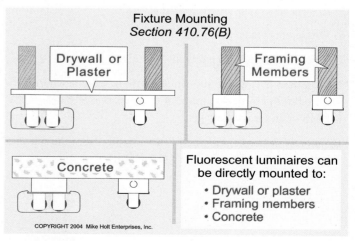

Figure 410–33

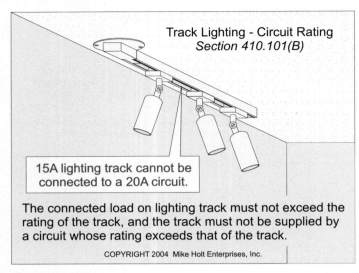

Figure 410–34

Decorative lighting, which can be used for a period of up to 90 days [590.3(B)], must be listed for the purpose.

Figure 410–35

wiring. Lampholders for track lighting are designed for lamps only, and a receptacle adapter isn't permitted [410.47].

(B) Circuit Rating. The connected load on lighting track must not exceed the rating of the track, and a circuit whose rating exceeds that of the track cannot supply the track.

> **Author's Comment:** This means 15A lighting track cannot be connected to a 20A circuit. **Figure 410–34**

(C) Locations not Permitted. Track lighting cannot be installed:

(1) Where it is likely to be subjected to physical damage.

(2) In wet or damp locations.

(3) Where subject to corrosive vapors.

(4) In storage battery rooms.

(5) In hazardous (classified) locations.

(6) Where concealed.

(7) Where extended through walls, partitions, or floors.

(8) Less than 5 ft above the finished floor, except where protected from physical damage or the track operates below 30V RMS open-circuit voltage.

(9) Within 3 ft horizontally and 8 ft vertically from the top of a bathtub rim or shower space [410.4(D)].

410.104 Fastening. Track lighting must be securely mounted to support the weight of the luminaires. A single track section 4 ft or shorter in length must have two supports, and, where installed in a continuous row, each individual track section of not more than 4 ft in length must have one additional support.

PART XVI. DECORATIVE LIGHTING AND SIMILAR ACCESSORIES

410.110. Listing of Decorative Lighting. Decorative lighting, which can be used for a period of up to 90 days [590.3(B)], must be listed for the purpose. **Figure 410–35**

> **Author's Comment:** Faulty or inadequate decorative lighting causes the majority of Christmas tree fires and UL recently developed UL 588, *Standard for Seasonal and Holiday Decorative Products,* to address this problem.

1. Article 410 covers luminaires, lampholders, pendants, and _____, and the wiring and equipment forming part of such products and lighting installations.

    (a) decorative lighting products
    (b) lighting accessories for temporary seasonal and holiday use
    (c) portable flexible lighting products
    (d) all of these

2. Luminaires using a _____ lamp, that are subject to physical damage and installed in playing and spectator seating areas of indoor sports, mixed-use, or all-purpose facilities, must be of the type that protects the lamp with a glass or plastic lens. Such luminaires are permitted to have an additional guard.

    (a) mercury vapor   (b) metal halide   (c) fluorescent   (d) a or b

3. Electric-discharge luminaires supported independently of the outlet box must be connected to the branch circuit through _____.

    (a) raceways   (b) Type MC, AC, MI, NM, or NMC cable
    (c) flexible cords   (d) a, b, or c

4. Metal poles used to support luminaires must be bonded to a(n) _____.

    (a) grounding electrode   (b) grounded neutral conductor
    (c) equipment grounding (bonding) conductor   (d) any of these

5. The flexible cord used to connect a luminaire may be terminated _____.

    (a) in a grounding-type attachment plug cap
    (b) as a part of a listed assembly incorporating a manufactured wiring system connector
    (c) as a part of a listed luminaire assembly with a strain relief and canopy
    (d) all of these

6. Recessed incandescent luminaires must have _____ protection and must be identified as thermally protected.

    (a) physical   (b) corrosion   (c) thermal   (d) all of these

7. The raceway or cable for tap conductors to recessed luminaires must have a minimum length of _____

    (a) 6 in.   (b) 12 in.   (c) 18 in.   (d) 24 in.

8. Lighting track is a manufactured assembly and its length may not be altered by the addition or subtraction of sections of track.

    (a) True   (b) False

9. Lighting track must be securely mounted so each fastening will be suitable to support the maximum weight of _____.

    (a) 35 lbs   (b) 50 lbs
    (c) luminaires that can be installed   (d) none of these

Notes

411 Lighting Systems Operating at 30V or Less

Introduction

Article 411 provides the requirements for 30V lighting systems which are found in such applications as landscaping, jewelry stores, and museums. Don't let the half-page size of Article 411 give you the impression that 30V lighting isn't something you need to be concerned about. Its applications are widespread and becoming more so.

Many of these systems now use LEDs, and 30V halogen lamps are also fairly common. All 30V lighting systems have an ungrounded secondary circuit supplied by an isolating transformer. These systems have restrictions that affect where they can be located, and they can have a maximum supply breaker size of 20A.

411.1 Scope. Article 411 covers the installation of lighting systems that operate at 30V or less and their associated components.

411.2 Definition.

Lighting Systems Operating at 30 Volts or Less. A 30V lighting system consisting of an isolating power source, which limits each secondary circuit to 25A. Figure 411–1

411.3 Listing Required. Lighting systems must be listed for the purpose.

411.4 Locations Not Permitted. Conductors for lighting systems operating at 30V or less must not be:

(A) Concealed Spaces. Installed in a concealed space or pass through a building wall, except as permitted in (1) or (2).

(1) Lighting system conductors installed within a Chapter 3 wiring method.

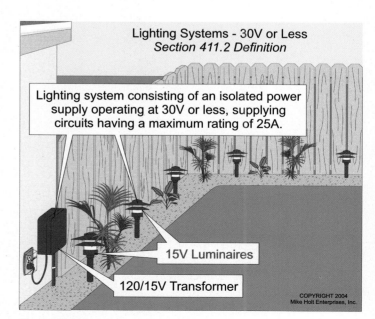

Lighting Systems - 30V or Less
Section 411.2 Definition

Lighting system consisting of an isolated power supply operating at 30V or less, supplying circuits having a maximum rating of 25A.

15V Luminaires

120/15V Transformer

COPYRIGHT 2004
Mike Holt Enterprises, Inc.

Figure 411–1

(2) Lighting system conductors supplied by a listed Class 2 power source and installed in accordance with 725.52.

(B) Pools, Spas, and Hot Tubs. Within 10 ft of permanently installed pools, storable pools, outdoor spas, outdoor hot tubs or fountains, or similar locations. **Figure 411–2**

411.5 Secondary Circuits.

(B) Isolation. The secondary circuit must be insulated from the branch circuit by an isolating transformer.

(C) Bare Conductors. Exposed bare conductors and current-carrying parts must not be installed less than 7 ft above the finished floor, unless listed for the height.

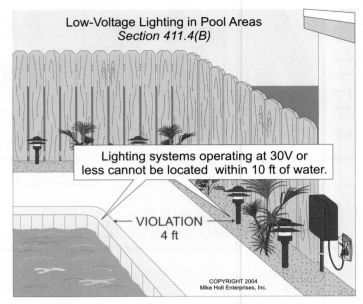

Low-Voltage Lighting in Pool Areas
Section 411.4(B)

Lighting systems operating at 30V or less cannot be located within 10 ft of water.

VIOLATION
4 ft

COPYRIGHT 2004
Mike Holt Enterprises, Inc.

Figure 411–2

Article 411 Questions

1. Lighting systems operating at 30V or less need not be listed for the purpose.

   (a) True   (b) False

2. Lighting systems operating at 30V or less are allowed to be concealed or extended through a building wall without regard to the wiring method used.

   (a) True   (b) False

3. Lighting systems operating at 30V or less are allowed to be concealed or extended through a building wall using _____.

   (a) any of the wiring methods specified in Chapter 3
   (b) wiring supplied by a listed class 2 power source installed in accordance with 725.52
   (c) both a and b
   (d) metal raceways only

4. Lighting systems operating at 30V or less must not be installed within 10 ft of pools, spas, fountains, or similar locations except as permitted by Article 680.

   (a) True   (b) False

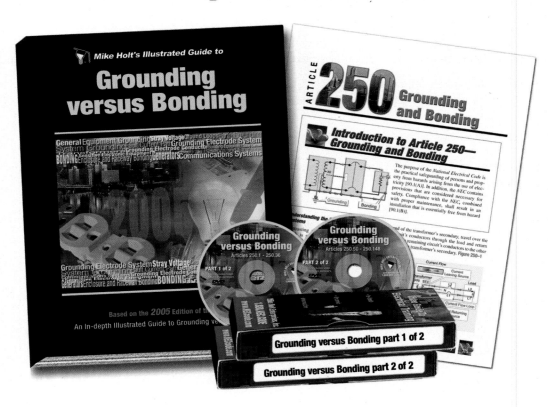

ARTICLE 422 Appliances

Introduction

Article 422 covers electric appliances used in any occupancy. The meat of what you need to know is in Parts II and III. Parts IV and V are primarily for manufacturers, but you should examine appliances for conformance before installing. If the appliance has a label from a recognized labeling authority (for example, UL), it conforms.

Two concepts drive the requirements of Article 422: On the one hand, an appliance should not overload the circuit supplying it; on the other hand, an appliance should not be supplied with more current than it should reasonably draw. The first concept is why 422.10 specifies the minimum circuit protection. The second concept is why 422.11 specifies the maximum circuit protection. As you read through Article 422 requirements, try to think of how each one relates to these two concepts.

Interestingly, the *NEC* doesn't include "Fixed Electric Space-Heating Equipment" in the scope of Article 422, but instead provides Article 424 to address fixed electrical equipment used for space heating.

PART I. GENERAL

422.1 Scope. The scope of Article 422 includes appliances that are fastened in place, permanently connected, or cord-and-plug connected in any occupancy. **Figure 422–1**

> **Author's Comment: Definition of** Appliance [Article 100]. Utilization equipment, generally other than industrial, that is normally built in standardized sizes or types and is installed or connected as a unit to perform one or more functions, such as clothes washing, air conditioning, food mixing, deep frying, and so forth.

422.3 Other Articles.

Appliances used in hazardous (classified) locations must comply with Articles 500 through 517. Motor-operated appliances must comply with Article 430, and appliances containing hermetic refrigerant motor compressor(s) must comply with Article 440.

> **Author's Comment:** Room air-conditioning equipment must be installed in accordance with Part VII of Article 440.

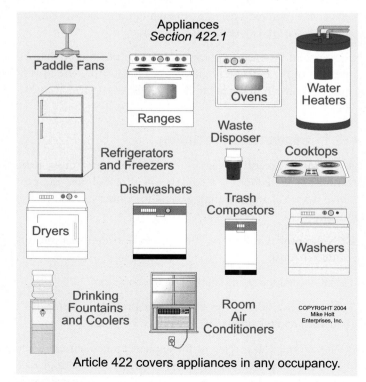

Appliances
Section 422.1

Paddle Fans · Ranges · Ovens · Water Heaters · Refrigerators and Freezers · Waste Disposer · Cooktops · Dishwashers · Trash Compactors · Dryers · Washers · Drinking Fountains and Coolers · Room Air Conditioners

COPYRIGHT 2004
Mike Holt
Enterprises, Inc.

Article 422 covers appliances in any occupancy.

Figure 422–1

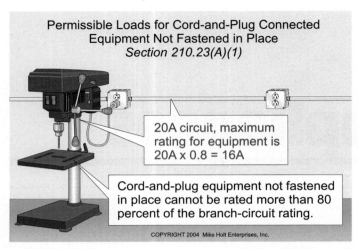

Permissible Loads for Cord-and-Plug Connected
Equipment Not Fastened in Place
Section 210.23(A)(1)

20A circuit, maximum rating for equipment is 20A x 0.8 = 16A

Cord-and-plug equipment not fastened in place cannot be rated more than 80 percent of the branch-circuit rating.

COPYRIGHT 2004 Mike Holt Enterprises, Inc.

Figure 422–2

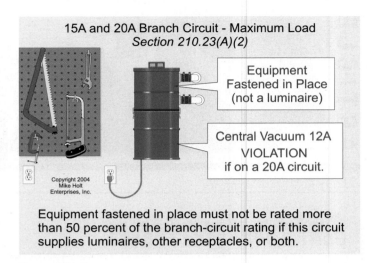

15A and 20A Branch Circuit - Maximum Load
Section 210.23(A)(2)

Equipment Fastened in Place (not a luminaire)

Central Vacuum 12A VIOLATION if on a 20A circuit.

Copyright 2004 Mike Holt Enterprises, Inc.

Equipment fastened in place must not be rated more than 50 percent of the branch-circuit rating if this circuit supplies luminaires, other receptacles, or both.

Figure 422–3

PART II. BRANCH-CIRCUIT REQUIREMENTS

422.10 Branch-Circuit Rating.

(A) Individual Circuits. The branch-circuit ampere rating for an individual appliance must not be less than the branch-circuit rating marked on the appliance [110.3(B)].

(B) Circuits Supplying Two or More Loads. Branch circuits supplying appliance and other loads must be sized in accordance with 210.23(A).

- Cord-and-plug connected equipment must not be rated more than 80 percent of the branch-circuit ampere rating [210.23(A)(1)]. **Figure 422–2**

- Equipment fastened in place must not be rated more than 50 percent of the branch-circuit ampere rating, if the circuit supplies both luminaires and receptacles [210.23(A)(2)]. **Figure 422–3**

422.11 Overcurrent Protection.

(A) Branch-Circuit. Branch-circuit conductors must have overcurrent protection in accordance with 240.4, and the protective device rating must not exceed the rating marked on the appliance.

(E) Nonmotor Appliances. The appliance overcurrent protection device must:

(1) Not exceed the rating marked on the appliance.

(2) Not exceed 20A if the overcurrent protection rating isn't marked and the appliance is rated 13.3A or less, or

(3) Not exceed 150 percent of the appliance rated current if the overcurrent protection rating isn't marked and the appliance is rated over 13.3A. Where 150 percent of the appliance

rating doesn't correspond to a standard overcurrent device ampere rating listed in 240.6(A), the next higher standard rating is permitted.

Question: What is the maximum size overcurrent protection for a 4,500W, 240V water heater? **Figure 422–4**

(a) 20A   *(b) 30A*   *(c) 40A*   *(d) 50A*

Answer: (b) 30A

4,500W/240V = 18.75A x 1.5 = 28A, next size up, 30A [240.6(A)]

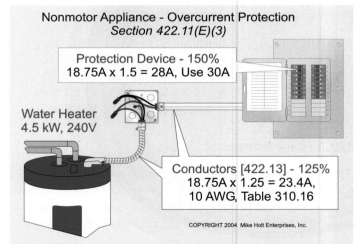

Nonmotor Appliance - Overcurrent Protection
Section 422.11(E)(3)

Protection Device - 150% 18.75A x 1.5 = 28A, Use 30A

Water Heater 4.5 kW, 240V

Conductors [422.13] - 125% 18.75A x 1.25 = 23.4A, 10 AWG, Table 310.16

COPYRIGHT 2004 Mike Holt Enterprises, Inc.

Note: Storage type water heaters with a capacity of 120 gallons or less are considered continuous loads [422.13]. 210.19(A)(1) requires conductors to be sized at 125%.

Figure 422–4

422.12 Fossil Fuel Heating Equipment (Furnaces).

An individual branch circuit must supply central heating equipment, such as gas, oil, or coal furnaces.

> **Author's Comment:** This rule isn't intended to apply to a listed wood-burning fireplace with a fan.

Exception 1: Auxiliary equipment to the fossil fuel heating equipment, such as pumps, valves, humidifiers, and electrostatic air cleaners, can be connected to the central heater circuit.

> **Author's Comment:** Electric space-heating equipment must be installed in accordance with Article 424 Electric Space-Heating Equipment.

Exception 2: Permanently connected air-conditioning equipment can be connected to the individual branch circuit that supplies central heating equipment.

422.13 Storage Water Heaters.

An electric water heater having a capacity of no more than 120 gallons is considered a continuous load.

> **Author's Comment:** Circuit conductors and overcurrent protection devices must have an ampacity of not less than 125 percent of the ampere rating of the appliance. **Figure 422–5**

- Branch-circuit conductors [210.19(A)(1)], overcurrent protection devices [210.20(A)]
- Feeder conductors [215.2(A)(1)], overcurrent protection devices [215.3]
- Service conductors [230.42(A)(1)]

Question: What is the calculated continuous load for a 4,500W, 230V water heater?

(a) 15A   (b) 20A   (c) 25A   (d) 30A

Answer: *(c) 25A*

I = P/E
P = 4,500W
E = 230V
I = 4,500W/230V
I = 20A

Calculated continuous load for
conductor sizing and protection = 20A x 1.25 = 25A

422.15 Central Vacuum.

(A) Circuit Loading. A separate circuit is not required for a central vacuum outlet, if the rating of the equipment does not exceed 50 percent of the ampere rating of the circuit.

> **Author's Comments:**
>
> - A separate 15A circuit is required for a central vacuum receptacle outlet if the rating of the central vacuum exceeds 7.5A, and a separate 20A circuit is required for a central vacuum receptacle outlet if the rating of the central vacuum exceeds 10A, but not 16A [210.23(A)(2)]. **Figure 422–6**
> - 210.23(A)(2) specifies that equipment fastened in place, other than luminaires, must not be rated more than 50 percent of the branch-circuit ampere rating if this circuit supplies both luminaires and receptacles.

Water Heater
Continuous Load
Section 422.13

A water heater with a capacity of 120 gallons or less is considered a continuous load.

Conductors and protection devices are calculated at 125% of the rated load.

120 Gallon Water Heater

COPYRIGHT 2004 Mike Holt Enterprises, Inc.

Figure 422–5

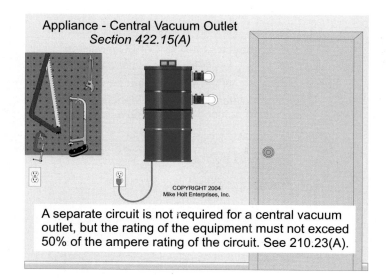

Appliance - Central Vacuum Outlet
Section 422.15(A)

COPYRIGHT 2004
Mike Holt Enterprises, Inc.

A separate circuit is not required for a central vacuum outlet, but the rating of the equipment must not exceed 50% of the ampere rating of the circuit. See 210.23(A).

Figure 422–6

422.16 Flexible Cords.

(A) General. Flexible cords are permitted for the connection of appliances if the appliance is identified for flexible cord use and it meets one of the following conditions [400.7(A)]:

(1) Facilitate frequent interchange or to prevent the transmission of noise and vibration.

(2) Facilitate the removal of appliances fastened in place, where the fastening means and mechanical connections are specifically designed to permit ready removal.

> **Author's Comment:** Flexible cords cannot be used for the connection of water heaters, furnaces, and other appliances fastened in place, unless the appliances are specifically identified to be used with a flexible cord. **Figure 422–7**

(B) Specific Appliances.

(1) Waste (Garbage) Disposal. A flexible cord is permitted for a waste disposal if:

   (1) The cord has a grounding-type attachment plug.

   (2) The cord length is at least 18 in. and not longer than 3 ft.

   (3) The waste disposer receptacle is located to avoid damage to the cord.

   (4) The waste disposer receptacle is accessible.

(2) Dishwasher and Trash Compactor. A cord is permitted for a dishwasher or trash compactor if:

   (1) The cord has a grounding-type attachment plug.

   (2) The cord length is at least 3 ft and not longer than 4 ft, measured from the rear plane of the appliance. **Figure 422–8**

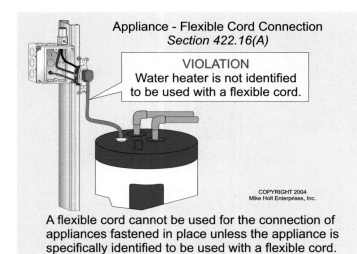

Appliance - Flexible Cord Connection
Section 422.16(A)

VIOLATION
Water heater is not identified to be used with a flexible cord.

COPYRIGHT 2004
Mike Holt Enterprises, Inc.

A flexible cord cannot be used for the connection of appliances fastened in place unless the appliance is specifically identified to be used with a flexible cord.

Figure 422–7

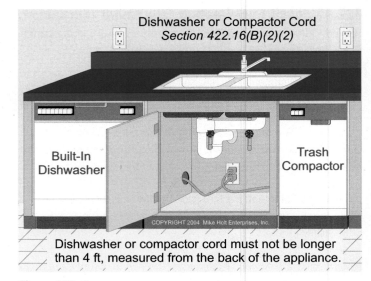

Dishwasher or Compactor Cord
Section 422.16(B)(2)(2)

Built-In Dishwasher

Trash Compactor

COPYRIGHT 2004 Mike Holt Enterprises, Inc.

Dishwasher or compactor cord must not be longer than 4 ft, measured from the back of the appliance.

Figure 422–8

> **Author's Comment:** This means that if 2 ft of cord is required to make the connections inside the appliance, the minimum cord length for a dishwasher is between 5 and 6 ft.

   (3) The appliance receptacle is located to avoid damage to the cord.

   (4) The receptacle is located in the space occupied by the appliance or in the space adjacent to the appliance.

   (5) The receptacle is accessible.

> **Author's Comment:** According to an article in the International Association of Electrical Inspectors magazine (*IAEI News*), a cord run through a cabinet for an appliance isn't considered as being run through a wall.

(3) Wall-Mounted Ovens and Counter-Mounted Cooking Units. Wall-mounted ovens and counter-mounted cooking units can be cord-and-plug connected.

(4) Range Hoods. Range hoods can be cord-and-plug connected, where all of the following conditions are met. **Figure 422–9**

   (1) The flexible cord terminates with a grounding-type attachment plug.

   (2) The length of the cord must not be less than 18 in. or longer than 36 in.

   (3) The range hood receptacle must be located to avoid physical damage to the flexible cord.

   (4) The range hood receptacle must be accessible.

   (5) The range hood receptacle must be supplied by an individual branch circuit.

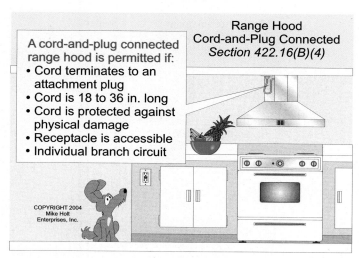

A cord-and-plug connected range hood is permitted if:
- Cord terminates to an attachment plug
- Cord is 18 to 36 in. long
- Cord is protected against physical damage
- Receptacle is accessible
- Individual branch circuit

Range Hood
Cord-and-Plug Connected
Section 422.16(B)(4)

COPYRIGHT 2004
Mike Holt
Enterprises, Inc.

Figure 422–9

422.18 Support of Ceiling-Suspended (Paddle) Fans.

Ceiling-suspended (paddle) fans must be supported by a listed fan outlet box, or outlet box system, in accordance with 314.27(D).

Author's Comment: An outlet box identified for use with a ceiling paddle fan is permitted for a ceiling paddle fan that doesn't weigh more than 70 lbs. An outlet box for ceiling paddle fans that weigh more than 35 lbs, and up to 70 lbs, must have the maximum weight marked on the box. Where the maximum isn't marked on the box, and the fan weighs over 35 lbs, the fan must be supported independently of the outlet box. All ceiling paddle fans over 70 lbs must be supported independently of the outlet box [314.27(D)]. **Figure 422–10**

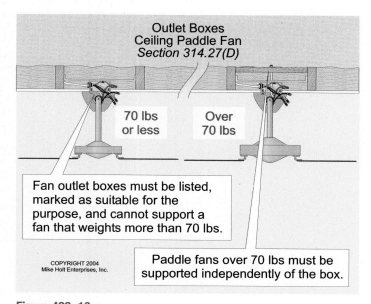

Outlet Boxes
Ceiling Paddle Fan
Section 314.27(D)

70 lbs or less

Over 70 lbs

Fan outlet boxes must be listed, marked as suitable for the purpose, and cannot support a fan that weights more than 70 lbs.

COPYRIGHT 2004
Mike Holt Enterprises, Inc.

Paddle fans over 70 lbs must be supported independently of the box.

Figure 422–10

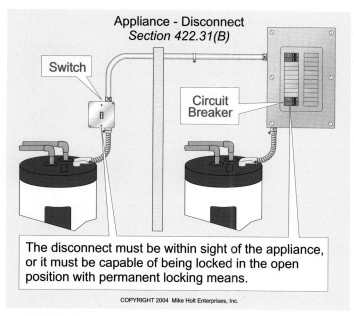

Appliance - Disconnect
Section 422.31(B)

Switch

Circuit Breaker

The disconnect must be within sight of the appliance, or it must be capable of being locked in the open position with permanent locking means.

COPYRIGHT 2004 Mike Holt Enterprises, Inc.

Figure 422–11

PART III. DISCONNECT

422.31 Permanently Connected Appliance Disconnect.

(A) Appliance Rated at Not Over 300 VA or ⅛ Horsepower. The branch-circuit overcurrent device, such as a plug fuse or circuit breaker, can serve as the appliance disconnect.

(B) Appliance Rated Over 300 VA or ⅛ Horsepower. A switch or circuit breaker located within sight from the appliance can serve as the appliance disconnecting means. If the switch or circuit breaker is capable of being locked in the open position, it does not need to be within sight if the provision for locking or adding a lock is on the disconnecting means. **Figure 422–11**

Author's Comment: "Within sight" is visible and not more than 50 ft from one to the other [Article 100].

422.33 Cord-and-Plug Connected Appliance Disconnect.

(A) Attachment Plug and Receptacle. An accessible plug and receptacle can serve as the disconnecting means for a cord-and-plug connected appliance. **Figure 422–12**

(B) Cord-and-Plug Connected Range. The plug and receptacle of a cord-and-plug connected household electric range can serve as the range disconnecting means if the plug is accessible from the front of the range by the removal of a drawer. **Figure 422–13**

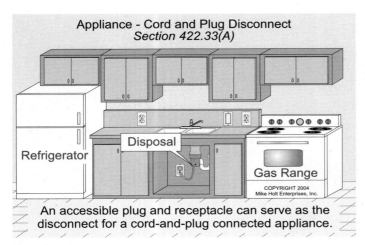

Appliance - Cord and Plug Disconnect
Section 422.33(A)

An accessible plug and receptacle can serve as the disconnect for a cord-and-plug connected appliance.

Figure 422–12

422.34 Unit Switch as Disconnect.

A unit switch with a marked "off" position that is a part of the appliance can serve as the appliance disconnect, if it disconnects all ungrounded conductors. **Figure 422–14**

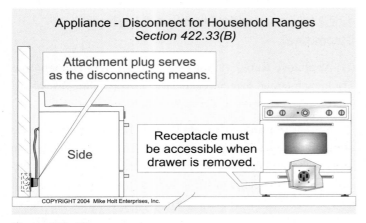

Appliance - Disconnect for Household Ranges
Section 422.33(B)

Attachment plug serves as the disconnecting means.

Receptacle must be accessible when drawer is removed.

Side

Figure 422–13

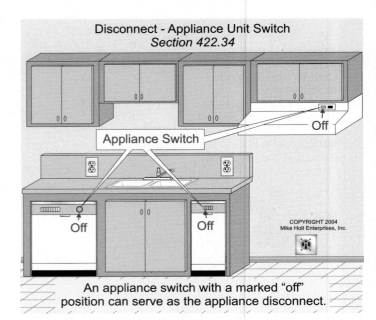

Disconnect - Appliance Unit Switch
Section 422.34

Appliance Switch

Off

Off   Off

An appliance switch with a marked "off" position can serve as the appliance disconnect.

Figure 422–14

422.51 Cord-and-Plug Connected Vending Machines.

Cord-and-plug connected vending machines manufactured or remanufactured on or after January 1, 2005 must include a GFCI as an integral part of the attachment plug. Cord-and-plug connected vending machines not incorporating integral GFCI protection must be connected to a GFCI-protected outlet.

Author's Comment: This change was driven by three electrocutions since 1995; two of the deaths were children and occurred when they touched the energized metal parts of a vending machine. Because electric vending machines are often located in damp or wet locations in public places, and are used by people standing on the ground, reliance on an equipment grounding (bonding) conductor for protection against electrocution is insufficient.

1. In general, branch-circuit conductors to individual appliances must not be sized _____ than required by the appliance markings or instructions.

    (a) larger   (b) smaller

2. The rating or setting of an overcurrent protection device for a 16.3A single nonmotor-operated appliance must not exceed _____.

    (a) 15A   (b) 35A   (c) 25A   (d) 45A

3. A waste disposal can be cord-and-plug connected, but the cord must not be less than 18 in. or more than _____ in length and must be protected from physical damage.

    (a) 30 in.   (b) 36 in.   (c) 42 in.   (d) 48 in.

4. Ceiling-suspended (paddle) fans must be supported independently of an outlet box or by outlet boxes or outlet box systems _____ for the application and installed in accordance with 314.27(D).

    (a) satisfactory   (b) approved   (c) strong enough   (d) identified

5. For cord-and-plug connected appliances, an accessible separable connector or _____ plug and receptacle is permitted to serve as the disconnecting means.

    (a) a labeled   (b) an accessible   (c) a metal enclosed   (d) none of these

Notes

Introduction

Many people are surprised to see how many pages are in Article 424. This is a nine-part article on fixed electric space-heaters. Why so much text for what seems to be a simple application? The answer is that Article 424 covers a variety of applications—heaters come in various configurations for various uses. Not all of these Parts are for the electrician in the field—the requirements in Part IV are for manufacturers.

Most electricians should focus on Part III, Part V, and Part VI. Fixed space heaters (wall-mounted, ceiling-mounted, or free-standing) are common in many utility buildings and other small structures, as well as in some larger structures. When used to heat floors, space-heating cables address the thermal layering problem typical of forced-air systems—so it's likely you will encounter them. Duct heaters are very common in large office and educational buildings. These provide a distributed heating scheme. Locating the heater in the ductwork, but close to the occupied space, eliminates the waste of transporting heated air through sheet metal routed in unheated spaces. So, it's likely you will encounter those as well.

PART I. GENERAL

424.1 Scope. Article 424 contains the installation requirements for fixed electrical equipment used for space heating, such as heating cable, unit heaters, boilers, or central systems.

> **Author's Comment:** Wiring for fossil-fuel heating equipment, such as gas, oil, or coal central furnaces, must be installed in accordance with Article 422, specifically 422.12.

424.3 Branch Circuits.

(B) Branch-Circuit Sizing. For the purpose of sizing branch-circuit conductors, fixed electric space-heating equipment is considered a continuous load.

> **Author's Comment:** The branch-circuit conductor and overcurrent protection devices for fixed electric space-heating equipment must have an ampacity not less than 125 percent of the total heating load [210.19(A)(1) and 210.20(A)].

> **Question:** What size conductor and protection device with 75°C terminals is required for a 10 kW, 240V fixed electric space heater that has a 3A blower motor? **Figure 424–1**

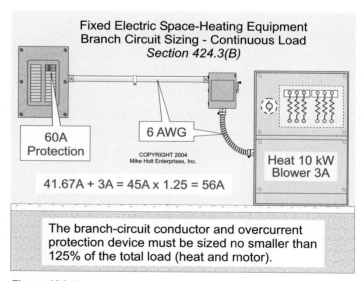

Fixed Electric Space-Heating Equipment
Branch Circuit Sizing - Continuous Load
Section 424.3(B)

60A Protection

6 AWG

COPYRIGHT 2004
Mike Holt Enterprises, Inc.

Heat 10 kW
Blower 3A

41.67A + 3A = 45A x 1.25 = 56A

The branch-circuit conductor and overcurrent protection device must be sized no smaller than 125% of the total load (heat and motor).

Figure 424–1

(a) 10 AWG, 30A   (b) 8 AWG, 40A
(c) 6 AWG, 60A   (d) 4 AWG, 80A

Answer: *(c) 6 AWG, 60A protection*

Step 1. Determine the total load

> $I = VA/E$
>
> $I = 10,000 VA/240V$
>
> $I = 41.67A + 3A = 44.67A$, round to 45A [220.5(B)]

Step 2. Size the conductors at 125 percent of the total current load [110.14(C), 210.19(A)(1), and Table 310.16]

> Conductor = 45A x 1.25
>
> Conductor = 56A, 6 AWG, rated 65A at 75°C

Step 3. Size the protection device at 125 percent of the total current load [210.20(A), 240.4(B) and 240.6(A)]

> Overcurrent Protection = 45A x 1.25
>
> Overcurrent Protection = 56A, next size up is 60A

424.9 Permanently Installed Electric Baseboard Heaters with Receptacles.
If a permanently installed electric baseboard heater has factory-installed receptacle outlets, the receptacles cannot be connected to the heater circuits.

> **FPN:** Listed baseboard heaters include instructions that prohibit their installation below receptacle outlets.

PART III. ELECTRIC SPACE-HEATING EQUIPMENT

424.19 Disconnecting Means.
Means must be provided to disconnect the heater, motor controller, and supplementary overcurrent protective devices of all fixed electric space-heating equipment from all ungrounded conductors.

The disconnecting means must be within sight from the equipment, or it must be capable of being locked in the open position [424.19(A)(1) and 424.19(A)(2)]. **Figure 424–2**

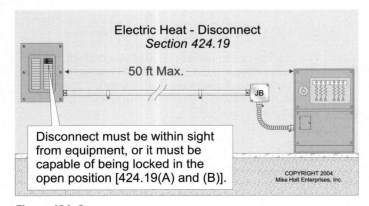

Figure 424–2

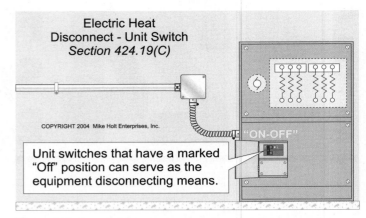

Figure 424–3

> **Author's Comment:** "Within sight" means that the specific equipment is visible and not more than 50 ft from one to the other [Article 100].

(C) Unit Switch as Disconnect. A unit switch with a marked "off" position that is an integral part of the equipment can serve as the heater disconnecting means if it disconnects all ungrounded conductors of the circuit. **Figure 424–3**

PART V. ELECTRIC SPACE-HEATING CABLES

424.44 Installation of Cables in Concrete or Poured Masonry Floors.

(G) GFCI Protection. GFCI protection is required for electric space-heating cables embedded in concrete or poured masonry floors of bathrooms and hydromassage bathtub locations. **Figure 424–4**

> **Author's Comment:** See 680.28(C)(3) for restrictions on the installation of radiant heating cables for spas and hot tubs installed outdoors.

PART VI. DUCT HEATERS

424.65 Disconnect for Electric Duct Heater Controller.
Means must be provided to disconnect the heater, motor controller, and supplementary overcurrent protective devices from all ungrounded conductors of the circuit. The disconnecting means must be within sight from the equipment, or it must be capable of being locked in the open position [424.19(A)]. **Figure 424–5**

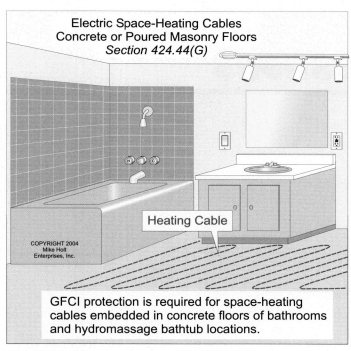

Electric Space-Heating Cables
Concrete or Poured Masonry Floors
Section 424.44(G)

Heating Cable

COPYRIGHT 2004
Mike Holt
Enterprises, Inc.

GFCI protection is required for space-heating
cables embedded in concrete floors of bathrooms
and hydromassage bathtub locations.

Figure 424–4

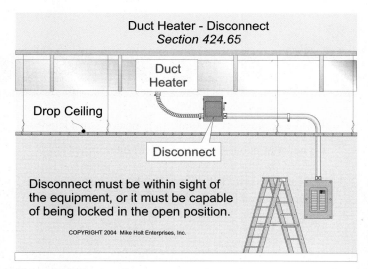

Duct Heater - Disconnect
Section 424.65

Duct
Heater

Drop Ceiling

Disconnect

Disconnect must be within sight of
the equipment, or it must be capable
of being locked in the open position.

COPYRIGHT 2004 Mike Holt Enterprises, Inc.

Figure 424–5

Author's Comment: The disconnecting means for a duct heater
isn't required to be readily accessible. Therefore, it can be
located within a suspended ceiling area adjacent to the duct
heater as long as it's accessible by portable means
[240.24(A)(4) and 404.8(A) Exception 2].

1. Fixed electric space-heating equipment is considered a(n) _____ load.

   (a) noncontinuous   (b) intermittent   (c) continuous   (d) none of these

2. Permanently installed electric baseboard heaters equipped with factory-installed receptacle outlets are permitted to be used as the outlets required by 210.50(B).

   (a) True   (b) False

3. If the disconnect is not within sight of the fixed electric space heater which includes a motor rated over ⅛ hp (without supplementary overcurrent protection devices), it must be capable of being _____.

   (a) locked   (b) locked in the closed position
   (c) locked in the open position   (d) within sight

4. A unit switch with a marked "off" position that is part of a fixed space heater, and disconnects all ungrounded conductors, is permitted as the disconnecting means required by Article 424 for one-family dwellings.

   (a) True   (b) False

5. Duct heater controller equipment must have a disconnecting means installed within _____ the controller.

   (a) 25 ft of   (b) sight from   (c) the side of   (d) none of these

430 Motors, Motor Circuits, and Controllers

Introduction

Article 430 contains the specific rules for conductor sizing, overcurrent protection, control circuit conductors, controllers, and disconnecting means for electric motors. The installation requirements for motor control centers are covered in Part VIII, and air-conditioning and refrigerating equipment are covered in Article 440.

Article 430 is by far the longest article in the *NEC*. It's also the most complex. But then, motors are complex. They are electrical *and* mechanical devices, but what makes motor applications complex is the fact that they are also inductive loads with a high-current demand at startup that is typically six, or more, times the running current. This makes circuit protection and motor protection necessarily different. So don't confuse circuit protection with motor protection—you must calculate and apply them separately. If you remember that as you study this Article, you will find it much easier to understand and apply.

PART I. GENERAL

430.1 Scope. Article 430 covers motors, motor branch-circuit and feeder conductors and their protection, motor overload protection, motor control circuits, motor controllers, and motor control centers. This article is divided into many parts, the most important include: **Figure 430–1**

- Branch Short-Circuit and Ground-Fault Protection — Part IV
- Circuit Conductors — Part II
- Control Circuits — Part VI
- Controllers — Part VII
- Disconnecting Means — Part IX
- Feeder Short-Circuit and Ground-Fault Protection — Part V
- General — Part I
- Overload Protection — Part III
- Motor Control Centers — Part VIII

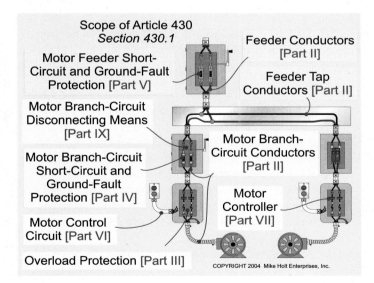

Figure 430–1

FPN 1: Article 440 contains the installation requirements for electrically driven air-conditioning and refrigeration equipment with hermetic refrigerant motor-compressors [440.1]. Also see 110.26(F) for dedicated space requirements for motor control centers.

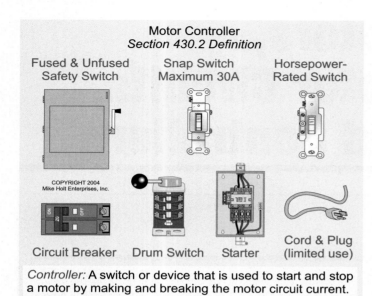

Motor Controller
Section 430.2 Definition

Fused & Unfused Safety Switch   Snap Switch Maximum 30A   Horsepower-Rated Switch

Circuit Breaker   Drum Switch   Starter   Cord & Plug (limited use)

Controller: A switch or device that is used to start and stop a motor by making and breaking the motor circuit current.

Figure 430–2

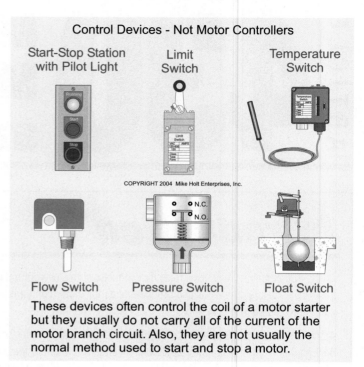

Control Devices - Not Motor Controllers

Start-Stop Station with Pilot Light   Limit Switch   Temperature Switch

Flow Switch   Pressure Switch   Float Switch

These devices often control the coil of a motor starter but they usually do not carry all of the current of the motor branch circuit. Also, they are not usually the normal method used to start and stop a motor.

Figure 430–3

430.2 Definitions.

Adjustable-Speed Drive. A combination of the power converter, motor, and motor mounted auxiliary devices such as encoders, tachometers, thermal switches and detectors, air blowers, heaters, and vibration sensors.

Adjustable-Speed Drive System. An interconnected combination of equipment that provides a means of adjusting the speed of a mechanical load coupled to a motor. A drive system typically consists of an adjustable-speed drive and auxiliary electrical apparatus.

Controller: A switch or device that is used to start and stop a motor by making and breaking the motor circuit current. Figure 430–2

> **Author's Comment:** A controller could be a horsepower-rated switch, snap switch, or circuit breaker. A pushbutton that operates an electromechanical relay isn't a controller, because it doesn't meet the controller rating requirements of 430.83. Devices such as start-stop stations and pressure switches are control devices, not motor controllers. Figure 430–3

Motor Control Circuit. The circuit that carries the electric signals directing the performance of the controller. Figure 430–4

430.6 Table FLC Versus Motor Nameplate Current Rating.

(A) General Requirements.

(1) Table Full-Load Current (FLC). The motor full-load current ratings listed in Tables 430.247, 430.248, and 430.250 are

used to determine the conductor ampacity [430.22], the branch-circuit short-circuit and ground-fault protection device size [430.52 and 62], and the ampere rating of disconnecting switches [430.110].

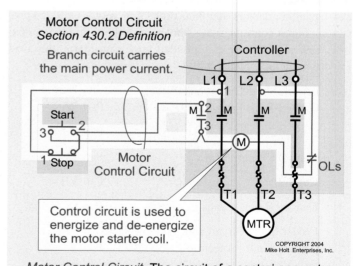

Motor Control Circuit
Section 430.2 Definition

Branch circuit carries the main power current.

Controller

L1   L2   L3

Start

Stop

Motor Control Circuit

Control circuit is used to energize and de-energize the motor starter coil.

OLs

T1   T2   T3

MTR

Motor Control Circuit. The circuit of a control apparatus or system that carries the electric signals directing the performance of the controller, but does not carry the main power current.

Figure 430–4

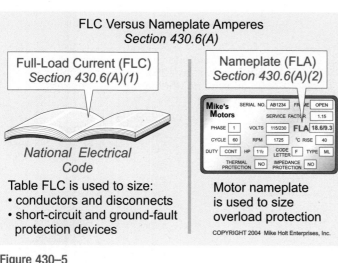

Figure 430–5

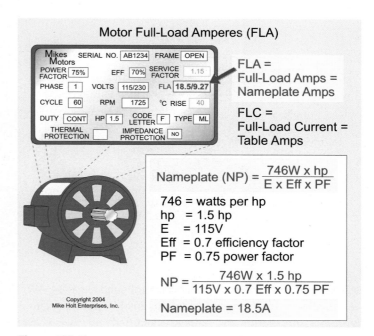

Figure 430–6

Author's Comment: The actual current rating on the motor nameplate full-load amperes (FLA) [430.6(A)(2)] is not permitted to be used to determine the conductor ampacity, the branch-circuit short-circuit and ground-fault protection device size, nor the ampere rating of disconnecting switches. **Figure 430–5**

Motors built to operate at less than 1,200 RPM or that have high torques may have higher full-load currents, and multispeed motors will have full-load current varying with speed, in which case the nameplate current ratings must be used.

Exception 3: For a listed motor-operated appliance, the motor full-load current marked on the nameplate of the appliance must be used instead of the horsepower rating on the appliance nameplate to determine the ampacity or rating of the disconnecting means, the branch-circuit conductors, the controller, and the branch-circuit short-circuit and ground-fault protection.

(2) Motor Nameplate Current Rating (FLA). Overload devices must be sized based on the motor nameplate current rating in accordance with 430.31.

Author's Comment: The motor nameplate full-load ampere rating is identified as full-load amperes (FLA). The FLA rating is the current in amperes that the motor draws while producing its rated horsepower load at its rated voltage, based on its rated efficiency and power factor. **Figure 430–6**

The actual current drawn by the motor depends upon the load on the motor and on the actual operating voltage at the motor terminals. That is, if the load increases, the current also increases, or if the motor operates at a voltage below its nameplate rating, the operating current will increase.

CAUTION: *To prevent damage to motor windings from excessive heat (caused by excessive current), never load a motor above its horsepower rating and/or be sure the voltage source matches the motor's voltage rating.*

430.8 Marking on Controllers. A controller must be marked with the manufacturer's name or identification, the voltage, the current or horsepower rating, the short-circuit current rating, and other necessary data to properly indicate the applications for which it's suitable.

Exception 1: The short-circuit current rating isn't required for controllers applied in accordance with 430.81(A), 430.81(B), or 430.83(C).

Exception 2: The short-circuit rating isn't required on the controller when the short-circuit current rating of the controller is marked elsewhere on the assembly.

Exception 3: The short-circuit rating isn't required on the controller when the assembly into which it's installed has a marked short-circuit current rating.

Exception 4: A short-circuit rating isn't required on controllers rated less than 2 hp, at 300V or less, and listed for general-purpose branch circuits.

430.9 Motor Controller Terminal Requirements.

(B) Copper Conductors. Motor controllers and terminals of control circuit devices must be connected with copper conductors.

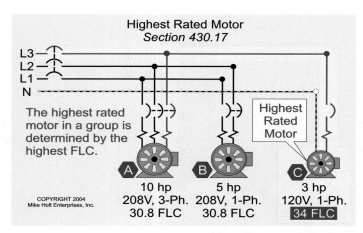

Highest Rated Motor
Section 430.17

Figure 430–7

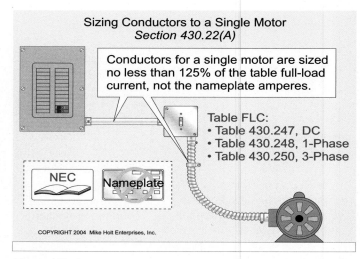

Sizing Conductors to a Single Motor
Section 430.22(A)

Figure 430–8

(C) Torque Requirements. Motor control conductors 14 AWG and smaller must be torqued at a minimum of 7 lbs-in. for screw-type pressure terminals, unless identified otherwise. See 110.3(B) and 110.14 FPN.

430.14 Location of Motors.

(A) Ventilation and Maintenance. Motors must be located so that adequate ventilation is provided and maintenance can be readily accomplished.

430.17 The Highest-Rated Motors. When sizing motor circuit conductors, the highest-rated motor is the motor with the highest-rated full-load current rating (FLC).

> *Question:* Which of the following motors has the highest FLC rating? **Figure 430–7**
>
> (a) 10 hp, three-phase, 208V   (b) 5 hp, single-phase, 208V
> (c) 3 hp, single-phase, 120V   (d) none of these
>
> *Answer:* (c) 3 hp, single-phase, 120V
>
> 10 hp = 30.8A [Table 430.250]
> 5 hp = 30.8A [Table 430.248]
> 3 hp = 34.0A [Table 430.248]

PART II. CONDUCTOR SIZE

430.22 Single Motor—Conductor Size.

(A) Conductor Size. Conductors to a single motor must be sized not smaller than 125 percent of the motor FLC rating as listed in: Figure 430–8

- Table 430.247 Direct Current Motors
- Table 430.248 Single-Phase Motors
- Table 430.250 Three-phase motors

> *Question:* What size branch-circuit conductor is required for a 7½ hp, 230V three-phase motor? **Figure 430–9**
>
> (a) 14 AWG   (b) 12 AWG
> (c) 10 AWG   (d) none of these
>
> *Answer:* (c) 10 AWG
>
> Motor FLC = 22A [Table 430.250]
> Conductor's Size = 22A x 1.25
> Conductor's Size = 27.5A, 10 AWG, rated 30A at 75°C
>   [Table 310.16]
>
> Note: The branch-circuit short-circuit and ground-fault protection device using an inverse-time breaker is sized at:
>   22A x 2.5 = 55A, next size up = 60A [240.6(A) and
>   430.52(C)(1) Exception 1].

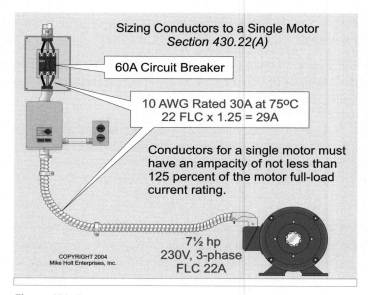

Sizing Conductors to a Single Motor
Section 430.22(A)

Figure 430–9

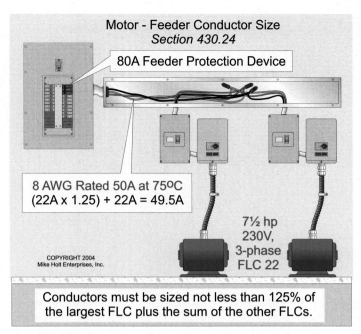

Motor - Feeder Conductor Size
Section 430.24

80A Feeder Protection Device

8 AWG Rated 50A at 75ºC
(22A x 1.25) + 22A = 49.5A

7½ hp
230V,
3-phase
FLC 22

COPYRIGHT 2004
Mike Holt Enterprises, Inc.

Conductors must be sized not less than 125% of
the largest FLC plus the sum of the other FLCs.

Figure 430–10

430.24 Several Motors—Conductor Size. Circuit conductors that supply several motors, typically feeders, must not be sized smaller than 125 percent of the largest motor FLC, plus the sum of the FLCs of the other motors.

Question: What size feeder conductor is required for two 7½ hp, 230V three-phase motors, if the terminals are rated for 75°C? **Figure 430–10**

(a) 14 AWG   (b) 12 AWG   (c) 10 AWG   (d) 8 AWG

Answer: (d) 8 AWG

Motor FLC = 22A [Table 430.250]
Motor Feeder Conductor = (22A x 1.25) + 22A
Motor Feeder Conductor = 49.5A, 8 AWG rated 50A at 75°C
[Table 310.16]

Author's Comment: The feeder protection device (inverse-time circuit breaker) must comply with 430.62 as follows:

Step 1. Determine the largest branch-circuit protection device rating [240.6(A) and 430.52(C)(1) Exception 1].

22A x 2.5 = 55A, next size up 60A

Step 2. Size the feeder protection device in accordance with 240.6(A) and 430.62.

Feeder Inverse-Time Breaker: 60A + 22A = 82A, next size down, 80A

430.28 Motor Feeder Taps. Motor circuit conductors tapped from a feeder must have an ampacity in accordance with 430.22(A), and the tap conductors must terminate in a branch-circuit protective device sized in accordance with 430.52. In addition, one of the following requirements must be met:

(1) 10 ft Tap. Tap conductors not over 10 ft long must have an ampacity not less than one-tenth the rating of the feeder protection device.

(2) 25 ft Tap. Tap conductors over 10 ft, but not over 25 ft, must have an ampacity not less than one-third the ampacity of the feeder conductor.

(3) Tap conductors must have an ampacity not less than the feeder conductors.

PART III. OVERLOAD PROTECTION

Part III contains the requirements for overload devices intended to protect motors, motor control apparatus, and motor branch-circuit conductors against excessive heating due to motor overloads and failure to start.

Overload is the operation of equipment in excess of the normal, full-load current rating, which, if it persists for a sufficient amount of time, would cause damage or dangerous overheating of the apparatus.

Author's Comment: Because of the difference between starting and running current, the overcurrent protection for motors is generally accomplished by having the overload device separate from the motor's short-circuit and ground-fault protection device.

Motor-Starting Current. When voltage is first applied to the field winding of an induction motor, only the conductor resistance opposes the flow of current through the motor winding. Because the conductor resistance is so low, the motor will have a very large inrush current (a minimum of six times the full-load ampere rating). **Figure 430–11A**

Motor-Running Current. However, once the rotor begins turning, the rotor bars (winding) will be increasingly cut by the stationary magnetic field, resulting in an increasing counter-electromotive force. **Figure 430–11B**

Counter-electromotive force (CEMF) opposes the applied voltage, resulting in an increased opposition to current flow within the conductor. This is called inductive reactance. The increase in inductive reactance, because of self-induction, causes the impedance of the conductor winding to increase and this results in a reduction of current flow.

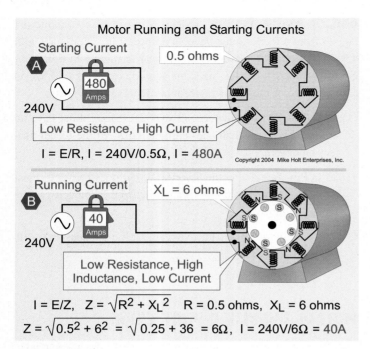

Motor Running and Starting Currents

Starting Current

A

0.5 ohms

240V 480 Amps

Low Resistance, High Current

$I = E/R$, $I = 240V/0.5\Omega$, $I = 480A$

Copyright 2004 Mike Holt Enterprises, Inc.

Running Current

B

$X_L = 6$ ohms

240V 40 Amps

Low Resistance, High Inductance, Low Current

$I = E/Z$, $Z = \sqrt{R^2 + X_L^2}$ $R = 0.5$ ohms, $X_L = 6$ ohms

$Z = \sqrt{0.5^2 + 6^2} = \sqrt{0.25 + 36} = 6\Omega$, $I = 240V/6\Omega = 40A$

Figure 430–11

Motor Locked-Rotor Current (LRC). If the rotating part of the motor winding (armature) becomes jammed so that it cannot rotate, no CEMF will be produced in the motor winding. This results in a decrease in conductor impedance to the point that it's effectively a short circuit. Result—the motor operates at locked-rotor current (LRC), which is usually about six times the running current value, and this will cause the motor winding to overheat to the point that it will be destroyed if the current isn't quickly stopped.

Author's Comment: *The National Electrical Code requires that most motors be provided with overcurrent protection to prevent damage to the motor winding because of locked-rotor current.*

Motor Overload Protection. Motors must be protected against excessive winding heat. Motors must not be overloaded; they must operate near their nameplate voltage, and measures must be taken to prevent the motor from jamming (LRC). If a motor is overloaded, or if it operates at a voltage below its rating, the operating current can increase to a value above the motor full-load amperes (FLA) rating. The excessive operating current may damage the motor winding from excessive heat.

Figure 430–12 shows a type of overload device called a melting-alloy type of overload. There are other types of overload devices that provide motor overload protection, but they do it in different ways. These include the dashpot type, the bimetallic type, and solid-state overload relays.

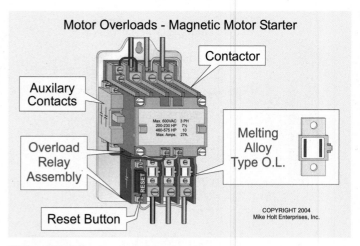

Motor Overloads - Magnetic Motor Starter

Contactor

Auxilary Contacts

Overload Relay Assembly

Melting Alloy Type O.L.

Reset Button

COPYRIGHT 2004 Mike Holt Enterprises, Inc.

Figure 430–12

Author's Comment: Motors are designed to operate with an inrush current of six to eight times the motor-rated FLA for short periods of time without damage to the motor windings. **Figure 430–13**

However, if a motor operates at LRC for a prolonged period of time, the motor insulation and lubrication can be destroyed by excessive heat. Most motors can operate at 600 percent of motor FLA satisfactorily for a period of less than one minute, or 300 percent of the motor FLA for not more than three minutes.

430.31 Overload. Overload protection devices, sometimes called "heaters," are intended to provide overload protection, and come in a variety of configurations; they can be conventional or electronic. In addition, a fuse sized in accordance with 430.32 can be used for circuit overload protection [430.55]. **Figure 430–14**

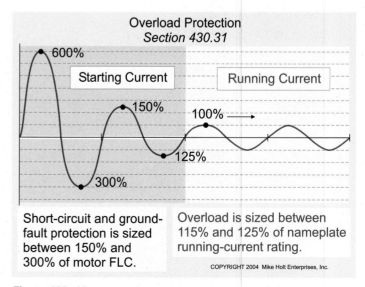

Overload Protection
Section 430.31

600%

Starting Current   Running Current

150%

100%

125%

300%

Short-circuit and ground-fault protection is sized between 150% and 300% of motor FLC.

Overload is sized between 115% and 125% of nameplate running-current rating.

COPYRIGHT 2004 Mike Holt Enterprises, Inc.

Figure 430–13

Overload Types
Section 430.31

Overloads protect the motor, conductors, and associated equipment from excessive heat due to motor overloads. They are not intended to protect against short circuits and ground faults.

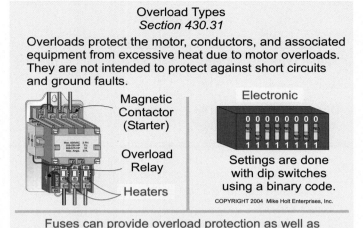

Magnetic Contactor (Starter)

Overload Relay

Heaters

Electronic

Settings are done with dip switches using a binary code.

COPYRIGHT 2004 Mike Holt Enterprises, Inc.

Fuses can provide overload protection as well as short-circuit and ground-fault protection. See 430.55.

Figure 430–14

Author's Comments:

- In reality, motor overload protection sizing is accomplished by simply installing or setting the overload protection device in accordance with the controller's instructions, based on the motor nameplate current rating.

- The intended level of protection required in Article 430 Part III is for overload and failure-to-start protection only, in order to protect against the motor from becoming a fire hazard. Part III isn't intended to provide specific protection requirements against voltage imbalance, loss of phase, or phase reversal conditions.

430.32 Overload Sizing—Continuous-Duty Motors.

(A) Motors Rated More Than One Horsepower. Motors rated more than 1 hp, used in a continuous-duty application without integral thermal protection, must have an overload device sized no more than 115 percent of the motor "nameplate current rating," except for the following: **Figure 430–15**

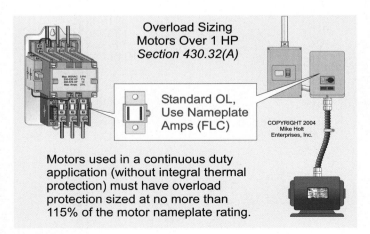

Overload Sizing
Motors Over 1 HP
Section 430.32(A)

Standard OL, Use Nameplate Amps (FLC)

COPYRIGHT 2004 Mike Holt Enterprises, Inc.

Motors used in a continuous duty application (without integral thermal protection) must have overload protection sized at no more than 115% of the motor nameplate rating.

Figure 430–15

Overload Sizing
Service Factor and Temperature Rise
Section 430.32

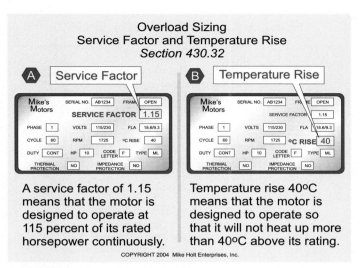

A Service Factor

B Temperature Rise

A service factor of 1.15 means that the motor is designed to operate at 115 percent of its rated horsepower continuously.

Temperature rise 40°C means that the motor is designed to operate so that it will not heat up more than 40°C above its rating.

COPYRIGHT 2004 Mike Holt Enterprises, Inc.

Figure 430–16

Author's Comment: Overload protection isn't required for a continuous-duty motor used in a noncontinuous application.

Service Factor. Motors with a marked service factor (SF) of 1.15 or more on the nameplate must have the overload device sized no more than 125 percent of the motor nameplate current rating. **Figure 430–16A**

Author's Comment: A service factor of 1.15 means that the motor is designed to operate continuously at 115 percent of its rated horsepower.

Temperature Rise. Motors with a nameplate temperature rise of 40°C or less must have the overload device sized no more than 125 percent of motor nameplate current rating. **Figure 430–16B**

Author's Comment: A motor with a nameplate temperature rise of 40°C means that the motor is designed to operate so it will not heat up more than 40°C above its rated ambient temperature when operated at rated load and voltage. Studies have shown that when the operating temperature of a motor is increased 10°C, the motor winding insulating material's anticipated life is reduced by 50 percent.

430.36 Use of Fuses for Overload Protection. Where fuses are used for motor overload protection, one must be used for each ungrounded conductor.

Author's Comment: If remote control isn't required for a motor, considerable savings can be achieved by using dual-element fuses (eliminate a motor controller) sized in accordance with 430.32 to protect the motor and the circuit conductors against overcurrent, which includes overload, short circuit, and ground faults. See 430.55 for additional details.

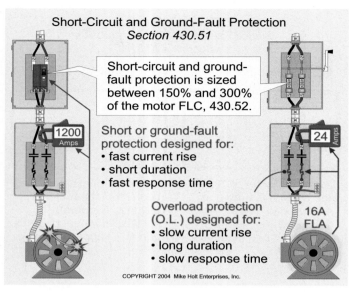

Figure 430–17

430.37 Number of Overload Devices. An overload protection device must be installed in each ungrounded conductor.

PART IV. BRANCH-CIRCUIT SHORT-CIRCUIT AND GROUND-FAULT PROTECTION

430.51 General. A branch-circuit short-circuit and ground-fault protection device protects the motor, the motor control apparatus, and the conductors against short circuits or ground faults, but not against overload. **Figure 430–17**

Author's Comment: Protection against heating from overloads must comply with 430.32.

430.52 Branch-Circuit Short-Circuit and Ground-Fault Protection.

(A) General. The motor branch-circuit short-circuit and ground-fault protective device must comply with 430.52(B) and 430.52(C).

(B) All Motors. A motor branch-circuit short-circuit and ground-fault protective device must be capable of carrying the motor's starting current.

(C) Rating or Setting.

(1) Table 430.52. Each motor branch circuit must be protected against short circuit and ground faults by a protection device sized no greater than the following percentages listed in Table 430.52.

Table 430.52

Circuit Motor Type	Nontime Delay	Dual-Element Fuse	Inverse-Time Breaker
Wound Rotor	150%	150%	150%
Direct Current	150%	150%	150%
All Other Motors	300%	175%	250%

Question: *What size conductor and inverse-time circuit breaker are required for a 2 hp, 230V single-phase motor?* **Figure 430–18**

(a) 14 AWG, 30A breaker   *(b) 14 AWG, 35A breaker*
(c) 14 AWG, 40A breaker   *(d) none of these*

Answer: *(a) 14 AWG, 30A breaker*

Step 1. Branch-Circuit Conductor [Table 310.16, 430.22(A), and Table 430.248]

   12A x 1.25 = 15A, 14 AWG, rated 20A at 75°C [Table 310.16]

Step 2. Branch-Circuit Protection [240.6(A), 430.52(C)(1), and Table 430.248]

   12A x 2.5 = 30A

Author's Comment: I know it bothers many in the electrical industry to see a 14 AWG conductor protected by a 30A circuit breaker, but branch-circuit conductors are protected against overloads sized between 115 and 125 percent of the motor nameplate current rating [430.32]. See 240.4(A) and 240.4(G) for details.

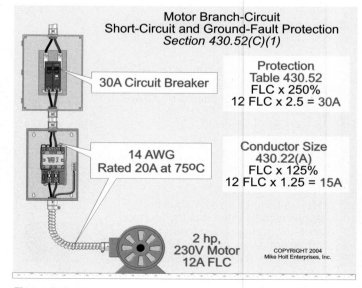

Figure 430–18

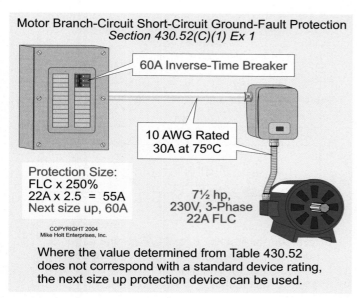

Motor Branch-Circuit Short-Circuit Ground-Fault Protection
Section 430.52(C)(1) Ex 1

60A Inverse-Time Breaker

10 AWG Rated
30A at 75°C

Protection Size:
FLC x 250%
22A x 2.5 = 55A
Next size up, 60A

COPYRIGHT 2004
Mike Holt Enterprises, Inc.

7½ hp,
230V, 3-Phase
22A FLC

Where the value determined from Table 430.52
does not correspond with a standard device rating,
the next size up protection device can be used.

Figure 430–19

Exception 1: *Where the motor short-circuit and ground-fault protection device values derived from Table 430.52 don't correspond with the standard overcurrent device ratings listed in 240.6(A), the next higher protection device rating can be used.*

Question: *What size conductor and inverse-time circuit breaker are required for a 7½ hp, 230V three-phase motor?* **Figure 430–19**

(a) 10 AWG, 50A breaker　　(b) 10 AWG, 60A breaker
(c) a or b　　(d) none of these

Answer. (c) 10 AWG, 50 or 60A breaker

Step 1. Branch-Circuit Conductor [Table 310.16, 430.22(A), and Table 430.250]

　　22A x 1.25 = 27.5A, 10 AWG, rated 30A at 75°C [Table 310.16]

Step 2. Branch-Circuit Protection [240.6(A), 430.52(C)(1) Exception 1, and Table 430.250]

　　22A x 2.5 = 55A, next size up = 60A

430.55 Single Overcurrent Protective Device. A motor can be protected against overload, short circuit, and ground faults by a single protection device sized in accordance with the overload requirements contained in 430.32.

Question: *What size dual-element fuse is permitted to protect a 5 hp, 230V single-phase motor with a service factor of 1.2 and a nameplate current rating of 28A?* **Figure 430–20**

(a) 20A　　　(b) 25A　　　(c) 30A　　　(d) 35A

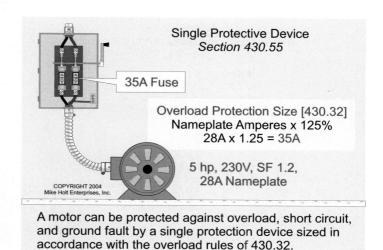

Single Protective Device
Section 430.55

35A Fuse

Overload Protection Size [430.32]
Nameplate Amperes x 125%
28A x 1.25 = 35A

5 hp, 230V, SF 1.2,
28A Nameplate

COPYRIGHT 2004
Mike Holt Enterprises, Inc.

A motor can be protected against overload, short circuit, and ground fault by a single protection device sized in accordance with the overload rules of 430.32.

Figure 430–20

Answer: (d) 35A

Overload Protection [430.32(A)(1)]
28A x 1.25 = 35A

PART V. FEEDER SHORT-CIRCUIT AND GROUND-FAULT PROTECTION

430.62 Feeder Protection.

(A) Motors Only. Feeder conductors must be protected against short circuits and ground faults by a protection device sized not greater than the largest rating of the branch-circuit short-circuit and ground-fault protective device for any motor, plus the sum of the full-load currents of the other motors in the group.

Question: *What size feeder protection (inverse-time breakers with 75°C terminals) and conductor are required for the following two motors?* **Figure 430–21**

Motor 1—20 hp, 460V three-phase = 27A

Motor 2—10 hp, 460V three-phase = 14A

(a) 8 AWG, 70A breaker　　(b) 8 AWG, 80A breaker
(c) 8 AWG, 90A breaker　　(d) none of these

Answer: (b) 8 AWG, 80A breaker

Step 1. Feeder Conductor Size [430.24]

　　(27A x 1.25) + 14A = 48A
　　8 AWG rated 50A at 75°C [110.14(C) and Table 310.16]

Step 2. Feeder Protection [430.62(A)] not greater than the largest branch-circuit protection device plus other motor FLC

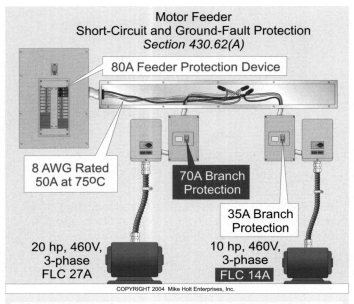

Motor Feeder
Short-Circuit and Ground-Fault Protection
Section 430.62(A)

80A Feeder Protection Device

8 AWG Rated
50A at 75°C

70A Branch
Protection

35A Branch
Protection

20 hp, 460V,
3-phase
FLC 27A

10 hp, 460V,
3-phase
FLC 14A

COPYRIGHT 2004 Mike Holt Enterprises, Inc.

Feeder protection device not to be larger than 70A + 14A

Figure 430–21

Largest Branch-Circuit Protection Device [430.52(C)(1) Exception]

20 hp Motor = 27A x 2.5 = 68, next size up = 70A
10 hp Motor = 14A x 2.5 = 35A

Step 4. Size Feeder Protection

Not greater than 70A + 14A, = 84A,
next size down = 80A [240.6(A)]

Author's Comment: The "next size up protection" rule for branch circuits [430.52(C)(1) Exception 1] doesn't apply to motor feeder protection device sizing.

PART VI. MOTOR CONTROL CIRCUITS

430.72 Overcurrent Protection for Control Circuits.

(A) Class 1 Control Conductors. Motor control conductors not tapped from the branch-circuit protection device are classified as a Class 1 remote-control circuit and they must have overcurrent protection in accordance with 725.23.

Author's Comment: 725.23 states that overcurrent protection for conductors 14 AWG and larger must comply with the conductor ampacity Table 310.16. Overcurrent protection for 18 AWG must not exceed 7A, and a 10A device must protect 16 AWG conductors. **Figure 430–22**

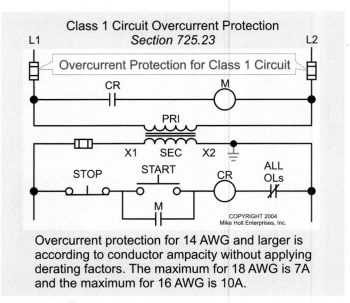

Class 1 Circuit Overcurrent Protection
Section 725.23

Overcurrent Protection for Class 1 Circuit

COPYRIGHT 2004
Mike Holt Enterprises, Inc.

Overcurrent protection for 14 AWG and larger is according to conductor ampacity without applying derating factors. The maximum for 18 AWG is 7A and the maximum for 16 AWG is 10A.

Figure 430–22

(B) Motor Control Conductors.

(2) Branch-Circuit Overcurrent Protective Device. Motor control circuit conductors tapped from the motor branch-circuit protection device that extend beyond the tap enclosure must have overcurrent protection as follows:

Conductor	Protection
18 AWG	7A
16 AWG	10A
14 AWG	45A
12 AWG	60A
10 AWG	90A

Author's Comment: The above limitations don't apply to the internal wiring of industrial control panels listed to UL 508.

(C) Control Circuit Transformer Protection. Transformers for motor control circuit conductors must have overcurrent protection on the primary side in accordance with 430.72(C)(1) through (5).

Author's Comment: Many control transformers have very small iron cores, which result in very high inrush (excitation) current when the coil is energized. This high inrush current might cause standard fuses to blow, so you should only use the fuses recommended by the control transformer manufacturer.

430.74 Disconnect for Control Circuit.

(A) Control Circuit Disconnect. Motor control circuit conductors must have a disconnecting means that opens all sources of

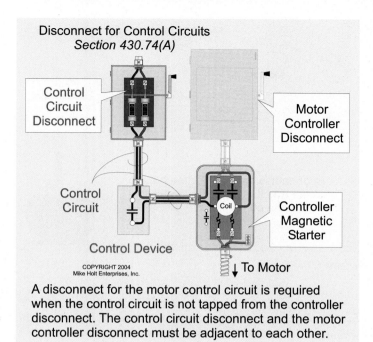

Disconnect for Control Circuits
Section 430.74(A)

Control Circuit Disconnect

Motor Controller Disconnect

Control Circuit

Coil

Controller Magnetic Starter

Control Device

↓ To Motor

COPYRIGHT 2004
Mike Holt Enterprises, Inc.

A disconnect for the motor control circuit is required when the control circuit is not tapped from the controller disconnect. The control circuit disconnect and the motor controller disconnect must be adjacent to each other.

Figure 430–23

supply when the disconnecting means is in the open position. If the control circuit conductors are tapped from the controller disconnect, the controller disconnecting means can serve as the disconnecting means for the control circuit conductors [430.102(A)].

If the control circuit conductors aren't tapped from the controller disconnect, a separate disconnecting means is required for the control circuit conductors, and it must be located adjacent to the controller disconnect. **Figure 430–23**

PART VII. MOTOR CONTROLLERS

430.83 Controller Rating.

(A) General. The controller must have one of the following ratings:

(1) Horsepower Rating. Controllers other than circuit breakers and molded case switches must have a horsepower rating not less than that of the motor.

(2) Circuit Breakers. A circuit breaker can serve as a motor controller [430.111].

> **Author's Comment:** Circuit breakers aren't required to be horsepower rated.

(3) Molded Case Switch. A molded case switch, rated in amperes, can serve as a motor controller.

> **Author's Comment:** A molded case switch isn't required to be horsepower rated.

(C) Stationary Motors of Two Horsepower or Less. For stationary motors rated at 2 hp or less, the controller can be one of the following:

(1) A general-use switch, having an ampere rating not less than twice the full-load current rating of the motor.

> **Author's Comment:** See Article 100 for the definition of "General-use Switch."

(2) A general-use ac snap switch, where the motor full-load current rating isn't more than 80 percent of the ampere rating of the switch.

> **Author's Comment:** A general-use snap switch is a general-use switch constructed for installation in device boxes or on box covers, or otherwise used in conjunction with wiring systems recognized by this *Code*.

(E) Voltage Rating. A controller with a straight voltage rating (for example, 240V or 480V), can be used in a circuit in which the nominal voltage between any two conductors doesn't exceed the controller's voltage rating.

A controller with a slash rating (for example, 240/120 volts or 480Y/277 volts), can only be used in a solidly grounded system, where the nominal voltage from any conductor to ground doesn't exceed the lower of the two values of the controller's voltage rating, and the nominal voltage between any two conductors doesn't exceed the higher value of the controller's voltage rating.

> **CAUTION:** *A controller with a slash rating, such as 120/240V, can only be used on a circuit where the nominal voltage of any one ungrounded conductor to metal case doesn't exceed the lower of the two values. A 120/240V slash-rated controller is not permitted on the high-leg of a solidly grounded 120/240V delta system, because the line-to-ground voltage can be as much as 208V, which exceeds the 120V line-to-ground voltage rating of the controller.* Figure 430–24

430.84 Need Not Open All Conductors of the Circuit.
The motor controller can open only as many conductors of the circuit as necessary to start and stop the motor.

> **Author's Comment:** The controller is only required to start and stop the motor; it isn't a disconnecting means. See the disconnecting means requirement in 430.103 for additional details.

430.87 Controller for Each Motor. Each motor must have its own individual controller.

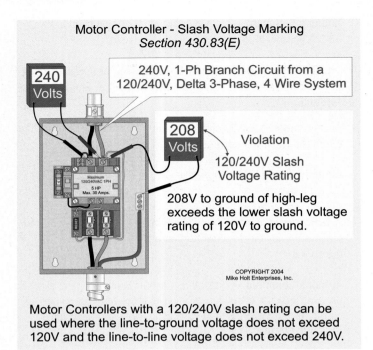

Motor Controller - Slash Voltage Marking
Section 430.83(E)

240V, 1-Ph Branch Circuit from a
120/240V, Delta 3-Phase, 4 Wire System

240 Volts

208 Volts

Violation
120/240V Slash
Voltage Rating

208V to ground of high-leg
exceeds the lower slash voltage
rating of 120V to ground.

COPYRIGHT 2004
Mike Holt Enterprises, Inc.

Motor Controllers with a 120/240V slash rating can be
used where the line-to-ground voltage does not exceed
120V and the line-to-line voltage does not exceed 240V.

Figure 430–24

430.91 Motor Controller Enclosure Types. Motor controllers are installed in a variety of environments: rain, snow, sleet, corrosive agents, submersion, dirt, liquids, dust, oil, and so on. The motor controller enclosure must be suitable for the environment, in accordance with Table 430.91.

PART IX. DISCONNECTING MEANS

430.102 Disconnect Requirement.

(A) Controller Disconnect. A disconnecting means is required for each motor controller and it must be located within sight from the controller. **Figure 430–25**

> **Author's Comment:** "Within sight" is visible and not more than 50 ft from each other [Article 100].

The controller disconnecting means must simultaneously open all ungrounded conductors of the circuit [430.103]. **Figure 430–26**

> **Author's Comment:** The controller disconnecting means can serve as the disconnecting means for the motor control circuit conductors [430.74], as well as for the motor [430.102(B)].

(B) Motor Disconnect. A disconnecting means is required for each motor and it must be located in sight from the motor location and the driven machinery location. **Figure 430–27**

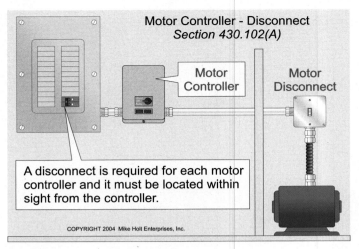

Motor Controller - Disconnect
Section 430.102(A)

Motor Controller

Motor Disconnect

A disconnect is required for each motor
controller and it must be located within
sight from the controller.

COPYRIGHT 2004 Mike Holt Enterprises, Inc.

Figure 430–25

Exception: A motor disconnecting means isn't required to be within sight from the motor under either condition (a) or (b), if the controller disconnecting means [430.102(A)] is capable of being locked in the open position and the provision for locking or adding a lock is permanently installed on the controller disconnect. **Figure 430–28**

(a) Where locating the disconnecting means is impracticable or introduces additional or increased hazards to persons or property.

(b) In industrial installations, with written safety procedures, where conditions of maintenance and supervision ensure that only qualified persons will service the equipment.

> **Author's Comment:** See Article 100 for the definition of "Within Sight."

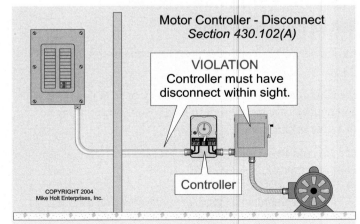

Motor Controller - Disconnect
Section 430.102(A)

VIOLATION
Controller must have
disconnect within sight.

Controller

COPYRIGHT 2004
Mike Holt Enterprises, Inc.

The controller disconnect must disconnect all circuit
conductors of the controller simultaneously [430.103].

Figure 430–26

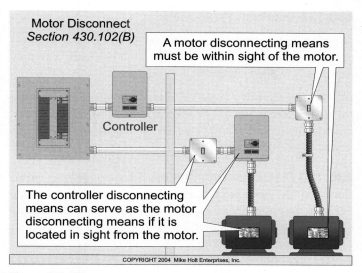

Figure 430–27

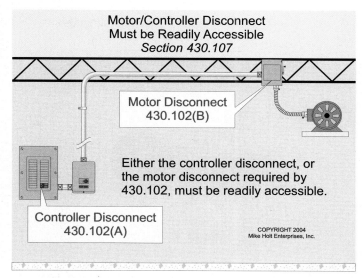

Figure 430–29

FPN 2: For information on lockout/tagout procedures, see NFPA 70E, *Standard for Electrical Safety in the Workplace.*

430.103 Disconnect Opens All Conductors. The disconnecting means for the motor controller and the motor must open all ungrounded supply conductors simultaneously.

430.104 Marking and Mounting. The controller and motor disconnecting means must indicate whether they are in the "on" or "off" position.

Author's Comment: The disconnecting means must be legibly marked to identify its intended purpose [110.22 and 408.4], and when operated vertically, the "up" position must be the "on" position [240.81 and 404.6(C)].

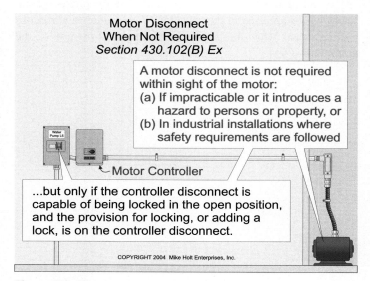

Figure 430–28

430.107 Readily Accessible. Either the controller disconnecting means or motor disconnecting means required by 430.102 must be readily accessible. Figure 430–29

430.109 Disconnecting Means Rating.

(A) General. The disconnecting means for the motor controller and/or the motor must be a:

(1) Motor-Circuit Switch. A listed horsepower-rated motor-circuit switch.

(2) Molded Case Circuit Breaker. A listed molded case circuit breaker.

(3) Molded Case Switch. A listed molded case switch.

(6) Manual Motor Controller. Listed manual motor controllers marked "Suitable as Motor Disconnect."

(C) Stationary Motors of Two Horsepower or Less.

(2) General-Use Snap Switch. A general-use ac snap switch is permitted as the motor disconnecting means for ac motors rated 2 hp or less, and 300V or less [430.83(C)]. Figure 430–30

(F) Cord-and-Plug Connected Motors. For a cord-and-plug connected motor, a horsepower-rated attachment plug and receptacle having ratings no less than the motor ratings is permitted to serve as the disconnecting means.

430.111 Combination Controller-Disconnect. A circuit breaker or horsepower-rated switch or circuit breaker can serve as a motor controller and disconnecting means if it opens all ungrounded conductors to the motor as required by 430.103.

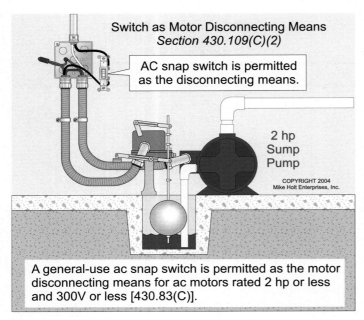

Switch as Motor Disconnecting Means
Section 430.109(C)(2)

AC snap switch is permitted
as the disconnecting means.

2 hp
Sump
Pump

COPYRIGHT 2004
Mike Holt Enterprises, Inc.

A general-use ac snap switch is permitted as the motor disconnecting means for ac motors rated 2 hp or less and 300V or less [430.83(C)].

Figure 430–30

PART X. ADJUSTABLE-SPEED DRIVE SYSTEMS

430.120 General. The installation provisions of Part I through Part IX apply unless modified or supplemented by this Part.

430.122 Conductors—Minimum Size and Ampacity.

(A) Branch/Feeder Circuit Conductors. Circuit conductors must have an ampacity not less than 125 percent of the rated input to power conversion equipment.

(B) Bypass Device. The ampacity of circuit conductors supplying power conversion equipment that utilizes a bypass device must be the larger of:

(1) 125 percent of the rated input to the power conversion equipment, or

(2) 125 percent of the motor full-load current rating as determined by 430.6.

430.124 Overload Protection.

(A) Included in Power Conversion Equipment. Where the power conversion equipment contains motor overload protection, additional overload protection is not required.

(B) Bypass Circuits. Adjustable-speed drive systems with a bypass that allows motor operation at rated full-load speed must provide motor overload protection in the bypass circuit, as described in Article 430 Part III.

430.128 Disconnecting Means. The disconnecting means can be in the incoming line to the conversion equipment and must have a rating not less than 115 percent of the rated input current of the conversion unit.

PART XIV. TABLES

Table 430.248 Full-Load Current, Single-Phase Motors. Table 430.248 lists the full-load current for single-phase alternating-current motors. These values are used to determine motor conductor sizing, ampere ratings of disconnects, controller rating, and branch-circuit and feeder protection, but not overload protection [430.6(A)(1)].

Table 430.250 Full-Load Current, Three-Phase Motors. Table 430.250 lists the full-load current for three-phase alternating-current motors. The values are used to determine motor conductor sizing, ampere ratings of disconnects, controller rating, and branch-circuit and feeder protection, but not overload protection [430.6(A)(1)].

Table 430.251 Locked-Rotor Currents. Table 251(A) contains locked-rotor current for single-phase motors, and Table 251(B) contains the locked-rotor current for three-phase motors. These values are used in the selection of controllers and disconnecting means when the horsepower rating isn't marked on the motor nameplate.

Article 430 Questions

1. For general motor applications, the motor branch-circuit short-circuit and ground-fault protection device must be sized based on the _____ amperes.

    (a) motor nameplate   (b) NEMA standard   (c) *NEC* Table   (d) Factory Mutual

2. Branch-circuit conductors supplying a single continuous-duty motor must have an ampacity not less than _____.

    (a) 125 percent of the motor's nameplate current rating
    (b) 125 percent of the motor's full-load current as determined by 430.6(A)(1)
    (c) 125 percent of the motor's full locked-rotor rating
    (d) 80 percent of the motor's full-load current rating

3. An overload device used to protect continuous-duty motors (rated more than 1 hp) must be selected to trip, or be rated, at no more than _____ percent of the motor nameplate full-load current rating for motors with a marked service factor of 1.15 or greater.

    (a) 110   (b) 115   (c) 120   (d) 125

4. A motor can be provided with combined overcurrent protection using a single protective device to provide branch-circuit _____ where the rating of the device provides the necessary overload protection specified in 430.32.

    (a) short-circuit protection   (b) ground-fault protection   (c) motor-overload protection   (d) all of these

5. The branch-circuit protective device is permitted to serve as the controller for a stationary motor rated at _____ or less that is normally left running and cannot be damaged by overload or failure to start.

    (a) ⅛ hp   (b) ¼ hp   (c) ⅜ hp   (d) ½ hp

6. Each motor must be provided with an individual controller.

    (a) True   (b) False

7. The disconnecting means for the controller and motor must open all ungrounded supply conductors and must be designed so that no pole can be operated independently.

    (a) True   (b) False

8. A branch-circuit overcurrent protection device such as a plug fuse may serve as the disconnecting means for a stationary motor of ⅛ hp or less.

    (a) True   (b) False

Notes

Mike Holt Enterprises, Inc. • www.NECcode.com • 1.888.NEC.Code

440 Air-Conditioning and Refrigeration Equipment

Introduction

This article applies to electrically driven air-conditioning and refrigeration equipment with a hermetic refrigerant motor-compressor. The rules in this article add to, or amend, the rules in Article 430 and other articles.

Why the special treatment? Several reasons. First, hermetic motors aren't general purpose. They are sized, fitted, and engineered for specific applications. You may have seen the phrase "hermetically sealed" on food containers. Something that is hermetically sealed is airtight—no gas can enter it or escape it. These motors are sealed airtight, and that affects what you can expect of the motor in terms of performance and heat dissipation.

Each equipment manufacturer has the motors for a given air-conditioning unit built to its own specifications. Cooling and other characteristics are different from those of nonhermetic motors. For each motor, the manufacturer has worked out all of the details and supplied the correct protection, conductor sizing, and other information on the nameplate. Thus, Article 440 requires you to use the nameplate circuit protection rather than what you might develop from applying Article 430.

The application itself—with the compressor motor often on the other side of an exterior building wall from the normal power sources so it can exchange heat with free air—poses additional problems, which the *NEC* addresses in Article 440.

Three key points to remember so you apply Article 440 correctly are:

1. Hermetic motors are different. Article 440 supercedes Article 430 in regard to these motors.
2. Use the nameplate information.
3. Be careful where you install your disconnects. For motors that fall under Article 440, there are no exceptions to the rules.

PART I. GENERAL

440.1 Scope. Article 440 applies to electrically driven air-conditioning and refrigeration equipment with a hermetic refrigerant motor-compressor.

440.2 Definitions.

Branch-Circuit Selection Current. The amperes used instead of the rated-load current, wherever the running overload protective device permits a sustained current greater than the specified percentage of the rated-load current. The branch-circuit selection current will always be equal to, or greater than, the marked rated-load current.

Hermetic Refrigerant Motor-Compressor. A compressor and motor enclosed in the same housing, operating in the refrigerant.

Rated-Load Current. The rated-load current resulting when the motor-compressor is operated at the rated load, rated voltage, and rated frequency.

440.3 Other Articles.

(B) Equipment with No Hermetic Motor-Compressors. Air-conditioning and refrigeration equipment that does not have hermetic refrigerant motor-compressors, such as furnaces with evaporator coils, must comply with Article 422 for appliances, Article 424 for electric space-heating, and in some cases, Article 430 for motors.

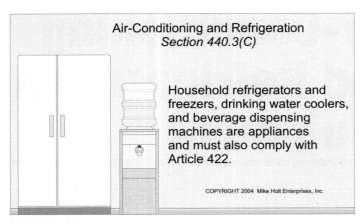

Figure 440–1

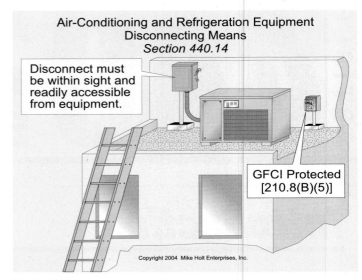

Figure 440–2

(C) Household Refrigerant Motor-Compressor Appliances. Household refrigerators and freezers, drinking water coolers, and beverage dispensing machines are each listed as an appliance, and their installation must comply with Article 422 for appliances. Figure 440–1

440.4 Marking on Hermetic Refrigerant Motor-Compressors and Equipment.

(B) Multimotor and Combination-Load Equipment. Multimotor and combination-load equipment must have a visible nameplate containing the maker's name, the rating in volts, frequency and number of phases, minimum supply circuit conductor ampacity, the maximum rating of the branch-circuit short-circuit and ground-fault protective device, and the short-circuit current rating of the motor controllers.

The circuit ampacity must be calculated by using Part IV, and the branch-circuit short-circuit and ground-fault protective device rating must not exceed the value calculated by using Part III.

Exception 3: Multimotor and combination-load equipment used in one- and two-family dwellings, cord-and-plug connected equipment, or equipment supplied from a branch circuit protected at 60A or less, aren't required to be marked with a short-circuit current rating.

440.6 Ampacity and Rating. The ampacity of conductors and protection devices must be determined according to (A) and (B).

(A) Hermetic Refrigerant Motor-Compressor. For a hermetic refrigerant motor-compressor, the rated-load current marked on the nameplate of the equipment or compressor must be used for all calculations.

Exception 1: Where so marked, the branch-circuit selection current must be used instead of the rated-load current.

PART II. DISCONNECTING MEANS

440.14 Location. The disconnecting means for air-conditioning or refrigerating equipment must be located within sight from and readily accessible from the equipment. Figures 440–2 and 440-3

> **Author's Comment:** "Within Sight" is visible and not more than 50 ft from each other [Article 100].

The disconnecting means can be mounted on or within the air-conditioning equipment, but it must not be located on panels that are designed to allow access to the equipment. Figure 440–4

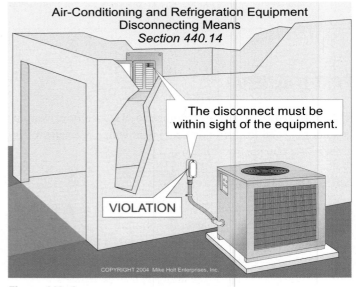

Figure 440–3

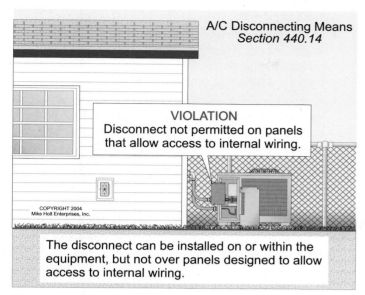

A/C Disconnecting Means
Section 440.14

VIOLATION
Disconnect not permitted on panels
that allow access to internal wiring.

COPYRIGHT 2004
Mike Holt Enterprises, Inc.

The disconnect can be installed on or within the
equipment, but not over panels designed to allow
access to internal wiring.

Figure 440–4

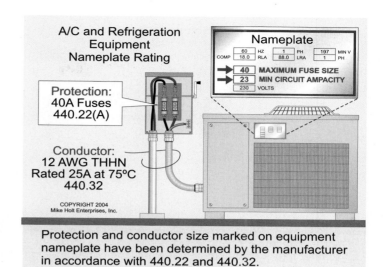

A/C and Refrigeration
Equipment
Nameplate Rating

Protection:
40A Fuses
440.22(A)

Conductor:
12 AWG THHN
Rated 25A at 75°C
440.32

Nameplate

| COMP | 60 | HZ | 1 | PH | 197 | MIN V |
| | 18.0 | RLA | 88.0 | LRA | 1 | PH |

➡ **40** MAXIMUM FUSE SIZE
➡ **23** MIN CIRCUIT AMPACITY
230 VOLTS

COPYRIGHT 2004
Mike Holt Enterprises, Inc.

Protection and conductor size marked on equipment
nameplate have been determined by the manufacturer
in accordance with 440.22 and 440.32.

Figure 440–5

Author's Comment: It would also be a good idea not to mount the air-conditioning disconnect over any machine data tags.

Exception 1: A disconnecting means isn't required to be within sight from the equipment, if the disconnecting means is capable of being individually locked in the open position, and if the equipment is essential to an industrial process in a facility that has written safety procedures, and where the conditions of maintenance and supervision ensure that only qualified persons service the equipment. The provision for locking or adding a lock to the disconnecting means must be permanently installed on or at the switch or circuit breaker used as the disconnecting means.

Exception 2: An accessible attachment plug and receptacle can serve as the disconnecting means.

Author's Comment: The receptacle for the attachment plug isn't required to be readily accessible.

PART III. CIRCUIT PROTECTION

440.21 General. The branch-circuit conductors, control apparatus, and circuits supplying hermetic refrigerant motor-compressors must be protected against short circuits and ground faults in accordance with 440.22.

Author's Comment: If the equipment nameplate specifies "Maximum Fuse Size," then a one-time or dual-element fuse must be used. If the nameplate specifies "HACR Circuit Breaker," then the equipment must be protected by an HACR-rated circuit breaker [110.3(B)].

440.22 Short-Circuit and Ground-Fault Protection Device Size.
Short-circuit and ground-fault protection for air-conditioning and refrigeration equipment must be sized no larger than identified on the equipment's nameplate. **Figure 440–5**

If the equipment doesn't have a nameplate specifying the size and type of protection device, then the protection device must be sized in accordance with (A) or (B).

(A) One Motor-Compressor. The short-circuit and ground-fault protection device must not be greater than 175 percent of the motor-compressor current rating. If the protection device sized at 175 percent isn't capable of carrying the starting current of the motor-compressor, the next size larger protection device can be used, but in no case can it exceed 225 percent of the motor-compressor current rating.

Question: What size conductor and protection is required for a 24A motor-compressor connected to a 240V circuit? **Figure 440–6**

(a) 10 AWG, 40A   *(b) 10 AWG, 60A*
(c) a or b   *(d) none of these*

Answer: *(a) 10 AWG, 40A*

Step 1. Branch-Circuit Conductor [Table 310.16 and 440.32]

> *24A x 1.25 = 30A, 10 AWG, rated 30A at 75°C [Table 310.16]*

Step 2. Branch-Circuit Protection [240.6(A) and 440.22(A)]

> *24A x 1.75 = 42A, next size down = 40A*

If the 40A protection device isn't capable of carrying the starting current, then the protection device can be sized up to 225 percent of the equipment load current rating. 24A x 2.25 = 54A, next size down 50A

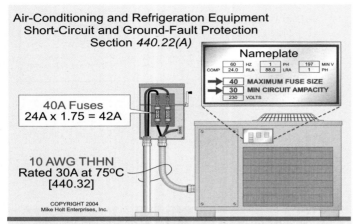

Figure 440–6

(B) Rating for Equipment. Where the equipment incorporates more than one hermetic refrigerant motor-compressor, or a hermetic refrigerant motor-compressor and other motors or other loads, the equipment short-circuit and ground-fault protection must be sized as follows:

(1) Motor-Compressor Largest Load. The rating of the branch-circuit short-circuit and ground-fault protective device must not be greater than the largest motor-compressor short-circuit ground-fault protection device, plus the sum of the rated-load currents of the other compressors.

> **Author's Comment:** The branch-circuit conductors are sized at 125 percent of the larger motor-compressor current, plus the sum of the rated-load currents of the other compressors [440.33].

PART IV. CONDUCTOR SIZING

> **Author's Comment:** The branch-circuit conductors for air-conditioning and refrigeration equipment must be sized not smaller than identified on the equipment's nameplate. If the equipment doesn't have a nameplate specifying the branch-circuit conductors, the conductors must be sized in accordance with 440.32. See Figure 440–5.

440.32 Conductor Size—One Motor-Compressor. If
equipment doesn't have a nameplate identifying the minimum circuit ampacity, the branch-circuit conductors to a single motor-compressor must have an ampacity not less than 125 percent of the motor-compressor rated-load current or the branch-circuit selection current, whichever is greater.

Author's Comments:

- In accordance with 440.22(A), branch-circuit conductors must have branch-circuit protection sized between 175 percent and 225 percent of the rated-load current to provide protection against short circuits and ground faults.

- In accordance with 440.22(A), branch-circuit conductors must have branch-circuit protection sized between 175 percent and 225 percent of the rated-load current to provide protection against short circuits and ground faults.

Question: *What size conductor and protection device is required for an 18A motor compressor?* **Figure 440–7**

(a) 12 AWG, 30A   *(b) 10 AWG, 50A*
(c) a or b   *(d) none of these*

Answer: *(a) 12 AWG, 30A*

Step 1. Branch-Circuit Conductor [Table 310.16 and 440.32]

> *18A x 1.25 = 22.5A, 12 AWG, rated 25A at 75°C [Table 310.16]*

Step 2. Branch-Circuit Protection [240.6(A) and 440.22(A)]

> *18A x 1.75 = 31.5A, next size down = 30A*

> *If the 30A protection device isn't capable of carrying the starting current, then the protection device can be sized up to 225 percent of the equipment load current rating.*
> *18A x 2.25 = 40.5A, next size down 40A*

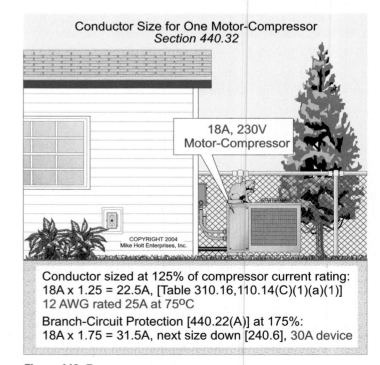

Conductor sized at 125% of compressor current rating:
18A x 1.25 = 22.5A, [Table 310.16,110.14(C)(1)(a)(1)]
12 AWG rated 25A at 75°C

Branch-Circuit Protection [440.22(A)] at 175%:
18A x 1.75 = 31.5A, next size down [240.6], 30A device

Figure 440–7

Author's Comment: A 30A or 40A protection device is permitted to protect a 12 AWG conductor. See 240.4(A) and 240.4(G) for details.

440.33 Conductor Size—Several Motor-Compressors.

Conductors that supply several motor-compressors must have an ampacity not less than 125 percent of the highest-rated motor-compressor current of the group, plus the sum of the rated-load or branch-circuit selection current ratings of the other compressors.

Author's Comment: These conductors must be protected against short circuits and ground faults in accordance with 440.22(B)(1).

PART VII. ROOM AIR CONDITIONERS

The requirements in this Part apply to a cord-and-plug connected room air conditioner of the window or in-wall type that incorporates a hermetic refrigerant motor-compressor rated not over 40A, 250V single-phase [440.60].

440.62 Branch-Circuit Requirements.

(A) Sizing Conductor and Protection. Branch-circuit conductors for a cord-and-plug connected room air conditioner must have an ampacity not less than 125 percent of the rated-load currents [440.32].

(B) Separate Circuit. Where the room air conditioner is the only load on a circuit, the marked rating of the air conditioner must not exceed 80 percent of the rating of the circuit [210.3].

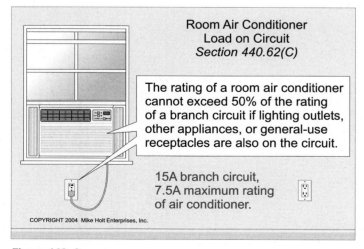

Room Air Conditioner
Load on Circuit
Section 440.62(C)

The rating of a room air conditioner cannot exceed 50% of the rating of a branch circuit if lighting outlets, other appliances, or general-use receptacles are also on the circuit.

15A branch circuit, 7.5A maximum rating of air conditioner.

COPYRIGHT 2004 Mike Holt Enterprises, Inc.

Figure 440–8

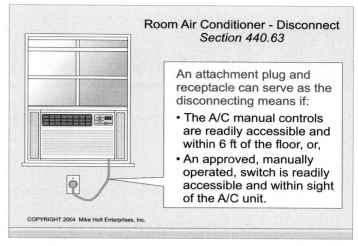

Room Air Conditioner - Disconnect
Section 440.63

An attachment plug and receptacle can serve as the disconnecting means if:
- The A/C manual controls are readily accessible and within 6 ft of the floor, or,
- An approved, manually operated, switch is readily accessible and within sight of the A/C unit.

COPYRIGHT 2004 Mike Holt Enterprises, Inc.

Figure 440–9

(C) Other Loads on Circuit. The total rating of a cord-and-plug connected room air conditioner must not exceed 50 percent of the rating of a branch circuit where lighting outlets, other appliances, or general-use receptacles are also supplied. **Figure 440–8**

440.63 Disconnecting Means. An attachment plug and receptacle can serve as the disconnecting means for a room air conditioner, provided: **Figure 440–9**

(1) The manual controls on the room air conditioner are readily accessible and within 6 ft of the floor, or

(2) A readily accessible disconnecting means is within sight from the room air conditioner.

Author's Comment: "Within sight" is visible and not more than 50 ft from each other [Article 100].

440.64 Supply Cord. Where a flexible cord is used to supply a room air conditioner, the cord must not exceed 10 ft for 120V units, or 6 ft for 208V through 240V units.

440.65 Leakage Current Detection and Interruption, and Arc-Fault Circuit Interrupter. Single-phase cord-and-plug connected room air conditioners must be provided with a factory-installed leakage current detector or with arc-fault circuit-interrupter (AFCI) protection.

1. Article 440 applies to electric motor-driven air-conditioning and refrigerating equipment that has a hermetic refrigerant motor-compressor.

    (a) True   (b) False

2. A disconnecting means that serves a hermetic refrigerant motor-compressor must be selected on the basis of the nameplate rated-load current or branch-circuit selection current, whichever is greater. It must have an ampere rating of at least _____ percent of the nameplate rated-load current or branch-circuit selection current, whichever is greater.

    (a) 125   (b) 80   (c) 100   (d) 115

3. Disconnecting means must be located within sight from and readily accessible from the air-conditioning or refrigerating equipment. The disconnecting means is permitted to be installed _____ the air-conditioning or refrigerating equipment, but not on panels that are designed to allow access to the air-conditioning or refrigeration equipment.

    (a) on   (b) within   (c) a or b   (d) none of these

4. Branch-circuit conductors supplying a single motor-compressor must have an ampacity not less than _____ percent of either the motor-compressor rated-load current or the branch-circuit selection current, whichever is greater.

    (a) 125   (b) 100   (c) 250   (d) 80

5. The rating of the attachment plug and receptacle must not exceed _____ at 250V for a cord-and-plug connected air conditioner.

    (a) 15A   (b) 20A   (c) 30A   (d) 40A

ARTICLE 445 Generators

Introduction

This article contains the electrical installation requirements for generators. These requirements include such things as where generators can be installed, nameplate markings, conductor ampacity, and disconnecting means.

Generators are basically motors that operate in reverse—they produce electricity when rotated, instead of rotating when supplied with electricity. Article 430, which covers motors, is the longest article in the *NEC*. Article 445, which covers generators, is one of the shortest. At first, this might not seem to make sense. But you don't need to size and protect conductors to a generator. You do need to size and protect them to a motor.

Generators need overload protection, and it's necessary to size the conductors that come from the generator. But these considerations are much more straightforward than the equivalent considerations for motors. Before you study Article 445, take a moment to read the definition of "Separately Derived System" in Article 100.

445.1 Scope. Article 445 covers the installation of generators.

Author's Comment: Generators, associated wiring, and equipment must be installed in accordance with the additional requirements depending on its use:

- Article 695, Fire Pumps
- Article 700, Emergency Systems
- Article 701, Legally Required Standby Systems
- Article 702, Optional Standby Systems

445.11 Marking. Each generator must be provided with a nameplate indicating the manufacturer's name, rated frequency, power factor, number of phases, rating in kilowatts or kilovolt amperes, volts and amperes corresponding to the rating, RPM, insulation class and rated ambient temperature or rated temperature rise, and time rating.

445.12 Overcurrent Protection.

(A) Constant-Voltage Generators. Constant-voltage generators must be protected from overloads by inherent design, circuit breakers, fuses, or other suitable acceptable overcurrent protective means.

445.13 Ampacity of Conductors. The ampacity of the conductors from the generator to distribution devices containing

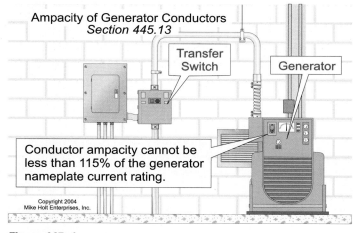

Ampacity of Generator Conductors
Section 445.13

Transfer Switch

Generator

Conductor ampacity cannot be less than 115% of the generator nameplate current rating.

Copyright 2004
Mike Holt Enterprises, Inc.

Figure 445–1

overcurrent protection must not be less than 115 percent of the nameplate current rating of the generator. **Figure 445–1**

Generators that aren't a separately derived system must have the grounded neutral conductor sized to: **Figure 445–2**

- Carry the maximum unbalanced current as determined by 220.61.
- Serve as the low-impedance fault-current path in accordance with 250.24(C).

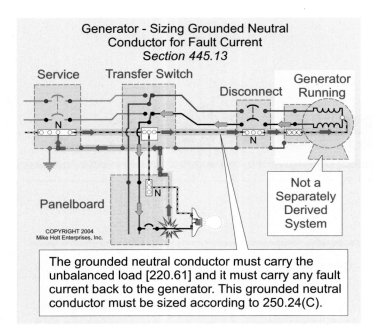

Figure 445–2

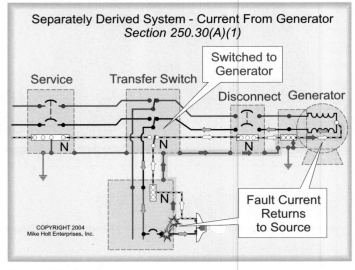

Figure 445–4

Author's Comment: Where the feeder conductors from the generator terminate in a transfer switch that doesn't open the grounded neutral conductor, the generator isn't considered a separately derived system [240.20(D)]. A neutral-to-case bond isn't permitted at the generator. Under this condition, the grounded neutral conductor from the normal power to the transfer switch, and the grounded neutral conductor from the generator to the transfer switch, are required to provide the low-impedance fault-current path back to the power source. **Figure 445–3**

Separately derived system generators must have the grounded neutral conductor sized not smaller than required to carry the maximum unbalanced current as determined by 220.61.

Author's Comment: Where the feeder conductors from the generator terminate in a transfer switch that opens the grounded neutral conductor, the generator is considered a separately derived system [Article100]. A neutral-to-case bond is required at the generator [250.30(A)(1)] to provide a low-impedance fault-current path back to the power source. **Figure 445–4**

445.18 Disconnecting Means. Generators must have one or more disconnecting means that disconnects all power, except where: **Figure 445–5**

(1) The driving means for the generator can be readily shut down, and

(2) The generator isn't arranged to operate in parallel with another generator or other source of voltage.

CAUTION: *If one generator is used to supply emergency, legally required, as well as optional standby power, then there must be at least two transfer switches; one for emergency power and another for legally required as well as optional stand-by power [700.6(D)].*

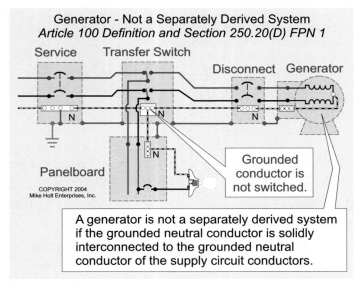

Figure 445–3

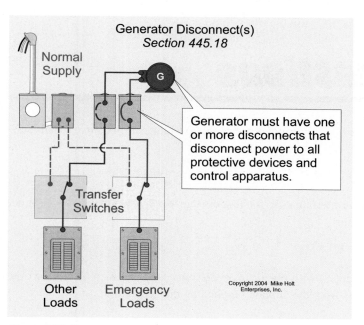

Figure 445–5

Article 445 Questions

1. Constant-voltage generators must be protected from overloads by _____ or other acceptable overcurrent protective means suitable for the conditions of use.

   (a) inherent design   (b) circuit breakers   (c) fuses   (d) any of these

2. The ampacity of ungrounded (phase) conductors from the generator terminals to the first overcurrent protection devices must not be less than _____ percent of the nameplate rating of the generator.

   (a) 75   (b) 115   (c) 125   (d) 140

3. Unless two restrictive conditions exist, a generator must be equipped with one or more disconnecting means to disconnect the generator, its protective devices, and all control apparatus entirely from the circuits supplied by the generator.

   (a) True   (b) False

Mike Holt Enterprises, Inc. • www.NECcode.com • 1.888.NEC.Code

Transformers and Transformer Vaults

Introduction

Article 450 opens by saying, "This article covers the installation of all transformers." Then it lists eight exceptions. So what does Article 450 really cover? Essentially, it covers power transformers, transformer vaults, and most kinds of lighting transformers.

One of the main concerns with transformers is the prevention of overheating. The *NEC* doesn't completely address this issue. Article 90 explains that the *NEC* isn't a design manual and that it assumes the person using the *Code* has a certain level of expertise. Proper transformer selection is an important part of preventing transformer overheating.

The *NEC* assumes you have already selected a transformer suitable to the load characteristics. For the *Code* to tell you how to do that would push it into the realm of a design manual. Article 450 then takes you to the next logical step—providing overcurrent protection and the proper connections. But Article 450 doesn't stop there; 450.9 provides ventilation requirements.

Part I of Article 450 contains the general requirements such as guarding, marking, and accessibility. Part II contains the requirements for different types of transformers, and Part III provides requirements for transformer vaults.

PART I. GENERAL

450.1 Scope. Article 450 covers the installation requirements of transformers and transformer vaults. **Figure 450–1**

> **Author's Comment:** Article 450 also applies to secondary systems whose voltage exceeds 600V; however, this textbook only addresses secondary systems operating at 600V or less.

450.3 Overcurrent Protection. Overcurrent protection of the primary winding of a transformer not exceeding 600V must comply with (B).

> **FPN 2:** 4-wire three-phase 120/208V or 277/480V systems that supply nonlinear line-to-neutral loads can overheat because of triplen harmonic currents (3rd, 9th, 15th, 21st, etc.) [450.9 FPN 2]. **Figure 450–2**

> **Author's Comment:** For more information on this subject, visit www.NECCode.com.

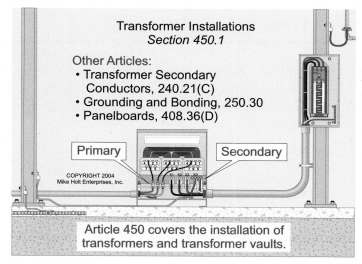

Transformer Installations
Section 450.1

Other Articles:
• Transformer Secondary Conductors, 240.21(C)
• Grounding and Bonding, 250.30
• Panelboards, 408.36(D)

Primary   Secondary

COPYRIGHT 2004
Mike Holt Enterprises, Inc.

Article 450 covers the installation of transformers and transformer vaults.

Figure 450–1

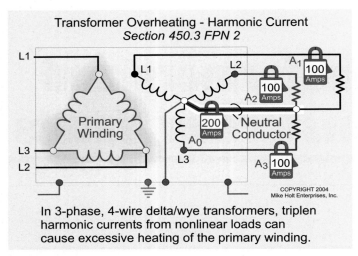

Transformer Overheating - Harmonic Current
Section 450.3 FPN 2

In 3-phase, 4-wire delta/wye transformers, triplen harmonic currents from nonlinear loads can cause excessive heating of the primary winding.

Figure 450–2

(B) Overcurrent Protection for Transformers Not Over 600V. The primary winding of a transformer must be protected against overcurrent in accordance with the percentages listed in Table 450.3(B) and all applicable notes.

Table 450.3(B) Primary Protection only

Primary Current Rating	Maximum Protection
9A or More	125%, see Note 1
Less Than 9A	167%
Less Than 2A	300%

Note 1. Where 125 percent of the primary current doesn't correspond to a standard rating of a fuse or nonadjustable circuit breaker, the next higher rating is permitted [240.6(A)].

Question: *What is the primary protection device rating and conductor size required for a 45 kVA, three-phase, 480V— 120/208V transformer that is fully loaded? Terminals are rated 75°C.* **Figure 450–3**

(a) 8 AWG, 40A   *(b) 6 AWG, 50A*
(c) 6 AWG, 60A   *(d) 4 AWG, 70A*

Answer: *(d) 70A*

Step 1. Primary current

$$I = VA/(E \times 1.732)$$
$$I = 45,000 \ VA/(480V \times 1.732)$$
$$I = 54A$$

Step 2. The primary protection device rating [240.6(A)]

    54A x 1.25 = 68A, next size up 70A, Note 1

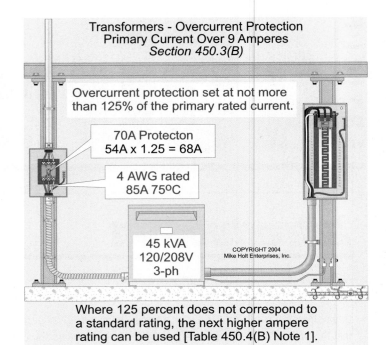

Transformers - Overcurrent Protection
Primary Current Over 9 Amperes
Section 450.3(B)

Overcurrent protection set at not more than 125% of the primary rated current.

70A Protecton
54A x 1.25 = 68A

4 AWG rated
85A 75ºC

45 kVA
120/208V
3-ph

Where 125 percent does not correspond to a standard rating, the next higher ampere rating can be used [Table 450.4(B) Note 1].

Figure 450–3

Step 3. The primary conductor must be sized to carry 54A continuously (54A x 1.25 = 68A) [215.2(A)(1)] and be protected by a 70A protection device [240.4(A)]. A 4 AWG conductor rated 85A at 75°C meets all of the requirements [110.14(C)(1) and 310.16].

Step 4. Secondary current

$$I = VA/(E \times 1.732)$$
$$I = 45,000 \ VA/ (208V \times 1.732), = 125A$$

Author's Comment: Secondary conductors having a maximum length of 25 ft that terminate in an overcurrent protection device that doesn't exceed the ampacity of the conductors, must be sized at 125 percent of the continuous load [215.2(A)(1) and 240.21(C)(6)].

Question: *What is the secondary conductor size required for a 45 kVA, three-phase, 480V—120/208V transformer, that supplies a 200A lighting and appliance panelboard that is fully loaded? Terminals are rated 75°C.* **Figure 450–4**

(a) 1 AWG   *(b) 1/0 AWG*   *(c) 2/0 AWG*   *(d) 3/0 AWG*

Answer: *(c) 2/0 AWG*

Step 1. Secondary current

$$I = VA/(E \times 1.732)$$
$$I = 45,000 \ VA/(208V \times 1.732)$$
$$I = 125A$$

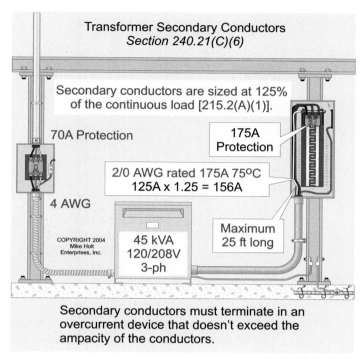

Transformer Secondary Conductors
Section 240.21(C)(6)

Secondary conductors are sized at 125% of the continuous load [215.2(A)(1)].

70A Protection

175A Protection

2/0 AWG rated 175A 75ºC
125A x 1.25 = 156A

4 AWG

COPYRIGHT 2004
Mike Holt
Enterprises, Inc.

45 kVA
120/208V
3-ph

Maximum
25 ft long

Secondary conductors must terminate in an overcurrent device that doesn't exceed the ampacity of the conductors.

Figure 450–4

Step 2. Secondary conductors must be sized to 125 percent of the continuous load [215.2(A)(1)].

125A x 1.25 = 156A, 2/0 AWG rated 175A at 75°C termination

Secondary conductors must terminate in an overcurrent protection device that doesn't exceed the ampacity of the conductors [240.21(C)(6)]. 2/0 AWG, rated 175A at 75°C, terminating on a 175A protection device meets this requirement.

450.9 Ventilation. Transformers must be installed in accordance with the manufacturer's instructions, and their ventilating openings must not be blocked [110.3(B)].

> FPN 2: Transformers can become excessively heated above their rating because of triplen harmonic currents (3rd, 9th, 15th, 21st, etc.) [450.3 FPN].

Author's Comments:

• The heating from harmonic currents is proportional to the square of the harmonic current. This means the third (3rd) harmonic currents (180 Hz) cause heat at nine times the rate of 60 Hz current. Figure 450–5

• A transformer supplying no harmonic loads can be loaded to its full-load capacity without overheating. However, because harmonic loads heat at the "square of the harmonic," the load

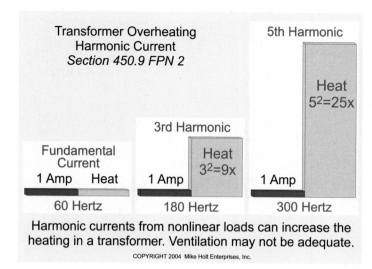

Transformer Overheating Harmonic Current
Section 450.9 FPN 2

Fundamental Current
1 Amp   Heat
60 Hertz

3rd Harmonic
Heat
$3^2=9x$
1 Amp
180 Hertz

5th Harmonic
Heat
$5^2=25x$
1 Amp
300 Hertz

Harmonic currents from nonlinear loads can increase the heating in a transformer. Ventilation may not be adequate.

COPYRIGHT 2004 Mike Holt Enterprises, Inc.

Figure 450–5

on a transformer would have to be limited, or the transformer must be "K-Factor" rated for the expected harmonic content. This topic is beyond the scope of this textbook, but for more information, visit the Technical link at www.MikeHolt.com.

450.11 Marking. Transformers must be provided with a nameplate identifying the manufacturer of the transformer and indicating the transformer's rated kVA, primary and secondary voltage, impedance if 25 kVA or larger, and required clearances for transformers with ventilating openings.

450.13 Transformer Accessibility. Transformers must be readily accessible to qualified personnel for inspection and maintenance, except as permitted by (A) or (B).

(A) Open Installations. Dry-type transformers can be located in the open on walls, columns, or structures. Figure 450–6

(B) Suspended Ceilings. Dry-type transformers, rated not more than 50 kVA, are permitted above suspended ceilings or in hollow spaces of buildings, if not permanently closed in by the structure. Figure 450–7

> Author's Comment: Dry-type transformers with a metal enclosure can be installed in a suspended-ceiling space for environmental air [300.22(C)(2)].

PART III. TRANSFORMER VAULTS

450.41 Location. Transformer vaults must be ventilated to the outside air, without using flues or ducts where practicable.

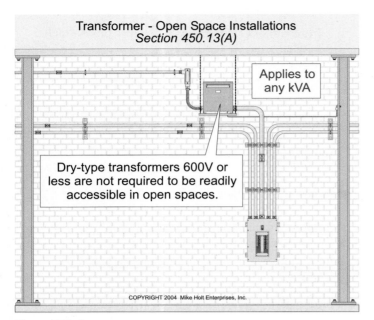

Transformer - Open Space Installations
Section 450.13(A)

Applies to any kVA

Dry-type transformers 600V or less are not required to be readily accessible in open spaces.

COPYRIGHT 2004 Mike Holt Enterprises, Inc.

Figure 450–6

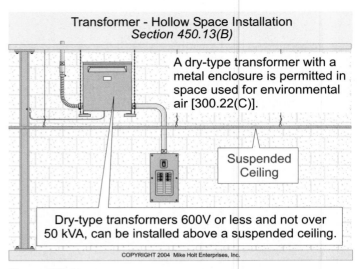

Transformer - Hollow Space Installation
Section 450.13(B)

A dry-type transformer with a metal enclosure is permitted in space used for environmental air [300.22(C)].

Suspended Ceiling

Dry-type transformers 600V or less and not over 50 kVA, can be installed above a suspended ceiling.

COPYRIGHT 2004 Mike Holt Enterprises, Inc.

Figure 450–7

450.42 Walls, Roofs, and Floors. The floors, walls, ceilings, and roofs of vaults must have adequate structural strength with a minimum fire resistance of 3 hours, such as 6 in. thick reinforced concrete.

Exception: Where the transformer is automatic sprinkler protected, a 1-hour fire-resistance rating is permitted.

450.43 Doorways.

(A) Type of Door. Each vault doorway must be provided with a tight-fitting door with a minimum fire-resistance rating of 3 hours.

Exception: Where an automatic sprinkler protects the transformer, a 1-hour fire-resistance rating is permitted.

(C) Locks. Vault doors must swing out, be equipped with panic bars or pressure plates so the door can open under simple pressure, and must be provided with locks that are only accessible to qualified persons.

450.45 Ventilation Openings.

(A) Location. Ventilation openings must be as far as possible from doors, windows, and combustible material.

(B) Arrangement. A vault ventilated by natural circulation must have no more than 50 percent of the total opening area near the floor, with the remainder of the opening area in the roof or sidewalls near the roof.

(C) Size. For a vault ventilated by natural circulation, the total opening area must not be less than 3 sq in. per kVA capacity. In no case can the area be less than 1 sq ft for 50 kVA or less.

(D) Covering. Ventilation openings must be covered to avoid unsafe conditions.

(E) Dampers. All indoor ventilation openings must be provided with automatic closing fire dampers rated not less than 1½ hours that close in response to a vault fire.

(F) Ducts. Ventilating ducts must be of fire-resistant material.

450.47 Water Pipes and Accessories. Only piping or other facilities provided for vault fire protection or cooling can be installed in the transformer vault.

450.48 Storage in Vaults. Nothing can be stored in a transformer vault.

Article 450 Questions

1. According to Article 450, a transformer rated 600V, nominal, or less, and whose primary current rating is 9A or more, is protected against overcurrent only when _____.

   (a) an individual overcurrent device on the primary side is set at not more than 125 percent of the rated primary current of the transformer.

   (b) a secondary overcurrent device is set at not more than 125 percent of the rated secondary current of the transformer, and a primary overcurrent device is set at not more than 250 percent of the rated primary current of the transformer.

   (c) a or b

   (d) none of these

2. For a transformer rated 600V, nominal, or less, if the primary overcurrent protection device is sized at 250 percent of the primary current, what size secondary overcurrent protection device is required if the secondary current is 42A?

   (a) 40A   (b) 70A   (c) 60A   (d) 90A

3. What size "primary only" overcurrent protection is required for a 600 volt, 45 kVA transformer that has a primary current rating of 54A?

   (a) 70   (b) 80   (c) 90   (d) 100

4. Transformers with ventilating openings must be installed so that the ventilating openings _____.

   (a) are a minimum 18 in. above the floor   (b) are not blocked by walls or obstructions

   (c) are aesthetically located   (d) are vented to the exterior of the building

5. Indoor transformers of greater than _____ rating must be installed in a transformer room of fire-resistant construction.

   (a) 35,000 kVA   (b) 87½ kVA   (c) 112½ kVA   (d) 75 kVA

6. Personnel doors for transformer vaults must _____ and be equipped with panic bars, pressure plates, or other devices that are normally latched but open under simple pressure.

   (a) be clearly identified   (b) swing out   (c) a and b   (d) a or b

Notes

ARTICLE 460 Capacitors

Introduction

This article covers the installation of capacitors, including those in hazardous (classified) locations as described by Articles 501 through 503.

Capacitors store energy. Thus, simply disconnecting capacitors doesn't de-energize them. Capacitors have requirements for ampacity, overcurrent protection, disconnecting means, and marking. The requirements for capacitors under 600V are less stringent than for those over 600V.

460.1 Scope. Article 460 covers the installation requirements for capacitors.

460.2 Enclosing and Guarding.

(B) Accidental Contact. Where accessible to unauthorized and unqualified persons, capacitors must be enclosed, located, or guarded to prevent contact with exposed energized parts, terminals, or buses.

PART I. 600V, NOMINAL, AND UNDER

460.8 Conductors.

(A) Ampacity. Capacitor circuit conductors must have an ampacity not less than 135 percent of the rated current of the capacitor. Conductors that connect a capacitor to a motor circuit must have an ampacity not less than one-third the ampacity of the motor circuit conductors, but not less than 135 percent of the rated current of the capacitor.

(B) Overcurrent Protection. The rating of the overcurrent protection device for a capacitor bank must be as low as practicable.

Exception: Separate overcurrent protection isn't required for a capacitor connected to the load side of a motor overload protective device.

(C) Disconnecting Means. A disconnecting means must be provided for each capacitor bank and must:

(1) Open all ungrounded conductors simultaneously.

(3) Not be rated less than 135 percent of the rated current of the capacitor.

Exception: A separate disconnecting means isn't required for a capacitor connected on the load side of a motor controller.

460.9 Rating or Setting of Motor Overload Device.
Where a capacitor is connected on the load side of the motor overload device, the rating or setting of the motor overload device must be based on the improved power factor of the motor circuit.

The effect of reduced current because of the capacitor must be disregarded when sizing motor circuit conductors in accordance with 430.22(A).

1. Capacitors must be _____ so that persons cannot come into accidental contact or bring conducting materials into accidental contact with exposed energized parts, terminals, or buses associated with them.

    (a) enclosed   (b) located   (c) guarded   (d) any of these

2. An overcurrent device must be provided in each ungrounded conductor for each capacitor bank. The rating or setting of the overcurrent device must be _____.

    (a) 20A   (b) as low as practicable
    (c) 400% of conductor ampacity   (d) 100A

3. A disconnect must be provided in each ungrounded conductor for each capacitor bank, and must _____.

    (a) open all ungrounded conductors simultaneously
    (b) be permitted to disconnect the capacitor from the line as a regular operating procedure
    (c) be rated no less than 135 percent of the rated current of the capacitor
    (d) all of these

4. Where a motor installation includes a capacitor connected on the load side of the motor overload device, the rating or setting of the motor overload device must be based on the improved power factor of the motor circuit.

    (a) True   (b) False

5. Where a motor installation includes a capacitor connected on the load side of the motor overload device, the effect of a capacitor must be disregarded in determining the motor circuit conductor size in accordance with 430.22.

    (a) True   (b) False

Index

Notes

Notes

Notes

Mike Holt Enterprises, Inc. • www.NECcode.com • 1.888.NEC.Code

Notes

What You Get With Mike's Products

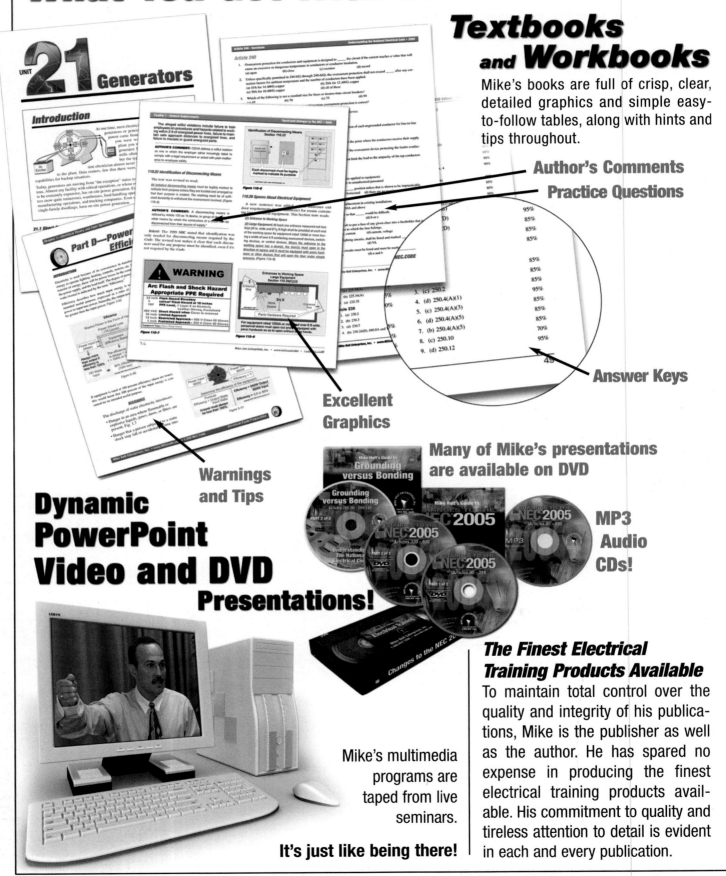

Textbooks and Workbooks

Mike's books are full of crisp, clear, detailed graphics and simple easy-to-follow tables, along with hints and tips throughout.

Author's Comments

Practice Questions

Answer Keys

Excellent Graphics

Warnings and Tips

Dynamic PowerPoint Video and DVD Presentations!

Many of Mike's presentations are available on DVD

MP3 Audio CDs!

Mike's multimedia programs are taped from live seminars.

It's just like being there!

The Finest Electrical Training Products Available

To maintain total control over the quality and integrity of his publications, Mike is the publisher as well as the author. He has spared no expense in producing the finest electrical training products available. His commitment to quality and tireless attention to detail is evident in each and every publication.

Exam Preparation Library

Printed in full color, the Electrical *NEC* Exam Preparation book covers Theory, *Code*, and Calculations in great detail. Clear colorful graphics guide you step-by-step through all the material you'll need to pass your exam the first time. Includes 480 pages, 350 graphics, hundreds of *NEC* questions and answer keys.

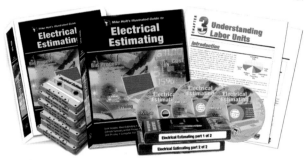

Supreme Estimating Library

Estimate like the experts. This program will show you how to take-off a job, determine material cost, accurately estimate labor cost, and how to determine your overhead, profit, and break-even point. This course also gives tips on selling and marketing your business. You'll learn how to estimate and project manage residential, commercial, and industrial projects. Learn how to make more money on every job by proper estimating and project management.

Grounding versus Bonding Textbook

Grounding and bonding problems are at epidemic levels. Surveys repeatedly show that 90 percent of power quality problems are due to poor grounding and bonding. Electrical theory has been applied to this difficult to understand Article, making it easier for students to grasp the concepts of grounding versus bonding. Additionally, the graphics are color-coded so you can easily differentiate between grounding and bonding.